Applied Mathematical Sciences

Volume 154

Springer
New York
Berlin
Heidelberg
Barcelona
Hong Kong
London
Milan
Paris
Singapore
Tokyo

Applied Mathematical Sciences

1. *John:* Partial Differential Equations, 4th ed.
2. *Sirovich:* Techniques of Asymptotic Analysis.
3. *Hale:* Theory of Functional Differential Equations, 2nd ed.
4. *Percus:* Combinatorial Methods.
5. *von Mises/Friedrichs:* Fluid Dynamics.
6. *Freiberger/Grenander:* A Short Course in Computational Probability and Statistics.
7. *Pipkin:* Lectures on Viscoelasticity Theory.
8. *Giacaglia:* Perturbation Methods in Non-linear Systems.
9. *Friedrichs:* Spectral Theory of Operators in Hilbert Space.
10. *Stroud:* Numerical Quadrature and Solution of Ordinary Differential Equations.
11. *Wolovich:* Linear Multivariable Systems.
12. *Berkovitz:* Optimal Control Theory.
13. *Bluman/Cole:* Similarity Methods for Differential Equations.
14. *Yoshizawa:* Stability Theory and the Existence of Periodic Solution and Almost Periodic Solutions.
15. *Braun:* Differential Equations and Their Applications, 3rd ed.
16. *Lefschetz:* Applications of Algebraic Topology.
17. *Collatz/Wetterling:* Optimization Problems.
18. *Grenander:* Pattern Synthesis: Lectures in Pattern Theory, Vol. I.
19. *Marsden/McCracken:* Hopf Bifurcation and Its Applications.
20. *Driver:* Ordinary and Delay Differential Equations.
21. *Courant/Friedrichs:* Supersonic Flow and Shock Waves.
22. *Rouche/Habets/Laloy:* Stability Theory by Liapunov's Direct Method.
23. *Lamperti:* Stochastic Processes: A Survey of the Mathematical Theory.
24. *Grenander:* Pattern Analysis: Lectures in Pattern Theory, Vol. II.
25. *Davies:* Integral Transforms and Their Applications, 2nd ed.
26. *Kushner/Clark:* Stochastic Approximation Methods for Constrained and Unconstrained Systems.
27. *de Boor:* A Practical Guide to Splines: Revised Edition.
28. *Keilson:* Markov Chain Models—Rarity and Exponentiality.
29. *de Veubeke:* A Course in Elasticity.
30. *Sniatycki:* Geometric Quantization and Quantum Mechanics.
31. *Reid:* Sturmian Theory for Ordinary Differential Equations.
32. *Meis/Markowitz:* Numerical Solution of Partial Differential Equations.
33. *Grenander:* Regular Structures: Lectures in Pattern Theory, Vol. III.
34. *Kevorkian/Cole:* Perturbation Methods in Applied Mathematics.
35. *Carr:* Applications of Centre Manifold Theory.
36. *Bengtsson/Ghil/Källén:* Dynamic Meteorology: Data Assimilation Methods.
37. *Saperstone:* Semidynamical Systems in Infinite Dimensional Spaces.
38. *Lichtenberg/Lieberman:* Regular and Chaotic Dynamics, 2nd ed.
39. *Piccini/Stampacchia/Vidossich:* Ordinary Differential Equations in $\mathbf{R}^n$.
40. *Naylor/Sell:* Linear Operator Theory in Engineering and Science.
41. *Sparrow:* The Lorenz Equations: Bifurcations, Chaos, and Strange Attractors.
42. *Guckenheimer/Holmes:* Nonlinear Oscillations, Dynamical Systems, and Bifurcations of Vector Fields.
43. *Ockendon/Taylor:* Inviscid Fluid Flows.
44. *Pazy:* Semigroups of Linear Operators and Applications to Partial Differential Equations.
45. *Glashoff/Gustafson:* Linear Operations and Approximation: An Introduction to the Theoretical Analysis and Numerical Treatment of Semi-Infinite Programs.
46. *Wilcox:* Scattering Theory for Diffraction Gratings.
47. *Hale/Magalhães/Oliva:* Dynamics in Infinite Dimensions, 2nd ed.
48. *Murray:* Asymptotic Analysis.
49. *Ladyzhenskaya:* The Boundary-Value Problems of Mathematical Physics.
50. *Wilcox:* Sound Propagation in Stratified Fluids.
51. *Golubitsky/Schaeffer:* Bifurcation and Groups in Bifurcation Theory, Vol. I.
52. *Chipot:* Variational Inequalities and Flow in Porous Media.
53. *Majda:* Compressible Fluid Flow and System of Conservation Laws in Several Space Variables.
54. *Wasow:* Linear Turning Point Theory.
55. *Yosida:* Operational Calculus: A Theory of Hyperfunctions.
56. *Chang/Howes:* Nonlinear Singular Perturbation Phenomena: Theory and Applications.
57. *Reinhardt:* Analysis of Approximation Methods for Differential and Integral Equations.
58. *Dwoyer/Hussaini/Voigt (eds):* Theoretical Approaches to Turbulence.
59. *Sanders/Verhulst:* Averaging Methods in Nonlinear Dynamical Systems.

(continued following index)

George W. Bluman Stephen C. Anco

Symmetry and Integration Methods for Differential Equations

With 18 Illustrations

Springer

George W. Bluman
Department of Mathematics
The University of British Columbia
Vancouver, British Columbia V6T 1Z2
Canada
bluman@math.ubc.ca

Stephen C. Anco
Department of Mathematics
Brock University
St. Catharines, Ontario L2S 3A1
Canada
sanco@brocku.ca

Mathematics Subject Classification (2000): 22E25, 22E65, 22E70, 34-01, 34A05, 35K05, 35L05, 35Q20, 58F35, 58F37, 58G35, 58G37, 70H35

Library of Congress Cataloging-in-Publication Data
Bluman, George W., 1943–
Symmetry and integration methods for differential equations/George W. Bluman, Stephen C. Anco.
p. cm. — (Applied mathematical sciences; v. 154)
Includes bibliographical references and index.
ISBN 0-387-98654-5 (alk. paper)
1. Differential equations—Numerical solutions. 2. Differential equations, Partial—Numerical solutions. 3. Lie groups. I. Bluman, George W., 1943– Symmetries and differential equations. II. Title. III. Applied mathematical sciences (Springer-Verlag New York, Inc.); v.154.
QA1 .A647 no. 154 2002
[QA372]
510 s—dc21
[515′.35] 2001054908

ISBN 0-387-98654-5 Printed on acid-free paper.

Printed in the United States of America.

9 8 7 6 5 4 3 2 1 SPIN 10696756

Typesetting: Pages created by the authors using Microsoft Word.

www.springer-ny.com

Springer-Verlag New York Berlin Heidelberg
A member of BertelsmannSpringer Science+Business Media GmbH

Preface

This book is a significant update of the first four chapters of *Symmetries and Differential Equations* (1989; reprinted with corrections, 1996), by George W. Bluman and Sukeyuki Kumei. Since 1989 there have been considerable developments in symmetry methods (group methods) for differential equations as evidenced by the number of research papers, books, and new symbolic manipulation software devoted to the subject. This is, no doubt, due to the inherent applicability of the methods to nonlinear differential equations. Symmetry methods for differential equations, originally developed by Sophus Lie in the latter half of the nineteenth century, are highly algorithmic and hence amenable to symbolic computation. These methods systematically unify and extend well-known ad hoc techniques to construct explicit solutions for differential equations, especially for nonlinear differential equations. Often ingenious tricks for solving particular differential equations arise transparently from the symmetry point of view, and thus it remains somewhat surprising that symmetry methods are not more widely known. Nowadays it is essential to learn the methods presented in this book to understand existing symbolic manipulation software for obtaining analytical results for differential equations. For ordinary differential equations (ODEs), these include reduction of order through group invariance or integrating factors. For partial differential equations (PDEs), these include the construction of special solutions such as similarity solutions or nonclassical solutions, finding conservation laws, equivalence mappings, and linearizations.

A large portion of this book discusses work that has appeared since the above-mentioned book, especially connected with finding first integrals for higher-order ODEs and using higher-order symmetries to reduce the order of an ODE. Also novel is a comparison of various complementary symmetry and integration methods for an ODE.

The present book includes a comprehensive treatment of dimensional analysis. There is a full discussion of aspects of Lie groups of point transformations (point symmetries), contact symmetries, and higher-order symmetries that are essential for finding solutions of differential equations. No knowledge of group theory is assumed. Emphasis is placed on explicit algorithms to discover symmetries and integrating factors admitted by a given differential equation and to construct solutions and first integrals resulting from such symmetries and integrating factors.

This book should be particularly suitable for applied mathematicians, engineers, and scientists interested in how to find systematically explicit solutions of differential equations. Almost all examples are taken from physical and engineering problems including those concerned with heat conduction, wave propagation, and fluid flow.

Chapter 1 includes a thorough treatment of dimensional analysis. The well-known Buckingham Pi-theorem is presented in a manner that introduces the reader concretely to the notion of invariance. This is shown to naturally lead to generalizations through invariance of boundary value problems under scalings of variables. This prepares the reader to consider the still more general invariance of differential equations under Lie groups of transformations in the third and fourth chapters. Basically, the first chapter gives the reader an intuitive grasp of some of the subject matter of the book in an elementary setting.

Chapter 2 develops the basic concepts of Lie groups of transformations and Lie algebras that are necessary in the following two chapters. By considering a Lie group of point transformations through its infinitesimal generator from the point of view of mapping functions into functions with their independent variables held fixed, we show how one is able to consider naturally other local transformations such as contact transformations and higher-order transformations. Moreover, this allows us to prepare the foundation for consideration of integrating factors for differential equations.

Chapter 3 is concerned with ODEs. A reduction algorithm is presented that reduces an nth-order ODE, admitting a solvable r-parameter Lie group of point transformations (point symmetries), to an $(n - r)$th-order differential equation and r quadratures. We show how to find admitted point, contact, and higher-order symmetries. We also show how to extend the reduction algorithm to incorporate such symmetries. It is shown how to find admitted first integrals through corresponding integrating factors and to obtain reductions of order using first integrals. We show how this simplifies and significantly extends the classical Noether's Theorem for finding conservation laws (first integrals) to any ODE (not just one admitting a variational principle). In particular, we show how to calculate integrating factors by various algorithmic procedures analogous to those for calculating symmetries in characteristic form where only the dependent variable undergoes a transformation. We also compare the distinct methods of reducing order through admitted local symmetries and through admitted integrating factors. We show how to use invariance under point symmetries to solve boundary value problems. We derive an algorithm to construct special solutions (invariant solutions) resulting from admitted symmetries. By studying their topological nature, we show that invariant solutions include separatrices and singular envelope solutions.

Chapter 4 is concerned with PDEs. It is shown how to find admitted point symmetries and how to construct related invariant solutions. There is a full discussion of the applicability to boundary value problems with numerous examples involving scalar PDEs and systems of PDEs.

Chapters 2 to 4 can be read independently of the first chapter. Moreover, a reader interested in PDEs can skip the third chapter.

Every topic is illustrated by examples. All sections have many exercises. It is essential to do the exercises to obtain a working knowledge of the material. The Discussion section at the end of each chapter puts its contents into perspective by summarizing major results, by referring to related works, and by introducing related material.

Within each section and subsection of a given chapter, we number separately, and consecutively, definitions, theorems, lemmas, and corollaries. For example, Definition 2.3.3-1 refers to the first definition and Theorem 2.3.3-1 to the first theorem in Section 2.3.3. Exercises appear at the conclusion of each section; Exercise 2.4-2 refers to the second problem of Exercises 2.4.

We thank Benny Bluman for the illustrations and Cecile Gauthier for typing several drafts of Sections 3.5 to 3.8.

Vancouver, British Columbia, Canada — George W. Bluman
St. Catharines, Ontario, Canada — Stephen C. Anco

Contents

Introduction

In the latter part of the nineteenth century, Sophus Lie introduced the notion of continuous groups, now known as Lie groups, in order to unify and extend various specialized methods for solving ordinary differential equations (ODEs). Lie was inspired by the lectures of Sylow given at Christiania (present-day Oslo) on Galois theory and Abel's related works. [In 1881 Sylow and Lie collaborated in a careful editing of Abel's complete works.] Lie showed that the order of an ODE could be reduced by one, constructively, if it is invariant under a one-parameter Lie group of point transformations.

Lie's work systematically related a miscellany of topics in ODEs including: integrating factor, separable equation, homogeneous equation, reduction of order, the methods of undetermined coefficients and variation of parameters for linear equations, solution of the Euler equation, and the use of the Laplace transform. Lie (1881) also indicated that for linear partial differential equations (PDEs), invariance under a Lie group leads directly to superpositions of solutions in terms of transforms.

A *symmetry* of a system of differential equations is a transformation that maps any solution to another solution of the system. In Lie's framework such transformations are groups that depend on continuous parameters and consist of either point transformations (*point symmetries*), acting on the system's space of independent and dependent variables, or, more generally, contact transformations (*contact symmetries*), acting on the space of independent and dependent variables as well as on all first derivatives of the dependent variables. Elementary examples of Lie groups include translations, rotations, and scalings. An autonomous system of first-order ODEs, i.e., a stationary flow, essentially defines a one-parameter Lie group of point transformations. Lie showed that for a given differential equation (linear or nonlinear), the admitted continuous group of point transformations, acting on the space of its independent and dependent variables, can be determined by an explicit computational algorithm (*Lie's algorithm*).

In this book, the applications of continuous groups to differential equations make no use of the global aspects of Lie groups. These applications use connected local Lie groups of transformations. Lie's fundamental theorems show that such groups are completely characterized by their *infinitesimal generators*. In turn, these form a *Lie algebra* determined by structure constants.

Lie groups, and hence their infinitesimal generators, can be naturally *extended* or *"prolonged"* to act on the space of independent variables, dependent variables, and derivatives of the dependent variables up to any finite order. As a consequence, the seemingly intractable nonlinear conditions for group invariance of a given system of differential equations reduce to linear homogeneous equations determining the infinitesimal generators of the group. Since these *determining equations* form an overdetermined system of linear homogeneous PDEs, one can usually determine the infinitesimal generators in explicit form. For a given system of differential equations, the setting up of the determining equations is entirely routine. Symbolic manipulation programs exist to set up the determining equations and in some cases explicitly solve them [Schwarz (1985, 1988); Kersten (1987); Head (1992); Champagne, Hereman, and Winternitz (1991); Wolf and Brand (1992); Hereman (1996); Reid (1990, 1991); Mansfield (1996); Mansfield and Clarkson (1997); Wolf (2002a)].

One can generalize Lie's work to find and use *higher-order symmetries* admitted by differential equations. The possibility of the existence of higher-order symmetries appears to have been first considered by Noether (1918). Such symmetries are characterized by infinitesimal generators that act only on dependent variables, with coefficients of the generators depending on independent variables, dependent variables and their derivatives to some finite order. Here, unlike the case for point symmetries or contact symmetries, any extension of the corresponding global transformation is not closed on any finite-dimensional space of independent variables, dependent variables and their derivatives to some finite order. In particular, globally, such transformations act on the infinite-dimensional space of independent variables, dependent variables, and their derivatives to all orders. Nonetheless, a natural extension of Lie's algorithm can be used to find such transformations for a given differential equation.

For a first-order ODE, Lie showed that invariance of the ODE under a point symmetry is equivalent to the existence of a first integral for the ODE. In this situation a *first integral* yields a conserved quantity that is constant for each solution of the ODE. Local existence theory for an nth-order ODE shows that there always exist n functionally independent first integrals of the ODE, which are quadratures relating the independent variable, dependent variable and its derivatives to order $n-1$. Correspondingly, an nth order ODE admits n essential conserved quantities. Moreover, it is a long-known result that any first integral arises from an *integrating factor*, given by a function of the independent variable, dependent variable and its derivatives to some order, which multiplies the ODE to transform it into an exact (total derivative) form.

For a higher-order ODE, a correspondence between first integrals and invariance under point symmetries holds only when the ODE has a variational principle (*Lagrangian*). In particular, Noether's work showed that invariance of such an ODE under a point symmetry, a contact symmetry, or a higher-order symmetry is equivalent to the existence of a first integral for the ODE if the symmetry leaves invariant the variational principle of the ODE (*variational symmetry*). Here it is essential to view a symmetry in its *characteristic form* where the coefficient of its infinitesimal generator acts only on the dependent variable (and its derivatives) in the ODE. The determining equation for symmetries is then given by the linearization (Frèchet derivative) of the ODE holding for *all* solutions of the ODE. The condition for a symmetry to be a variational symmetry is expressed by augmenting the linearization of the ODE through extra determining equations. Integrating factors are solutions of the resulting augmented system of determining equations.

For an ODE with no variational principle, we show that integrating factors are related to *adjoint-symmetries* defined as solutions of the adjoint equation of the linearization (Frèchet derivative) of the ODE, holding for all solutions of the ODE. In particular, there are necessary and sufficient extra determining equations for an adjoint-symmetry to be an integrating factor. This generalizes the equivalence between first integrals and variational symmetries in the case of an ODE with a variational principle, to an equivalence between first integrals and adjoint-symmetries that satisfy extra *adjoint invariance conditions* in the case of an ODE with no variational principle.

As a consequence, adjoint-symmetries play a central role in the study of first integrals of ODEs. Most important, an obvious extension of the calculational algorithm for solving the symmetry-determining equation can be used to solve the determining equation for adjoint-symmetries and the augmented system of determining equations for integrating factors.

Integrating factors provide another method for constructively reducing the order of an ODE through finding a first integral. This reduction of order method is complementary to, and independent of, Lie's reduction method for second- and higher-order ODEs. In particular, the integrating factor method is just as algorithmic and no more computationally complex than Lie's algorithm. Moreover, with the integrating factor approach one obtains a reduction of order in terms of the given variables in the original ODE, unlike reduction through point symmetries where the reduced ODE involves derived independent and dependent variables (and usually remains of the same order as the given ODE if expressed in the original variables).

If a system of PDEs is invariant under a Lie group of point transformations, one can find, constructively, special solutions, called *similarity solutions* or *invariant solutions,* that are invariant under a subgroup of the full group admitted by the system. These solutions result from solving a reduced system of differential equations with fewer independent variables. This application of Lie groups was discovered by Lie but first came to prominence in the late 1950s through the work of the Soviet group at Novosibirsk, led by Ovsiannikov (1962, 1982). Invariant solutions can also be constructed for specific boundary value problems. Here one seeks a subgroup of the full group of a given PDE that leaves invariant the boundary curves and the conditions imposed on them [Bluman and Cole (1974)]. Such solutions include *self-similar* (*automodel*) *solutions* that can be obtained through *dimensional analysis* or, more generally, from invariance under groups of scalings. Connections between invariant solutions and separation of variables have been studied extensively by Miller (1977) and coworkers. For ODEs, invariant solutions have particularly nice geometrical properties and include separatrices and envelope solutions [Bluman (1990c); Dresner (1999)].

1

Dimensional Analysis, Modeling, and Invariance

1.1 INTRODUCTION

In this chapter, we introduce the ideas behind invariance concretely through a thorough treatment of dimensional analysis. We show how dimensional analysis is connected to modeling and the construction of solutions obtained through invariance for boundary value problems for PDEs.

Often, for a quantity of interest, one knows at most the independent quantities it depends upon, say n in total, and the dimensions of all these $n + 1$ quantities. The application of dimensional analysis usually reduces the number of essential independent quantities. This is the starting point of modeling where the objective is to reduce significantly the number of necessary experimental measurements. In the following sections we will show that dimensional analysis can lead to a reduction in the number of independent variables appearing in a boundary value problem for a PDE. Most important, we show that for PDEs the reduction of the number of independent variables through dimensional analysis is a special case of reduction from invariance under groups of scaling (stretching) transformations.

1.2 DIMENSIONAL ANALYSIS: BUCKINGHAM PI-THEOREM

The basic theorem of dimensional analysis is the so-called *Buckingham Pi-theorem,* attributed to the American engineering scientist Buckingham (1914, 1915a,b). General references on the subject include those of Bridgman (1931), Barenblatt (1979, 1987, 1996), Sedov (1982), and Bluman (1983a). An historical perspective is given by Görtler (1975). For a detailed mathematical perspective, see Curtis, Logan, and Parker (1982).

The following assumptions and conclusions of dimensional analysis constitute the Buckingham Pi-theorem.

1.2.1 ASSUMPTIONS BEHIND DIMENSIONAL ANALYSIS

Essentially, no real problem violates the following assumptions:

(i) A quantity u is to be determined in terms of n *measurable quantities* (variables and parameters) $W_1, W_2, \ldots, W_n$:

$$u = f(W_1, W_2, \ldots, W_n), \tag{1.1}$$

where f is some function of $W_1, W_2, \ldots, W_n$.

(ii) The quantities $u, W_1, W_2, \ldots, W_n$ are measured in terms of *m fundamental dimensions* labeled by $L_1, L_2, \ldots, L_m$. For example, in a mechanical problem these are usually the mechanical fundamental dimensions, L_1 = length, L_2 = mass, and L_3 = time.

(iii) Let Z represent any of $u, W_1, W_2, \ldots, W_n$. Then the *dimension of* Z, denoted by $[Z]$, is a product of powers of the fundamental dimensions, in particular,

$$[Z] = L_1^{\alpha_1} L_2^{\alpha_2} \cdots L_m^{\alpha_m}, \tag{1.2}$$

for some real numbers $\alpha_1, \alpha_2, \ldots, \alpha_m$, usually rational, which are called the dimension exponents of Z. The *dimension vector* of Z is the column vector

$$\boldsymbol{\alpha} = \begin{bmatrix} \alpha_1 \\ \alpha_2 \\ \vdots \\ \alpha_m \end{bmatrix}. \tag{1.3}$$

A quantity Z is said to be *dimensionless* if and only if $[Z] = 1$, i.e., if and only if all of its dimension exponents are zero. For example, in terms of the mechanical fundamental dimensions, the dimension vector of the energy E is given by

$$\boldsymbol{\alpha}(E) = \begin{bmatrix} 2 \\ 1 \\ -2 \end{bmatrix}.$$

Let

$$\mathbf{b}_i = \begin{bmatrix} b_{1i} \\ b_{2i} \\ \vdots \\ b_{mi} \end{bmatrix} \tag{1.4}$$

be the dimension vector of W_i, $i = 1,2,\ldots,n$, and let

$$\mathrm{B} = \begin{bmatrix} b_{11} & b_{12} & \cdots & b_{1n} \\ b_{21} & b_{22} & \cdots & b_{2n} \\ \vdots & \vdots & \vdots & \vdots \\ b_{m1} & b_{m2} & \cdots & b_{mn} \end{bmatrix} \tag{1.5}$$

be the $m \times n$ dimension matrix of the given problem.

(iv) For any set of fundamental dimensions, one can choose a *system of units* for measuring the value of any quantity Z. A change from one system of units to another involves a positive *scaling* of each fundamental dimension that in turn induces a scaling of each quantity Z. For example, for the mechanical fundamental dimensions, the common systems of units are mks (meter-kilogram-second), cgs (centimeter-gram-second), or British foot-pounds. In changing from cgs to mks units, L_1 is scaled by 10^{-2}, L_2 is scaled by 10^{-3}, L_3 is unchanged, and hence the value of the energy E is scaled by 10^{-7}. Under a change of system of units, the value of a dimensionless quantity is unchanged, i.e., its value is *invariant* under an arbitrary scaling of any fundamental dimension. Hence, it is meaningful to deem dimensionless quantities as large or small. The last assumption of dimensional analysis is that formula (1.1) acts as a dimensionless equation in the sense that (1.1) is invariant under an arbitrary scaling of any fundamental dimension, i.e., (1.1) is independent of the choice of system of units.

1.2.2 CONCLUSIONS FROM DIMENSIONAL ANALYSIS

The assumptions of the Buckingham Pi-theorem stated in Section 1.2.1 lead to the following conclusions:

(i) Formula (1.1) can be expressed in terms of dimensionless quantities.

(ii) The number of dimensionless quantities is $k + 1 = n + 1 - r(\mathrm{B})$, where $r(\mathrm{B})$ is the rank of matrix B. Precisely k of these dimensionless quantities depend on the measurable quantities $W_1, W_2, \ldots, W_n$.

(iii) Let

$$\mathbf{x}^{(i)} = \begin{bmatrix} x_{1i} \\ x_{2i} \\ \vdots \\ x_{ni} \end{bmatrix}, \quad i = 1,2,\ldots,k, \tag{1.6}$$

represent the $k = n - r(\mathrm{B})$ linearly independent solutions $\mathbf{x}$ of the system

$$\mathrm{B}\mathbf{x} = 0. \tag{1.7}$$

Let

$$\mathbf{a} = \begin{bmatrix} a_1 \\ a_2 \\ \vdots \\ a_m \end{bmatrix} \tag{1.8}$$

be the dimension vector of u, and let

$$\mathbf{y} = \begin{bmatrix} y_1 \\ y_2 \\ \vdots \\ y_n \end{bmatrix} \tag{1.9}$$

represent a solution of the system

$$\mathbf{B}\mathbf{y} = -\mathbf{a}. \tag{1.10}$$

Then formula (1.1) simplifies to

$$\pi = g(\pi_1, \pi_2, \ldots, \pi_k), \tag{1.11}$$

where π, π_i are dimensionless quantities given by

$$\pi = uW_1^{y_1}W_2^{y_2}\cdots W_n^{y_n}, \tag{1.12a}$$

$$\pi_i = W_1^{x_{1i}}W_2^{x_{2i}}\cdots W_n^{x_{ni}}, \quad i = 1,2,\ldots,k, \tag{1.12b}$$

and g is some function of its arguments. In particular, (1.1) becomes

$$u = W_1^{-y_1}W_2^{-y_2}\cdots W_n^{-y_n}g(\pi_1, \pi_2, \ldots, \pi_k). \tag{1.13}$$

[In terms of experimental modeling, formula (1.13) is "cheaper" than formula (1.1) by $r(\mathrm{B})$ orders of magnitude.]

1.2.3 PROOF OF THE BUCKINGHAM PI-THEOREM

First of all,

$$[u] = L_1^{a_1}L_2^{a_2}\cdots L_m^{a_m}, \tag{1.14a}$$

$$[W_i] = L_1^{b_{1i}}L_2^{b_{2i}}\cdots L_m^{b_{mi}}, \quad i = 1,2,\ldots,n. \tag{1.14b}$$

Next, we use assumption (iv), and consider the invariance of (1.1) under arbitrary scalings of the fundamental dimensions by taking each fundamental dimension in turn. We first scale L_1 by letting

$$L^*_1 = e^{\varepsilon}L_1, \quad \varepsilon \in \mathbf{R}. \tag{1.15}$$

In turn, this induces the following scalings of the measurable quantities:

$$u^* = e^{\varepsilon a_1}u, \tag{1.16a}$$

$$W^*_i = e^{\varepsilon b_{1i}}W_i, \quad i = 1,2,\ldots,n. \tag{1.16b}$$

Equations (1.16a,b) happen to define a one-parameter (ε) Lie group of scaling transformations of the $n + 1$ quantities $u, W_1, W_2, \ldots, W_n$, with $\varepsilon = 0$ corresponding to the identity transformation. This group is induced by the one-parameter group of scalings (1.15) of the fundamental dimension L_1. [It is not necessary to be familiar with Lie groups to read the rest of this chapter.]

From assumption (iv), formula (1.1) holds if and only if

$$u^* = f(W^*_1, W^*_2, \ldots, W^*_n),$$

i.e.,

$$e^{\varepsilon a_1}u = f(e^{\varepsilon b_{11}}W_1, e^{\varepsilon b_{12}}W_2, \ldots, e^{\varepsilon b_{1n}}W_n) \quad \text{for all } \varepsilon \in \mathbf{R}. \tag{1.17}$$

Then two cases need to be distinguished:

Case I. $b_{11} = b_{12} = \cdots = b_{1n} = a_1 = 0$.
Here, L_1 is not a fundamental dimension of the problem, or in other words, formula (1.1) is *dimensionless with respect to* L_1.

Case II. $b_{11} = b_{12} = \cdots = b_{1n} = 0$, $a_1 \neq 0$.
Here, it follows that $u \equiv 0$, a trivial situation.

Hence, it follows that $b_{1i} \neq 0$ for some $i = 1,2,\ldots,n$. Without loss of generality, we assume $b_{11} \neq 0$. We define new measurable quantities

$$X_{i-1} = W_i W_1^{-b_{1i}/b_{11}}, \quad i = 2,3,\ldots,n, \tag{1.18}$$

and let

$$X_n = W_1. \tag{1.19}$$

We choose as the new unknown

$$v = uW_1^{-a_1/b_{11}}. \tag{1.20}$$

The transformation given by (1.18)–(1.20) defines a one-to-one mapping of the quantities $W_1, W_2, \ldots, W_n$ to the quantities $X_1, X_2, \ldots, X_n$, and a one-to-one mapping of the quantities $u, W_1, W_2, \ldots, W_n$ to the quantities $v, X_1, X_2, \ldots, X_n$. Consequently, formula (1.1) is equivalent to

$$v = F(X_1, X_2, \ldots, X_n), \tag{1.21}$$

where F is some function of $X_1, X_2, \ldots, X_n$. Thus, the group of transformations (1.16a,b) becomes

$$v^* = v, \tag{1.22a}$$

$$X^*_i = X_i, \quad i = 1,2,\ldots,n-1, \tag{1.22b}$$

$$X^*_n = e^{\varepsilon b_{11}}X_n, \tag{1.22c}$$

so that $v, X_1, X_2, \ldots, X_{n-1}$ are *invariants* of (1.16a,b). Moreover, the quantities $v, X_1, X_2, \ldots, X_n$ satisfy assumption (iii), and formula (1.21) satisfies assumption (iv). Hence,

$$v = F(X_1, X_2, \ldots, X_{n-1}, e^{\varepsilon b_{11}}X_n), \tag{1.23}$$

for all $\varepsilon \in \mathbf{R}$. Consequently, F is independent of the quantity X_n. Moreover, the measurable quantities $X_1, X_2, \ldots, X_{n-1}$ are products of powers of $W_1, W_2, \ldots, W_n$, and v is a product of u and powers of $W_1, W_2, \ldots, W_n$. Thus, formula (1.1) reduces to

$$v = G(X_1, X_2, \dots, X_{n-1}), \tag{1.24}$$

where $v, X_1, X_2, \dots, X_{n-1}$ are dimensionless with respect to L_1 and G is some function of its $n-1$ arguments.

Continuing in turn with the other $m-1$ fundamental dimensions, we reduce formula (1.1) to a dimensionless formula

$$\pi = g(\pi_1, \pi_2, \dots, \pi_k), \tag{1.25}$$

where $[\pi] = [\pi_i] = 1$, and g is some function of $\pi_1, \pi_2, \dots, \pi_k$:

$$\pi = uW_1^{y_1} W_2^{y_2} \cdots W_n^{y_n}, \tag{1.26a}$$

and

$$\pi_i = W_1^{x_{1i}} W_2^{x_{2i}} \cdots W_n^{x_{ni}}, \quad i = 1, 2, \dots, k, \tag{1.26b}$$

for some real numbers y_j, x_{ji}, $i = 1,2,\dots,k$; $j = 1,2,\dots,n$.

Next, we show that the number of measurable dimensionless quantities is $k = n - r(\mathrm{B})$. This follows immediately since

$$[W_1^{x_1} W_1^{x_2} \cdots W_n^{x_n}] = 1$$

if and only if

$$\mathbf{x} = \begin{bmatrix} x_1 \\ x_2 \\ \vdots \\ x_n \end{bmatrix}$$

satisfies (1.7). Equation (1.7) has $k = n - r(\mathrm{B})$ linearly independent solutions $\mathbf{x}^{(i)}$ given by (1.6). The real numbers

$$\mathbf{y} = \begin{bmatrix} y_1 \\ y_2 \\ \vdots \\ y_n \end{bmatrix}$$

follow from setting

$$[uW_1^{y_1} W_1^{y_2} \cdots W_n^{y_n}] = 1,$$

leading to $\mathbf{y}$ satisfying (1.10).

Note that the proof of the Buckingham Pi-theorem makes no assumption about the continuity of the function f, and hence of g, with respect to any of their arguments.

1.2.4 EXAMPLES

(1) *The Atomic Explosion of 1945*
Sir Geoffrey Taylor (1950) deduced the approximate energy released by the first atomic explosion that took place in New Mexico in 1945 from the motion picture records of J.E. Mack that were declassified in 1947. But the amount of energy released by the blast was still classified in 1947! [Taylor carried out the analysis necessary for his deduction in 1941.] A dimensional analysis argument of Taylor's deduction follows.

An atomic explosion is approximated by the release of a large amount of energy E from a "point." A consequence is an expanding spherical fireball whose edge corresponds to a powerful shock wave. Let $u = R$ be the radius of the shock wave. We treat R as the unknown and assume that

$$R = f(W_1, W_2, W_3, W_4), \tag{1.27}$$

where

$W_1 = E$ is the energy released by the explosion,
$W_2 = t$ is the elapsed time after the explosion takes place,
$W_3 = \rho_0$ is the initial or ambient air density,
$W_4 = P_0$ is the initial or ambient air pressure.

For this problem, we use the mechanical fundamental dimensions. The corresponding dimension matrix is given by

$$\mathrm{B} = \begin{bmatrix} 2 & 0 & -3 & -1 \\ 1 & 0 & 1 & 1 \\ -2 & 1 & 0 & -2 \end{bmatrix}. \tag{1.28}$$

Clearly, $r(\mathrm{B}) = 3$, and hence, $k = n - r(\mathrm{B}) = 4 - 3 = 1$. The general solution of $\mathrm{B}\mathbf{x} = 0$ is $x_1 = -\frac{2}{5}x_4$, $x_2 = \frac{6}{5}x_4$, $x_3 = -\frac{3}{5}x_4$, where x_4 is arbitrary. Setting $x_4 = 1$, we get the measurable dimensionless quantity

$$\pi_1 = P_0\left[\frac{t^6}{E^2(\rho_0)^3}\right]^{1/5}. \tag{1.29}$$

The dimension vector of R is

$$\mathbf{a} = \begin{bmatrix} 1 \\ 0 \\ 0 \end{bmatrix}. \tag{1.30}$$

The general solution of $\mathrm{B}\mathbf{y} = -\,\mathbf{a}$ is

$$\mathbf{y} = \frac{1}{5}\begin{bmatrix} -1 \\ -2 \\ 1 \\ 0 \end{bmatrix} + \mathbf{x}, \tag{1.31}$$

where $\mathbf{x}$ is the general solution of $\mathrm{B}\mathbf{x} = 0$. Setting $\mathbf{x} = 0$ in (1.31), we obtain the dimensionless unknown

$$\pi = R\left[\frac{Et^2}{\rho_0}\right]^{-1/5}. \tag{1.32}$$

Thus, from dimensional analysis, we get

$$R = \left[\frac{Et^2}{\rho_0}\right]^{1/5} g(\pi_1), \tag{1.33}$$

where g is some function of π_1.

Equivalently,

$$R = \left[\frac{Et^2}{\rho_0}\right]^{1/5} (P_0)^q \left[\frac{t^6}{E^2(\rho_0)^3}\right]^{q/5} h_q(\pi_1),$$

for some function $h_q(\pi_1) = g(\pi_1)\pi_1^{-q}$. Now we assume that for some $q = Q$, $h_Q(0) \neq 0$ and that $h_Q(\pi_1)$ is continuous at $\pi_1 = 0$, i.e., we are essentially assuming that $R \propto t^{(2+6q)/5}$ near $t = 0$ for some $q = Q$. Then

$$R = E^{(1-2Q)/5}(\rho_0)^{-(1+3Q)/5}(P_0)^Q t^{(2+6Q)/5} h_Q(\pi_1),$$

and

$$\log R = \frac{2+6Q}{5}\log t + C,$$

for some constant

$$C = \frac{1-2Q}{5}\log E - \frac{1+3Q}{5}\log \rho_0 + Q\log P_0 + \log h_Q(0),$$

near $\pi_1 = 0$. Plotting $\log R$ versus $\log t$ for a light explosives experiment, one can determine that $Q \cong 0$ and that $g(0) = h_0(0) \cong 1$. This leads to Taylor's approximation formula

$$R = At^{2/5}, \tag{1.34}$$

where

$$A = \left(\frac{E}{\rho_0}\right)^{1/5} g(0). \tag{1.35}$$

Using Mack's motion picture for the first atomic explosion, Taylor plotted $\frac{5}{2}\log_{10}R$ versus $\log_{10} t$ with R and t measured in terms of cgs units. [See Figure 1.1 where the motion picture data is indicated by +.] This led to an accurate estimation of the classified energy E of the explosion!

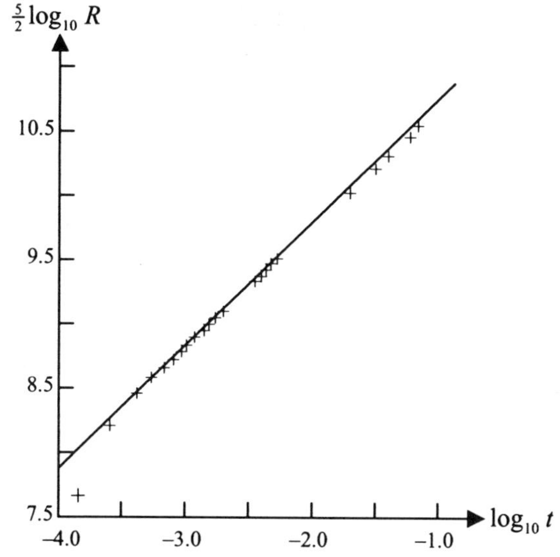

Figure 1.1

(2) *An Example in Heat Conduction Illustrating the Choice of Fundamental Dimensions* Consider the standard problem of one-dimensional heat conduction in an "infinite" bar with constant thermal properties, initially heated by a point source of heat. Let u be the temperature at any point of the bar. We assume that

$$u = f(W_1, W_2, W_3, W_4, W_5, W_6), \tag{1.36}$$

where

$W_1 = x$ is the distance along the bar from the point source of heat,

$W_2 = t$ is the elapsed time after the initial heating,

$W_3 = \rho$ is the mass density of the bar,

$W_4 = c$ is the specific heat of the bar,

$W_5 = K$ is the thermal conductivity of the bar,

$W_6 = Q$ is the strength of the heat source measured in energy units per (length units)^2.

It is interesting to consider the effects of dimensional analysis in simplifying (1.36) with two different choices of fundamental dimensions.

Choice I (Dynamical Units). Here, we let L_1 = length, L_2 = mass, L_3 = time, and L_4 = temperature. Correspondingly, the dimension matrix is given by

$$\mathrm{B_I} = \begin{bmatrix} 1 & 0 & -3 & 2 & 1 & 0 \\ 0 & 0 & 1 & 0 & 1 & 1 \\ 0 & 1 & 0 & -2 & -3 & -2 \\ 0 & 0 & 0 & -1 & -1 & 0 \end{bmatrix}. \tag{1.37}$$

Here, $r(\mathrm{B_I}) = 4$, and hence, the number of measurable dimensionless quantities is $k = 6 - 4 = 2$. One can choose two linearly independent solutions $\mathbf{x}^{(1)}$ and $\mathbf{x}^{(2)}$ of $\mathrm{B_I}\mathbf{x} = 0$ such that π_1 is linear in x and independent of t, and π_2 is linear in t and independent of x. Then

$$\pi_1 = \xi = \frac{\rho c^2 Q}{K^2} x, \tag{1.38a}$$

$$\pi_2 = \tau = \frac{\rho c^3 Q^2}{K^3} t. \tag{1.38b}$$

For the dimensionless quantity π, it is convenient to choose a solution of

$$\mathrm{B_I}\mathbf{y} = -\mathbf{a} = \begin{bmatrix} 0 \\ 0 \\ 0 \\ -1 \end{bmatrix},$$

where $y_1 = y_2 = 0$, so that π is independent of x and t. Consequently,

$$\pi = \frac{K^2}{Q^2 c} u. \tag{1.39}$$

Hence, dimensional analysis with dynamical units reduces (1.36) to

$$u = \frac{Q^2 c}{K^2} F(\xi, \tau), \tag{1.40}$$

where F is some function of ξ and τ.

Choice II (Thermal Units). Motivated by the implicit assumption that in the posed problem there is no conversion of heat energy to mechanical energy, we refine the dynamical units by introducing a thermal unit L_5 = "calories." The corresponding dimension matrix is given by

$$B_{II} = \begin{bmatrix} 1 & 0 & -3 & 0 & -1 & -2 \\ 0 & 0 & 1 & -1 & 0 & 0 \\ 0 & 1 & 0 & 0 & -1 & 0 \\ 0 & 0 & 0 & -1 & -1 & 0 \\ 0 & 0 & 0 & 1 & 1 & 1 \end{bmatrix}. \tag{1.41}$$

Here, $r(B_{II}) = 5$, and hence, there is only one measurable dimensionless quantity. For dimensionless quantities, it is convenient to choose

$$\pi_1 = \eta = \frac{x}{\sqrt{\kappa t}} \quad \text{where } \kappa = \frac{K}{\rho c}, \tag{1.42a}$$

and

$$\pi = \frac{\sqrt{\rho c K t}}{Q} u. \tag{1.42b}$$

Thus, dimensional analysis with thermal units reduces (1.36) to

$$u = \frac{Q}{\sqrt{\rho c K t}} G(\eta), \tag{1.43}$$

where G is some function of η.

Note that (1.43) is a special case of (1.40) with

$$\eta = \frac{\xi}{\sqrt{\tau}} \quad \text{and} \quad F(\xi, \tau) = \frac{1}{\sqrt{\tau}} G\left(\frac{\xi}{\sqrt{\tau}}\right).$$

[In terms of thermal units, each of the quantities ξ, τ, $K^2 u / Q^2 c$ is not dimensionless.]

Obviously, if it is correct, expression (1.43) is a significant simplification of (1.40). By conducting experiments or associating a properly posed boundary value problem to determine u, one can show that thermal units are justified. In turn, thermal units can then be used for other heat (diffusion) problems where the governing equations are not completely known.

EXERCISES 1.2

1. Use dimensional analysis to prove the Pythagoras theorem. [*Hint*: Drop a perpendicular to the hypotenuse of a right-angle triangle and consider the areas of the resulting three similar triangles.]

2. How would you use dimensional analysis and experimental modeling to find the time of flight T of a body dropped vertically from a height h?
   (a) *Model* I: Assume that T depends on h, the mass m of the body, the acceleration g due to gravity, and the shape s of the body.

(b) *Model* II: Now take into account a resistance force proportional to the velocity v of the body as it falls. Let k be the constant of proportionality. How does the extra dimensionless quantity depend on h and m? How important is the constant k as the values of h and m change?

3. Given that in cgs units $\rho_0 = 1.3\times10^{-3}$, and $P_0 = 1.0\times10^{-6}$, use the data from Figure 1.1 to estimate the domain of π_1 and E.

4. *Cooking a turkey.* Assume that a turkey is composed of a uniform material with specific heat c, mass density ρ, thermal conductivity K, and weight m. Assume that the cooking temperature is T. Let t be the time to cook the turkey.
   (a) Choose, as fundamental dimensions: length, mass, time, and temperature. Use dimensional analysis to find t in terms of c, ρ, K, m, T, and the shape of the turkey.
   (b) Repeat as for (a) and determine t with heat as an added fifth fundamental dimension. How can one justify introducing this fifth fundamental dimension? Is this extra fundamental dimension helpful?
   (c) Interpret your answer for t in (b) in terms of the surface area of the turkey.
   (d) You should find that t is proportional to $m^{2/3}$.
   (e) Suppose one assumes that t is proportional to m^p for some constant p. Use cookbook data to determine p. How good is the crude "dimensional analysis" estimate of $p = 2/3$?
   (f) How would stuffing affect the answer?

1.3 APPLICATION OF DIMENSIONAL ANALYSIS TO PDEs

Consider the use of dimensional analysis where the quantities $u, W_1, W_2, \ldots, W_n$ arise in a boundary value problem for a PDE which has a unique solution. Then the unknown u (the *dependent variable* of the PDE) is the solution of the boundary value problem, and $W_1, W_2, \ldots, W_n$ denote all *independent variables* and *constants* appearing in the boundary value problem. From the Buckingham Pi-theorem it follows that such a boundary value problem can always be re-expressed in dimensionless form where π is a dimensionless dependent variable and $\pi_1, \pi_2, \ldots, \pi_k$ are dimensionless independent variables and dimensionless constants.

Suppose $W_1, W_2, \ldots, W_\ell$ are the ℓ independent variables and $W_{\ell+1}, W_{\ell+2}, \ldots, W_n$ are the $n-\ell$ constants appearing in the boundary value problem. Let

$$\mathrm{B}_1 = \begin{bmatrix} b_{11} & b_{12} & \cdots & b_{1\ell} \\ b_{21} & b_{22} & \cdots & b_{2\ell} \\ \vdots & \vdots & \vdots & \vdots \\ b_{m1} & b_{m2} & \cdots & b_{m\ell} \end{bmatrix} \tag{1.44a}$$

be the dimension matrix of the independent variables, and let

$$\mathrm{B}_2 = \begin{bmatrix} b_{1,\ell+1} & b_{1,\ell+2} & \cdots & b_{1n} \\ b_{2,\ell+1} & b_{2,\ell+2} & \cdots & b_{2n} \\ \vdots & \vdots & \vdots & \vdots \\ b_{m,\ell+1} & b_{m,\ell+2} & \cdots & b_{mn} \end{bmatrix} \tag{1.44b}$$

be the dimension matrix of the constants. The dimension matrix of the boundary value problem is given by

$$\mathrm{B} = [\mathrm{B}_1 \mid \mathrm{B}_2]. \tag{1.45}$$

A dimensionless π_i quantity is called a *dimensionless constant* if and only if it does not depend on the variables $W_1, W_2, \ldots, W_\ell$, i.e., in (1.26b), $x_{ji} = 0$, $j = 1,2,\ldots,\ell$. A dimensionless π_i quantity is a *dimensionless variable* if $x_{ji} \neq 0$ for some $j = 1,2,\ldots,\ell$. An important objective in applying dimensional analysis to a boundary value problem is to reduce the number of independent variables. The rank of B_2, i.e., $r(\mathrm{B}_2)$, represents the reduction in the number of constants through dimensional analysis. Consequently, the reduction in the number of independent variables is $\rho = r(\mathrm{B}) - r(\mathrm{B}_2)$. In particular, the number of dimensionless measurable quantities is $k = n - r(\mathrm{B}) = [\ell - \rho] + [(n - \ell - r(\mathrm{B}_2)]$, where $\ell - \rho$ of the quantities are dimensionless independent variables and $n - \ell - r(\mathrm{B}_2)$ are dimensionless constants.

If $r(\mathrm{B}) = r(\mathrm{B}_2)$, then dimensional analysis reduces the given boundary value problem to a dimensionless boundary value problem with $(n - \ell) - r(\mathrm{B}_2)$ dimensionless constants. In this case the number of independent variables is not reduced. Nonetheless, this is useful as a starting point for perturbation analysis if any dimensionless constant is small.

If $\ell \geq 2$, $\ell - \rho = 1$, then the resulting solution of the boundary value problem is called a *self-similar* or *automodel solution.*

1.3.1 EXAMPLES

(1) *Source Problem for Heat Conduction*

Consider the unknown temperature u of the heat conduction problem of Section 1.2.4 as the solution $u(x, t)$ of the boundary value problem

$$\rho c \frac{\partial u}{\partial t} - K \frac{\partial^2 u}{\partial x^2} = 0, \quad -\infty < x < \infty, \ t > 0, \tag{1.46a}$$

$$u(x,0) = \frac{Q}{\rho c} \delta(x), \tag{1.46b}$$

$$\lim_{x \to \pm\infty} u(x,t) = 0. \tag{1.46c}$$

In (1.46b), $\delta(x)$ is the Dirac delta function.

The use of dimensional analysis with dynamical units reduces (1.46a–c) to

$$\frac{\partial F}{\partial \tau} - \frac{\partial^2 F}{\partial \xi^2} = 0, \quad -\infty < \xi < \infty, \ t > 0, \tag{1.47a}$$

$$F(\xi, 0) = \delta(\xi), \tag{1.47b}$$

$$\lim_{\xi \to \pm\infty} F(\xi, \tau) = 0, \tag{1.47c}$$

with u defined in terms of $F(\xi, \tau)$ by (1.40) and ξ, τ given by (1.38a,b). Consequently, there is no essential progress in solving the boundary value problem (1.46a–c).

We now justify the use of dimensional analysis with thermal units to solve (1.46a–c) as follows: First, note that from (1.47a,c) we have

$$\frac{\partial}{\partial \tau} \int_{-\infty}^{\infty} F(\xi, \tau)\, d\xi = \int_{-\infty}^{\infty} \frac{\partial^2 F}{\partial \xi^2}(\xi, \tau)\, d\xi = 0.$$

Then, from this equation and (1.47b), we get the conservation law

$$\int_{-\infty}^{\infty} F(\xi, \tau)\, d\xi = 1 \quad \text{valid for all } \tau > 0.$$

Consequently, the substitution $F(\xi, \tau) = (1/\sqrt{\tau})G(\xi/\sqrt{\tau})$, which results from using dimensional analysis with thermal units [cf. Section 1.2.4], reduces (1.47a–c), and hence (1.46a–c), to a boundary value problem for an ODE with independent variable $\eta = \xi/\sqrt{\tau}$ and dependent variable $G(\eta)$:

$$2\frac{d^2 G}{d\eta^2} + \eta \frac{dG}{d\eta} + G = 0, \quad -\infty < \eta < \infty, \tag{1.48a}$$

$$\int_{-\infty}^{\infty} G(\eta)\, d\eta = 1, \tag{1.48b}$$

$$G(\pm\infty) = 0. \tag{1.48c}$$

This reduction of (1.46a–c) to a boundary value problem for an ODE is obtained much more naturally and easily in Section 1.4 from the invariance of (1.46a–c) under a one-parameter group of scalings of its variables.

(2) *Prandtl–Blasius Problem for a Flat Plate*
Consider the Prandtl boundary layer equations for flow past a semi-infinite flat plate:

$$u\frac{\partial u}{\partial x} + v\frac{\partial u}{\partial y} = \kappa \frac{\partial^2 u}{\partial y^2}, \tag{1.49a}$$

$$\frac{\partial u}{\partial x} + \frac{\partial v}{\partial y} = 0, \tag{1.49b}$$

$0 < x < \infty$, $0 < y < \infty$, with boundary conditions

$$u(x, 0) = 0, \tag{1.49c}$$

$$v(x,0) = 0, \tag{1.49d}$$

$$u(x,\infty) = U, \tag{1.49e}$$

$$u(0,y) = U. \tag{1.49f}$$

In the boundary value problem (1.49a–f), x is the distance along the plate surface from its edge (tangential coordinate), y is the distance from the plate surface (normal coordinate), u is the x-component of velocity, v is the y-component of velocity, κ is the kinematic viscosity, and U is the velocity of the incident flow [Figure 1.2].

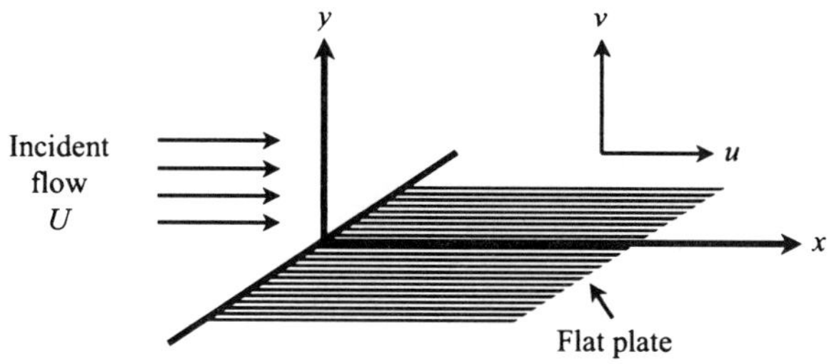

Figure 1.2

Our aim is to calculate the shear at the plate (skin friction), $u_y(x, 0)$, which in turn leads to the determination of the viscous drag on the plate.

We look at the problem of determining $u_y(x, 0)$ as defined through the boundary value problem (1.49a–f) from three analytical perspectives:

(i) *Dimensional Analysis.* From (1.49a–f), it follows that

$$\frac{\partial u}{\partial y}(x, 0) = f(x, U, \kappa), \tag{1.50}$$

with the unknown function f to be determined in terms of measurable quantities x, U, κ. The fundamental dimensions are L = length and T = time. Then, with respect to these fundamental dimensions, one has

$$\left[\frac{\partial u}{\partial y}(x, 0)\right] = T^{-1}, \tag{1.51a}$$

$$[x] = L, \tag{1.51b}$$

$$[U] = LT^{-1}, \tag{1.51c}$$

$$[\kappa] = L^2T^{-1}. \tag{1.51d}$$

Consequently, $r(\mathrm{B}) = 2$. Dimensionless quantities are given by

$$\pi_1 = \frac{Ux}{\kappa}, \tag{1.52a}$$

and

$$\pi = \frac{\kappa}{U^2}\frac{\partial u}{\partial y}(x, 0). \tag{1.52b}$$

Hence, dimensional analysis leads to

$$\frac{\partial u}{\partial y}(x, 0) = \frac{U^2}{\kappa} g\left(\frac{Ux}{\kappa}\right), \tag{1.53}$$

where g is an unknown function of Ux/κ.

(ii) *Scalinqs of Quantities Followed by Dimensional Analysis.* Consider a linear transformation of the variables of the boundary value problem (1.49a–f) given by $x = aX$, $y = bY$, $u = UQ$, $v = cR$, where a, b, c are undetermined positive constants, U is the velocity of the incident flow, and X, Y, Q, R represent new (dimensional) independent and dependent variables: $Q = Q(X, Y)$, $R = R(X, Y)$;

$$u(x, y) = UQ(X, Y) = UQ\left(\frac{x}{a}, \frac{y}{b}\right), \quad v(x, y) = cR(X, Y) = cR\left(\frac{x}{a}, \frac{y}{b}\right).$$

Consequently,

$$\frac{\partial u}{\partial y}(x, 0) = \frac{U}{b}\frac{\partial Q}{\partial Y}(X, 0), \tag{1.54}$$

and the boundary value problem (1.49a–f) transforms to

$$\frac{U}{a}Q\frac{\partial Q}{\partial X} + \frac{c}{b}R\frac{\partial Q}{\partial Y} = \frac{\kappa}{b^2}\frac{\partial^2 Q}{\partial Y^2}, \tag{1.55a}$$

$$\frac{U}{a}\frac{\partial Q}{\partial X} + \frac{c}{b}\frac{\partial R}{\partial Y} = 0, \tag{1.55b}$$

$0 < X < \infty$, $0 < Y < \infty$, with

$$Q(X, 0) = 0, \tag{1.55c}$$

$$R(X, 0) = 0, \tag{1.55d}$$

$$Q(X, \infty) = 1, \tag{1.55e}$$

$$Q(0, Y) = 1. \tag{1.55f}$$

From the form of (1.55a,b), it is convenient to choose a, b, c so that

$$\frac{U}{a} = \frac{c}{b} = \frac{\kappa}{b^2}.$$

Hence, we set $c = 1, b = \kappa, a = U\kappa$. As a result, (1.55a,b) become cleared of constants:

$$Q\frac{\partial Q}{\partial X} + R\frac{\partial Q}{\partial Y} = \frac{\partial^2 Q}{\partial Y^2}, \tag{1.56a}$$

$$\frac{\partial Q}{\partial X} + \frac{\partial R}{\partial Y} = 0, \tag{1.56b}$$

$0 < X < \infty, 0 < Y < \infty$. Moreover,

$$\frac{\partial u}{\partial y}(x, 0) = \frac{U}{\kappa}\frac{\partial Q}{\partial Y}(X, 0) = \frac{U}{\kappa}\frac{\partial Q}{\partial Y}\left(\frac{x}{U\kappa}, 0\right). \tag{1.57}$$

Since $Q(X, Y)$ results from the solution of (1.56a,b), (1.55c–f), we have

$$\frac{\partial Q}{\partial Y}(X, 0) = h(X), \tag{1.58}$$

for some function $h(X)$. To determine $h(X)$, we apply dimensional analysis to (1.58):

$$\left[\frac{\partial Q}{\partial Y}\right] = LT^{-1}, \tag{1.59a}$$

$$[X] = L^{-2}T^{2}. \tag{1.59b}$$

Hence, it is easy to see that (1.58) reduces to

$$h(X) = \sigma X^{-1/2}, \tag{1.60}$$

for some fixed dimensionless constant σ to be determined. Thus, (1.53) simplifies further to $g(Ux/\kappa) = \sigma(Ux/\kappa)^{-1/2}$, so that

$$\frac{\partial u}{\partial y}(x, 0) = \sigma\left(\frac{U^3}{x\kappa}\right)^{1/2}. \tag{1.61}$$

(iii) *Further Use of Dimensional Analysis on the Full Boundary Value Problem.* We now apply dimensional analysis to the boundary value problem (1.56a,b), (1.55c-f), to reduce it to a boundary value problem for an ODE. It is convenient (but not necessary) to introduce a *potential* (*stream function*) $\psi(X, Y)$ from the form of (1.56b).

Let $Q = \partial\psi/\partial Y, R = -\partial\psi/\partial X$. Then in terms of the single dependent variable ψ, the boundary value problem (1.56a,b), (1.55c–f), becomes

$$\frac{\partial \psi}{\partial Y}\frac{\partial^2 \psi}{\partial X \partial Y} - \frac{\partial \psi}{\partial X}\frac{\partial^2 \psi}{\partial Y^2} = \frac{\partial^3 \psi}{\partial Y^3}, \tag{1.62a}$$

$0 < X < \infty, 0 < Y < \infty$, with

$$\frac{\partial \psi}{\partial Y}(X, 0) = 0, \tag{1.62b}$$

$$\frac{\partial\psi}{\partial X}(X,0)=0, \tag{1.62c}$$

$$\frac{\partial\psi}{\partial Y}(X,\infty)=1, \tag{1.62d}$$

$$\frac{\partial\psi}{\partial Y}(0,Y)=1. \tag{1.62e}$$

Moreover, from (1.58) and (1.60), we get

$$\frac{\partial Q}{\partial Y}(X,0)=\frac{\partial^2\psi}{\partial Y^2}(X,0)=\sigma X^{-1/2}. \tag{1.63}$$

We now use dimensional analysis to simplify $\psi(X, Y)$. Since the boundary value problem (1.62a–e) has no constants, we have

$$\psi = F(X, Y), \tag{1.64}$$

for some unknown function F. We see that

$$[\psi]=[Y]=L^{-1}T, \tag{1.65a}$$

$$[X]=L^{-2}T^2. \tag{1.65b}$$

Consequently, there is only one measurable dimensionless quantity. It is convenient to choose as dimensionless quantities

$$\pi_1=\eta=\frac{Y}{\sqrt{X}}, \tag{1.66a}$$

and

$$\pi=\frac{\psi}{\sqrt{X}}. \tag{1.66b}$$

Hence,

$$\psi(X,Y)=\sqrt{X}G(\eta), \tag{1.67}$$

where $G(\eta)$ solves a boundary value problem for an ODE that is obtained by substituting (1.67) into (1.62a–e). Moreover, from (1.67) and (1.63), it follows that

$$\sigma = G''(0). \tag{1.68}$$

[A prime denotes differentiation with respect to η.] Note that

$$\frac{\partial\psi}{\partial Y}=G'(\eta),\quad \frac{\partial\psi}{\partial X}=\tfrac{1}{2}X^{-1/2}[G-\eta G'],$$
$$\frac{\partial^2\psi}{\partial Y^2}=X^{-1/2}G'',\quad \frac{\partial^3\psi}{\partial Y^3}=X^{-1}G''',\quad \frac{\partial^2\psi}{\partial X\,\partial Y}=\tfrac{1}{2}X^{-1}[-\eta G''],$$

$0<X<\infty$, $0<Y<\infty$ leads to $0<\eta<\infty$; $Y=0$ leads to $\eta=0$; $Y\to\infty$ leads to $\eta\to\infty$;

and $X = 0$ leads to $\eta \to \infty$. Correspondingly, the boundary value problem (1.62a–e) reduces to solving a third-order ODE, known as the *Blasius equation*, for $G(\eta)$:

$$2\frac{d^3G}{d\eta^3} + G\frac{d^2G}{d\eta^2} = 0, \quad 0 < \eta < \infty, \tag{1.69a}$$

with boundary conditions

$$G(0) = G'(0) = 0, \quad G'(\infty) = 1. \tag{1.69b}$$

The aim is to find $\sigma = G''(0)$.

A numerical procedure for solving the boundary value problem (1.69a,b) is the shooting method where one considers the auxiliary initial value problem

$$2\frac{d^3H}{dz^3} + H\frac{d^2H}{dz^2} = 0, \qquad 0 < z < \infty, \tag{1.70a}$$

$$H(0) = H'(0) = 0, \qquad H''(0) = A, \tag{1.70b}$$

for some initial guess A. One integrates out the initial value problem (1.70a,b) and determines that $H'(\infty) = B$, for some number $B = B(A)$. One continues this process with different values of A until B is close enough to 1.

We now show that the invariance of (1.70a) and the initial conditions $H(0) = H'(0) = 0$ under a one-parameter family of scalings (one-parameter Lie group of scaling transformations) lead to the determination of σ with only one shooting.

The transformation

$$z = \frac{\eta}{\alpha}, \tag{1.71a}$$

$$H(z) = \alpha G(\eta), \tag{1.71b}$$

where $\alpha > 0$ is an arbitrary constant, maps (1.70a,b) to (1.69a) with initial conditions

$$G(0) = G'(0) = 0, \qquad G''(0) = \frac{A}{\alpha^3}. \tag{1.72}$$

Moreover, $H'(\infty) = B$ implies that

$$G'(\infty) = \frac{B}{\alpha^2}. \tag{1.73}$$

Hence, we pick α so that $\alpha^2 = B$, i.e., $\alpha = \sqrt{B}$. Then

$$\sigma = G''(0) = \frac{A}{B^{3/2}}. \tag{1.74}$$

From the numerical solution of the initial value problem (1.70a,b) for any particular value of A, one can show that

$$\sigma = 0.332\ldots. \tag{1.75}$$

EXERCISES 1.3

1. For the heat conduction problem (1.46a–c), show that $r(B_2) = 4$ for both dynamical and thermal units.
2. Derive (1.47a–c).
3. Derive (1.48a–c).
4. The boundary value problem (1.46a–c), in effect, has only two constants: $\kappa = K/\rho c$ (diffusivity) and $\lambda = Q/\rho c$. Use dimensional analysis with dynamical units to reduce (1.46a–c), where now $W_1 = x$, $W_2 = t$, $W_3 = \kappa$, $W_4 = \lambda$.
5. Consider the *Rayleigh flow problem* [see Schlichting (1955)], where an infinite flat plate is immersed in an incompressible fluid at rest. The plate is instantaneously accelerated so that it moves parallel to itself with constant velocity U. Let u be the fluid velocity in the direction of U (x-direction). Let the y-direction be the direction normal to the plate. The situation is illustrated in Figure 1.3.

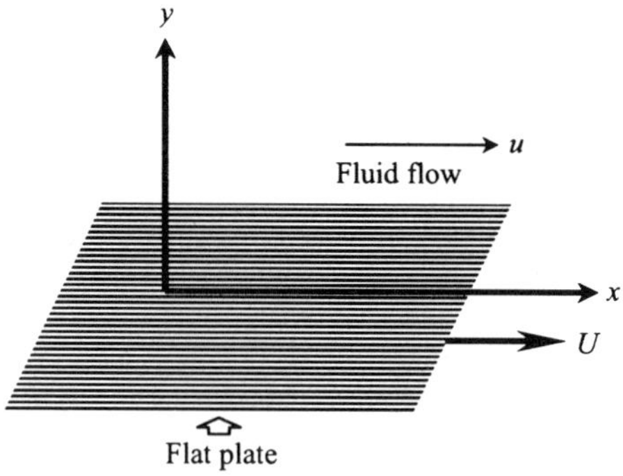

Figure 1.3

From symmetry considerations, one can show that the Navier–Stokes equations governing this problem reduce to the viscous diffusion equation

$$\frac{\partial u}{\partial t} = \nu \frac{\partial^2 u}{\partial y^2}, \quad 0 < t < \infty, 0 < y < \infty, \nu = \text{const}, \tag{1.76a}$$

with boundary conditions

$$u(y, 0) = 0, \tag{1.76b}$$

$$u(0, t) = U, \tag{1.76c}$$

$$u(\infty, t) = 0. \tag{1.76d}$$

(a) Use dimensional analysis to simplify the boundary value problem (1.76a–d).
(b) Use scalings of quantities followed by dimensional analysis to further simplify (1.76a–d). Find the explicit self-similar solution $u(y, t)$ of (1.76a–d).

1.4 GENERALIZATION OF DIMENSIONAL ANALYSIS: INVARIANCE OF PDEs UNDER SCALINGS OF VARIABLES

In both examples of Section 1.3.1, the use of dimensional analysis to reduce a boundary value problem for a PDE to a boundary value problem for an ODE is rather cumbersome. For the heat conduction problem, the use of dimensional analysis depends on either making the right choice of fundamental dimensions (thermal units) or combining effectively the constants before using dynamical units [cf. Exercise 1.3-4]. For the Prandtl–Blasius problem we used scaled variables before applying dimensional analysis.

A much easier way to accomplish such a reduction for a boundary value problem is to consider the invariance property of the boundary value problem under a one-parameter family of scalings (one-parameter Lie group of scaling transformations) where its variables are scaled but the constants of the boundary value problem are not scaled. If the boundary value problem is invariant under such a family of scaling transformations, then the number of independent variables is reduced constructively by one. We show that if, for some choice of fundamental dimensions, dimensional analysis leads to a reduction of the number of independent variables of a boundary value problem, then such a reduction is always possible through invariance of the boundary value problem under scalings applied strictly to its variables. [Recall that dimensional analysis involves scalings of *both* variables and constants.] Moreover, as will be shown, there exist boundary value problems for which the number of independent variables is reduced from invariance under a one-parameter family of scalings of their variables but the number of independent variables is not reduced from the use of dimensional analysis for any known choice of fundamental dimensions. [One could argue that this is a way of discovering new sets of fundamental dimensions!] Hence, for the purpose of reducing the number of independent variables of a boundary value problem, the invariance of a boundary value problem under a one-parameter family of scalings of its variables is a generalization of dimensional analysis.

Zel'dovich (1956) [see also Barenblatt and Zel'dovich (1972) and Barenblatt (1979, 1987, 1996)] calls a *self-similar solution of the first kind* a solution of a boundary value problem obtained by reduction through dimensional analysis, and calls a *self-similar solution of the second kind* a solution to a boundary value problem obtained by reduction through invariance under scalings of the variables when this reduction is not possible through dimensional analysis. The two examples of Section 1.3.1 show that these distinctions are somewhat blurred.

Before proving a general theorem relating dimensional analysis and invariance under scalings of variables, we consider the invariance property of the heat conduction problem (1.46a–c) under scalings of its variables.

Consider the family of scaling transformations

$$x^* = \alpha x, \tag{1.77a}$$

$$t^* = \beta t, \tag{1.77b}$$

$$u^* = \gamma u, \tag{1.77c}$$

where α, β, γ are arbitrary positive constants.

Definition 1.4-1. A transformation of the form (1.77a–c) *leaves invariant* the boundary value problem (1.46a–c) (*is admitted by* the boundary value problem (1.46a–c)) if and only if for any solution $u = \Theta(x,t)$ of (1.46a–c), it follows that

$$v(x^*, t^*) = u^* = \gamma u = \gamma\Theta(x, t) \tag{1.78}$$

solves the boundary value problem

$$\rho c \frac{\partial v}{\partial t^*} - K\frac{\partial^2 v}{\partial x^{*2}} = 0, \quad -\infty < x^* < \infty,\ t^* > 0, \tag{1.79a}$$

$$v(x^*, 0) = \frac{Q}{\rho c}\delta(x^*), \tag{1.79b}$$

$$\lim_{x^* \to \pm\infty} v(x^*, t^*) = 0. \tag{1.79c}$$

Clearly, the domain $-\infty < x^* < \infty, t^* > 0$, corresponds to the domain $-\infty < x < \infty, t > 0$; $t^* = 0$ corresponds to $t = 0$; and $x^* \to \pm\infty$ corresponds to $x \to \pm\infty$. Hence, (1.77a–c) leaves invariant the boundary of the boundary value problem (1.46a–c).

Lemma 1.4-1. *If a scaling* (1.77a–c) *leaves invariant the boundary value problem* (1.46a–c) *and* $u = \Theta(x,t)$ *solves* (1.46a–c), *then* $u = \gamma\Theta(x/\alpha, t/\beta)$ *also solves* (1.46a–c).

Proof. Left to Exercise 1.4-1. □

In order that (1.77a–c) leaves invariant the boundary value problem (1.46a-c), it is sufficient that each of the three equations of (1.46a–c) is separately invariant. Invariance of (1.46a) means that $u = \Theta(x, t)$ solves (1.46a) if and only if $v = \gamma\Theta(x, t)$ solves (1.79a). This leads to $\beta = \alpha^2$. Invariance of (1.46b,c) similarly leads to $\gamma = 1/\alpha$. Hence, the one-parameter ($\alpha > 0$) family of scaling transformations

$$x^* = \alpha x, \tag{1.80a}$$

$$t^* = \alpha^2 t, \tag{1.80b}$$

$$u^* = \alpha^{-1} u, \tag{1.80c}$$

is admitted by (1.46a–c).

Clearly, if $u = \Theta(x, t)$ solves (1.46a–c), then

$$v(x^*, t^*) = \Theta(x^*, t^*) = \Theta(\alpha x, \alpha^2 t) \tag{1.81}$$

solves (1.79a–c). Hence, the transformation (1.80a–c) maps any solution $v = \Theta(x^*, t^*)$ of (1.79a–c) to the solution

$$v = \alpha^{-1}\Theta(x, t) = \alpha^{-1}\Theta(\alpha^{-1}x^*, \alpha^{-2}t^*)$$

of (1.79a–c) or, equivalently, maps any solution $u = \Theta(x, t)$ of (1.46a–c) to the solution $u = \alpha^{-1}\Theta(\alpha^{-1}x, \alpha^{-2}t)$ of (1.46a–c).

The solution of (1.46a–c), and hence of (1.79a–c), is unique. Consequently, from this uniqueness property, the solution $u = \Theta(x, t)$ of (1.46a–c) must satisfy the functional equation

$$\Theta(x^*, t^*) = \alpha^{-1}\Theta(x, t). \tag{1.82}$$

Such a solution of a PDE, arising from invariance under a one-parameter Lie group of transformations, is called a *similarity* or *invariant solution.* The functional equation (1.82), satisfied by the invariant solution, is called the *invariant surface condition.* An invariant solution arising from invariance under a one-parameter Lie group of scalings such as (1.80a–c) is also called a *self-similar* or *automodel solution.*

From (1.80a,b), the invariant surface condition (1.82) satisfied by $\Theta(x, t)$ is given by

$$\Theta(\alpha x, \alpha^2 t) = \alpha^{-1}\Theta(x, t). \tag{1.83}$$

In order to solve (1.83), let $z = x/\sqrt{t}$ and $\Theta(x, t) = (1/\sqrt{t})\phi(z,t)$. Then in terms of $z,\ t,\ \phi(z, t)$, (1.83) becomes

$$\alpha\Theta(\alpha x, \alpha^2 t) = \Theta(x,t) = \frac{1}{\sqrt{t}}\phi(z,t) = \frac{\alpha}{\sqrt{\alpha^2 t}}\phi(z,\alpha^2 t) = \frac{\phi(z,\alpha^2 t)}{\sqrt{t}}.$$

Hence, $\phi(z, t)$ satisfies the functional equation

$$\phi(z, t) = \phi(z, \alpha^2 t) \quad \text{for any } \alpha > 0. \tag{1.84}$$

Thus, $\phi(z, t)$ does not depend on t. This leads to the *invariant form* (*similarity form*)

$$u = \Theta(x, t) = \frac{1}{\sqrt{t}}F(z) \tag{1.85}$$

for the solution of the boundary value problem (1.46a–c); z is called the *similarity variable.* The substitution of (1.85) into (1.46a–c) leads to a boundary value problem for an ordinary differential equation with unknown $F(z)$. The details are left to Exercise 1.4–2.

Now consider the following theorem that connects dimensional analysis and invariance under scalings of variables:

Theorem 1.4-1. *If the number of independent variables appearing in a boundary value problem for a PDE can be reduced by ρ through dimensional analysis, then the number of variables can be reduced by ρ through invariance of the boundary value problem under a ρ-parameter family of scaling transformations of its variables.*

Proof. Consider the dimension matrices B, B_1, and B_2 defined by (1.44a,b) and (1.45). Through dimensional analysis the number of independent variables of the given boundary value problem is reduced by $\rho = r(B) - r(B_2)$.

An arbitrary scaling of any fundamental dimension is represented by the *m*-

parameter family of scaling transformations

$$L^*_j = e^{\varepsilon_j} L_j, \quad j = 1,2,\ldots,m, \tag{1.86}$$

where $\varepsilon_1, \varepsilon_2, \ldots, \varepsilon_m$ are arbitrary real numbers. Let $\boldsymbol{\varepsilon}$ be the row vector

$$\boldsymbol{\varepsilon} = [\varepsilon_1, \varepsilon_2, \ldots, \varepsilon_m]. \tag{1.87}$$

The scaling (1.86) induces a scaling of the value of each measurable quantity W_i:

$$W^*_i = e^{\sum_{j=1}^m \varepsilon_j b_{ji}} W_i = e^{(\boldsymbol{\varepsilon}\mathrm{B})_i} W_i, \quad i = 1,2,\ldots,n, \tag{1.88}$$

where $(\boldsymbol{\varepsilon}\mathrm{B})_i$ is the ith component of the n-component row vector $\boldsymbol{\varepsilon}\mathrm{B}$. The value of u scales to

$$u^* = e^{\sum_{j=1}^m \varepsilon_j a_j} u. \tag{1.89}$$

From assumption (iv) of the Buckingham Pi-theorem, the family of scaling transformations (1.88), (1.89), induced by the m-parameter family of scalings of the fundamental dimensions (1.86), leaves invariant the given boundary value problem. Our aim is to find the number of essential parameters in the subfamily of transformations of the form (1.88), (1.89) for which the constants are all invariant, i.e., we aim to find the dimension of the vector space of all vectors $\boldsymbol{\varepsilon} = [\varepsilon_1, \varepsilon_2, \ldots, \varepsilon_m]$ such that

$$W^*_i = W_i, \quad i = l+1, l+2, \ldots, n, \tag{1.90a}$$

and

$$W^*_j \neq W_j \quad \text{for some } j = 1,2,\ldots,l. \tag{1.90b}$$

Equation (1.90a) holds if and only if

$$\boldsymbol{\varepsilon}\mathrm{B}_2 = 0. \tag{1.91}$$

The number of essential parameters is the number of linearly independent solutions $\boldsymbol{\varepsilon}$ of (1.91) such that $\boldsymbol{\varepsilon}\mathrm{B}_1 \neq 0$.

It is helpful to introduce a few definitions and some notation:

Let A be a matrix linear transformation acting on vector space V such that if $\mathbf{v} \in V$ is a row vector, then $\mathbf{v}\mathrm{A}$ is the action of A on $\mathbf{v}$. The *null space* of A is the vector space $V_{(\mathrm{A})_N} = \{\boldsymbol{\varepsilon} \in V : \boldsymbol{\varepsilon}\mathrm{A} = 0\}$, the *range space* of A is the vector space $V_{(\mathrm{A})_R} = \{z : z = \boldsymbol{\varepsilon}\mathrm{A} \text{ for some } \boldsymbol{\varepsilon} \in V\}$, and dim V is the dimension of the vector space V. It follows that

$$\dim V = \dim V_{(\mathrm{A})_R} + \dim V_{(\mathrm{A})_N}.$$

Consider the matrices B, B_1, and B_2 defined by (1.44a,b), (1.45). Let V be $\mathbf{R}^m$, where m is the number of rows of each of these three matrices, so that dim $V = m$. Then $\dim V_{(\mathrm{B})_N}$ is the number of linearly independent solutions $\boldsymbol{\varepsilon}$ of the set of equations $\boldsymbol{\varepsilon}\mathrm{B} = 0$, and $\dim V_{(\mathrm{B}_2)_N}$ is the number of linearly independent solutions ε of $\varepsilon\mathrm{B}_2 = 0$. It follows that

$$\dim V_{(\mathrm{B}_2)_N} = m - r(\mathrm{B}_2), \quad \dim V_{(\mathrm{B})_N} m - r(\mathrm{B}) = m - r(\mathrm{B}_2) - \rho.$$

Since $V_{(\mathrm{B}_2)_N(\mathrm{B}_1)_N} = V_{(\mathrm{B})_N}$, it follows that

$$\dim V_{(\mathrm{B}_2)_N} = \dim V_{(\mathrm{B}_2)_N(\mathrm{B}_1)_N} + \dim V_{(\mathrm{B}_2)_N(\mathrm{B}_1)_R} = \dim V_{(\mathrm{B})_N} + \dim V_{(\mathrm{B}_2)_N(\mathrm{B}_1)_R}.$$

Hence, $\dim V_{(\mathrm{B}_2)_N(\mathrm{B}_1)_R} = \rho$. But $\dim V_{(\mathrm{B}_2)_N(\mathrm{B}_1)_R}$ is the number of linearly independent solutions ε of the system $\varepsilon\mathrm{B}_2 = 0$ such that $\varepsilon\mathrm{B}_1 \neq 0$. Hence, the number of essential parameters is ρ, completing the proof of the theorem. □

EXERCISES 1.4

1. Prove Lemma 1.4-1.
2. Set up the boundary value problem for $F(z)$ as defined by (1.85). Put this boundary value problem in dimensionless form using:
   (a) dynamical units; and
   (b) thermal units. Explain.
3. Consider diffusion in a half-space with a concentration-dependent diffusion coefficient which is directly proportional to the concentration of a substance $C(x, t)$. Initially, and far from the front face $x = 0$, the concentration is assumed to be zero. The concentration is fixed on the front face. The aim is to find the concentration flux on the front face, $C_x(0,t)$. In special units, $C(x, t)$ satisfies the boundary value problem

$$\frac{\partial C}{\partial t} = \frac{\partial}{\partial x}\left(C\frac{\partial C}{\partial x}\right), \quad 0 < x < \infty, 0 < t < \infty, \tag{1.92a}$$

   where

$$C(x, 0) = C(\infty, t) = 0, \quad C(0, t) = A. \tag{1.92b}$$

   (a) Exploit similarity to determine $C_x(0,t)$ as effectively as possible.
   (b) Use scaling invariance to reduce the boundary value problem (1.92a,b) to a boundary value problem for an ODE.
   (c) Discuss a numerical procedure to determine $C_x(0,t)$ based on the scaling property of the reduced boundary value problem derived in (b).
4. For boundary layer flow over a semi-infinite wedge at zero angle of attack, the governing PDEs are given by

$$u\frac{\partial u}{\partial x}+v\frac{\partial u}{\partial y}-U(x)\frac{dU}{dx}=\nu\frac{\partial^2 u}{\partial y^2},$$

$$\frac{\partial u}{\partial x}+\frac{\partial v}{\partial y}=0, \quad 0<x<\infty,\ 0<y<\infty,$$

with boundary conditions $u(x, 0) = v(x, 0) = 0$, $\lim_{y\to\infty} u(x, y) = U(x)$, $U(x) = Ax$, where A, l are constants with $l = \beta/(2-\beta)$ corresponding to the opening angle $\pi\beta$ of the semi-infinite wedge. In this problem, x is the distance from the leading edge on the wedge surface and y is the distance from the wedge surface [Figure 1.4].

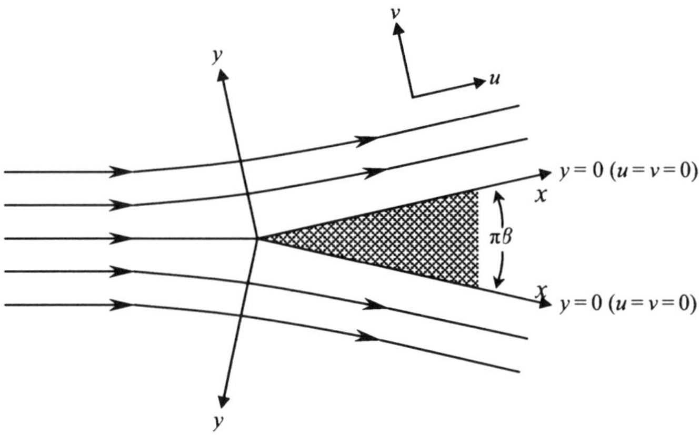

Figure 1.4

As for the Prandtl boundary layer equations (1.49a,b), introduce a stream function $\psi(x, y)$. Use scaling invariance to reduce the given problem to a boundary value problem for an ODE. Choose coordinates so that the Blasius equation arises if $l = 0$.

5. The following boundary value problem for a nonlinear diffusion equation arises from a biphasic continuum model of soft tissue [Holmes (1984)]:

$$\frac{\partial^2 u}{\partial x^2}-K\left(\frac{\partial u}{\partial x}\right)\frac{\partial u}{\partial t}=0, \quad 0<x<\infty,\ 0<t<\infty,$$

where K is a function of $\partial u/\partial x$, with boundary conditions $\frac{\partial u}{\partial x}(0,t)=-1$, $u(\infty, t) = u(x, 0) = 0$. Reduce this problem to a boundary value problem for an ODE.

6. Use invariance under scalings of the variables to solve the Rayleigh flow problem (1.76a–d).

7. Consider again the source problem for heat conduction in terms of the dimensionless form arising from dynamical units:

$$\frac{\partial u}{\partial t} - \frac{\partial^2 u}{\partial x^2} = 0, \quad -\infty < x < \infty, \ t > 0,$$

$$u(x,0) = \delta(x),$$

$$\lim_{x \to \pm\infty} u(x,t) = 0.$$

The use of scaling invariance with respect to the variables (1.80a–c) leads to the similarity form for the solution, $u = (1/\sqrt{t})G(x/\sqrt{t})$.

(a) Show that this problem is invariant under the one-parameter (β) family of transformations

$$x^* = x - \beta t, \quad t^* = t, \quad u^* = ue^{(1/2)\beta x - (1/4)\beta^2 t}, \tag{1.93}$$

for any constant $\beta, -\infty < \beta < \infty$.

(b) Check that t and $ue^{x^2/4t}$ are invariants of these transformations.

(c) Show that these transformations lead to the similarity form

$$u(x,t) = H(t) = e^{-x^2/4t}. \tag{1.94}$$

Hence show that invariance under the scalings (1.80a–c) and the transformations (1.93) lead to the well-known fundamental solution

$$u(x,t) = \frac{1}{\sqrt{4\pi t}} e^{-x^2/4t}.$$

1.5 DISCUSSION

Dimensional analysis is necessary for ascertaining fundamental dimensions and the consequent essential quantities that arise in a real problem in order to design proper model experiments. If a given problem can be described in terms of a boundary value problem for a system of PDEs then dimensional analysis may lead to a reduction in the number of independent variables. Moreover, if such a reduction exists, it can always be accomplished by considering the invariance properties of the boundary value problem under scaling transformations applied only to its variables.

As will be seen in Chapter 4, the invariance properties of PDEs (or, more particularly, boundary value problems) under scalings of variables can be generalized to the study of the invariance properties of PDEs under arbitrary one-parameter Lie groups of point transformations of their variables. Moreover, for a given differential equation, such transformations can be found algorithmically. [For example, one can easily deduce transformations (1.93) and (1.94).] This follows from the properties of such transformations and, in particular, their characterization by infinitesimal generators [see Chapter 2].

References on dimensional analysis specific to various fields include: de Jong (1967) [economics]; Sedov (1982), Birkhoff (1950), Barenblatt (1979, 1987, 1996), and Zierep (1971) [mechanics, elasticity, and hydrodynamics]; Venikov (1969) [electrical engineering]; Taylor (1974) [mechanical engineering]; Becker (1976) [chemical engineering]; Haynes (1982) [geography]; Kurth (1972) [astrophysics]; Murota (1985) [systems analysis]; Schepartz (1980) and Barenblatt (1987) [biomedical sciences]. Examples of dimensional analysis and scaling invariance applied to boundary value problems appear in Sedov (1982), Birkhoff (1950), Barenblatt (1979, 1996), Dresner (1983, 1999), Hansen (1964), Zierep (1971), and Seshadri and Na (1985). Examples which use scalings to convert boundary value problems to initial value problems for ODEs appear in Klamkin (1962), Na (1967, 1979), Dresner (1983, 1999), and Seshadri and Na (1985). Fractals are connected with self-similarity [Mandelbrot (1977, 1982)]. There are important connections between self-similarity, asymptotics, and renormalization groups [Barenblatt (1996); Goldenfeld (1992); Cole and Wagner (1996)].

2

Lie Groups of Transformations and Infinitesimal Transformations

2.1 INTRODUCTION

In dimensional analysis, the scalings of the fundamental dimensions (1.86), the induced scalings of the measurable quantities (1.88), the induced scalings of all quantities (1.88), (1.89), and the induced scalings preserving all constants (1.88), (1.91), are all examples of Lie groups of transformations. From the point of view of finding solutions to partial differential equations (PDEs), a general theory of Lie groups of transformations is unnecessary if transformations are restricted to scalings, translations, or rotations. However, it turns out that much wider classes of transformations can leave invariant PDEs. For the use and discovery of such transformations, the infinitesimal characterization of a Lie group of transformations is crucial.

Sophus Lie introduced the notion of a continuous group of transformations to put order to the hodgepodge of techniques for solving ordinary differential equations (ODEs). He was motivated by the lectures of his fellow Norwegian, Sylow, on the works of Abel and Galois on solving algebraic equations.

Lie groups of transformations are characterized by infinitesimal generators. Lie gave an algorithm to find all infinitesimal generators of point transformations and, more generally, contact transformations admitted by a given differential equation. Significantly, for a given differential equation, the basic applications of Lie groups of transformations only require knowledge of the admitted infinitesimal generators.

A point transformation acts on the space of independent and dependent variables of a differential equation. A Lie group of point transformations extends naturally to act on a space that includes the derivatives of dependent variables to any finite order. The functions appearing in the infinitesimal generator of a Lie group of transformations satisfy an overdetermined system of linear differential equations. These functions only depend on independent and dependent variables in the case of point transformations and include dependence on first derivatives of dependent variables in the case of contact transformations. More generally, the method of calculation, as well as many applications of infinitesimal generators for point and contact transformations, extend to infinitesimal generators of higher-order local transformations which allow the functions in their generators to depend on a finite number of higher-order derivatives.

A Lie group of transformations admitted by a differential equation corresponds to a mapping of each of its solutions to another solution of the same differential equation. There are an infinite number of ways of representing such a mapping by allowing an arbitrary change of independent variables. The representation is unique if the independent variables are kept fixed. This point of view is essential when one extends Lie's algorithm to the computation and use of higher-order local transformations admitted by differential equations as well as when one extends Lie's work on integrating factors for first-order ODEs to higher-order ODEs.

2.2 LIE GROUP OF TRANSFORMATIONS

We start with the definition of a group, then consider a group of transformations and, more specifically, a one-parameter Lie group of transformations. Here, the transformations act on $\mathbf{R}^n$.

2.2.1 GROUPS

Definition 2.2.1-1. A *group* G is a set of elements with a law of composition ϕ between elements satisfying the following axioms:

(i) *Closure property.* For any elements a and b of G, $\phi(a, b)$ is an element of G.

(ii) *Associative property.* For any elements a, b, c of G:

$$\phi(a, \phi(b, c)) = \phi(\phi(a, b), c).$$

(iii) *Identity element.* There exists a unique identity element e of G such that for any element a of G:

$$\phi(a, e) = \phi(e, a) = a.$$

(iv) *Inverse element.* For any element a of G there exists a unique inverse element a^{-1} in G such that

$$\phi(a, a^{-1}) = \phi(a^{-1}, a) = e.$$

Definition 2.2.1-2. A group G is *Abelian* if $\phi(a, b) = \phi(b, a)$ holds for all elements a and b in G.

Definition 2.2.1-3. A *subgroup* of G is a group formed by a subset of elements of G with the same law of composition ϕ.

2.2.2 EXAMPLES OF GROUPS

(1) G is the set of all integers with $\phi(a, b) = a + b$. Here, $e = 0$ and $a^{-1} = -a$.

(2) G is the set of all positive reals with $\phi(a, b) = a \cdot b$. Here, $e = 1$ and $a^{-1} = 1/a$.

(3) G is the set of symmetries (transformations) which leave invariant an equilateral triangle ABC with both faces painted in the same color [Figure 2.1].

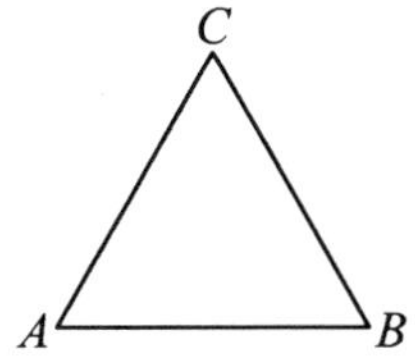

Figure 2.1

Here, the group can be represented by all permutations of the vertices A, B, C. The *identity element* $e = (1,2,3)$ corresponds to vertex 1 located at A, vertex 2 at B, and vertex 3 at C [Figure 2.2(a)]. The *rotation element* $R = (3,1,2)$ corresponds to a counterclockwise rotation of $2\pi/3$ radians of the configuration of Figure 2.1 about an axis coming out of the page through the center of the triangle [Figure 2.2(b)]. As a consequence vertex 3 is located at A, vertex 1 at B, and vertex 2 at C. The *flip element* $r = (3,2,1)$ represents the rotation of the configuration of Figure 2.1 about the indicated perpendicular by π radians. As a consequence, vertex 3 is located at A, vertex 2 at B, and vertex 1 at C [Figure 2.2(c)].

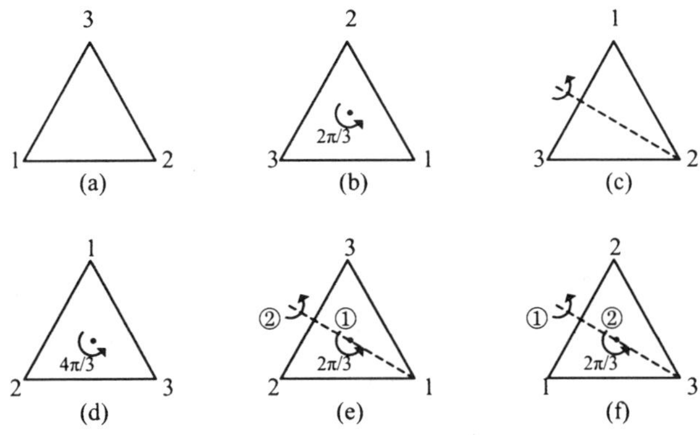

Figure 2.2. Symmetry group of an equilateral triangle: (a) identity, e; (b) rotation by $2\pi/3$, R; (c) flip, r; (d) rotation by $4\pi/3$, $\phi(R,R)$; (e) rotation by $2\pi/3$ followed by flip, $\phi(R,r)$; and (f) flip followed by rotation by $2\pi/3$, $\phi(r,R)$.

The element $\phi(R,R) = R = (2,3,1)$ represents a counterclockwise rotation of $4\pi/3$ radians of the configuration in Figure 2.1 [Figure 2.2(d)]. It is the composition of a counterclockwise rotation of $2\pi/3$ radians followed by another counterclockwise rotation of $2\pi/3$ radians. The composition element $\phi(R,r) = rR = (2,1,3)$ represents a counterclockwise rotation of $2\pi/3$ radians followed by a flip [Figure 2.2(e)]. The composition element $\phi(r,R) = Rr = (1,3,2)$ represents a flip followed by a counterclockwise rotation of $2\pi/3$ radians [Figure 2.2(f)]. It is left to Exercise 2.2–1 to prove that the symmetries of an equilateral triangle form a group with six elements. Note that this group is not *Abelian* since $\phi(r,R) \neq \phi(R,r)$.

2.2.3 GROUP OF TRANSFORMATIONS

Definition 2.2.3-1. Let $\mathbf{x} = (x_1, x_2, \ldots, x_n)$ lie in region $D \subset \mathbf{R}^n$. The set of transformations

$$\mathbf{x}^* = \mathbf{X}(\mathbf{x}; \varepsilon), \tag{2.1}$$

defined for each $\mathbf{x}$ in D and parameter ε in set $S \subset \mathbf{R}$, with $\phi(\varepsilon, \delta)$ defining a law of composition of parameters ε and δ in S, forms a *one-parameter group of transformations* on D if the following hold:

(i) For each ε in S the transformations are one-to-one onto D. [Hence, $\mathbf{x}^*$ lies in D.]

(ii) S with the law of composition ϕ forms a group G.

(iii) For each $\mathbf{x}$ in D, $\mathbf{x}^* = \mathbf{x}$ when $\varepsilon = \varepsilon_0$ corresponds to the identity e, i.e.,

$$\mathbf{X}(\mathbf{x}; \varepsilon_0) = \mathbf{x}.$$

(iv) If $\mathbf{x}^* = \mathbf{X}(\mathbf{x}; \varepsilon)$, $\mathbf{x}^{**} = \mathbf{X}(\mathbf{x}^*; \delta)$, then

$$\mathbf{x}^{**} = \mathbf{X}(\mathbf{x}; \phi(\varepsilon, \delta)).$$

2.2.4 ONE-PARAMETER LIE GROUP OF TRANSFORMATIONS

Definition 2.2.4-1. A one-parameter group of transformations defines a *one-parameter Lie group of transformations* if, in addition to satisfying axioms (i)–(iv) of Definition 2.2.3-1, the following hold:

(v) ε is a continuous parameter, i.e., S is an interval in $\mathbf{R}$. Without loss of generality, $\varepsilon = 0$ corresponds to the identity element e.

(vi) $\mathbf{X}$ is infinitely differentiable with respect to $\mathbf{x}$ in D and an analytic function of ε in S.

(vii) $\phi(\varepsilon, \delta)$ is an analytic function of ε and δ, $\varepsilon \in S$, $\delta \in S$.

If one thinks of ε as a time variable and $\mathbf{x}$ as spatial variables, then a one-parameter Lie group of transformations, in effect, defines a stationary flow. This will be shown in Section 2.3.1 but can now be partially seen as follows: Let

$$\mathbf{X}(\mathbf{x}; \varepsilon) \tag{2.2}$$

define the evolution of $\mathbf{x}$ over all elements $\varepsilon \in S$. This defines a curve γ_1 [Figure 2.3(a)].

Now let $\mathbf{y} = \mathbf{X}(\mathbf{x}; \varepsilon)$ represent a point on γ_1. Then $\mathbf{x}^{**} = \mathbf{X}(\mathbf{y}; \delta)$ $= \mathbf{X}(\mathbf{x}; \phi(\varepsilon, \delta))$ must lie on γ_1. Note that the self-intersecting curve γ_2 [Figure 2.3(b)] cannot represent the evolution defined by (2.2) [Exercise 2.2-2].

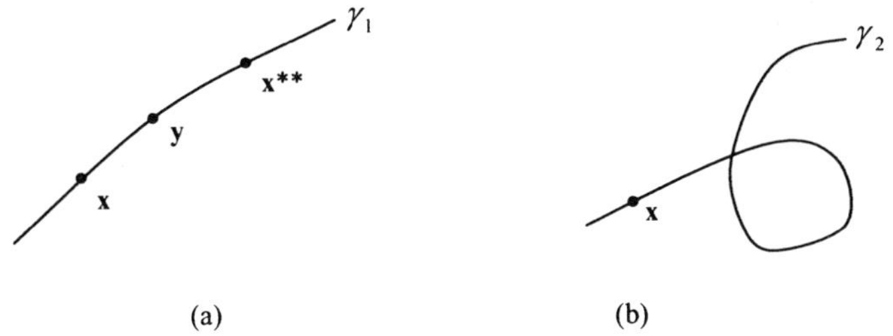

Figure 2.3

2.2.5 EXAMPLES OF ONE-PARAMETER LIE GROUPS OF TRANSFORMATIONS

(1) *Group of Translations in the Plane*
Consider the group of translations

$$x^* = x + \varepsilon,$$
$$y^* = y, \quad \varepsilon \in \mathbf{R}.$$

Here, $\phi(\varepsilon, \delta) = \varepsilon + \delta$. This group corresponds to motions parallel to the x-axis. [In Figure 2.4, the curve γ represents the evolution $\mathbf{X}(\mathbf{x}_0; \varepsilon)$.]

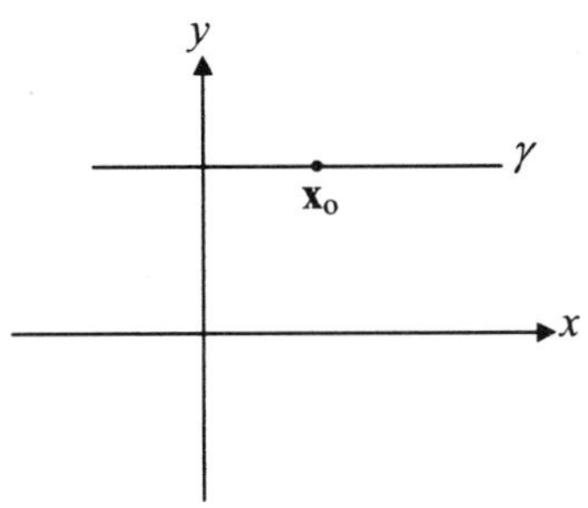

Figure 2.4

(2) *Group of Scalings in the Plane*
Consider the group of scalings

$$x^* = \alpha x,$$
$$y^* = \alpha^2 y, \quad 0 < \alpha < \infty.$$

Here $\phi(\alpha, \beta) = \alpha\beta$, and the identity element corresponds to $\alpha = 1$. This group of transformations can also be reparametrized in terms of $\varepsilon = \alpha - 1$:

$$x^* = (1 + \varepsilon)x, \tag{2.3a}$$

$$y^* = (1 + \varepsilon)^2 y, \quad -1 < \varepsilon < \infty, \tag{2.3b}$$

so that the identity element corresponds to $\varepsilon = 0$ with the law of composition of parameters given by

$$\phi(\varepsilon,\delta) = \varepsilon + \delta + \varepsilon\delta. \tag{2.4}$$

EXERCISES 2.2

1. Show that the symmetries of an equilateral triangle, with both faces painted in the same color, form a group with six elements. What happens if the faces are painted in different colors?

2. Show that the curve γ_2 of Figure 2.3(b) cannot represent the transformations (2.2).

3. Show that the transformations

$$x^* = x + 2\varepsilon, \tag{2.5a}$$

$$y^* = y + 3\varepsilon, \quad \varepsilon \in \mathbf{R}, \ (x, y) \in \mathbf{R}^2, \tag{2.5b}$$

define a Lie group of transformations. Trace out the evolution curves of the points (0, 0) and (1, 0) under this group. Explain the geometrical situation of the resulting curves.

4. Show that the set $S = \{\varepsilon : -1 < \varepsilon < \infty\}$ with the law of composition $\phi(\varepsilon, \delta) = \varepsilon + \delta + \varepsilon\delta$ defines a group.

5. Trace out the evolution curves of the points (1, 0), (1, 1), and (0, 0) for the Lie group of transformations (2.3a,b).

6. Show that the transformations (1.93) define a one-parameter Lie group of transformations acting on:
(a) (x, t)-space; and
(b) (x, t, u)-space.

7. Show whether or not each of the following one-parameter (ε) families of transformations of the plane defines a Lie group of transformations:
(a) $x^* = x - \varepsilon y$, $y^* = y + \varepsilon x$;
(b) $y^* = x + \varepsilon^2$, $y^* = y$; and
(c) $x^* = x + \varepsilon$, $y^* = \dfrac{xy}{x + \varepsilon}$.

2.3 INFINITESIMAL TRANSFORMATIONS

Consider a one-parameter (ε) Lie group of transformations

$$\mathbf{X}^* = \mathbf{X}(\mathbf{x}; \varepsilon) \tag{2.6}$$

with the identity $\varepsilon = 0$ and law of composition ϕ. Expanding (2.6) about $\varepsilon = 0$, in some neighborhood of $\varepsilon = 0$, we get

$$x^* = x + \varepsilon\left(\left.\frac{\partial X(x;\varepsilon)}{\partial \varepsilon}\right|_{\varepsilon=0}\right) + \tfrac{1}{2}\varepsilon^2\left(\left.\frac{\partial^2 X(x;\varepsilon)}{\partial \varepsilon^2}\right|_{\varepsilon=0}\right) + \cdots = x + \varepsilon\left(\left.\frac{\partial X(x;\varepsilon)}{\partial \varepsilon}\right|_{\varepsilon=0}\right) + O(\varepsilon^2). \tag{2.7}$$

Let

$$\xi(\mathbf{x}) = \left.\frac{\partial \mathbf{X}(\mathbf{x};\varepsilon)}{\partial \varepsilon}\right|_{\varepsilon=0}. \tag{2.8}$$

The transformation $\mathbf{x} + \varepsilon\xi(\mathbf{x})$ is called the *infinitesimal transformation* of the Lie group of transformations (2.6). The components of $\xi(\mathbf{x})$ are called the *infinitesimals* of (2.6).

2.3.1 FIRST FUNDAMENTAL THEOREM OF LIE

The following lemma is useful:

Lemma 2.3.1-1. *The one-parameter* (ε) *Lie group of transformations* (2.6) *satisfies the relation*

$$\mathbf{X}(\mathbf{x};\varepsilon+\Delta\varepsilon) = \mathbf{X}(\mathbf{X}(\mathbf{x};\varepsilon);\phi(\varepsilon^{-1},\varepsilon+\Delta\varepsilon)). \tag{2.9}$$

Proof.

$$\begin{aligned}\mathbf{X}(\mathbf{X}(\mathbf{x};\varepsilon);\phi(\varepsilon^{-1},\varepsilon+\Delta\varepsilon)) &= \mathbf{X}(\mathbf{x};\phi(\varepsilon,\phi(\varepsilon^{-1},\varepsilon+\Delta\varepsilon)))\\ &= \mathbf{X}(\mathbf{x};\phi(\phi(\varepsilon,\varepsilon^{-1}),\varepsilon+\Delta\varepsilon))\\ &= \mathbf{X}(\mathbf{x};\phi(0,\varepsilon+\Delta\varepsilon))\\ &= \mathbf{X}(\mathbf{x};\varepsilon+\Delta\varepsilon).\end{aligned}$$

□

Theorem 2.3.1-1 (First Fundamental Theorem of Lie). *There exists a parametrization* $\tau(\varepsilon)$ *such that the Lie group of transformations* (2.6) *is equivalent to the solution of an initial value problem for a system of first-order ODEs given by*

$$\frac{d\mathbf{x}^*}{d\tau} = \xi(\mathbf{x}^*), \tag{2.10a}$$

with

$$\mathbf{x}^* = \mathbf{x} \quad \textit{when} \quad \tau = 0. \tag{2.10b}$$

In particular,

$$\tau(\varepsilon) = \int_0^{\varepsilon} \Gamma(\varepsilon')\,d\varepsilon', \tag{2.11}$$

where

$$\Gamma(\varepsilon)=\left.\frac{\partial\phi(a,b)}{\partial b}\right|_{(a,b)=(\varepsilon^{-1},\varepsilon)} \tag{2.12}$$

and

$$\Gamma(0)=1. \tag{2.13}$$

[ε^{-1} *denotes the inverse element to* ε.]

Proof. First we show that (2.6) leads to (2.10a,b), (2.11), and (2.12). Expand the left-hand side of (2.9) in a power series in $\Delta\varepsilon$ about $\Delta\varepsilon=0$, so that

$$\mathbf{X}(\mathbf{x};\,\varepsilon+\Delta\varepsilon)=\mathbf{x}^*+\frac{\partial\mathbf{X}(\mathbf{x};\varepsilon)}{\partial\varepsilon}\Delta\varepsilon+O((\Delta\varepsilon)^2), \tag{2.14}$$

where $\mathbf{x}^*$ is given by (2.6). Then, expanding $\phi(\varepsilon^{-1},\varepsilon+\Delta\varepsilon)$ in a power series in $\Delta\varepsilon$ about $\Delta\varepsilon=0$, we have

$$\phi(\varepsilon^{-1},\varepsilon+\Delta\varepsilon)=\phi(\varepsilon^{-1},\varepsilon)+\Gamma(\varepsilon)\Delta\varepsilon+O((\Delta\varepsilon)^2)=\Gamma(\varepsilon)\Delta\varepsilon+O((\Delta\varepsilon)^2), \tag{2.15}$$

where $\Gamma(\varepsilon)$ is defined by (2.12). Consequently, after expanding the right-hand side of (2.9) in a power series in $\Delta\varepsilon$ about $\Delta\varepsilon=0$, we obtain

$$\begin{aligned}\mathbf{X}(\mathbf{x};\,\varepsilon+\Delta\varepsilon)&=\mathbf{X}(\mathbf{x}^*;\,\phi(\varepsilon^{-1},\varepsilon+\Delta\varepsilon))=\mathbf{X}(\mathbf{x}^*;\,\Gamma(\varepsilon)\Delta\varepsilon+O((\Delta\varepsilon)^2))\\&=\mathbf{X}(\mathbf{x}^*;0)+\Delta\varepsilon\Gamma(\varepsilon)\left(\left.\frac{\partial\mathbf{X}(\mathbf{x}^*;\delta)}{\partial\delta}\right|_{\delta=0}\right)+O((\Delta\varepsilon)^2)\\&=\mathbf{x}^*+\Gamma(\varepsilon)\xi(\mathbf{x}^*)\,\Delta\varepsilon+O((\Delta\varepsilon)^2).\end{aligned} \tag{2.16}$$

Equating (2.14) and (2.16), we see that $\mathbf{x}^*=\mathbf{X}(\mathbf{x};\varepsilon)$ satisfies the initial value problem for the system of differential equations given by

$$\frac{d\mathbf{x}^*}{d\varepsilon}=\Gamma(\varepsilon)\xi(\mathbf{x}^*) \tag{2.17a}$$

with

$$\mathbf{x}^*=\mathbf{x}\quad\text{at}\ \varepsilon=0. \tag{2.17b}$$

From (2.7), it follows that $\Gamma(0)=1$. The parametrization $\tau(\varepsilon)=\int_0^\varepsilon\Gamma(\varepsilon')\,d\varepsilon'$ leads to (2.10a,b).

Since $\partial\xi(\mathbf{x})/\partial x_i$ is continuous, $i=1,2,\ldots,n,$ then from the existence and uniqueness theorem for an initial value problem for a system of first-order differential equations [Coddington (1961)], it follows that the solution of (2.10a,b), and hence (2.17a,b), exists and is unique. This solution must be (2.6), completing the proof of Lie's First Fundamental Theorem. □

Lie's First Fundamental Theorem shows that the infinitesimal transformation contains the essential information determining a one-parameter Lie group of transformations. Since the system of first-order ODEs (2.10a) is invariant under translations in τ, one can always reparameterize a given group in terms of a parameter τ such that for parameter values τ_1 and τ_2, the law of composition becomes $\phi(\tau_1, \tau_2) = \tau_1 + \tau_2$. Lie's First Fundamental Theorem also shows that a one-parameter Lie group of transformations (2.6) defines a stationary flow given by (2.10a,b) and, moreover, that any stationary flow (2.10a,b) defines a one-parameter Lie group of transformations.

2.3.2 EXAMPLES ILLUSTRATING LIE'S FIRST FUNDAMENTAL THEOREM

(1) *Group of Translations in the Plane*
For the group of translations

$$x^* = x + \varepsilon, \tag{2.18a}$$

$$y^* = y, \tag{2.18b}$$

the law of composition is given by $\phi(a, b) = a + b$, and $\varepsilon^{-1} = -\varepsilon$. Then $\partial\phi(a,b)/\partial b = 1$, and hence, $\Gamma(\varepsilon) \equiv 1$.

Let $\mathbf{x} = (x, y)$. Then the group (2.18a,b) becomes $\mathbf{X}(\mathbf{x};\varepsilon) = (x+\varepsilon, y)$. Thus, $\partial\mathbf{X}(\mathbf{x};\varepsilon)/\partial\varepsilon = (1, 0)$. Hence,

$$\xi(\mathbf{x}) = \left.\frac{\partial \mathbf{X}(\mathbf{x};\varepsilon)}{\partial\varepsilon}\right|_{\varepsilon=0} = (1, 0).$$

Consequently, (2.17a,b) becomes

$$\frac{dx^*}{d\varepsilon} = 1, \quad \frac{dy^*}{d\varepsilon} = 0, \tag{2.19a}$$

with

$$x^* = x, \quad y^* = y \quad \text{at } \varepsilon = 0. \tag{2.19b}$$

The solution of the initial value problem (2.19a,b) is easily seen to be given by (2.18a,b).

(2) *Group of Scalings*
For the group of scalings

$$x^* = (1+\varepsilon)x, \tag{2.20a}$$

$$y^* = (1+\varepsilon)^2 y, \quad -1 < \varepsilon < \infty, \tag{2.20b}$$

the law of composition is given by $\phi(a, b) = a + b + ab$, and $\varepsilon^{-1} = -\varepsilon/(1+\varepsilon)$. Here, $\partial\phi(a,b)/\partial b = 1 + a$, and hence,

$$\Gamma(\varepsilon) = \left.\frac{\partial\phi(a,b)}{\partial b}\right|_{(a,b)=(\varepsilon^{-1},\varepsilon)} = 1 + \varepsilon^{-1} = \frac{1}{1+\varepsilon}.$$

Let $\mathbf{x} = (x, y)$. Then the group (2.20a,b) becomes $\mathbf{X}(\mathbf{x};\varepsilon) = ((1+\varepsilon)x, (1+\varepsilon)^2 y)$. Thus, $\partial\mathbf{X}(\mathbf{x};\varepsilon)/\partial\varepsilon = (x, 2(1+\varepsilon)y)$, and

$$\xi(\mathbf{x}) = \left.\frac{\partial \mathbf{X}(\mathbf{x};\varepsilon)}{\partial\varepsilon}\right|_{\varepsilon=0} = (x, 2y).$$

As a result, (2.17a,b) here becomes

$$\frac{dx^*}{d\varepsilon} = \frac{x^*}{1+\varepsilon}, \quad \frac{dy^*}{d\varepsilon} = \frac{2y^*}{1+\varepsilon}, \tag{2.21a}$$

with

$$x^* = x, \quad y^* = y \quad \text{at } \varepsilon = 0. \tag{2.21b}$$

The solution of the initial value problem (2.21a,b) is, of course, given by (2.20a,b).

In terms of the parameterization

$$\tau = \int_0^{\varepsilon} \Gamma(\varepsilon')\, d\varepsilon' = \int_0^{\varepsilon} \frac{1}{1+\varepsilon'}\, d\varepsilon' = \log(1+\varepsilon),$$

the group (2.20a,b) becomes

$$x^* = e^{\tau} x, \tag{2.22a}$$

$$y^* = e^{2\tau} y, \quad -\infty < \tau < \infty, \tag{2.22b}$$

with the law of composition $\phi(\tau_1, \tau_2) = \tau_1 + \tau_2$.

2.3.3 INFINITESIMAL GENERATORS

In view of Lie's First Fundamental Theorem, from now on, without loss of generality, we assume that a one-parameter (ε) Lie group of transformations is parameterized such that its law of composition is given by $\phi(a, b) = a + b$, so that $\varepsilon^{-1} = -\varepsilon$ and $\Gamma(\varepsilon) \equiv 1$. Thus, in terms of its infinitesimals $\xi(\mathbf{x})$, the one-parameter Lie group of transformations (2.6) becomes

$$\frac{d\mathbf{x}^*}{d\varepsilon} = \xi(\mathbf{x}^*), \tag{2.23a}$$

with

$$\mathbf{x}^* = \mathbf{x} \quad \text{at } \varepsilon = 0. \tag{2.23b}$$

Definition 2.3.3-1. The *infinitesimal generator* of the one-parameter Lie group of transformations (2.6) is the operator

$$\mathrm{X} = \mathrm{X}(\mathbf{x}) = \xi(\mathbf{x}) \cdot \nabla = \sum_{i=1}^{n} \xi_i(\mathbf{x}) \frac{\partial}{\partial x_i}, \tag{2.24}$$

where ∇ is the gradient operator

$$\nabla = \left(\frac{\partial}{\partial x_1}, \frac{\partial}{\partial x_2}, \ldots, \frac{\partial}{\partial x_n} \right). \tag{2.25}$$

For any differentiable function $F(\mathbf{x}) = F(x_1, x_2, \ldots, x_n)$, one has

$$\mathrm{X}F(\mathbf{x}) = \xi(\mathbf{x}) \cdot \nabla F(\mathbf{x}) = \sum_{i=1}^{n} \xi_i(\mathbf{x}) \frac{\partial F(\mathbf{x})}{\partial x_i}.$$

Note that $\mathrm{X}\mathbf{x} = \xi(\mathbf{x})$.

It follows that a one-parameter Lie group of transformations, which from Lie's First Fundamental Theorem is determined by its infinitesimal transformation, is also determined by its infinitesimal generator. The following theorem shows that use of the infinitesimal generator (2.24) leads to an algorithm to find the explicit solution of the initial value problem (2.23a,b).

Theorem 2.3.3-1. *The one-parameter Lie group of transformations* (2.6) *is equivalent to*

$$\mathbf{x}^* = e^{\varepsilon \mathrm{X}}\mathbf{x} = \mathbf{x} + \varepsilon \mathrm{X}\mathbf{x} + \tfrac{1}{2}\varepsilon^2 \mathrm{X}^2 \mathbf{x} + \cdots = [1 + \varepsilon \mathrm{X} + \tfrac{1}{2}\varepsilon^2 \mathrm{X}^2 + \cdots]\mathbf{x} = \sum_{k=0}^{\infty} \frac{\varepsilon^k}{k!} \mathrm{X}^k \mathbf{x}, \tag{2.26}$$

where the operator $\mathrm{X} = \mathrm{X}(\mathbf{x})$ *is defined by* (2.24) *and the operator* $\mathrm{X}^k = \mathrm{X}^k(\mathbf{x})$ *is given by* $\mathrm{X}^k = \mathrm{X}\mathrm{X}^{k-1}$, $k = 1,2,\ldots$ *. In particular,* $\mathrm{X}^k F(\mathbf{x})$ *is the function obtained by applying the operator* X *to the function* $\mathrm{X}^{k-1}F(\mathbf{x})$, $k = 1,2,\ldots$, *with* $\mathrm{X}^0 F(\mathbf{x}) \equiv F(\mathbf{x})$.

Proof. Let

$$\mathrm{X} = \mathrm{X}(\mathbf{x}) = \sum_{i=1}^{n} \xi_i(\mathbf{x}) \frac{\partial}{\partial x_i} \tag{2.27a}$$

and

$$\mathrm{X}(\mathbf{x}^*) = \sum_{i=1}^{n} \xi_i(\mathbf{x}^*) \frac{\partial}{\partial x^*_i}, \tag{2.27b}$$

where

$$\mathbf{x}^* = \mathbf{X}(\mathbf{x}; \varepsilon) \tag{2.28}$$

is the Lie group of transformations (2.6). From Taylor's theorem, expanding (2.28) about $\varepsilon = 0$, we have

$$\mathbf{x}^* = \sum_{k=0}^{\infty} \frac{\varepsilon^k}{k!} \left(\left. \frac{\partial^k \mathbf{X}(\mathbf{x}; \varepsilon)}{\partial \varepsilon^k} \right|_{\varepsilon=0} \right) = \sum_{k=0}^{\infty} \frac{\varepsilon^k}{k!} \left(\left. \frac{d^k \mathbf{x}^*}{d\varepsilon^k} \right|_{\varepsilon=0} \right). \tag{2.29}$$

For any differentiable function $F(\mathbf{x})$, we obtain

$$\frac{d}{d\varepsilon}F(\mathbf{x}^*)=\sum_{i=1}^{n}\frac{\partial F(\mathbf{x}^*)}{\partial x^*_i}\frac{dx^*_i}{d\varepsilon}=\sum_{i=1}^{n}\xi_i(\mathbf{x}^*)\frac{\partial F(\mathbf{x}^*)}{\partial x^*_i}=\mathrm{X}(\mathbf{x}^*)F(\mathbf{x}^*). \tag{2.30}$$

Hence, it follows that

$$\frac{d\mathbf{x}^*}{d\varepsilon}=\mathrm{X}(\mathbf{x}^*)\mathbf{x}^*, \tag{2.31a}$$

$$\frac{d^2\mathbf{x}^*}{d\varepsilon^2}=\frac{d}{d\varepsilon}\left(\frac{d\mathbf{x}^*}{d\varepsilon}\right)=\mathrm{X}(\mathbf{x}^*)\mathrm{X}(\mathbf{x}^*)\mathbf{x}^*=\mathrm{X}^2(\mathbf{x}^*)\mathbf{x}^*, \tag{2.31b}$$

and, in general,

$$\frac{d^k\mathbf{x}^*}{d\varepsilon^k}=\mathrm{X}^k(\mathbf{x}^*)\mathbf{x}^*,\quad k=1,2,\ldots. \tag{2.31c}$$

Consequently, we have

$$\left.\frac{d^k\mathbf{x}^*}{d\varepsilon^k}\right|_{\varepsilon=0}=\mathrm{X}^k(\mathbf{x})\mathbf{x}=\mathrm{X}^k\mathbf{x},\quad k=1,2,\ldots, \tag{2.32}$$

which leads to (2.26). □

Thus, the Taylor series expansion about $\varepsilon=0$ of a function $\mathbf{X}(\mathbf{x};\varepsilon)$ that defines a Lie group of transformations (2.6), with law of composition $\phi(a,b)=a+b$, is determined by the coefficient of its $O(\varepsilon)$ term, which is the infinitesimal

$$\left.\frac{\partial\mathbf{X}(\mathbf{x};\varepsilon)}{\partial\varepsilon}\right|_{\varepsilon=0}=\boldsymbol{\xi}(\mathbf{x}).$$

In summary, there are two ways to find explicitly a one-parameter Lie group of transformations from its infinitesimal transformation:

(i) Express the group in terms of a power series (2.26), called a *Lie series,* that is developed from the infinitesimal generator (2.24) corresponding to the infinitesimal transformation; or

(ii) Solve the initial value problem (2.23a,b) through explicitly finding the general solution of the system of first-order differential equations (2.23a).

The following corollary results from Theorem 2.3.3-1:

Corollary 2.3.3-1. *If* $F(\mathbf{x})$ *is infinitely differentiable, then for a one-parameter Lie group of transformations* (2.6) *with infinitesimal generator* (2.27a), *we have*

$$F(\mathbf{x}^*)=F(e^{\varepsilon\mathrm{X}}\mathbf{x})=e^{\varepsilon\mathrm{X}}F(\mathbf{x}). \tag{2.33}$$

Proof.

$$F(e^{\varepsilon \mathrm{X}}\mathbf{x}) = F(\mathbf{x}^*) = \sum_{k=0}^{\infty} \frac{\varepsilon^k}{k!}\left(\left. \frac{d^k F(\mathbf{x}^*)}{d\varepsilon^k} \right|_{\varepsilon=0} \right).$$

From (2.30) we see that $\dfrac{d^2 F(\mathbf{x}^*)}{d\varepsilon^2} = \mathrm{X}^2(\mathbf{x}^*)F(\mathbf{x}^*)$ and, hence, $\dfrac{d^k F(\mathbf{x}^*)}{d\varepsilon^k} = \mathrm{X}^k(\mathbf{x}^*)F(\mathbf{x}^*)$.

Thus, $\left.\dfrac{d^k F(\mathbf{x}^*)}{d\varepsilon^k}\right|_{\varepsilon=0} = \mathrm{X}^k(\mathbf{x})F(\mathbf{x})$. Consequently,

$$F(\mathbf{x}^*) = F(e^{\varepsilon \mathrm{X}}\mathbf{x}) = \left(\sum_{k=0}^{\infty} \frac{\varepsilon^k}{k!} \mathrm{X}^k(\mathbf{x}) \right) F(\mathbf{x}) = e^{\varepsilon \mathrm{X}} F(\mathbf{x}). \qquad \square$$

As an example, consider the rotation group

$$x^* = x\cos\varepsilon + y\sin\varepsilon, \tag{2.34a}$$

$$y^* = -x\sin\varepsilon + y\cos\varepsilon. \tag{2.34b}$$

The infinitesimal for (2.34a,b) is given by

$$\boldsymbol{\xi}(\mathbf{x}) = (\xi_1(x,y), \xi_2(x,y)) = \left(\left.\frac{dx^*}{d\varepsilon}\right|_{\varepsilon=0}, \left.\frac{dy^*}{d\varepsilon}\right|_{\varepsilon=0} \right) = (y, -x). \tag{2.35}$$

The infinitesimal generator for (2.34a,b) is

$$\mathrm{X} = \xi_1(x,y)\frac{\partial}{\partial x} + \xi_2(x,y)\frac{\partial}{\partial y} = y\frac{\partial}{\partial x} - x\frac{\partial}{\partial y}. \tag{2.36}$$

The Lie series corresponding to (2.36) is obtained as follows:

$$(x^*, y^*) = (e^{\varepsilon \mathrm{X}}x, e^{\varepsilon \mathrm{X}}y),$$

where

$$\mathrm{X}x = y, \quad \mathrm{X}^2 x = \mathrm{X}y = -x.$$

Hence,

$$\mathrm{X}^{4n}x = x, \quad \mathrm{X}^{4n-1}x = -y, \quad \mathrm{X}^{4n-2}x = -x, \quad \mathrm{X}^{4n-3}x = y, \quad n = 1, 2, \ldots,$$

$$\mathrm{X}^{4n}y = y, \quad \mathrm{X}^{4n-1}y = x, \quad \mathrm{X}^{4n-2}y = -y, \quad \mathrm{X}^{4n-3}y = -x, \quad n = 1, 2, \ldots.$$

Consequently,

$$x^* = e^{\varepsilon \mathrm{X}}x = \sum_{k=0}^{\infty} \frac{\varepsilon^k}{k!}\mathrm{X}^k x = \left(1 - \frac{\varepsilon^2}{2!} + \frac{\varepsilon^4}{4!} + \cdots\right)x + \left(\varepsilon - \frac{\varepsilon^3}{3!} + \frac{\varepsilon^5}{5!} + \cdots\right)y$$

$$= x\cos\varepsilon + y\sin\varepsilon.$$

Similarly,

$$y^* = e^{\varepsilon X} y = -x \sin\varepsilon + y \cos\varepsilon.$$

2.3.4 INVARIANT FUNCTIONS

Definition 2.3.4-1. An infinitely differentiable function $F(\mathbf{x})$ is an *invariant function* of the Lie group of transformations (2.6) if and only if, for any group transformation (2.6),

$$F(\mathbf{x}^*) \equiv F(\mathbf{x}). \tag{2.37}$$

If $F(\mathbf{x})$ is an invariant function of (2.6), then $F(\mathbf{x})$ is called an *invariant* of (2.6) and $F(\mathbf{x})$ is said to be *invariant under* (2.6).

Theorem 2.3.4-1. *$F(\mathbf{x})$ is invariant under a Lie group of transformations* (2.6) *if and only if*

$$XF(\mathbf{x}) \equiv 0. \tag{2.38}$$

Proof.

$$F(\mathbf{x}^*) \equiv e^{\varepsilon X} F(\mathbf{x}) \equiv \sum_{k=0}^{\infty} \frac{\varepsilon^k}{k!} X^k F(\mathbf{x}) \equiv F(\mathbf{x}) + \varepsilon XF(\mathbf{x}) + \tfrac{1}{2}\varepsilon^2 X^2 F(\mathbf{x}) + \cdots. \tag{2.39}$$

Suppose $F(\mathbf{x}^*) \equiv F(\mathbf{x})$. Then $XF(\mathbf{x}) \equiv 0$ follows from (2.39).

Conversely, let $F(\mathbf{x})$ satisfy $XF(\mathbf{x}) \equiv 0$. Then $X^n F(\mathbf{x}) \equiv 0$, $n = 1,2,\ldots.$ Hence, from (2.39), we have $F(\mathbf{x}^*) \equiv F(\mathbf{x})$. □

Theorem 2.3.4-2. *For a Lie group of transformations* (2.6), *the identity*

$$F(\mathbf{x}^*) \equiv F(\mathbf{x}) + \varepsilon \tag{2.40}$$

holds if and only if $F(\mathbf{x})$ is such that

$$XF(\mathbf{x}) \equiv 1. \tag{2.41}$$

Proof. Let $F(\mathbf{x})$ satisfy (2.40). Then

$$F(\mathbf{x}) + \varepsilon \equiv F(\mathbf{x}) + \varepsilon XF(\mathbf{x}) + \tfrac{1}{2}\varepsilon^2 X^2 F(\mathbf{x}) + \cdots.$$

Hence, $XF(\mathbf{x}) \equiv 1$.

Conversely, let $F(\mathbf{x})$ satisfy (2.41). Then $X^n F(\mathbf{x}) \equiv 0$, $n = 2,3,\ldots.$ Hence,

$$F(\mathbf{x}^*) \equiv e^{\varepsilon X} F(\mathbf{x}) \equiv F(\mathbf{x}) + \varepsilon XF(\mathbf{x}) \equiv F(\mathbf{x}) + \varepsilon.$$ □

2.3.5 CANONICAL COORDINATES

Suppose one makes a change of coordinates (one-to-one and continuously differentiable in some appropriate domain)

$$\mathbf{y} = \mathbf{Y}(\mathbf{x}) = (y_1(\mathbf{x}), y_2(\mathbf{x}), \ldots, y_n(\mathbf{x})). \tag{2.42}$$

For the one-parameter Lie group of transformations (2.6), the infinitesimal generator $\mathrm{X} = \sum_{i=1}^{n} \xi_i(\mathbf{x}) \frac{\partial}{\partial x_i}$ with respect to the coordinates $\mathbf{x} = (x_1, x_2, \ldots, x_n)$, becomes the infinitesimal generator

$$\mathrm{Y} = \sum_{i=1}^{n} \eta_i(\mathbf{y}) \frac{\partial}{\partial y_i}, \tag{2.43}$$

with respect to the coordinates $\mathbf{y} = (y_1, y_2, \ldots, y_n)$ defined by (2.42). Then it is necessary that Y = X in order to have the same group action. The infinitesimal, with respect to the coordinates $\mathbf{y}$, is given by

$$\boldsymbol{\eta}(\mathbf{y}) = (\eta_1(\mathbf{y}), \eta_2(\mathbf{y}), \ldots, \eta_n(\mathbf{y})) = \mathrm{Y}\mathbf{y}. \tag{2.44}$$

Note that, using the chain rule, we have

$$\mathrm{X} = \sum_{i=1}^{n} \xi_i(\mathbf{x}) \frac{\partial}{\partial x_i} = \sum_{i,j=1}^{n} \xi_i(\mathbf{x}) \frac{\partial y_j(\mathbf{x})}{\partial x_i} \frac{\partial}{\partial y_j} = \sum_{j=1}^{n} \eta_j(\mathbf{y}) \frac{\partial}{\partial y_j} = \mathrm{Y}.$$

Hence, in order to have Y = X, it is necessary that

$$\eta_j(\mathbf{y}) = \sum_{i=1}^{n} \xi_i(\mathbf{x}) \frac{\partial y_j(\mathbf{x})}{\partial x_i} = \mathrm{X} y_j, \quad j = 1, 2, \ldots, n. \tag{2.45}$$

Theorem 2.3.5-1. *With respect to the coordinates* $\mathbf{y}$ *given by* (2.42), *the one-parameter Lie group of transformations* (2.6) *becomes*

$$\mathbf{y}^* = e^{\varepsilon \mathrm{Y}} \mathbf{y}. \tag{2.46}$$

Proof. From (2.33) and (2.42), we obtain

$$\mathbf{y}^* = \mathbf{Y}(\mathbf{x}^*) = e^{\varepsilon \mathrm{X}} \mathbf{Y}(\mathbf{x}) = e^{\varepsilon \mathrm{X}} \mathbf{Y} = e^{\varepsilon \mathrm{Y}} \mathbf{y}.$$ □

Definition 2.3.5-1. A change of coordinates (2.42) defines a set of *canonical coordinates* for the one-parameter Lie group of transformations (2.6) if, in terms of such coordinates, the group (2.6) becomes

$$y^*_i = y_i, \quad i = 1, 2, \ldots, n-1, \tag{2.47a}$$

$$y^*_n = y_n + \varepsilon. \tag{2.47b}$$

Theorem 2.3.5-2. *For any Lie group of transformations* (2.6), *there exists a set of canonical coordinates* $\mathbf{y} = (y_1, y_2, \ldots, y_n)$ *such that* (2.6) *is equivalent to* (2.47a,b).

Proof. From Theorem 2.3.4-1, we have

$$y^*_i = y_i(\mathbf{x}^*) = y_i(\mathbf{x})$$

if and only if

$$\mathrm{X}y_i(\mathbf{x}) \equiv 0, \quad i = 1,2,\ldots,n-1. \tag{2.48}$$

The homogeneous first-order linear PDE

$$\mathrm{X}u(\mathbf{x}) = \xi_1(\mathbf{x})\frac{\partial u}{\partial x_1} + \xi_2(\mathbf{x})\frac{\partial u}{\partial x_2} + \cdots + \xi_n(\mathbf{x})\frac{\partial u}{\partial x_n} = 0 \tag{2.49}$$

has $n-1$ functionally independent solutions for $u(\mathbf{x})$. These solutions are $n-1$ essential constants $y_1(\mathbf{x}), y_2(\mathbf{x}), \ldots, y_{n-1}(\mathbf{x})$ appearing in the general solution of the system of n first-order ODEs

$$\frac{d\mathbf{x}}{dt} = \boldsymbol{\xi}(\mathbf{x}), \tag{2.50}$$

resulting from the characteristic equations

$$\frac{dx_1}{\xi_1(\mathbf{x})} = \frac{dx_2}{\xi_2(\mathbf{x})} = \cdots = \frac{dx_n}{\xi_n(\mathbf{x})}.$$

This yields the $n-1$ coordinates satisfying (2.47a). □

From Theorem 2.3.4-2, it follows that

$$y^*_n = y_n(\mathbf{x}^*) = y_n(\mathbf{x}) + \varepsilon$$

if and only if

$$\mathrm{X}y_n(\mathbf{x}) \equiv 1. \tag{2.51}$$

Hence, $y_n(\mathbf{x})$ is given by any particular solution $v(\mathbf{x})$ of the nonhomogeneous first-order linear PDE

$$\mathrm{X}v(\mathbf{x}) = \xi_1(\mathbf{x})\frac{\partial v}{\partial x_1} + \xi_2(\mathbf{x})\frac{\partial v}{\partial x_2} + \cdots + \xi_n(\mathbf{x})\frac{\partial v}{\partial x_n} = 1 \tag{2.52}$$

that is solved by determining a particular solution of the corresponding characteristic system of $n+1$ first-order ODEs

$$\frac{dv}{dt} = 1, \tag{2.53a}$$

$$\frac{d\mathbf{x}}{dt} = \xi(\mathbf{x}). \tag{2.53b}$$

Theorem 2.3.5-3. *In terms of any set of canonical coordinates* $\mathbf{y} = (y_1(\mathbf{x}), y_2(\mathbf{x}), \ldots, y_{n-1}(\mathbf{x}))$, *the infinitesimal generator of the one-parameter Lie group of transformations* (2.6) *becomes*

$$\mathrm{Y} = \frac{\partial}{\partial y_n}. \tag{2.54}$$

Proof. We have $\mathrm{Y} = \sum_{i=1}^{n} \eta_i(\mathbf{y}) \frac{\partial}{\partial y_i}$. In terms of canonical coordinates, from (2.48) and (2.51) it follows that

$$\eta_i(\mathbf{y}) = \mathrm{X}y_i = 0, \quad i = 1,2,\ldots,n-1,$$
$$\eta_n(\mathbf{y}) = \mathrm{X}y_n = 1.$$

Hence, we obtain (2.54). □

2.3.6 EXAMPLES OF SETS OF CANONICAL COORDINATES

In $\mathbf{R}^2$, we set $x_1 = x$, $x_2 = y$, and let canonical coordinates be denoted by $y_1 = r$, $y_2 = s$, so that a one-parameter Lie group of transformations becomes

$$r^* = r, \tag{2.55a}$$

$$s^* = s + \varepsilon, \tag{2.55b}$$

with infinitesimal generator

$$\mathrm{Y} = \frac{\partial}{\partial s}.$$

(1) *Group of Scalings*
For the group of scalings

$$x^* = e^{\varepsilon} x, \tag{2.56a}$$

$$y^* = e^{2\varepsilon} y, \tag{2.56b}$$

the infinitesimal generator is given by $\mathrm{X} = x\frac{\partial}{\partial x} + 2y\frac{\partial}{\partial y}$. The canonical coordinate $r(x, y)$ satisfies

$$\mathrm{X}r = x\frac{\partial r}{\partial x} + 2y\frac{\partial r}{\partial y} = 0. \tag{2.57}$$

The corresponding characteristic differential equations reduce to

$$\frac{dy}{dx} = \frac{2y}{x} \tag{2.58}$$

with the general solution given by

$$r(x, y) = \frac{y}{x^2} = \text{const.} \tag{2.59}$$

The canonical coordinate $s(x, y)$ satisfies

$$\mathrm{X}s = x\frac{\partial s}{\partial x} + 2y\frac{\partial s}{\partial y} = 1. \tag{2.60}$$

A particular solution of (2.60) is given by $s(x, y) = s(x)$ satisfying

$$\frac{ds}{dx} = \frac{1}{x}. \tag{2.61}$$

Thus,

$$s(x, y) = \log x, \tag{2.62}$$

and hence, (2.56a,b) has canonical coordinates $(r, s) = (y/x^2, \log x)$.

(2) *Group of Rotations*
For the group of rotations

$$x^* = x\cos\varepsilon - y\sin\varepsilon, \tag{2.63a}$$

$$y^* = x\sin\varepsilon + y\cos\varepsilon, \tag{2.63b}$$

the infinitesimal generator is given by $\mathrm{X} = x\frac{\partial}{\partial y} - y\frac{\partial}{\partial x}$. Correspondingly,

$$r(x, y) = \text{const}$$

is the general solution of

$$\frac{dy}{dx} = -\frac{x}{y} \tag{2.64}$$

so that

$$r = \sqrt{x^2 + y^2}. \tag{2.65}$$

Then a particular solution for $s(x, y)$ is given by $s(x, y) = s(y)$ satisfying

$$\frac{ds}{dy} = \frac{1}{x} = \frac{1}{\sqrt{r^2 - y^2}}. \tag{2.66}$$

Thus,

$$s = \theta = \sin^{-1}\frac{y}{r}. \tag{2.67}$$

Canonical coordinates are the polar coordinates

$$(r, s) = (r, \theta) = \left(\sqrt{x^2 + y^2},\ \sin^{-1}\frac{y}{r}\right), \tag{2.68}$$

in terms of which the rotation group (2.63a,b) is expressed in the usual form

$$r^* = r, \tag{2.69a}$$

$$\theta^* = \theta + \varepsilon. \tag{2.69b}$$

EXERCISES 2.3

1. Consider the rotation group

$$x^* = \sqrt{1-\varepsilon^2}\,x - \varepsilon y, \tag{2.70a}$$

$$y^* = \varepsilon x + \sqrt{1-\varepsilon^2}\,y. \tag{2.70b}$$

   (a) Show that (2.70a,b) defines a one-parameter Lie group of transformations in some neighborhood of $\varepsilon = 0$. In particular, find the law of composition $\phi(a, b)$ and ε^{-1}.
   (b) Determine $\Gamma(\varepsilon)$ and the infinitesimal generator for (2.70a,b).
   (c) Integrate the initial value problem for the infinitesimals to obtain (2.70a,b).
   (d) Parametrize (2.70a,b) in terms of $\tau = \int_0^{\varepsilon}\Gamma(\varepsilon')\,d\varepsilon'$.

2. Formally, consider the one-parameter (ε) family of transformations

$$x^* = x + \varepsilon, \tag{2.71a}$$

$$y^* = \frac{xy}{x+\varepsilon}. \tag{2.71b}$$

   (a) Determine $\Gamma(\varepsilon), \boldsymbol{\xi}(\mathbf{x})$, and explicitly integrate the initial value problem for the infinitesimals to obtain (2.71a,b).
   (b) Find canonical coordinates for (2.71a,b).
   (c) Determine (2.71a,b) in terms of its Lie series developed from $\boldsymbol{\xi}(\mathbf{x})$.

3. For the group of transformations (1.93), find the infinitesimal generator, explicitly integrate out the initial value problem for the infinitesimals, and find canonical coordinates:
   (a) in (x, t)-space; and
   (b) in (x, t, u)-space.

4. Find the one-parameter groups of transformations and canonical coordinates corresponding to the infinitesimal generators:
   (a) $X_1 = x\frac{\partial}{\partial x} + y\frac{\partial}{\partial y}$;
   (b) $X_2 = x\frac{\partial}{\partial x} - y\frac{\partial}{\partial y}$; and
   (c) $X_3 = x^2\frac{\partial}{\partial x} + y^2\frac{\partial}{\partial y}$.

5. Show that $X(\mathbf{x})\mathbf{x}^* = \xi(\mathbf{x}^*)$. Hence, show that $X(\mathbf{x}^*) \equiv X(\mathbf{x}) \equiv X$.

6. For the infinitesimal generator

$$X = x^2\frac{\partial}{\partial x} + xy\frac{\partial}{\partial y} - \left(\tfrac{1}{4}y^2 + \tfrac{1}{2}x\right)z\frac{\partial}{\partial z}:$$

   (a) find canonical coordinates; and
   (b) determine the corresponding one-parameter Lie group of transformations by
      (i) integrating the appropriate initial value problem; and
      (ii) developing it in terms of a Lie series.

2.4 POINT TRANSFORMATIONS AND EXTENDED TRANSFORMATIONS (PROLONGATIONS)

In later chapters, we will be interested in determining one-parameter Lie groups of point transformations admitted by a given system S of differential equations.

Definition 2.4-1. A *one-parameter* (ε) *Lie group of point transformations* is a group of transformations of the form

$$x^* = X(x, u; \varepsilon), \tag{2.72a}$$

$$u^* = U(x, u; \varepsilon), \tag{2.72b}$$

acting on the space of $n + m$ variables

$$x = (x_1, x_2, \ldots, x_n), \tag{2.73}$$

$$u = (u^1, u^2, \ldots, u^m); \tag{2.74}$$

x represents n independent variables and u represents m dependent variables.

A Lie group of point transformations (2.72a,b) admitted by S maps any solution $u = \Theta(x)$ of S into a one-parameter family of solutions $u = \phi(x;\varepsilon)$ of S. Equivalently, a Lie group of point transformations (2.72a,b) leaves S invariant in the sense that the form of S is unchanged in terms of the transformed variables (2.72a,b) for any solution

$u = \Theta(x)$ of S. The expression for $\phi(x;\varepsilon)$ is derived in Section 2.6.2.

Let ∂u denote the set of nm coordinates corresponding to all first order partial derivatives of u with respect to x:

$$\partial u = \left(\frac{\partial u^1}{\partial x_1}, \frac{\partial u^1}{\partial x_2}, \ldots, \frac{\partial u^1}{\partial x_n}, \frac{\partial u^2}{\partial x_1}, \frac{\partial u^2}{\partial x_2}, \ldots, \frac{\partial u^2}{\partial x_n}, \ldots, \frac{\partial u^m}{\partial x_1}, \frac{\partial u^m}{\partial x_2}, \ldots, \frac{\partial u^m}{\partial x_n}\right). \quad (2.75)$$

In general, for $k \geq 1$, let $\partial^k u$ denote the set of coordinates

$$u^\mu_{i_1 i_2 \cdots i_k} = \frac{\partial^k u^\mu}{\partial x_{i_1} \partial x_{i_2} \cdots \partial x_{i_k}} \quad \text{with } \mu = 1, 2, \ldots, m \text{ and } i_j = 1, 2, \ldots, n, \quad j = 1, 2, \ldots, k$$

corresponding to all kth-order partial derivatives of u with respect to x.

It turns out that the natural transformation of partial derivatives of the dependent variables leads successively to *extensions* (*prolongations*) of a one-parameter Lie group of transformations (2.72a,b) acting on (x,u)-space to one-parameter Lie groups of transformations acting on $(x, u, \partial u)$-space, $(x, u, \partial u, \partial^2 u)$-space,..., $(x, u, \partial u, \partial^2 u, \ldots, \partial^k u)$-space for any $k > 2$. [For a given system S of differential equations, k would be the order of the highest order derivative appearing in S.] Then the infinitesimal transformation of (2.72a,b) is naturally extended (prolonged) successively to infinitesimal transformations acting on $(x, u, \partial u, \ldots, \partial^\ell u)$-space, $\ell = 1,2,\ldots,k$.

In the following subsections, because of the importance of scalar differential equations, we consider separately the cases of one dependent ($m = 1$) and one independent variable ($n = 1$); and of one dependent ($m = 1$) and n independent variables. Key results will be stated for the general case of m dependent and n independent variables with proofs left as an exercise.

The motivation for introducing extended transformations is that we can formulate the problem of finding one-parameter Lie groups of transformations of the form (2.72a,b), admitted by a given system S of differential equations, in terms of infinitesimal generators admitted by S. This will be shown to be an algorithmic procedure.

2.4.1 EXTENDED GROUP OF POINT TRANSFORMATIONS: ONE DEPENDENT AND ONE INDEPENDENT VARIABLE

In studying the invariance properties of a kth-order ODE with independent variable x and dependent variable y, the aim is to find admitted one-parameter Lie groups of point transformations of the form

$$x^* = X(x, y; \varepsilon), \quad (2.76a)$$

$$y^* = Y(x, y; \varepsilon), \quad (2.76b)$$

where $y = y(x)$.

Let

$$y_k = y^{(k)} = \frac{d^k y}{dx^k}, \qquad k \geq 1. \tag{2.77}$$

We naturally extend (2.76a,b) to $(x, y, y', \ldots, y^{(k)})$-space, $k \geq 1$, by demanding that (2.76a,b) preserve the contact conditions relating the differentials dx, dy, dy_1, …, dy_k:

$$dy = y_1\, dx, \tag{2.78a}$$

and

$$dy_k = y_{k+1}\, dx, \quad k \geq 1. \tag{2.78b}$$

In particular, under the action of the group of transformations (2.76a,b), the transformed derivatives, y^*_k, $k \geq 1$, are defined successively by

$$dy^* = y^*_1\, dx^*, \tag{2.79a}$$

$$\vdots$$

$$dy^*_k = y^*_{k+1}\, dx^*, \tag{2.79b}$$

where x^* is defined by (2.76a) and y^* by (2.76b). Then

$$dy^* = dY(x, y; \varepsilon) = \frac{\partial Y(x,y;\varepsilon)}{\partial x} dx + \frac{\partial Y(x,y;\varepsilon)}{\partial y} dy, \tag{2.80a}$$

$$dx^* = dX(x, y; \varepsilon) = \frac{\partial X(x,y;\varepsilon)}{\partial x} dx + \frac{\partial X(x,y;\varepsilon)}{\partial y} dy. \tag{2.80b}$$

Consequently, from (2.79a) and (2.80a,b), it follows that y^*_1 satisfies

$$\frac{\partial Y(x,y;\varepsilon)}{\partial x} dx + \frac{\partial Y(x,y;\varepsilon)}{\partial y} dy = y^*_1 \left[\frac{\partial X(x,y;\varepsilon)}{\partial x} dx + \frac{\partial X(x,y;\varepsilon)}{\partial y} dy \right]. \tag{2.81}$$

Substituting (2.78a) into (2.81), we see that

$$y^*_1 = Y_1(x, y, y_1; \varepsilon) = \frac{\dfrac{\partial Y(x,y;\varepsilon)}{\partial x} + y_1 \dfrac{\partial Y(x,y;\varepsilon)}{\partial y}}{\dfrac{\partial X(x,y;\varepsilon)}{\partial x} + y_1 \dfrac{\partial X(x,y;\varepsilon)}{\partial y}}. \tag{2.82}$$

Theorem 2.4.1-1. *The one-parameter Lie group of point transformations* (2.76a,b) *acting on* (x, y)*-space (naturally) extends to the following one-parameter Lie group of transformations acting on* (x, y, y_1)*-space:*

$$x^* = X(x, y; \varepsilon), \tag{2.83a}$$

$$y^* = Y(x, y; \varepsilon), \tag{2.83b}$$

$$y^*_1 = Y_1(x, y, y_1; \varepsilon), \tag{2.83c}$$

where $Y_1(x, y, y_1; \varepsilon)$ *is given by* (2.82).

Proof. The proof is accomplished by showing that the closure property is preserved in this *first extension* of (2.76a,b) to (x, y, y_1)-space. The other properties of a one-parameter Lie group of transformations then follow immediately for this first extension.

Let $\phi(\varepsilon, \delta)$ define the law of composition of parameters ε and δ. Let

$$(x^{**}, y^{**}) = (X(x^*, y^*; \delta), Y(x^*, y^*; \delta)). \tag{2.84}$$

Then, from the closure property of the group (2.76a,b), it follows that

$$(x^{**}, y^{**}) = (X(x, y; \phi(\varepsilon, \delta)),\ Y(x, y; \phi(\varepsilon, \delta))).$$

But y^{**}_1 satisfies $dy^{**} = y^{**}_1\, dx^{**}$. Consequently,

$$y^{**}_1 = Y_1(x, y, y_1; \phi(\varepsilon, \delta)) = \frac{\dfrac{\partial Y(x, y; \phi(\varepsilon, \delta))}{\partial x} + y_1 \dfrac{\partial Y(x, y; \phi(\varepsilon, \delta))}{\partial y}}{\dfrac{\partial X(x, y; \phi(\varepsilon, \delta))}{\partial x} + y_1 \dfrac{\partial X(x, y; \phi(\varepsilon, \delta))}{\partial y}}. \qquad \square$$

Theorem 2.4.1-2. *The second extension of the one-parameter Lie group of point transformations* (2.76a,b) *is the following one-parameter Lie group of transformations acting on* (x, y, y_1, y_2)*-space*:

$$x^* = X(x, y; \varepsilon), \tag{2.85a}$$

$$y^* = Y(x, y; \varepsilon), \tag{2.85b}$$

$$y^*_1 = Y_1(x, y, y_1; \varepsilon), \tag{2.85c}$$

$$y^*_2 = Y_2(x, y, y_1, y_2; \varepsilon) = \frac{\dfrac{\partial Y_1}{\partial x} + y_1 \dfrac{\partial Y_1}{\partial y} + y_2 \dfrac{\partial Y_1}{\partial y_1}}{\dfrac{\partial X(x, y; \varepsilon)}{\partial x} + y_1 \dfrac{\partial X(x, y; \varepsilon)}{\partial y}}, \tag{2.85d}$$

where $Y_1 = Y_1(x, y, y_1; \varepsilon)$ *is defined by* (2.82).

The proof of Theorem 2.4.1-2 is left to Exercise 2.4-2. □

The proof of the next theorem follows by induction:

Theorem 2.4.1-3. *The kth extension of the one-parameter Lie group of point transformations* (2.76a,b), $k \geq 2$, *is the following one-parameter Lie group of transformations acting on* $(x, y, y_1, \ldots, y_k)$*-space*:

$$x^* = X(x, y; \varepsilon), \tag{2.86a}$$

$$y^* = Y(x, y; \varepsilon), \tag{2.86b}$$

$$y^*_1 = Y_1(x, y, y_1; \varepsilon), \tag{2.86c}$$

$$\vdots$$

$$y^*_k = Y_k(x, y, y_1, \ldots, y_k; \varepsilon) = \frac{\dfrac{\partial Y_{k-1}}{\partial x} + y_1 \dfrac{\partial Y_{k-1}}{\partial y} + \cdots + y_k \dfrac{\partial Y_{k-1}}{\partial y_{k-1}}}{\dfrac{\partial X(x, y; \varepsilon)}{\partial x} + y_1 \dfrac{\partial X(x, y; \varepsilon)}{\partial y}}, \tag{2.86d}$$

where $Y_1 = Y_1(x, y, y_1; \varepsilon)$ *is defined by* (2.82), *and* $Y_i = Y_i(x, y, y_1, \ldots, y_i; \varepsilon),\ i = 1, 2, \ldots, k.$

Note that we can extend any set of one-to-one transformations (not necessarily a group of transformations)

$$x^\dagger = X(x, y), \tag{2.87a}$$

$$y^\dagger = Y(x, y), \tag{2.87b}$$

from some domain D in (x, y)-space to another domain $D^\dagger$ in $(x^\dagger, y^\dagger)$-space, where the functions $X(x, y)$ and $Y(x, y)$ are k times differentiable in D. One can naturally extend the transformations (2.87a,b) to $(x, y, y_1, \ldots, y_k)$-space so that the contact conditions (2.79a,b) are preserved, i.e.,

$$dy^\dagger = y_1^\dagger\, dx^\dagger, \tag{2.88a}$$

$$dy_k^\dagger = dy_{k+1}^\dagger\, dx^\dagger, \quad k \geq 1. \tag{2.88b}$$

Here the kth-extended transformation from $(x, y, y_1, \ldots, y_k)$-space to $(x^\dagger, y^\dagger, y_1^\dagger, \ldots, y_k^\dagger)$-space is given by

$$x^\dagger = X(x, y), \tag{2.89a}$$

$$y^\dagger = Y(x, y), \tag{2.89b}$$

$$y_1^\dagger = Y_1(x, y, y_1), \tag{2.89c}$$

$$\vdots$$

$$y_k^\dagger = Y_k(x, y, y_1, \ldots, y_k) = \frac{\dfrac{\partial Y_{k-1}}{\partial x} + y_1 \dfrac{\partial Y_{k-1}}{\partial y} + \cdots + y_k \dfrac{\partial Y_{k-1}}{\partial y_{k-1}}}{\dfrac{\partial X(x, y)}{\partial x} + y_1 \dfrac{\partial X(x, y)}{\partial y}}, \tag{2.89d}$$

where

$$Y_1 = Y_1(x, y, y_1) = \frac{\dfrac{\partial Y(x,y)}{\partial x} + y_1 \dfrac{\partial Y(x,y)}{\partial y}}{\dfrac{\partial X(x,y)}{\partial x} + y_1 \dfrac{\partial X(x,y)}{\partial y}},$$

and $Y_i = Y_i(x, y, y_1, \dots, y_i)$, $i = 1, 2, \dots, k-1$.

Now consider examples of extended group transformations.

(1) *Translation Group*
For the translation group

$$x^* = X = x + \varepsilon, \tag{2.90a}$$

$$y^* = Y = y, \tag{2.90b}$$

we have

$$y^*_1 = \left(\frac{dy}{dx}\right)^* = \frac{dy^*}{dx^*} = Y_1 = \frac{dy}{dx} = y_1,$$

and, in general,

$$y^*_k = \left(\frac{d^k y}{dx^k}\right)^* = \frac{d^k y^*}{dx^{*k}} = Y_k = \frac{d^k y}{dx^k} = y_k, \quad k \ge 1.$$

Then, for the translation group (2.90a,b), the kth-extended group is given by

$$x^* = x + \varepsilon, \tag{2.91a}$$

$$y^* = y, \tag{2.91b}$$

$$y^*_i = y_i, \quad i = 1, \dots, k. \tag{2.91c}$$

(2) *Scaling Group*
For the scaling group

$$x^* = X = e^{\varepsilon} x, \tag{2.92a}$$

$$y^* = Y = e^{2\varepsilon} y, \tag{2.92b}$$

we have

$$y^*_1 = \left(\frac{dy}{dx}\right)^* = \frac{dy^*}{dx^*} = Y_1 = \frac{y_1 \dfrac{\partial Y}{\partial y}}{\dfrac{\partial X}{\partial x}} = e^{\varepsilon} y_1,$$

and, in general,

$$y^*_k = \left(\frac{d^k y}{dx^k}\right)^* = \frac{d^k y^*}{dx^{*k}} = Y_k = \frac{y_k \dfrac{\partial Y_{k-1}}{\partial y_{k-1}}}{\dfrac{\partial X}{\partial x}} = e^{(2-k)\varepsilon} y_k, \qquad k \geq 1.$$

Here the kth-extended Lie group of transformations is given by

$$x^* = e^{\varepsilon} x, \tag{2.93a}$$

$$y^* = e^{2\varepsilon} y, \tag{2.93b}$$

$$y^*_i = e^{(2-i)\varepsilon} y_i, \quad i = 1, 2, \ldots, k. \tag{2.93c}$$

(3) *Rotation Group*
For the rotation group

$$x^* = X = x \cos\varepsilon + y \sin\varepsilon, \tag{2.94a}$$

$$y^* = Y = -x \sin\varepsilon + y \cos\varepsilon, \tag{2.94b}$$

we obtain

$$\frac{\partial X}{\partial x} = \cos\varepsilon, \quad \frac{\partial X}{\partial y} = \sin\varepsilon, \quad \frac{\partial Y}{\partial x} = -\sin\varepsilon, \quad \frac{\partial Y}{\partial y} = \cos\varepsilon.$$

Hence, from (2.83), we obtain

$$Y_1 = \frac{-\sin\varepsilon + y_1 \cos\varepsilon}{\cos\varepsilon + y_1 \sin\varepsilon}.$$

Then

$$\frac{\partial Y_1}{\partial x} = \frac{\partial Y_1}{\partial y} = 0, \quad \frac{\partial Y_1}{\partial y_1} = \frac{1}{(\cos\varepsilon + y_1 \sin\varepsilon)^2}.$$

Consequently, from (2.85d), we have

$$Y_2 = \frac{y_2}{(\cos\varepsilon + y_1 \sin\varepsilon)^3}.$$

Then

$$\frac{\partial Y_2}{\partial x} = \frac{\partial Y_2}{\partial y} = 0, \quad \frac{\partial Y_2}{\partial y_1} = \frac{-3(\sin\varepsilon) y_2}{(\cos\varepsilon + y_1 \sin\varepsilon)^4}, \quad \frac{\partial Y_2}{\partial y_2} = \frac{1}{(\cos\varepsilon + y_1 \sin\varepsilon)^3}.$$

As a result, from (2.86d), we get

$$Y_3 = \frac{(y_1 \sin\varepsilon + \cos\varepsilon) y_3 - 3(y_2)^2 \sin\varepsilon}{(\cos\varepsilon + y_1 \sin\varepsilon)^5}.$$

Thus, the third-extended Lie group of transformations corresponding to (2.94a,b) is given by

$$x^* = X = x\cos\varepsilon + y\sin\varepsilon, \tag{2.95a}$$

$$y^* = Y = -x\sin\varepsilon + y\cos\varepsilon, \tag{2.95b}$$

$$y^*_1 = \frac{-\sin\varepsilon + y_1\cos\varepsilon}{\cos\varepsilon + y_1\sin\varepsilon}, \tag{2.95c}$$

$$y^*_2 = \frac{y_2}{(\cos\varepsilon + y_1\sin\varepsilon)^3}, \tag{2.95d}$$

$$y^*_3 = \frac{(y_1\sin\varepsilon + \cos\varepsilon)y_3 - 3(y_2)^2\sin\varepsilon}{(\cos\varepsilon + y_1\sin\varepsilon)^5}. \tag{2.95e}$$

This is a one-parameter Lie group of transformations acting on (x, y, y_1, y_2, y_3)-space. Of course, one can extend this Lie group of transformations successively to $(x, y, y_1, y_2, y_3, \ldots, y_k)$-space, $k = 4,5,\ldots$, but the calculations quickly get more and more complicated as k increases.

From Section 2.3, we know that a one-parameter Lie group of point transformations is characterized by its infinitesimal generator. Since the kth extension of a one-parameter Lie group of point transformations is also a one-parameter Lie group of transformations, it follows that the study of extended Lie groups of transformations reduces to the study of extended infinitesimal transformations. In the next subsection we formulate Theorems 2.4.1-1 to 2.4.1-3 in terms of infinitesimal transformations. Consequently, we will have an explicit algorithm to determine the extended infinitesimal transformations (and the corresponding infinitesimal generators) of an infinitesimal transformation.

Before proceeding further, we introduce the following convenient notations: Let a subscript denote differentiation with respect to the corresponding coordinate, e.g., $F_x = \partial F/\partial x$, $F_y = \partial F/\partial y$.

Definition 2.4.1-1. The *total derivative operator* is defined by

$$\mathrm{D} = \frac{\partial}{\partial x} + y_1\frac{\partial}{\partial y} + y_2\frac{\partial}{\partial y_1} + \cdots + y_{n+1}\frac{\partial}{\partial y_n} + \cdots. \tag{2.96}$$

For a differentiable function $F(x, y, y_1, y_2, \ldots, y_\ell)$, its total derivative is then given by

$$\mathrm{D}F(x, y, y_1, y_2, \ldots, y_\ell) = F_x + y_1F_y + y_2F_{y_1} + \cdots + y_{\ell+1}F_{y_\ell}.$$

In terms of the total derivative operator (2.96), the kth extension of the one-parameter Lie group of point transformations (2.86a,b) is given by

$$x^* = X(x, y; \varepsilon), \tag{2.97a}$$

$$y^* = Y(x, y; \varepsilon), \tag{2.97b}$$

$$y^*_i = Y_i(x, y, y_1, \ldots, y_i; \varepsilon) = \frac{\mathrm{D}Y_{i-1}(x, y, y_1, \ldots, y_{i-1}; \varepsilon)}{\mathrm{D}X(x, y; \varepsilon)}, \quad i = 1,2,\ldots,k, \tag{2.97c}$$

where $Y_0 = Y(x, y; \varepsilon)$.

2.4.2 EXTENDED INFINITESIMAL TRANSFORMATIONS: ONE DEPENDENT AND ONE INDEPENDENT VARIABLE

The one-parameter Lie group of point transformations

$$x^* = X(x, y; \varepsilon) = x + \varepsilon \xi(x, y) + O(\varepsilon^2), \tag{2.98a}$$

$$y^* = Y(x, y; \varepsilon) = y + \varepsilon \eta(x, y) + O(\varepsilon^2), \tag{2.98b}$$

acting on (x, y)-space, has infinitesimals

$$\xi(x, y), \eta(x, y), \tag{2.98c}$$

with the corresponding infinitesimal generator

$$\mathrm{X} = \xi(x, y)\frac{\partial}{\partial x} + \eta(x, y)\frac{\partial}{\partial y}. \tag{2.98d}$$

The kth extension of (2.98a,b), given by

$$x^* = X(x, y; \varepsilon) = x + \varepsilon \xi(x, y) + O(\varepsilon^2), \tag{2.99a}$$

$$y^* = Y(x, y; \varepsilon) = y + \varepsilon \eta(x, y) + O(\varepsilon^2), \tag{2.99b}$$

$$y^*_1 = Y_1(x, y, y_1; \varepsilon) = y_1 + \varepsilon \eta^{(1)}(x, y, y_1) + O(\varepsilon^2), \tag{2.99c}$$

$$\vdots$$

$$y^*_k = Y_k(x, y, y_1, \ldots, y_k; \varepsilon) = y_k + \varepsilon \eta^{(k)}(x, y, y_1, \ldots, y_k) + O(\varepsilon^2), \tag{2.99d}$$

has *kth-extended infinitesimals*

$$\xi(x, y), \eta(x, y), \eta^{(1)}(x, y, y_1), \ldots, \eta^{(k)}(x, y, y_1, \ldots, y_k), \tag{2.99e}$$

with the corresponding *kth-extended infinitesimal generator*

$$\mathrm{X}^{(k)} = \xi(x, y)\frac{\partial}{\partial x} + \eta(x, y)\frac{\partial}{\partial y} + \eta^{(1)}(x, y, y_1)\frac{\partial}{\partial y_1} + \cdots + \eta^{(k)}(x, y, y_1, \ldots, y_k)\frac{\partial}{\partial y_k}, \tag{2.99f}$$

$k = 1,2,\ldots$. Explicit formulas for the extended infinitesimals $\eta^{(k)}$ result from the following theorem:

Theorem 2.4.2-1. *The extended infinitesimals* $\eta^{(k)}$ *satisfy the recursion relation*

$$\eta^{(k)}(x, y, y_1, \ldots, y_k) = \mathrm{D}\eta^{(k-1)}(x, y, y_1, \ldots, y_{k-1}) - y_k \mathrm{D}\xi, \quad k = 1,2,\ldots, \tag{2.100a}$$

where $\eta^{(0)} = \eta(x, y)$. *In particular,*

$$\eta^{(k)} = \mathrm{D}^k\eta - \sum_{j=1}^{k} \frac{k!}{(k-j)!\,j!} y_{k-j+1}\mathrm{D}^j\xi, \quad k \geq 1. \tag{2.100b}$$

Proof. From (2.97c), (2.99a–d), and (2.96), we have

$$\begin{aligned} Y_k(x, y, y_1, \ldots, y_k) &= \frac{\mathrm{D}Y_{k-1}}{\mathrm{D}X} = \frac{\mathrm{D}[y_{k-1} + \varepsilon\eta^{(k-1)} + O(\varepsilon^2)]}{\mathrm{D}[x + \varepsilon\xi + O(\varepsilon^2)]} = \frac{y_k + \varepsilon\mathrm{D}\eta^{(k-1)}}{1 + \varepsilon\mathrm{D}\xi} + O(\varepsilon^2) \\ &= y_k + \varepsilon[\mathrm{D}\eta^{(k-1)} - y_k\mathrm{D}\xi] + O(\varepsilon^2) = y_k + \varepsilon\eta^{(k)} + O(\varepsilon^2), \end{aligned}$$

leading to (2.100a). Then we obtain (2.100b) by finite induction on k. □

Explicit formulas for $\eta^{(k)}$ follow immediately from Theorem 2.4.2-1. In particular,

$$\eta^{(1)} = \eta_x + (\eta_y - \xi_x)y_1 - \xi_y(y_1)^2, \tag{2.101}$$

$$\begin{aligned} \eta^{(2)} &= \eta_{xx} + (2\eta_{xy} - \xi_{xx})y_1 + (\eta_{yy} - 2\xi_{xy})(y_1)^2 \\ &\quad - \xi_{yy}(y_1)^3 + (\eta_y - 2\xi_x)y_2 - 3\xi_y y_1 y_2, \end{aligned} \tag{2.102}$$

$$\begin{aligned} \eta^{(3)} &= \eta_{xxx} + (3\eta_{xxy} - \xi_{xxx})y_1 + 3(\eta_{xyy} - 2\xi_{xxy})(y_1)^2 \\ &\quad + (\eta_{yyy} - 3\xi_{xyy})(y_1)^3 - \xi_{yyy}(y_1)^4 + 3(\eta_{xy} - \xi_{xx})y_2 \\ &\quad + 3(\eta_{yy} - 3\xi_{xy})y_1 y_2 - 6\xi_{yy}(y_1)^2 y_2 \\ &\quad - 3\xi(y_2)^2 + (\eta_y - 3\xi_x)y_3 - 4\xi_y y_1 y_3. \end{aligned} \tag{2.103}$$

The following observations are important:

Theorem 2.4.2-2. *The extended infinitesimals* $\eta^{(k)}$ *have the following properties:*

(i) $\eta^{(k)}$ *is linear in* y_k *for* $k \geq 2$.

(ii) $\eta^{(k)}$ *is a polynomial in* $y_1, y_2, \ldots, y_k$, *whose coefficients are linear homogeneous in* $\xi(x, y)$, $\eta(x, y)$, *and their partial derivatives up to kth-order.*

Proof. Left to Exercise 2.4-5. □

We now find the extended infinitesimals $\eta^{(k)}$ for the examples of Section 2.4.1.

(1) *Translation Group* (2.90a,b)
Here

$$\eta^{(k)} = 0, \; k \geq 1.$$

(2) *Scaling Group* (2.92a,b)
From the form of (2.93c), it is immediately obvious that

$$\eta^{(k)} = (2-k)y_k, \quad k \geq 1.$$

(3) *Rotation Group* (2.94a,b)
Here $\xi(x,y) = y$, $\eta(x,y) = -x$. So $\xi_y = 1$, $\eta_x = -1$, $\xi_x = \eta_y = 0$. From (2.101)–(2.103), we see that

$$\begin{aligned} \eta^{(1)} &= -[1+(y_1)^2], \\ \eta^{(2)} &= -3y_1y_2, \\ \eta^{(3)} &= -[3(y_2)^2 + 4y_1y_3], \end{aligned}$$

From (2.100a), for $k \geq 4$, we have $\eta^{(k)} = \mathrm{D}\eta^{(k-1)} - y_k y_1$, so that

$$\begin{aligned} \eta^{(4)} &= -5[2y_2y_3 + y_1y_4], \\ \eta^{(5)} &= -[10(y_3)^2 + 15y_2y_4 + 6y_1y_5], \text{ etc.} \end{aligned}$$

2.4.3 EXTENDED TRANSFORMATIONS: ONE DEPENDENT AND *n* INDEPENDENT VARIABLES

In studying the invariance properties of a *k*th-order PDE with one dependent variable *u* and *n* independent variables $x = (x_1, x_2, \ldots, x_n)$, with $u = u(x)$, we are naturally led to the problem of finding the extensions of transformations on (x,u)-space to $(x,u,\partial u,\ldots,\partial^k u)$-space where $\partial^k u$ denotes the components of all *k*th-order partial derivatives of *u* with respect to x.

First we consider the extended transformations of a *point transformation*

$$x^{\dagger} = X(x,u), \tag{2.104a}$$

$$u^{\dagger} = U(x,u). \tag{2.104b}$$

The transformation (2.104a,b) is assumed to be one-to-one on some domain *D* in (*x*,*u*)-space with functions $X(x,u), U(x,u)$ that are *k* times differentiable in *D*. The transformation (2.104a,b) preserves the contact conditions, i.e.,

$$du = \partial u \; dx, \tag{2.105a}$$

$$\vdots$$

$$d\;\partial u^{k-1} = \partial u^k \; dx, \tag{2.105b}$$

in some domain *D* in $(x,u,\partial u,\ldots,\partial^k u)$-space if and only if

$$du^{\dagger} = \partial u^{\dagger} \; dx^{\dagger}, \tag{2.106a}$$

$$\vdots$$

$$d\,\partial^{k-1}u^\dagger = \partial^k u^\dagger\, dx^\dagger, \tag{2.106b}$$

in the corresponding domain $D^\dagger$ in $(x^\dagger, u^\dagger, \partial u^\dagger, \ldots, \partial^k u^\dagger)$-space.

In order to express the contact conditions in an explicit form, we let

$$u_i = \frac{\partial u}{\partial x_i}, \quad u_i^\dagger = \frac{\partial u^\dagger}{\partial x_i^\dagger} = \frac{\partial U}{\partial X_i}, \quad \text{etc.}$$

From now on, we assume summation over a repeated index. The conditions (2.105a,b) are given by the set of equations

$$du = u_j\, dx_j,$$
$$du_{i_1 i_2 \cdots i_{k-1}} = u_{i_1 i_2 \cdots i_{k-1} j} dx_j, \quad i_\ell = 1,2,\ldots,n \text{ for } \ell = 1,2,\ldots,k-1.$$

Similar representations hold for (2.106a,b).

We introduce the total derivative operators

$$\mathrm{D}_i = \frac{\partial}{\partial x_i} + u_i \frac{\partial}{\partial u} + u_{ij} \frac{\partial}{\partial u_j} + \cdots + u_{i i_1 i_2 \cdots i_n} \frac{\partial}{\partial u_{i_1 i_2 \cdots i_n}} + \cdots, \quad i = 1,2,\ldots,n. \tag{2.107}$$

For a given differentiable function $F(x, u, \partial u, \ldots, \partial^\ell u)$, we have

$$\mathrm{D}_i F(x, u, \partial u, \ldots, \partial^\ell u) = \frac{\partial F}{\partial x_i} + u_i \frac{\partial F}{\partial u} + u_{ij} \frac{\partial F}{\partial u_j} + \cdots + u_{i i_1 i_2 \cdots i_\ell} \frac{\partial F}{\partial u_{i_1 i_2 \cdots i_\ell}}, \quad i = 1,2,\ldots,n.$$

We now determine the extended transformation

$$u_j^\dagger = U_j(x, u, \partial u), \quad j = 1,2,\ldots,n. \tag{2.108}$$

From (2.104a,b), we obtain

$$du^\dagger = u_i^\dagger\, dx_i^\dagger = (\mathrm{D}_i U)\, dx_i,$$

and

$$dx_j^\dagger = (\mathrm{D}_i X_j)\, dx_i, \quad j = 1,2,\ldots,n,$$

where D_i is defined by (2.107), $i = 1,2,\ldots,n.$ Then

$$(\mathrm{D}_i X_j) u_j^\dagger = \mathrm{D}_i U, \quad i = 1,2,\ldots,n.$$

Let A be the $n \times n$ matrix

$$A = \begin{bmatrix} \mathrm{D}_1 X_1 & \cdots & \mathrm{D}_1 X_n \\ \vdots & & \vdots \\ \mathrm{D}_n X_1 & \cdots & \mathrm{D}_n X_n \end{bmatrix} \tag{2.109}$$

and assume that A^{-1} exists. Then

$$\begin{bmatrix} u_1^\dagger \\ u_2^\dagger \\ \vdots \\ u_n^\dagger \end{bmatrix} = \begin{bmatrix} U_1 \\ U_2 \\ \vdots \\ U_n \end{bmatrix} = A^{-1} \begin{bmatrix} \mathrm{D}_1 U \\ \mathrm{D}_2 U \\ \vdots \\ \mathrm{D}_n U \end{bmatrix}. \tag{2.110}$$

This leads to the extended transformation in $(x, u, \partial u)$-space given by

$$x^\dagger = X(x, u), \tag{2.111a}$$

$$u^\dagger = U(x, u), \tag{2.111b}$$

$$\partial u^\dagger = \partial U(x, u, \partial u). \tag{2.111c}$$

It is easy to show that the extension to $(x, u, \partial u, \ldots, \partial^k u)$-space is given by

$$x^\dagger = X(x, u), \tag{2.112a}$$

$$u^\dagger = U(x, u), \tag{2.112b}$$

$$\partial u^\dagger = \partial U(x, u, \partial u), \tag{2.112c}$$

$$\vdots$$

$$\partial^k u^\dagger = \partial^k U(x, u, \partial u, \ldots, \partial^k u), \tag{2.112d}$$

where the components of $\partial^k u^\dagger$ are determined by

$$\begin{bmatrix} u_{i_1 i_2 \cdots i_{k-1} 1}^\dagger \\ u_{i_1 i_2 \cdots i_{k-1} 2}^\dagger \\ \vdots \\ u_{i_1 i_2 \cdots i_{k-1} n}^\dagger \end{bmatrix} = \begin{bmatrix} U_{i_1 i_2 \cdots i_{k-1} 1} \\ U_{i_1 i_2 \cdots i_{k-1} 2} \\ \vdots \\ U_{i_1 i_2 \cdots i_{k-1} n} \end{bmatrix} = A^{-1} \begin{bmatrix} \mathrm{D}_1 U_{i_1 i_2 \cdots i_{k-1}} \\ \mathrm{D}_2 U_{i_1 i_2 \cdots i_{k-1}} \\ \vdots \\ \mathrm{D}_n U_{i_1 i_2 \cdots i_{k-1}} \end{bmatrix}, \tag{2.113}$$

$i_\ell = 1, 2, \ldots, n$ for $\ell = 1, 2, \ldots, k-1$, with $k \geq 2$; $\partial U(x, u, \partial u)$ is determined by (2.110), and A is the matrix (2.109).

Now we specialize to the case where (2.104a,b) defines a Lie group of point transformations

$$x^* = X(x, u; \varepsilon), \tag{2.114a}$$

$$u^* = U(x, u; \varepsilon), \tag{2.114b}$$

acting on (x, u)-space. Then it is easy to show (following the proofs of Theorems 2.4.1-1 to 2.4.1-3) that its kth extension to $(x, u, \partial u, \ldots, \partial^k u)$-space, given by

$$x^* = X(x, u; \varepsilon), \tag{2.115a}$$

$$u^* = U(x, u; \varepsilon), \tag{2.115b}$$

$$\partial u^* = \partial U(x, u, \partial u; \varepsilon), \tag{2.115c}$$

$$\vdots$$

$$\partial^k u^* = \partial^k U(x, u, \partial u, \ldots, \partial^k u; \varepsilon), \tag{2.115d}$$

defines a kth-extended one-parameter Lie group of transformations. In (2.115a–d),

$$\begin{bmatrix} u^*_1 \\ u^*_2 \\ \vdots \\ u^*_n \end{bmatrix} = \begin{bmatrix} U_1 \\ U_2 \\ \vdots \\ U_n \end{bmatrix} = A^{-1} \begin{bmatrix} \mathrm{D}_1 U \\ \mathrm{D}_2 U \\ \vdots \\ \mathrm{D}_n U \end{bmatrix}, \tag{2.116a}$$

$$\begin{bmatrix} u^*_{i_1 i_2 \cdots i_{k-1} 1} \\ u^*_{i_1 i_2 \cdots i_{k-1} 2} \\ \vdots \\ u^*_{i_1 i_2 \cdots i_{k-1} n} \end{bmatrix} = \begin{bmatrix} U_{i_1 i_2 \cdots i_{k-1} 1} \\ U_{i_1 i_2 \cdots i_{k-1} 2} \\ \vdots \\ U_{i_1 i_2 \cdots i_{k-1} n} \end{bmatrix} = A^{-1} \begin{bmatrix} \mathrm{D}_1 U_{i_1 i_2 \cdots i_{k-1}} \\ \mathrm{D}_2 U_{i_1 i_2 \cdots i_{k-1}} \\ \vdots \\ \mathrm{D}_n U_{i_1 i_2 \cdots i_{k-1}} \end{bmatrix}, \tag{2.116b}$$

where $u^*_i = U_i$ are the components of $\partial u^* = \partial U$ and $u^*_{i_1 i_2 \ldots i_{k-1} i} = U_{i_1 i_2 \ldots i_{k-1} i}$ are the components of $\partial^k u^* = \partial^k U$. In (2.116b), $i_\ell = 1,2,\ldots,n$ for $\ell = 1,2,\ldots,k-1$ with $k \geq 2$; the operators D_i are given by (2.107); and A^{-1} is the inverse of the matrix A given by (2.109) for X and U given by (2.115a,b).

2.4.4 EXTENDED INFINITESIMAL TRANSFORMATIONS: ONE DEPENDENT AND n INDEPENDENT VARIABLES

The one-parameter Lie group of point transformations

$$x^*_i = X_i(x, u; \varepsilon) = x_i + \varepsilon \xi_i(x, u) + O(\varepsilon^2), \tag{2.117a}$$

$$u^* = U(x, u; \varepsilon) = u + \varepsilon \eta(x, u) + O(\varepsilon^2), \tag{2.117b}$$

$i = 1,2,\ldots,n$, acting on (x, u)-space, has infinitesimal generator

$$\mathrm{X} = \xi_i(x, u)\frac{\partial}{\partial x_i} + \eta(x, u)\frac{\partial}{\partial u}. \tag{2.117c}$$

The *kth extension* of (2.117a,b) is given by

$$x^*_i = X_i(x, u; \varepsilon) = x_i + \varepsilon \xi_i(x, u) + O(\varepsilon^2), \tag{2.118a}$$

$$u^* = U(x, u; \varepsilon) = u + \varepsilon \eta(x, u) + O(\varepsilon^2), \tag{2.118b}$$

$$u^*_i = U_i(x, u, \partial u; \varepsilon) = u_i + \varepsilon \eta_i^{(1)}(x, u, \partial u) + O(\varepsilon^2), \tag{2.118c}$$

$$\vdots$$

$$u^*_{i_1 i_2 \cdots i_k} = U_{i_1 i_2 \cdots i_k}(x, u, \partial u, \ldots, \partial^k u; \varepsilon) = u_{i_1 i_2 \cdots i_k} + \varepsilon \eta^{(k)}_{i_1 i_2 \cdots i_k}(x, u, \partial u, \ldots, \partial^k u) + O(\varepsilon^2), \tag{2.118d}$$

where $i = 1,2,\ldots,n$ and $i_\ell = 1,2,\ldots,n$ for $\ell = 1,2,\ldots,k$ with $k \geq 1$. Its *kth-extended infinitesimals* are

$$\xi(x,u),\ \eta(x,u),\ \eta^{(1)}(x,u,\partial u),\ldots,\ \eta^{(k)}(x,u,\partial u,\ldots,\partial^k u), \tag{2.118e}$$

with the corresponding *kth-extended infinitesimal generator*

$$\mathrm{X}^{(k)} = \xi_i \frac{\partial}{\partial x_i} + \eta \frac{\partial}{\partial u} + \eta_i^{(1)} \frac{\partial}{\partial u_i} + \cdots + \eta_{i_1 i_2 \cdots i_k}^{(k)} \frac{\partial}{\partial u_{i_1 i_2 \cdots i_k}}, \quad k \geq 1. \tag{2.118f}$$

Explicit formulas for the extended infinitesimals $\eta^{(k)}$ result from the following theorem:

Theorem 2.4.4-1. *The extended infinitesimals satisfy the recursion relations*

$$\eta_i^{(1)} = \mathrm{D}_i \eta - (\mathrm{D}_i \xi_j) u_j, \quad i = 1,2,\ldots,n, \tag{2.119a}$$

$$\eta_{i_1 i_2 \cdots i_k}^{(k)} = \mathrm{D}_{i_k} \eta_{i_1 i_2 \cdots i_{k-1}}^{(k-1)} - (\mathrm{D}_{i_k} \xi_j) u_{i_1 i_2 \cdots i_{k-1} j}, \tag{2.119b}$$

$i_\ell = 1,2,\ldots,n$ *for* $\ell = 1,2,\ldots,k$ *with* $k \geq 2$.

Proof. From (2.109) and (2.118a), we have

$$A = \begin{bmatrix} \mathrm{D}_1(x_1 + \varepsilon\xi_1) & \mathrm{D}_1(x_2 + \varepsilon\xi_2) & \cdots & \mathrm{D}_1(x_n + \varepsilon\xi_n) \\ \mathrm{D}_2(x_1 + \varepsilon\xi_1) & \mathrm{D}_2(x_2 + \varepsilon\xi_2) & \cdots & \mathrm{D}_2(x_n + \varepsilon\xi_n) \\ \vdots & \vdots & & \vdots \\ \mathrm{D}_n(x_1 + \varepsilon\xi_1) & \mathrm{D}_n(x_2 + \varepsilon\xi_2) & \cdots & \mathrm{D}_n(x_n + \varepsilon\xi_n) \end{bmatrix} + O(\varepsilon^2) = I + \varepsilon B + O(\varepsilon^2),$$

where I is the $n \times n$ identity matrix and

$$B = \begin{bmatrix} \mathrm{D}_1\xi_1 & \mathrm{D}_1\xi_2 & \cdots & \mathrm{D}_1\xi_n \\ \mathrm{D}_2\xi_1 & \mathrm{D}_2\xi_2 & \cdots & \mathrm{D}_2\xi_n \\ \vdots & \vdots & & \vdots \\ \mathrm{D}_n\xi_1 & \mathrm{D}_n\xi_2 & \cdots & \mathrm{D}_n\xi_n \end{bmatrix}. \tag{2.120}$$

Then

$$A^{-1} = I - \varepsilon B + O(\varepsilon^2). \tag{2.121}$$

From (2.116a), (2.118b,c), (2.120), and (2.121), it follows that

$$\begin{bmatrix} u_1 + \varepsilon\eta_1^{(1)} \\ u_2 + \varepsilon\eta_2^{(1)} \\ \vdots \\ u_n + \varepsilon\eta_n^{(1)} \end{bmatrix} = [I - \varepsilon B] \begin{bmatrix} u_1 + \varepsilon \mathrm{D}_1\eta \\ u_2 + \varepsilon \mathrm{D}_2\eta \\ \vdots \\ u_n + \varepsilon \mathrm{D}_n\eta \end{bmatrix} + O(\varepsilon^2),$$

and, thus,

$$\begin{bmatrix} \eta_1^{(1)} \\ \eta_2^{(1)} \\ \vdots \\ \eta_n^{(1)} \end{bmatrix} = \begin{bmatrix} D_1\eta \\ D_2\eta \\ \vdots \\ D_n\eta \end{bmatrix} - B \begin{bmatrix} u_1 \\ u_2 \\ \vdots \\ u_n \end{bmatrix},$$

leading to (2.119a). Then, from (2.116b), (2.118c,d), (2.120), and (2.121), we get

$$\begin{bmatrix} u_{i_1i_2\cdots i_{k-1}1} + \varepsilon\eta^{(k)}_{i_1i_2\cdots i_{k-1}1} \\ u_{i_1i_2\cdots i_{k-1}2} + \varepsilon\eta^{(k)}_{i_1i_2\cdots i_{k-1}2} \\ \vdots \\ u_{i_1i_2\cdots i_{k-1}n} + \varepsilon\eta^{(k)}_{i_1i_2\cdots i_{k-1}n} \end{bmatrix} = [I - \varepsilon B] \begin{bmatrix} u_{i_1i_2\cdots i_{k-1}1} + \varepsilon D_1\eta^{(k-1)}_{i_1i_2\cdots i_{k-1}} \\ u_{i_1i_2\cdots i_{k-1}2} + \varepsilon D_2\eta^{(k-1)}_{i_1i_2\cdots i_{k-1}} \\ \vdots \\ u_{i_1i_2\cdots i_{k-1}n} + \varepsilon D_n\eta^{(k-1)}_{i_1i_2\cdots i_{k-1}} \end{bmatrix} + O(\varepsilon^2)$$

and hence,

$$\begin{bmatrix} \eta^{(k)}_{i_1i_2\cdots i_{k-1}1} \\ \eta^{(k)}_{i_1i_2\cdots i_{k-1}2} \\ \vdots \\ \eta^{(k)}_{i_1i_2\cdots i_{k-1}n} \end{bmatrix} = \begin{bmatrix} D_1\eta^{(k-1)}_{i_1i_2\cdots i_{k-1}} \\ D_2\eta^{(k-1)}_{i_1i_2\cdots i_{k-1}} \\ \vdots \\ D_n\eta^{(k-1)}_{i_1i_2\cdots i_{k-1}} \end{bmatrix} - B \begin{bmatrix} u_{i_1i_2\cdots i_{k-1}1} \\ u_{i_1i_2\cdots i_{k-1}2} \\ \vdots \\ u_{i_1i_2\cdots i_{k-1}n} \end{bmatrix},$$

$i_\ell = 1,2,\dots,n$ for $\ell = 1,2,\dots,k-1$ with $k \geq 2$, leading to (2.119b). □

Specializing Theorem 2.4.4-1 to the case of one dependent variable u and two independent variables x_1 and x_2, the extended one-parameter Lie group of transformations

$$x^*_i = X_i(x_1, x_2, u; \varepsilon) = x_i + \varepsilon\xi_i(x_1, x_2, u) + O(\varepsilon^2), \quad i = 1,2, \tag{2.122a}$$

$$u^* = U(x_1, x_2, u; \varepsilon) = u + \varepsilon\eta(x_1, x_2, u) + O(\varepsilon^2), \tag{2.122b}$$

$$u^*_i = U_i(x_1, x_2, u, u_1, u_2; \varepsilon) = u_i + \varepsilon\eta_i^{(1)}(x_1, x_2, u, u_1, u_2) + O(\varepsilon^2), \quad i = 1,2, \tag{2.122c}$$

$$\begin{aligned} u^*_{ij} &= U_{ij}(x_1, x_2, u, u_1, u_2, u_{11}, u_{12}, u_{22}; \varepsilon) \\ &= u_{ij} + \varepsilon\eta_{ij}^{(2)}(x_1, x_2, u, u_1, u_2, u_{11}, u_{12}, u_{22}) + O(\varepsilon^2), \quad i, j = 1,2, \end{aligned} \tag{2.122d}$$

etc., has its extended infinitesimals given by

$$\eta_1^{(1)} = \frac{\partial\eta}{\partial x_1} + \left[\frac{\partial\eta}{\partial u} - \frac{\partial\xi_1}{\partial x_1}\right]u_1 - \frac{\partial\xi_2}{\partial x_1}u_2 - \frac{\partial\xi_1}{\partial u}(u_1)^2 - \frac{\partial\xi_2}{\partial u}u_1u_2, \tag{2.123}$$

$$\eta_2^{(1)} = \frac{\partial\eta}{\partial x_2} + \left[\frac{\partial\eta}{\partial u} - \frac{\partial\xi_2}{\partial x_2}\right]u_2 - \frac{\partial\xi_1}{\partial x_2}u_1 - \frac{\partial\xi_2}{\partial u}(u_2)^2 - \frac{\partial\xi_1}{\partial u}u_1u_2, \tag{2.124}$$

$$\eta_{11}^{(2)} = \frac{\partial^2\eta}{\partial x_1{}^2} + \left[2\frac{\partial^2\eta}{\partial x_1\,\partial u} - \frac{\partial^2\xi_1}{\partial x_1{}^2}\right]u_1 - \frac{\partial^2\xi_2}{\partial x_1{}^2}u_2 + \left[\frac{\partial\eta}{\partial u} - 2\frac{\partial\xi_1}{\partial x_1}\right]u_{11} - 2\frac{\partial\xi_2}{\partial x_1}u_{12}$$
$$+\left[\frac{\partial^2\eta}{\partial u^2} - 2\frac{\partial^2\xi_1}{\partial x_1\,\partial u}\right](u_1)^2 - 2\frac{\partial^2\xi_2}{\partial x_1\,\partial u}u_1u_2 - \frac{\partial^2\xi_1}{\partial u^2}(u_1)^3 - \frac{\partial^2\xi_2}{\partial u^2}(u_1)^2u_2$$
$$-3\frac{\partial\xi_1}{\partial u}u_1u_{11} - \frac{\partial\xi_2}{\partial u}u_2u_{11} - 2\frac{\partial\xi_2}{\partial u}u_1u_{12}, \tag{2.125}$$

$$\eta_{12}^{(2)} = \eta_{21}^{(2)}$$
$$= \frac{\partial^2\eta}{\partial x_1\,\partial x_2} + \left[\frac{\partial^2\eta}{\partial x_1\,\partial u} - \frac{\partial^2\xi_2}{\partial x_1\,\partial x_2}\right]u_2 + \left[\frac{\partial^2\eta}{\partial x_2\,\partial u} - \frac{\partial^2\xi_1}{\partial x_1\,\partial x_2}\right]u_1 - \frac{\partial\xi_2}{\partial x_1}u_{22}$$
$$+\left[\frac{\partial\eta}{\partial u} - \frac{\partial\xi_1}{\partial x_1} - \frac{\partial\xi_2}{\partial x_2}\right]u_{12} - \frac{\partial\xi_1}{\partial x_2}u_{11} - \frac{\partial^2\xi_2}{\partial x_1\,\partial u}(u_2)^2$$
$$+\left[\frac{\partial^2\eta}{\partial u^2} - \frac{\partial^2\xi_1}{\partial x_1\,\partial u} - \frac{\partial^2\xi_2}{\partial x_2\,\partial u}\right]u_1u_2 - \frac{\partial^2\xi_1}{\partial x_2\,\partial u}(u_1)^2 - \frac{\partial^2\xi_2}{\partial u^2}u_1(u_2)^2 - \frac{\partial^2\xi_1}{\partial u^2}(u_1)^2u_2$$
$$-2\frac{\partial\xi_2}{\partial u}u_2u_{12} - 2\frac{\partial\xi_1}{\partial u}u_1u_{12} - \frac{\partial\xi_1}{\partial u}u_2u_{11} - \frac{\partial\xi_2}{\partial u}u_1u_{22}, \tag{2.126}$$

$$\eta_{22}^{(2)} = \frac{\partial^2\eta}{\partial x_2{}^2} + \left[2\frac{\partial^2\eta}{\partial x_2\,\partial u} - \frac{\partial^2\xi_2}{\partial x_2{}^2}\right]u_2 - \frac{\partial^2\xi_1}{\partial x_2{}^2}u_1 + \left[\frac{\partial\eta}{\partial u} - 2\frac{\partial\xi_2}{\partial x_2}\right]u_{22} - 2\frac{\partial\xi_1}{\partial x_2}u_{12}$$
$$+\left[\frac{\partial^2\eta}{\partial u^2} - 2\frac{\partial^2\xi_2}{\partial x_2\,\partial u}\right](u_2)^2 - 2\frac{\partial^2\xi_1}{\partial x_2\,\partial u}u_1u_2 - \frac{\partial^2\xi_2}{\partial u^2}(u_2)^3 - \frac{\partial^2\xi_2}{\partial u^2}u_1(u_2)^2$$
$$-3\frac{\partial\xi_2}{\partial u}u_2u_{22} - 3\frac{\partial\xi_1}{\partial u}u_1u_{22} - 2\frac{\partial\xi_1}{\partial u}u_2u_{12}, \tag{2.127}$$

etc.

2.4.5 EXTENDED TRANSFORMATIONS AND EXTENDED INFINITESIMAL TRANSFORMATIONS: m DEPENDENT AND n INDEPENDENT VARIABLES

The situation of m dependent variables $u = (u^1, u^2, \ldots, u^m)$ and n independent variables $x = (x_1, x_2, \ldots, x_n)$, $u = u(x)$, with $m \geq 2$, arises in studying systems of differential equations. This leads to consideration of extended transformations from (x, u)-space to $(x, u, \partial u, \ldots, \partial^k u)$-space where $\partial^k u$ denotes the components of all kth-order partial derivatives of u with respect to x. These extended transformations preserve the corresponding contact conditions.

We consider a point transformation

$$x^\dagger = X(x,u), \tag{2.128a}$$

$$u^\dagger = U(x,u). \tag{2.128b}$$

Let

$$u_i^\sigma = \frac{\partial u^\sigma}{\partial x_i}, \quad (u_i^\sigma)^\dagger = \frac{\partial (u^\sigma)^\dagger}{\partial x_i^\dagger} = \frac{\partial U^\sigma}{\partial X_i}, \quad \text{etc.,}$$

$$D_i = \frac{\partial}{\partial x_i} + u_i^\mu \frac{\partial}{\partial u^\mu} + u_{ij}^\mu \frac{\partial}{\partial u_j^\mu} + \cdots + u_{ii_1i_2\cdots i_n}^\mu \frac{\partial}{\partial u_{i_1i_2\cdots i_n}^\mu} + \cdots,$$

with summation over a repeated index. The *k*th-extended transformation of (2.128a,b) is given by [Exercise 2.4-12]

$$x^\dagger = X(x,u), \tag{2.129a}$$

$$u^\dagger = U(x,u), \tag{2.129b}$$

$$\partial u^\dagger = \partial U(x,u,\partial u), \tag{2.129c}$$

$$\vdots$$

$$\partial^k u^\dagger = \partial^k U(x,u,\partial u,\ldots,\partial^k u), \tag{2.129d}$$

where the components $(u_i^\mu)^\dagger$ of $\partial u^\dagger$ are determined by

$$\begin{bmatrix} (u_1^\mu)^\dagger \\ (u_2^\mu)^\dagger \\ \vdots \\ (u_n^\mu)^\dagger \end{bmatrix} = \begin{bmatrix} U_1^\mu \\ U_2^\mu \\ \vdots \\ U_n^\mu \end{bmatrix} = A^{-1} \begin{bmatrix} \mathrm{D}_1 U^\mu \\ \mathrm{D}_2 U^\mu \\ \vdots \\ \mathrm{D}_n U^\mu \end{bmatrix}, \tag{2.130}$$

A^{-1} is the inverse (assumed to exist) of the matrix

$$A = \begin{bmatrix} \mathrm{D}_1 X_1 & \mathrm{D}_1 X_2 & \cdots & \mathrm{D}_1 X_n \\ \mathrm{D}_2 X_1 & \mathrm{D}_2 X_2 & \cdots & \mathrm{D}_2 X_n \\ \vdots & \vdots & & \vdots \\ \mathrm{D}_n X_1 & \mathrm{D}_n X_2 & \cdots & \mathrm{D}_n X_n \end{bmatrix}, \tag{2.131}$$

and the components $(u_{i_1i_2\cdots i_k}^\mu)^\dagger$ of $\partial^k u^\dagger$ are determined by

$$\begin{bmatrix} (u_{i_1i_2\cdots i_{k-1}1}^\mu)^\dagger \\ (u_{i_1i_2\cdots i_{k-1}2}^\mu)^\dagger \\ \vdots \\ (u_{i_1i_2\cdots i_{k-1}n}^\mu)^\dagger \end{bmatrix} = \begin{bmatrix} U_{i_1i_2\cdots i_{k-1}1}^\mu \\ U_{i_1i_2\cdots i_{k-1}2}^\mu \\ \vdots \\ U_{i_1i_2\cdots i_{k-1}n}^\mu \end{bmatrix} = A^{-1} \begin{bmatrix} \mathrm{D}_1 U_{i_1i_2\cdots i_{k-1}}^\mu \\ \mathrm{D}_2 U_{i_1i_2\cdots i_{k-1}}^\mu \\ \vdots \\ \mathrm{D}_n U_{i_1i_2\cdots i_{k-1}}^\mu \end{bmatrix}, \quad k = 2,3,\ldots,n. \tag{2.132}$$

Now we specialize the point transformation (2.128a,b) to the case of a one-parameter Lie group of point transformations

$$X^* = X(x, u; \varepsilon), \tag{2.133a}$$

$$U^* = U(x, u; \varepsilon). \tag{2.133b}$$

Here, the kth-extended transformation (2.129)–(2.132), with † replaced by *, is a one-parameter Lie group of transformations acting on $(x, u, \partial u, \ldots, \partial^k u)$-space. Then we have

$$x^*_i = X_i(x, u; \varepsilon) = x_i + \varepsilon\xi_i(x, u) + O(\varepsilon^2), \tag{2.134a}$$

$$(u^\mu)^* = U^\mu(x, u; \varepsilon) = u^\mu + \varepsilon\eta^\mu(x, u) + O(\varepsilon^2), \tag{2.134b}$$

$$(u_i^\mu)^* = U_i^\mu(x, u, \partial u; \varepsilon) = u_i^\mu + \varepsilon\eta_i^{(1)\mu}(x, u, \partial u) + O(\varepsilon^2), \tag{2.134c}$$

$$\vdots$$

$$(u^\mu_{i_1 i_2 \cdots i_k})^* = U^\mu_{i_1 i_2 \cdots i_k}(x,u,\partial u,\ldots,\partial^k u; \varepsilon) = u^\mu_{i_1 i_2 \cdots i_k} + \varepsilon\eta^{(k)\mu}_{i_1 i_2 \cdots i_k}(x,u,\partial u,\ldots,\partial^k u) + O(\varepsilon^2), \tag{2.134d}$$

with the extended infinitesimals $\eta^{(k)\mu}_{i_1 i_2 \cdots i_k}$ given by

$$\eta_i^{(1)\mu} = \mathrm{D}_i\eta^\mu - (\mathrm{D}_i\xi_j)u_j^\mu, \tag{2.135}$$

and

$$\eta^{(k)\mu}_{i_1 i_2 \cdots i_k} = \mathrm{D}_{i_k}\eta^{(k-1)\mu}_{i_1 i_2 \cdots i_{k-1}} - (\mathrm{D}_{i_k}\xi_j)u^\mu_{i_1 i_2 \cdots i_{k-1} j}, \tag{2.136}$$

$i_\ell = 1,2,\ldots,n$ for $\ell = 1,2,\ldots,k$ with $k \geq 2$. Here, the kth-extended infinitesimal generator is given by

$$\mathrm{X}^{(k)} = \xi_i(x,u)\frac{\partial}{\partial x_i} + \eta^\mu(x,u)\frac{\partial}{\partial u^\mu} + \eta_i^{(1)\mu}(x,u,\partial u)\frac{\partial}{\partial u_i^\mu} + \cdots$$
$$+ \eta^{(k)\mu}_{i_1 i_2 \cdots i_k}(x,u,\partial u,\partial^2 u,\ldots,\partial^k u)\frac{\partial}{\partial u^\mu_{i_1 i_2 \cdots i_k}}, \quad k \geq 1. \tag{2.137}$$

EXERCISES 2.4

1. In Theorem 2.4.1-3, show that Y_k, $k \geq 2$, defined by (2.86d), is:
   (a) linear in y_k; and
   (b) a polynomial in $y_2, y_3, \ldots, y_k$ whose coefficients are functions of (x, y, y_1).
2. Prove Theorem 2.4.1-2.
3. For the rotation group (2.94a,b), determine $y^*_4 = Y_4$:
   (a) using Theorem 2.4.1-3; and
   (b) from its extended infinitesimals, i.e., using Theorem 2.4.2-1.
4. (a) Derive (2.101)–(2.103).

(b) Determine $\eta^{(4)}$.

5. Prove Theorem 2.4.2-2.

6. For the group

$$x^* = x + \varepsilon, \quad y^* = \frac{xy}{x+\varepsilon}, \quad \text{with } y = y(x),$$

determine:
(a) $\xi, \eta, \eta^{(1)}, \eta^{(2)}, \eta^{(3)}$; and
(b) $y^*_1 = Y_1, y^* = Y_2, y^*_3 = Y_3$.

7. For the rotation group (2.94a,b), find the invariants of its first- and second-extended group. Interpret geometrically.

8. Explain the geometrical significance of preserving the contact conditions (2.78a,b).

9. Show that each component of $\partial^k U, k \geq 2$, defined by (2.115d), (2.116a,b), is:
   (a) linear in the components of $\partial^k u$; and
   (b) a polynomial in the components of $\partial^2 u, \partial^3 u, \ldots, \partial^{k-1} u$, with coefficients that are functions of the components of $x, u, \partial u$.

10. State and prove the analog of Theorem 2.4.2-2 for the extended infinitesimals $\eta_{i_1 i_2 \cdots i_k}$ determined by Theorem 2.4.4-1.

11. Derive (2.123)–(2.127).

12. Derive (2.129a–d), (2.130)–(2.132).

13. Derive (2.135), (2.136).

14. For the following two examples (arising from study of the group properties of the heat equation), involving two independent variables (x, t) and one dependent variable $u = u(x, t)$, determine: (i) the extended infinitesimal generators $\mathrm{X}^{(1)}$ and $\mathrm{X}^{(2)}$; and (ii) the extended one-parameter Lie groups of transformations acting on $(x, u, \partial u)$ – space and $(x, u, \partial u, \partial^2 u)$ – space with:
   (a) $\mathrm{X} = 2t\dfrac{\partial}{\partial x} - xu\dfrac{\partial}{\partial u}$; and
   (b) $\mathrm{X} = 4xt\dfrac{\partial}{\partial x} + 4t^2\dfrac{\partial}{\partial t} - (x^2 + 2t)u\dfrac{\partial}{\partial u}$.

15. Consider the case of one independent variable x and one dependent variable $y = y(x)$. Assume that the point transformation

$$x^\dagger = X(x, y), \tag{2.138a}$$

$$y^\dagger = Y(x, y), \tag{2.138b}$$

preserves the contact conditions and can be inverted so that

$$x = X^{\dagger}(x^{\dagger}, y^{\dagger}),$$
$$y = Y^{\dagger}(x^{\dagger}, y^{\dagger}),$$

where $X^{\dagger}$, $Y^{\dagger}$ are known explicitly as functions of $x^{\dagger}$, $y^{\dagger}$. Express y_1 and y_2 as functions of $x^{\dagger}, y^{\dagger}, y_1^{\dagger}, y_2^{\dagger}$. Show how this simplifies in the situation when (2.138a,b) is a one-parameter Lie group of point transformations. Illustrate for the rotation group (2.94a,b).

16. Consider the situation of two independent variables (x, t) and one dependent variable $u = u(x, t)$. Assume that the point transformation

$$x^{\dagger} = X(x,t,u), \tag{2.139a}$$

$$t^{\dagger} = T(x,t,u), \tag{2.139b}$$

$$u^{\dagger} = U(x,t,u), \tag{2.139c}$$

preserves the contact conditions and can be inverted so that

$$x = X^{\dagger}(x^{\dagger}, t^{\dagger}, u^{\dagger}),$$
$$t = T^{\dagger}(x^{\dagger}, t^{\dagger}, u^{\dagger}),$$
$$u = U^{\dagger}(x^{\dagger}, t^{\dagger}, u^{\dagger}),$$

where $X^{\dagger}, T^{\dagger}, U^{\dagger}$ are known as explicit functions of $x^{\dagger}, t^{\dagger}, u^{\dagger}$. Express the components of ∂u and $\partial^2 u$ as functions of $x^{\dagger}, t^{\dagger}, u^{\dagger}$ and the components of $\partial u^{\dagger}$ and $\partial^2 u^{\dagger}$. Show how this simplifies in the situation when (2.139a–c) is a one-parameter Lie group of point transformations. Illustrate for the one-parameter Lie group of point transformations with the infinitesimal generator $\mathrm{X} = 2t\dfrac{\partial}{\partial x} - xu\dfrac{\partial}{\partial u}$.

17. For $X(x,u)$ defined by (2.104a) [(2.129a)], give criteria so that the corresponding matrix A defined by (2.109) [(2.131)] has an inverse.

2.5 MULTIPARAMETER LIE GROUPS OF TRANSFORMATIONS AND LIE ALGEBRAS

So far in this chapter, we have only considered one-parameter Lie groups of transformations. In Chapter 1, on dimensional analysis, we encountered invariance under multiparameter families of scalings. These are examples of multiparameter Lie groups of point transformations. In this section we summarize some key results pertaining to multiparameter Lie groups of transformations. We assume a finite number of parameters.

Each parameter of an r-parameter Lie group of transformations leads to an infinitesimal generator. The infinitesimal generators belong to an r-dimensional vector space on which there is an additional structure, called the *commutator.* This special vector space is called a *Lie algebra* (*r-dimensional Lie algebra*).

For our purposes, the study of an r-parameter Lie group of transformations is equivalent to the study of its infinitesimal generators and the structure of the corresponding Lie algebra. The exponentiation of any infinitesimal generator is a one-parameter Lie group of transformations that is a subgroup of the r-parameter Lie group of transformations. Most important, the discovery of multiparameter Lie groups of transformations admitted by a given differential equation requires one to consider only invariance of the differential equation under one-parameter Lie groups of transformations.

Special Lie algebras called *solvable Lie algebras* play an important role in the study of the invariance of ODEs of at least third-order under multiparameter Lie groups of transformations.

For further details of the material of this section, the reader is referred to the books of Cohn (1965), Eisenhart (1933), Gilmore (1974), and Ovsiannikov (1962, 1982).

2.5.1 r-PARAMETER LIE GROUPS OF TRANSFORMATIONS

Consider an *r-parameter Lie group of point transformations*

$$\mathbf{x}^* = \mathbf{X}(\mathbf{x};\boldsymbol{\varepsilon}), \tag{2.140}$$

with $\mathbf{x} = (x_1, x_2, \ldots, x_n)$ and parameters $\boldsymbol{\varepsilon} = (\varepsilon_1, \varepsilon_2, \ldots, \varepsilon_r)$. Let the law of composition of parameters be denoted by

$$\boldsymbol{\varphi}(\boldsymbol{\varepsilon},\boldsymbol{\delta}) = (\phi_1(\boldsymbol{\varepsilon},\boldsymbol{\delta}), \phi_2(\boldsymbol{\varepsilon},\boldsymbol{\delta}), \ldots, \phi_r(\boldsymbol{\varepsilon},\boldsymbol{\delta})),$$

with $\boldsymbol{\delta} = (\delta_1, \delta_2, \ldots, \delta_r)$, where $\boldsymbol{\varphi}(\boldsymbol{\varepsilon},\boldsymbol{\delta})$ satisfies the group axioms with $\boldsymbol{\varepsilon} = 0$ corresponding to the identity $\varepsilon_1 = \varepsilon_2 = \cdots = \varepsilon_r = 0$, and $\boldsymbol{\varphi}(\boldsymbol{\varepsilon},\boldsymbol{\delta})$ is assumed to be analytic in its domain of definition.

Let the *infinitesimal matrix* $\boldsymbol{\Xi}(\mathbf{x})$ be the $r \times n$ matrix with entries

$$\xi_{\alpha j}(\mathbf{x}) = \left.\frac{\partial x^*_j}{\partial \varepsilon_\alpha}\right|_{\boldsymbol{\varepsilon}=0} = \left.\frac{\partial X_j(\mathbf{x};\boldsymbol{\varepsilon})}{\partial \varepsilon_\alpha}\right|_{\boldsymbol{\varepsilon}=0}, \quad \alpha = 1,2,\ldots,r, \; j = 1,2,\ldots,n. \tag{2.141}$$

Let $\boldsymbol{\Theta}(\boldsymbol{\varepsilon})$ be the $r \times r$ matrix with entries

$$\Theta_{\alpha\beta}(\boldsymbol{\varepsilon}) = \left.\frac{\partial \phi_\beta(\boldsymbol{\varepsilon},\boldsymbol{\delta})}{\partial \delta_\alpha}\right|_{\boldsymbol{\delta}=0}, \tag{2.142}$$

and denote the inverse of the matrix $\boldsymbol{\Theta}(\boldsymbol{\varepsilon})$ by

$$\boldsymbol{\Psi}(\boldsymbol{\varepsilon}) = \boldsymbol{\Theta}^{-1}(\boldsymbol{\varepsilon}). \tag{2.143}$$

Then *Lie's First Fundamental Theorem* for an r-parameter Lie group of transformations states that in some neighborhood of $\boldsymbol{\varepsilon} = 0$, (2.140) is equivalent to the solution of the initial value problem for the system of nr first-order PDEs given by

$$\begin{bmatrix} \frac{\partial x^*_1}{\partial \varepsilon_1} & \frac{\partial x^*_2}{\partial \varepsilon_1} & \cdots & \frac{\partial x^*_n}{\partial \varepsilon_1} \\ \frac{\partial x^*_1}{\partial \varepsilon_2} & \frac{\partial x^*_2}{\partial \varepsilon_2} & \cdots & \frac{\partial x^*_n}{\partial \varepsilon_2} \\ \vdots & \vdots & \cdots & \vdots \\ \frac{\partial x^*_1}{\partial \varepsilon_r} & \frac{\partial x^*_2}{\partial \varepsilon_r} & \cdots & \frac{\partial x^*_n}{\partial \varepsilon_r} \end{bmatrix} = \mathbf{\Psi}(\boldsymbol{\varepsilon})\mathbf{\Xi}(\mathbf{x}^*), \tag{2.144a}$$

with

$$\mathbf{x}^* = \mathbf{x} \quad \text{at} \quad \boldsymbol{\varepsilon} = 0. \tag{2.144b}$$

Definition 2.5.1-1. The *infinitesimal generator* X_α, corresponding to the parameter ε_α of the r-parameter Lie group of transformations (2.140), is given by

$$\mathrm{X}_\alpha = \sum_{j=1}^{n} \xi_{\alpha j}(\mathbf{x}) \frac{\partial}{\partial x_j}, \quad \alpha = 1,2,\ldots,r. \tag{2.145}$$

One can show that the r-parameter Lie group of transformations (2.140) is equivalent to

$$\mathbf{x}^* = \prod_{\alpha=1}^{r} e^{\mu_\alpha \mathrm{X}_\alpha} \mathbf{x} = e^{\mu_1 \mathrm{X}_1} e^{\mu_2 \mathrm{X}_2} \cdots e^{\mu_r \mathrm{X}_r} \mathbf{x}, \tag{2.146}$$

where $\mu_1, \mu_2, \ldots, \mu_r$ are arbitrary real constants. [The order of the operations in (2.146) can be rearranged by renumbering the infinitesimal generators even though it is not necessarily true that $e^{\mu_\alpha \mathrm{X}_\alpha} e^{\mu_\beta \mathrm{X}_\beta} = e^{\mu_\beta \mathrm{X}_\beta} e^{\mu_\alpha \mathrm{X}_\alpha}$ for $\alpha \neq \beta$. A reordering would correspond to a different parameterization, i.e., $\mathbf{\Psi}(\boldsymbol{\varepsilon})$ would change. An r-parameter Lie group of transformations is equivalent to (2.140) if it can be expressed in the form (2.144a,b) with the same $\mathbf{\Xi}(\mathbf{x})$.]

One can also show that the one-parameter Lie group of transformations

$$\mathbf{x}^* = e^{\varepsilon \mathrm{X}} \mathbf{x} = e^{\varepsilon \sum_{\alpha=1}^{r} \sigma_\alpha \mathrm{X}_\alpha} \mathbf{x}, \tag{2.147}$$

obtained by exponentiating the infinitesimal generator

$$\mathrm{X} = \sum_{\alpha=1}^{r} \sigma_\alpha \mathrm{X}_\alpha = \sum_{j=1}^{n} \varsigma_j(\mathbf{x}) \frac{\partial}{\partial x_j}, \tag{2.148}$$

where

$$\varsigma_j(\mathbf{x}) = \sum_{\alpha=1}^{r} \sigma_\alpha \xi_{\alpha j}(\mathbf{x}), \quad j = 1,2,\ldots,n, \tag{2.149}$$

in terms of any fixed real constants $\sigma_1, \sigma_2, \ldots, \sigma_r$, defines a one-parameter (ε) subgroup of the r-parameter Lie group of transformations (2.140).

As an example, consider the two-parameter $[\boldsymbol{\varepsilon}=(\varepsilon_1,\varepsilon_2)]$ Lie group of transformations $[(x_1, x_2) = (x, y)]$ given by

$$x^* = e^{\varepsilon_1}x + \varepsilon_2, \tag{2.150a}$$

$$y^* = e^{2\varepsilon_1}y. \tag{2.150b}$$

Then

$$x^{**} = e^{\delta_1}x^* + \delta_2 = e^{\phi_1(\boldsymbol{\varepsilon},\boldsymbol{\delta})}x + \phi_2(\boldsymbol{\varepsilon},\boldsymbol{\delta}),$$
$$y^{**} = e^{2\delta_1}y^* = e^{2\phi_1(\boldsymbol{\varepsilon},\boldsymbol{\delta})}y,$$

with the law of composition given by

$$\boldsymbol{\varphi}(\boldsymbol{\varepsilon},\boldsymbol{\delta}) = (\phi_1(\boldsymbol{\varepsilon},\boldsymbol{\delta}),\phi_2(\boldsymbol{\varepsilon},\boldsymbol{\delta})) = (\varepsilon_1+\delta_1, e^{\delta_1}\varepsilon_2+\delta_2). \tag{2.151}$$

One can easily check that the two-parameter family of transformations (2.150a,b), with the law of composition (2.151), defines a two-parameter Lie group of transformations with $x^* = x, y^* = y$, when $\boldsymbol{\varepsilon} = (\varepsilon_1,\varepsilon_2) = 0$.

We now check that (2.144a,b) holds:

$$\frac{\partial x^*}{\partial \varepsilon_1} = e^{\varepsilon_1}x = x^* - \varepsilon_2, \quad \frac{\partial y^*}{\partial \varepsilon_1} = 2e^{2\varepsilon_1}y = 2y^*, \quad \frac{\partial x^*}{\partial \varepsilon_2} = 1, \quad \frac{\partial y^*}{\partial \varepsilon_2} = 0.$$

Hence,

$$\begin{bmatrix} \dfrac{\partial x^*}{\partial \varepsilon_1} & \dfrac{\partial y^*}{\partial \varepsilon_1} \\ \dfrac{\partial x^*}{\partial \varepsilon_2} & \dfrac{\partial y^*}{\partial \varepsilon_2} \end{bmatrix} = \begin{bmatrix} x^* - \varepsilon_2 & 2y^* \\ 1 & 0 \end{bmatrix}. \tag{2.152}$$

Then

$$\xi_{11}(\mathbf{x}) = \left.\frac{\partial x^*}{\partial \varepsilon_1}\right|_{\boldsymbol{\varepsilon}=0} = x, \quad \xi_{12}(\mathbf{x}) = \left.\frac{\partial y^*}{\partial \varepsilon_1}\right|_{\boldsymbol{\varepsilon}=0} = 2y,$$

$$\xi_{21}(\mathbf{x}) = \left.\frac{\partial x^*}{\partial \varepsilon_2}\right|_{\boldsymbol{\varepsilon}=0} = 1, \quad \xi_{22}(\mathbf{x}) = \left.\frac{\partial y^*}{\partial \varepsilon_2}\right|_{\boldsymbol{\varepsilon}=0} = 0.$$

Consequently, the infinitesimal matrix is given by

$$\boldsymbol{\Xi}(\mathbf{x}) = \begin{bmatrix} x & 2y \\ 1 & 0 \end{bmatrix}. \tag{2.153}$$

To determine $\boldsymbol{\Psi}(\boldsymbol{\varepsilon})$, we have

$$\frac{\partial \phi_1}{\partial \delta_1} = 1, \quad \frac{\partial \phi_2}{\partial \delta_1} = e^{\delta_1}\varepsilon_2, \quad \frac{\partial \phi_1}{\partial \delta_2} = 0, \quad \frac{\partial \phi_2}{\partial \delta_2} = 1,$$

and, hence,

$$\Theta_{11} = \left.\frac{\partial \phi_1}{\partial \delta_1}\right|_{\delta=0} = 1, \quad \Theta_{12} = \left.\frac{\partial \phi_2}{\partial \delta_1}\right|_{\delta=0} = \varepsilon_2, \quad \Theta_{21} = \left.\frac{\partial \phi_1}{\partial \delta_2}\right|_{\delta=0} = 0, \quad \Theta_{22} = \left.\frac{\partial \phi_2}{\partial \delta_2}\right|_{\delta=0} = 1.$$

Thus, we get

$$\boldsymbol{\Theta}(\boldsymbol{\varepsilon}) = \begin{bmatrix} 1 & \varepsilon_2 \\ 0 & 1 \end{bmatrix}, \tag{2.154}$$

and

$$\boldsymbol{\Psi}(\boldsymbol{\varepsilon}) = \boldsymbol{\Theta}^{-1}(\boldsymbol{\varepsilon}) = \begin{bmatrix} 1 & -\varepsilon_2 \\ 0 & 1 \end{bmatrix}. \tag{2.155}$$

Then it is easily seen that

$$\boldsymbol{\Psi}(\boldsymbol{\varepsilon})\boldsymbol{\Xi}(\mathbf{x}^*) = \begin{bmatrix} x^* - \varepsilon_2 & 2y^* \\ 1 & 0 \end{bmatrix},$$

which is the matrix (2.152), verifying (2.144a,b). It is left to Exercise 2.5-1 to solve the initial value problem for the system of PDEs

$$\frac{\partial x^*}{\partial \varepsilon_1} = x^* - \varepsilon_2, \tag{2.156a}$$

$$\frac{\partial y^*}{\partial \varepsilon_1} = 2y^*, \tag{2.156b}$$

$$\frac{\partial x^*}{\partial \varepsilon_2} = 1, \tag{2.156c}$$

$$\frac{\partial y^*}{\partial \varepsilon_2} = 0, \tag{2.156d}$$

with

$$x^* = x, \quad y^* = y, \quad \text{when} \quad \varepsilon_1 = 0, \varepsilon_2 = 0, \tag{2.156e}$$

to recover (2.150a,b).

For the two-parameter Lie group of transformations (2.150a,b), the corresponding infinitesimal generators are

$$\mathrm{X}_1 = x\frac{\partial}{\partial x} + 2y\frac{\partial}{\partial y}, \tag{2.157a}$$

$$\mathrm{X}_2 = \frac{\partial}{\partial x}. \tag{2.157b}$$

For any differentiable function $F(x, y)$, we have

$$e^{\varepsilon \mathrm{X}_1} F(x, y) = F(e^{\varepsilon} x, e^{2\varepsilon} y), \tag{2.158a}$$

$$e^{\varepsilon X_2}F(x,y)=F(x+\varepsilon,y). \tag{2.158b}$$

We now check that the representations (2.146) and (2.147) lead to (2.150a,b).

From (2.158a,b), it follows that for any real constants μ_1, μ_2, we have

$$e^{\mu_1 X_1}e^{\mu_2 X_2}(x,y)=e^{\mu_1 X_1}(x+\mu_2,y)=(e^{\mu_1}x+\mu_2,e^{2\mu_1}y), \tag{2.159}$$

and

$$e^{\mu_2 X_2}e^{\mu_1 X_1}(x,y)=e^{\mu_2 X_2}(e^{\mu_1}x,e^{2\mu_1}y)=(e^{\mu_1}(x+\mu_2),e^{2\mu_1}y). \tag{2.160}$$

Let $\tilde{x}=\lambda_1 x+\lambda_2$. Then

$$e^{\varepsilon(\lambda_1 X_1+\lambda_2 X_2)}(x,y)=e^{\varepsilon\lambda_1\tilde{x}\frac{\partial}{\partial\tilde{x}}+2\varepsilon\lambda_1 y\frac{\partial}{\partial y}}\left(\frac{\tilde{x}-\lambda_2}{\lambda_1},y\right)=\left(\frac{e^{\varepsilon\lambda_1}\tilde{x}-\lambda_2}{\lambda_1},e^{2\varepsilon\lambda_1}y\right)$$

$$=\left(e^{\varepsilon\lambda_1}x+\frac{\lambda_2}{\lambda_1}\left(e^{\varepsilon\lambda_1}-1\right),e^{2\varepsilon\lambda_1}y\right). \tag{2.161}$$

Thus, (2.159) is identical to (2.150a,b), with the same law of composition (2.151); (2.160) is equivalent to (2.150a,b) with the law of composition $\boldsymbol{\varphi}(\boldsymbol{\varepsilon},\boldsymbol{\delta})=(\varepsilon_1+\delta_1,\varepsilon_2+e^{-\delta_1}\delta_2)$; and (2.161) is equivalent to (2.150a,b) with the law of composition

$$\boldsymbol{\varphi}(\boldsymbol{\varepsilon},\boldsymbol{\delta})=\left(\varepsilon_1+\delta_1,\frac{\varepsilon_1+\delta_1}{e^{\varepsilon_1+\delta_1}-1}\left(e^{\delta_1}\left(\frac{\varepsilon_2}{\varepsilon_1}(e^{\varepsilon_1}-1)+\frac{\delta_2}{\delta_1}\right)-\frac{\delta_2}{\delta_1}\right)\right).$$

2.5.2 LIE ALGEBRAS

Definition 2.5.2-1. Consider an r-parameter Lie group of transformations (2.140) with infinitesimal generators X_α, $\alpha=1,2,\ldots,r$, defined by (2.141) and (2.145). The *commutator* (*Lie bracket*) of X_α and X_β is a first-order operator

$$[X_\alpha,X_\beta]=X_\alpha X_\beta-X_\beta X_\alpha=\sum_{i,j=1}^{n}\left[\left(\xi_{\alpha i}(\mathbf{x})\frac{\partial}{\partial x_i}\right)\left(\xi_{\beta j}(\mathbf{x})\frac{\partial}{\partial x_j}\right)-\left(\xi_{\beta i}(\mathbf{x})\frac{\partial}{\partial x_i}\right)\left(\xi_{\alpha j}(\mathbf{x})\frac{\partial}{\partial x_j}\right)\right]$$

$$=\sum_{j=1}^{n}\eta_j(\mathbf{x})\frac{\partial}{\partial x_j}, \tag{2.162a}$$

where

$$\eta_j(\mathbf{x})=\sum_{i=1}^{n}\left(\xi_{\alpha i}(\mathbf{x})\frac{\partial\xi_{\beta j}(\mathbf{x})}{\partial x_i}-\xi_{\beta i}(\mathbf{x})\frac{\partial\xi_{\alpha j}(\mathbf{x})}{\partial x_i}\right). \tag{2.162b}$$

It immediately follows that

$$[X_\alpha, X_\beta] = -[X_\beta, X_\alpha]. \tag{2.163}$$

Theorem 2.5.2-1 (Second Fundamental Theorem of Lie). *The commutator of any two infinitesimal generators of an r-parameter Lie group of transformations is also an infinitesimal generator. In particular,*

$$[\mathrm{X}_\alpha, \mathrm{X}_\beta] = \sum_{\gamma=1}^{r} C_{\alpha\beta}^{\gamma} \mathrm{X}_\gamma, \tag{2.164}$$

where the coefficients $C_{\alpha\beta}^{\gamma}$ *are constants called* **structure constants**, $\alpha, \beta, \gamma = 1,2,\ldots,r$.

Proof. The proof of this theorem essentially depends on the integrability conditions

$$\frac{\partial^2 x^*_i}{\partial \varepsilon_\alpha \, \partial \varepsilon_\beta} = \frac{\partial^2 x^*_i}{\partial \varepsilon_\beta \, \partial \varepsilon_\alpha}, \quad i = 1,2,\ldots,n, \quad \alpha, \beta = 1,2,\ldots,r, \tag{2.165}$$

applied to (2.144a). For complete details, see any of the earlier-mentioned references of this section. □

Definition 2.5.2-2. Equations (2.164) are called the *commutation relations* of the r-parameter Lie group of transformations (2.140) with the infinitesimal generators (2.145).

For any three infinitesimal generators X_α, X_β, X_γ, by direct computation one can show that *Jacobi's identity* holds:

$$[\mathrm{X}_\alpha, [\mathrm{X}_\beta, \mathrm{X}_\gamma]] + [\mathrm{X}_\beta, [\mathrm{X}_\gamma, \mathrm{X}_\alpha]] + [\mathrm{X}_\gamma, [\mathrm{X}_\alpha, \mathrm{X}_\beta]] = 0. \tag{2.166}$$

From (2.163), (2.164), and (2.166), the following theorem relating the structure constants is easily proved:

Theorem 2.5.2-2 (Third Fundamental Theorem of Lie). *The structure constants defined by the commutation relations* (2.164) *satisfy the relations*

$$C_{\alpha\beta}^{\gamma} = -C_{\beta\alpha}^{\gamma}, \tag{2.167a}$$

$$\sum_{\rho=1}^{r} [C_{\alpha\beta}^{\rho} C_{\rho\gamma}^{\delta} + C_{\beta\gamma}^{\rho} C_{\rho\alpha}^{\delta} + C_{\gamma\alpha}^{\rho} C_{\rho\beta}^{\delta}] = 0. \tag{2.167b}$$

In particular, (2.167a) *is equivalent to the commutator anti-symmetry property* (2.163), *and* (2.167b) *is equivalent to Jacobi's identity* (2.166).

Definition 2.5.2-3. A *Lie algebra* $\mathcal{L}$ is a vector space over **R** or **C** with a bilinear bracket operation (the *commutator*) satisfying the properties (2.163), (2.166) and, most important, (2.164). In particular, the set of infinitesimal generators $\{\mathrm{X}_\alpha\}$, $\alpha = 1,2,\ldots,r$, of an r-parameter Lie group of transformations (2.140) forms an r-dimensional Lie algebra over **R**.

One can motivate the definition of the commutator $[X_\alpha, X_\beta]$ by the following argument.

Let G^r denote the r-parameter Lie group of transformations (2.140). Any one-parameter (ε) subgroup of G^r has a corresponding infinitesimal generator in $\mathcal{L}^r$. For example, $X_\alpha \in \mathcal{L}^r$ corresponds to $e^{\varepsilon X_\alpha}\mathbf{x} \in G^r$, $\alpha = 1,2,\ldots,r$; $aX_\alpha + bX_\beta \in \mathcal{L}^r$ corresponds to both $e^{\varepsilon(aX_\alpha+bX_\beta)}\mathbf{x} \in G^r$, and $e^{\varepsilon aX_\alpha}e^{\varepsilon bX_\beta}\mathbf{x} \in G^r$. If $X_\alpha, X_\beta \in \mathcal{L}^r$, then both $e^{\varepsilon X_\alpha}\mathbf{x}$ and $e^{\varepsilon X_\beta}\mathbf{x}$ belong to G^r for any real ε. Consider the one-parameter (ε) commutator group transformations

$$e^{-\varepsilon X_\alpha}e^{-\varepsilon X_\beta}e^{\varepsilon X_\alpha}e^{\varepsilon X_\beta}\mathbf{x} = [e^{\varepsilon X_\alpha}]^{-1}[e^{\varepsilon X_\beta}]^{-1}e^{\varepsilon X_\alpha}e^{\varepsilon X_\beta}\mathbf{x} \in G^r.$$

Then

$$\begin{aligned}
e^{-\varepsilon X_\alpha}e^{-\varepsilon X_\beta}e^{\varepsilon X_\alpha}e^{\varepsilon X_\beta} &= (1-\varepsilon X_\alpha + \tfrac{1}{2}\varepsilon^2(X_\alpha)^2)(1-\varepsilon X_\beta + \tfrac{1}{2}\varepsilon^2(X_\beta)^2)\\
&\quad \times (1+\varepsilon X_\alpha + \tfrac{1}{2}\varepsilon^2(X_\alpha)^2)(1+\varepsilon X_\beta + \tfrac{1}{2}\varepsilon^2(X_\beta)^2) + O(\varepsilon^3)\\
&= (1-\varepsilon(X_\alpha + X_\beta) + \varepsilon^2(X_\alpha X_\beta + \tfrac{1}{2}(X_\alpha)^2 + \tfrac{1}{2}(X_\beta)^2)\\
&\quad \times (1+\varepsilon(X_\alpha + X_\beta) + \varepsilon^2(X_\alpha X_\beta + \tfrac{1}{2}(X_\alpha)^2 + \tfrac{1}{2}(X_\beta)^2) + O(\varepsilon^3)\\
&= 1 + \varepsilon^2(2X_\alpha X_\beta + (X_\alpha)^2 + (X_\beta)^2 - (X_\alpha + X_\beta)^2) + O(\varepsilon^3)\\
&= 1 + \varepsilon^2(X_\alpha X_\beta - X_\beta X_\alpha) + O(\varepsilon^3)\\
&= 1 + \varepsilon^2[X_\alpha, X_\beta] + O(\varepsilon^3).
\end{aligned}$$

Hence, $[X_\alpha, X_\beta] \in \mathcal{L}^r$.

One can show that $e^{\varepsilon X_\alpha}e^{\delta X_\beta} = e^{\delta X_\beta}e^{\varepsilon X_\alpha} = e^{\varepsilon X_\alpha + \delta X_\beta}$ if and only if $[X_\alpha, X_\beta] = 0$ [Exercise 2.5-10].

Theorem 2.5.2-3. *Let* $X_\alpha^{(k)}$, $X_\beta^{(k)}$ *be the kth-extended infinitesimal generators of the infinitesimal generators* X_α, X_β, *and let* $[X_\alpha, X_\beta]^{(k)}$ *be the kth-extended infinitesimal generator of the commutator* $[X_\alpha, X_\beta]$. *Then* $[X_\alpha, X_\beta]^{(k)} = [X_\alpha^{(k)}, X_\beta^{(k)}]$, $k \geq 1$. *Hence, if* $[X_\alpha, X_\beta] = X_\gamma$, *then* $[X_\alpha^{(k)}, X_\beta^{(k)}] = X_\gamma^{(k)}$, $k \geq 1$.

Proof. Left to Exercise 2.5-11 [Ovsiannikov (1962, 1982), Olver (1986)]. □

Definition 2.5.2-4. A subspace $\mathcal{J} \subset \mathcal{L}$ is called a *subalgebra* of the Lie algebra $\mathcal{L}$ if $[X_\alpha, X_\beta] \in \mathcal{J}$ for all $X_\alpha, X_\beta \in \mathcal{J}$.

2.5.3 EXAMPLES OF LIE ALGEBRAS

(1) *Eight-Parameter Lie Group of Projective Transformations in* $\mathbf{R}^2$

Projective transformations in $\mathbf{R}^2$ map straight lines into straight lines. In particular, they are defined by the eight-parameter Lie group of transformations

$$x^* = \frac{(1+\varepsilon_3)x + \varepsilon_4 y + \varepsilon_5}{\varepsilon_1 x + \varepsilon_2 y + 1}, \tag{2.168a}$$

$$y^* = \frac{\varepsilon_6 x + (1+\varepsilon_7)y + \varepsilon_8}{\varepsilon_1 x + \varepsilon_2 y + 1}, \tag{2.168b}$$

with parameters $\varepsilon_\ell \in \mathbf{R}$, $\ell = 1,2,\ldots,8$. The infinitesimal generators of the corresponding Lie algebra $\mathcal{L}^8$ are given by

$$\begin{gathered} \mathrm{X}_1 = x^2\frac{\partial}{\partial x} + xy\frac{\partial}{\partial y}, \quad \mathrm{X}_2 = xy\frac{\partial}{\partial x} + y^2\frac{\partial}{\partial y}, \quad \mathrm{X}_3 = x\frac{\partial}{\partial x}, \quad \mathrm{X}_4 = y\frac{\partial}{\partial x}, \\ \mathrm{X}_5 = \frac{\partial}{\partial x}, \quad \mathrm{X}_6 = x\frac{\partial}{\partial y}, \quad \mathrm{X}_7 = y\frac{\partial}{\partial y}, \quad \mathrm{X}_8 = \frac{\partial}{\partial y}. \end{gathered} \tag{2.169}$$

It is convenient to display the commutators of a Lie algebra through its *commutator table* whose (i, j)th entry is $[\mathrm{X}_i, \mathrm{X}_j]$. From (2.163), it follows that the table is antisymmetric with its diagonal elements all zero. The structure constants are easily read off from the commutator table.

For the infinitesimal generators (2.169), we have the following commutator table:

	X_1	X_2	X_3	X_4
X_1	0	0	$-\mathrm{X}_1$	$-\mathrm{X}_2$
X_2	0	0	0	0
X_3	X_1	0	0	$-\mathrm{X}_4$
X_4	X_2	0	X_4	0
X_5	$2\mathrm{X}_3+\mathrm{X}_7$	X_4	X_5	0
X_6	0	X_1	$-\mathrm{X}_6$	$\mathrm{X}_3-\mathrm{X}_7$
X_7	0	X_2	0	X_4
X_8	X_6	$\mathrm{X}_3+2\mathrm{X}_7$	0	X_5

	X_5	X_6	X_7	X_8
X_1	$-2\mathrm{X}_3-\mathrm{X}_7$	0	0	$-\mathrm{X}_6$
X_2	$-\mathrm{X}_4$	$-\mathrm{X}_1$	$-\mathrm{X}_2$	$-\mathrm{X}_3-2\mathrm{X}_7$
X_3	$-\mathrm{X}_5$	X_6	0	0
X_4	0	$\mathrm{X}_7-\mathrm{X}_3$	$-\mathrm{X}_4$	$-\mathrm{X}_5$
X_5	0	X_8	0	0
X_6	$-\mathrm{X}_8$	0	X_6	0
X_7	0	$-\mathrm{X}_6$	0	$-\mathrm{X}_8$
X_8	0	0	X_8	0

(2) *Group of Rigid Motions in* $\mathbf{R}^2$

The group of rigid motions in $\mathbf{R}^2$ preserves distances between any two points in $\mathbf{R}^2$. This group is the three-parameter Lie group of transformations of rotations and translations in $\mathbf{R}^2$ given by

$$x^* = x\cos\varepsilon_1 - y\sin\varepsilon_1 + \varepsilon_2, \tag{2.170a}$$

$$y^* = x\sin\varepsilon_1 + y\cos\varepsilon_1 + \varepsilon_3, \tag{2.170b}$$

The corresponding infinitesimal generators are given by

$$\mathrm{X}_1 = -y\frac{\partial}{\partial x} + x\frac{\partial}{\partial y}, \quad \mathrm{X}_2 = \frac{\partial}{\partial x}, \quad \mathrm{X}_3 = \frac{\partial}{\partial y}. \tag{2.171}$$

The commutator table of its Lie algebra follows:

	X_1	X_2	X_3
X_1	0	$-\mathrm{X}_3$	X_2
X_2	X_3	0	0
X_3	$-\mathrm{X}_2$	0	0

(3) *Similitude Group in* $\mathbf{R}^2$

The similitude group in $\mathbf{R}^2$ consists of uniform scalings and rigid motions in $\mathbf{R}^2$. It is the four-parameter Lie group of transformations given by

$$x^* = e^{\varepsilon_4}(x\cos\varepsilon_1 - y\sin\varepsilon_1) + \varepsilon_2, \tag{2.172a}$$

$$y^* = e^{\varepsilon_4}(x\sin\varepsilon_1 + y\cos\varepsilon_1) + \varepsilon_3. \tag{2.172b}$$

The infinitesimal generators are X_1, X_2, X_3 given by (2.171) and

$$\mathrm{X}_4 = x\frac{\partial}{\partial x} + y\frac{\partial}{\partial y}. \tag{2.173}$$

The corresponding commutator table is given by

	X_1	X_2	X_3	X_4
X_1	0	$-\mathrm{X}_3$	X_2	0
X_2	X_3	0	0	X_2
X_3	$-\mathrm{X}_2$	0	0	X_3
X_4	0	$-\mathrm{X}_2$	$-\mathrm{X}_3$	0

The group of rigid motions in $\mathbf{R}^2$ [(2.170a,b)] is a three-parameter subgroup of the similitude group in $\mathbf{R}^2$ [(2.172a,b)]. This also follows from noticing that the Lie algebra with infinitesimal generators (2.171) is a three-dimensional subalgebra of the four-dimensional Lie algebra with infinitesimal generators (2.171) and (2.173).

By comparing the infinitesimal generators of the Lie algebra for the projective group (2.168a,b) with those for the similitude group (2.172a,b), one can see that the similitude group is a four-parameter subgroup of the eight-parameter projective group.

The commutator can be most useful as an aid for finding additional symmetries. For example, if a problem in $\mathbf{R}^2$ is invariant under both rotational symmetry $X_1 = -y\frac{\partial}{\partial x} + x\frac{\partial}{\partial y}$ and translational symmetry in the x-direction, i.e., $X_2 = \frac{\partial}{\partial x}$, then it must also be invariant under the symmetry $[X_1, X_2] = -\frac{\partial}{\partial y} = -X_3$ [cf. (2.171)], i.e., translational symmetry in the y-direction.

2.5.4 SOLVABLE LIE ALGEBRAS

In the next chapter, we will consider nth-order ODEs admitting r-parameter Lie groups of transformations. We will show that if $r = 1$, then the order of an ODE can be reduced constructively by one. If $n \geq 2$ and $r = 2$, the order can be reduced constructively by two. But if $n \geq 2$ and $r > 2$, it will not necessarily follow that the order can be reduced by more than one. However, if the r-dimensional Lie algebra of infinitesimal generators of an admitted r-parameter group has a q-dimensional *solvable subalgebra*, then the order of the ODE can be reduced constructively by q.

Definition 2.5.4-1. A subalgebra $\mathcal{J} \subset \mathcal{L}$ is called an *ideal* or *normal subalgebra* of $\mathcal{L}$ if $[X, Y] \in \mathcal{J}$ for all $X \in \mathcal{J}$, $Y \in \mathcal{L}$.

Definition 2.5.4-2. $\mathcal{L}^q$ is a q-dimensional *solvable Lie algebra* if there exists a chain of subalgebras

$$\mathcal{L}^{(1)} \subset \mathcal{L}^{(2)} \subset \cdots \subset \mathcal{L}^{(q-1)} \subset \mathcal{L}^{(q)} = \mathcal{L}^q, \tag{2.174}$$

such that $\mathcal{L}^{(k)}$ is a k-dimensional Lie algebra and $\mathcal{L}^{(k-1)}$ is an ideal of $\mathcal{L}^{(k)}$, $k = 1,2,\ldots,q$. [$\mathcal{L}^{(0)}$ is the null ideal consisting of only the zero vector.]

Definition 2.5.4-3. $\mathcal{L}$ is called an *Abelian Lie algebra* if $[X_\alpha, X_\beta] = 0$, for all $X_\alpha, X_\beta \in \mathcal{L}$.

The proof of the following theorem is obvious and left to Exercise 2.5-12:

Theorem 2.5.4-1. *Every Abelian Lie algebra is solvable.*

The following theorem holds for any two-dimensional Lie algebra:

Theorem 2.5.4-2. *Every two-dimensional Lie algebra is solvable.*

Proof. Let $\mathcal{L}$ be a two-dimensional Lie algebra with infinitesimal generators X_1 and X_2 as basis vectors. Suppose $[X_1, X_2] = aX_1 + bX_2 = Y$. If $c_1X_1 + c_2X_2 \in \mathcal{L}$, for arbitrary constants c_1 and c_2, then

$$\begin{aligned}[Y, c_1X_1 + c_2X_2] &= c_1 [Y, X_1] + c_2[Y, X_2] \\ &= c_1b[X_2, X_1] + c_2a[X_1, X_2] \\ &= (c_2a - c_1b)Y.\end{aligned}$$

Hence, Y is a one-dimensional ideal of $\mathcal{L}$. [If $a = b = 0$, then $\mathcal{L}$ is an Abelian Lie algebra.] □

It turns out that a three-dimensional Lie algebra is not necessarily solvable. For example, the three-dimensional Lie algebra with infinitesimal generators

$$X_1 = \frac{\partial}{\partial x}, \quad X_2 = x\frac{\partial}{\partial x}, \quad X_3 = x^2\frac{\partial}{\partial x}, \tag{2.175}$$

is not solvable.

As an example of a three-dimensional solvable Lie algebra, consider the Lie algebra for the group of rigid motions (2.170a,b). The solvability of its Lie algebra follows from the chain

$$\mathcal{L}^{(1)} \subset \mathcal{L}^{(2)} \subset \mathcal{L}^{(3)} = \mathcal{L},$$

where $\mathcal{L}^{(3)}$ has basis vectors X_1, X_2, X_3 given by (2.171), $\mathcal{L}^{(2)}$ has basis vectors X_2, X_3, and $\mathcal{L}^{(1)}$ has basis vector X_2.

EXERCISES 2.5

1. Solve the initial value problem (2.156a-e) and recover (2.150a,b).

2. In the case of a one-parameter Lie group of transformations [$r = 1$], show that the law of composition $\phi(a, b)$ satisfies

$$\Gamma(\varepsilon) = \left.\frac{\partial \phi(a,b)}{\partial b}\right|_{(a,b)=(\varepsilon^{-1},\varepsilon)} = \left[\left.\frac{\partial \phi(\varepsilon,\delta)}{\partial \delta}\right|_{\delta=0}\right]^{-1}.$$

[*Hint*: Consider $\phi(\varepsilon^{-1}, \phi(\varepsilon,\delta))$ in some neighborhood of $\delta = 0$.]

3. Show that the set of conformal transformation $x^* = X(x, y)$, $y^* = Y(x, y)$, where $F(z) = X(x,y) + iY(x,y)$ is analytic in domain D, forms an infinite-parameter Lie group of transformations. Let $z = x + iy$. Characterize the infinitesimal generators of the group.

4. Consider the set of all conformal transformations that are one-to-one on the extended plane, i.e., the *bilinear* (*Möbius*) *transformations*,

$$z^* = \frac{az+b}{cz+d}, \quad ad - bc \neq 0, \tag{2.176}$$

where $a, b, c, d \in \mathbf{C}$ and $z = x + iy$.
(a) Show that (2.176) defines a six-parameter Lie group of transformations.
(b) Find the infinitesimal generators of the group.
(c) Establish the commutator table of the corresponding Lie algebra.
(d) Find the subalgebra of largest dimension that is identical to a subalgebra of the Lie algebra of the projective group (2.168a,b).
(e) Determine the subgroup of (2.176) with the largest number of parameters that is in common with a subgroup of the projective group (2.168a,b).

5. (a) Show that the infinitesimal generators X_3, X_4, X_6, X_7 of (2.169) form a four-dimensional Lie algebra.
(b) Find the corresponding four-parameter Lie group of transformations.

6. Show that the three-parameter family of transformations $x^* = ax + b$, $y^* = cx + y$, does not form a three-parameter Lie group of transformations:
(a) from the definition of a Lie group of transformations; or
(b) from the algebra of its infinitesimal generators.

7. Consider the three-parameter family of transformations

$$x^* = ax + b, \tag{2.177a}$$

$$y^* = cy. \tag{2.177b}$$

(a) Show that (2.177a,b) defines a three-parameter Lie group of transformations.
(b) Establish the commutator table of the corresponding infinitesimal generators.
(c) Show that the Lie algebra of (2.177a,b) is solvable.

8. In Chapter 1, it was shown that problem (1.46a–c) is invariant under the two-parameter family of transformations

$$x^* = \alpha(x - \beta t), \tag{2.178a}$$

$$t^* = \alpha^2 t, \tag{2.178b}$$

$$u^* = \frac{1}{\alpha} u e^{(1/2)\beta x - (1/4)\beta^2 t}. \tag{2.178c}$$

(a) Show that (2.178a–c) defines a two-parameter Lie group of point transformations.
(b) Establish the commutator table of its Lie algebra.

9. Check that the projective transformations (2.168a,b) map straight lines into straight lines. Also check this in terms of their infinitesimal generators (2.169).

10. Show that $e^{\varepsilon X_\alpha} e^{\delta X_\beta} = e^{\delta X_\beta} e^{\varepsilon X_\alpha} = e^{\varepsilon X_\alpha + \delta X_\beta}$ if and only if $[X_\alpha, X_\beta] = 0$.

11. Prove Theorem 2.5.2-3.

12. Prove Theorem 2.5.4-1.

13. (a) Show that the infinitesimal generators

$$\mathrm{X}_1 = (1+x^2)\frac{\partial}{\partial x} + xy\frac{\partial}{\partial y}, \quad \mathrm{X}_2 = xy\frac{\partial}{\partial x} + (1+y^2)\frac{\partial}{\partial y}, \quad \mathrm{X}_3 = y\frac{\partial}{\partial x} - x\frac{\partial}{\partial y},$$

form a three-dimensional Lie algebra $\mathcal{L}^{(3)}$.

(b) Show that $\mathcal{L}^{(3)}$ does not have a two-dimensional subalgebra and, hence, is not solvable.

2.6 MAPPINGS OF CURVES AND SURFACES

Under the action of a one-parameter Lie group of point transformations admitted by an ODE, each solution curve is mapped into a one-parameter family of solution curves of the same differential equation or is invariant under the action of the group. Corresponding remarks apply to solution surfaces of PDEs.

2.6.1 INVARIANT SURFACES, INVARIANT CURVES, INVARIANT POINTS

Definition 2.6.1-1. A surface $F(\mathbf{x}) = 0$ is an *invariant surface* for a one-parameter Lie group of transformations (2.6) if and only if $F(\mathbf{x}^*) = 0$ when $F(\mathbf{x}) = 0$.

Definition 2.6.1-2. A curve $F(x, y) = 0$ is an *invariant curve* for a one-parameter Lie group of transformations (2.98a,b) if and only if $F(x^*, y^*) = 0$ when $F(x, y) = 0$.

The proof of the following theorem is left to Exercise 2.6-3:

Theorem 2.6.1-1.

(i) *A surface* $F(\mathbf{x}) = 0$ *is an invariant surface for a one-parameter Lie group of transformations* (2.6) *if and only if*

$$\mathrm{X}F(\mathbf{x}) = 0 \quad \textit{when } F(\mathbf{x}) = 0, \tag{2.179}$$

where X *is the infinitesimal generator given by* (2.24).

(ii) *A curve* $F(x, y) = 0$ *is an invariant curve for a one-parameter Lie group of transformations* (2.98a,b) *if and only if*

$$\mathrm{X}F(x, y) = 0 \quad \textit{when } F(x, y) = 0, \tag{2.180}$$

where X *is the infinitesimal generator given by* (2.98d).

This theorem gives a means for finding the invariant surface of a given Lie group of transformations, namely, by solving (2.179).

A curve written in a solved form, $F(x,y) = y - f(x) = 0$, is an invariant curve for (2.98a,b) if and only if

$$\mathrm{X}F(x,y) = \eta(x,y) - \xi(x,y)f'(x) = 0$$

when $F(x,y) = y - f(x) = 0$, i.e., if and only if

$$\eta(x,f(x)) - \xi(x,f(x))f'(x) = 0. \tag{2.181}$$

As an example, consider the scaling group

$$x^* = e^{\varepsilon}x, \tag{2.182a}$$

$$y^* = e^{\varepsilon}y. \tag{2.182b}$$

The corresponding infinitesimal generator is given by

$$\mathrm{X} = x\frac{\partial}{\partial x} + y\frac{\partial}{\partial y}. \tag{2.183}$$

A ray $y - \lambda x = 0$, $x > 0$, $\lambda = \text{const}$, is an invariant curve for (2.182a,b) since $\mathrm{X}(y - \lambda x) = y - \lambda x = 0$ when $y - \lambda x = 0$; a parabola $y - \lambda x^2 = 0$, $\lambda = \text{const}$, is not an invariant curve for (2.182a,b) since $\mathrm{X}(y - \lambda x^2) = y - 2\lambda x^2 \neq 0$ when $y - \lambda x^2 = 0$.

To find all invariant curves $y - f(x) = 0$ for (2.182a,b), we first find the general solution $u(x,y)$ of the PDE

$$x\frac{\partial u}{\partial x} + y\frac{\partial u}{\partial y} = 0.$$

This yields

$$u(x,y) = F\left(\frac{y}{x}\right),$$

where F is an arbitrary function of y/x. Invariant curves then include the curves

$$y - \lambda x = 0, \quad \lambda = \text{const}, \quad x > 0 \text{ or } x < 0.$$

Definition 2.6.1-3. A point $\mathbf{x}$ is an *invariant point* for the Lie group of transformations (2.6) if and only if $\mathbf{x}^* \equiv \mathbf{x}$ under (2.6).

The proof of the following theorem is left to Exercise 2.6-5:

Theorem 2.6.1-2. *A point* $\mathbf{x}$ *is an invariant point for the Lie group of transformations* (2.6) *if and only if*

$$\xi(\mathbf{x}) = 0. \tag{2.184}$$

For the scaling group (2.182a,b), note that $\xi(x, y) = \eta(x, y) = 0$ if and only if

$x = y = 0$, so that the only invariant point is the origin $(0, 0)$.

Definition 2.6.1-4. The family of surfaces

$$\omega(\mathbf{x}) = \text{const} = c$$

is an *invariant family of surfaces* for the Lie group of transformations (2.6) if and only if

$$\omega(\mathbf{x}^*) = \text{const} = c^* \quad \text{when } \omega(\mathbf{x}) = c.$$

Definition 2.6.1-5. The family of curves

$$\omega(x, y) = \text{const} = c$$

is an *invariant family of curves* for the Lie group of transformations (2.98a,b) if and only if

$$\omega(x^*, y^*) = \text{const} = c^* \quad \text{when } \omega(x, y) = c.$$

From these definitions, it follows that

$$c^* = C(c; \varepsilon) \tag{2.185}$$

for some function C of the constant c and the group parameter ε.

Theorem 2.6.1-3.

(i) *A family of surfaces,* $\omega(\mathbf{x}) = \text{const} = c$, *is an invariant family of surfaces for the Lie group of transformations* (2.6) *if and only if*

$$X\omega = \Omega(\omega) \tag{2.186}$$

for some infinitely differentiable function $\Omega(\omega)$.

(ii) *A family of curves,* $\omega(x, y) = \text{const} = c$, *is an invariant family of curves for the Lie group of transformations* (2.98a,b) *if and only if*

$$X\omega = \xi(x, y)\frac{\partial \omega}{\partial x} + \eta(x, y)\frac{\partial \omega}{\partial y} = \Omega(\omega) \tag{2.187}$$

for some infinitely differentiable function $\Omega(\omega)$.

Proof. Let $\omega(\mathbf{x}) = c$ be an invariant family of surfaces for (2.6). Then

$$\omega(\mathbf{x}^*) = e^{\varepsilon X}\omega(\mathbf{x}) = \omega(\mathbf{x}) + \varepsilon X\omega(\mathbf{x}) + \frac{\varepsilon^2}{2}X^2\omega(\mathbf{x}) + \cdots = c^* = C(c; \varepsilon).$$

Hence, $X\omega(\mathbf{x}) = \Omega(\omega)$ for some function $\Omega(\omega)$ when $\omega(\mathbf{x}) = c$. It follows that $X^2\omega = \Omega'(\omega)\,X\omega = \Omega'(\omega)\,\Omega(\omega)$, etc.

Conversely, suppose $X\omega(\mathbf{x}) = \Omega(\omega)$ for some infinitely differentiable function $\Omega(\omega)$. Then $X^2\omega = \Omega'(\omega)\,\Omega(\omega)$, and $X^n\omega = f_n(\omega)$ for some function $f_n(\omega)$, $n = 1,2,\ldots$. Consequently, if $\omega(\mathbf{x}) = c$, then

$$\omega(\mathbf{x}^*) = e^{\varepsilon \mathrm{X}}\omega(\mathbf{x}) = \omega(\mathbf{x}) + \varepsilon \mathrm{X}\omega(\mathbf{x}) + \frac{\varepsilon^2}{2}\mathrm{X}^2\omega(\mathbf{x}) + \cdots$$
$$= \omega(\mathbf{x}) + \sum_{n=1}^{\infty}\frac{\varepsilon^n}{n!} f_n(\omega(\mathbf{x})) = c + \sum_{n=1}^{\infty}\frac{\varepsilon^n}{n!} f_n(c) = c^*. \qquad \square$$

There are two distinguished types of invariant families of surfaces (curves). The trivial type is where each surface (curve) in the family is itself invariant. This type is characterized by $\Omega(\omega) \equiv 0$. The nontrivial type is where no surface (curve) in the family is itself invariant, i.e., each surface (curve) is moved to a different surface (curve). This type is characterized by $\Omega(\omega) \equiv 1$. This follows from the fact that if $\omega(\mathbf{x}) = c$ is an invariant family of surfaces, then so is $F(\omega(\mathbf{x})) = F(c)$ for any function F; then $\mathrm{X}F(\omega(\mathbf{x})) = F'(\omega)\,\mathrm{X}\omega = F'(\omega)\,\Omega(\omega)$, so that setting $F'(\omega) = 1/\Omega(\omega)$ we have $\mathrm{X}F(\omega) \equiv 1$. [We assume that $\Omega(\omega) \neq 0$, for otherwise some surface in the invariant family of surfaces is itself an invariant surface for (2.6).]

As an example, consider again the scaling group (2.182a,b). The invariant family of curves $\omega(x, y) = c$ satisfies

$$\mathrm{X}\omega = x\frac{\partial \omega}{\partial x} + y\frac{\partial \omega}{\partial y} = 1.$$

The corresponding characteristic equations are given by

$$\frac{d\omega}{1} = \frac{dx}{x} = \frac{dy}{y},$$

with their general solution given by

$$\omega(x, y) = \log x + f\left(\frac{y}{x}\right)$$

for an arbitrary function f. Hence, any family of curves

$$F(\omega) = F\left(\log x + f\left(\frac{y}{x}\right)\right) = \text{const} = c \tag{2.188}$$

is an invariant family of curves for (2.182a,b) for any choice of F and f. In particular, the family of circles $x^2 + y^2 = \text{const} = r^2$ is an invariant family of curves for (2.182a,b) obtained by choosing $F(\omega) = e^{2\omega}$ and $f(z) = \frac{1}{2}\log(1 + z^2)$ in (2.188). The family of lines $x = \text{const}$ is invariant, corresponding to $F(\omega) = e^{\omega}$, $f(z) = 0$. The family of logarithmic spirals $r^2 e^{\theta} = \text{const}$ is invariant, corresponding to $F(\omega) = e^{2\omega}$ and $f(z) = \frac{1}{2}(\log(1 + z^2) + \arctan z)$.

2.6.2 MAPPINGS OF CURVES

Consider a one-parameter Lie group of transformations

$$x^* = X(x, y; \varepsilon) = e^{\varepsilon X} x, \tag{2.189a}$$

$$y^* = Y(x, y; \varepsilon) = e^{\varepsilon X} y, \tag{2.189b}$$

with infinitesimal generator

$$\mathrm{X} = \xi(x, y)\frac{\partial}{\partial x} + \eta(x, y)\frac{\partial}{\partial y}. \tag{2.189c}$$

Consider a curve $y = \Theta(x)$. The transformation (2.189a,b) maps a point (x, y) on the curve $y = \Theta(x)$ into the point (x^*, y^*) with

$$x^* = X(x, \Theta(x); \varepsilon), \tag{2.190a}$$

$$y^* = Y(x, \Theta(x); \varepsilon). \tag{2.190b}$$

For a fixed value of ε, (2.190a,b) defines a parametric representation of a new curve with x playing the role of a parameter [Figure 2.5]. One can eliminate x from (2.190a,b) by substitution through the inverse transformation of (2.189a,b), i.e., through substitution of

$$x = X(x^*, y^*; -\varepsilon) \tag{2.191}$$

into (2.190b). Then

$$y^* = Y(X(x^*, y^*; -\varepsilon), \Theta(X(x^*, y^*; -\varepsilon)); \varepsilon) = Y(e^{-\varepsilon X} x^*, \Theta(e^{-\varepsilon X} x^*); \varepsilon). \tag{2.192}$$

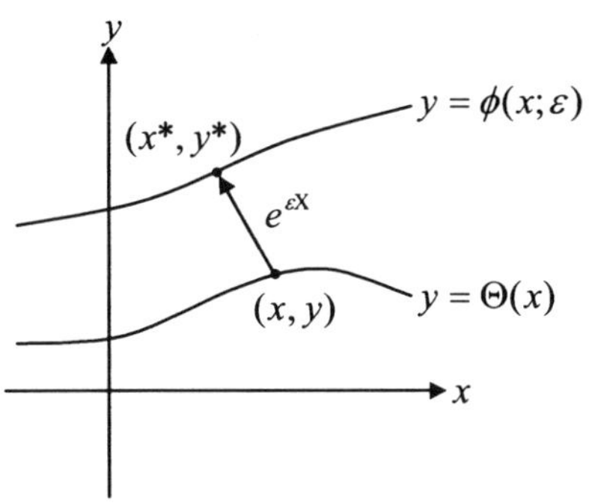

Figure 2.5. Mapping of a curve: A different curve $y = \phi(x; \varepsilon)$ corresponds to each parameter value ε.

Equation (2.192) yields the relationship between the x- and y-coordinates of the new curve denoted by $y = \phi(x; \varepsilon)$. After substituting in x and y for x^* and y^* and replacing ε

by $-\varepsilon$, we have the following:

Theorem 2.6.2-1. *Suppose* $y=\Theta(x)$ *is not an invariant curve of* (2.189a,b). *Then*

$$y = Y(e^{\varepsilon X}x, \Theta(e^{\varepsilon X}x); -\varepsilon) \tag{2.193a}$$

$$= Y(X(x,y;\varepsilon), \Theta(X(x,y;\varepsilon)); -\varepsilon) \tag{2.193b}$$

implicitly defines a one-parameter family of curves $y=\phi(x;\varepsilon)$.

2.6.3 EXAMPLES OF MAPPINGS OF CURVES

(1) *Scaling Group*
For the scaling group

$$x^* = X = e^{\varepsilon}x, \tag{2.194a}$$

$$y^* = Y = e^{2\varepsilon}y, \tag{2.194b}$$

we have

$$y^* = e^{2\varepsilon}\Theta(x) = e^{2\varepsilon}\Theta(e^{-\varepsilon}x^*),$$

and hence, $y=\Theta(x)$ maps into the family of curves

$$y = e^{2\varepsilon}\Theta(e^{-\varepsilon}x) = \phi(x;\varepsilon). \tag{2.195}$$

(2) *Projection Group*
For the projection group

$$x^* = \frac{x}{1-\varepsilon y}, \tag{2.196a}$$

$$y^* = \frac{y}{1-\varepsilon y}, \tag{2.196b}$$

we have

$$y^* = \frac{\Theta(x)}{1-\varepsilon\Theta(x)},$$

and

$$x = x^*(1-\varepsilon y) = \frac{x^*}{1+\varepsilon y^*}.$$

Hence,

$$y^* = \frac{\Theta\left(\dfrac{x^*}{1+\varepsilon y^*}\right)}{1-\varepsilon\Theta\left(\dfrac{x^*}{1+\varepsilon y^*}\right)}.$$

Consequently, the curve $y = \Theta(x)$ maps into the family of curves $y = \phi(x;\varepsilon)$ satisfying the implicit equation

$$\frac{y}{1+\varepsilon y} = \Theta\left(\frac{x}{1+\varepsilon y}\right). \tag{2.197}$$

2.6.4 MAPPINGS OF SURFACES

We derive a formula for families of surfaces analogous to formula (2.193a,b) for families of curves. Consider a one-parameter Lie group of transformations

$$x^* = X(x,u;\varepsilon) = e^{\varepsilon \mathrm{X}} x, \tag{2.198a}$$

$$u^* = U(x,u;\varepsilon) = e^{\varepsilon \mathrm{X}} u, \tag{2.198b}$$

with infinitesimal generator

$$\mathrm{X} = \sum_{i=1}^{n} \xi_i(x,u)\frac{\partial}{\partial x_i} + \eta(x,u)\frac{\partial}{\partial u}.$$

Consider a surface $u = \Theta(x)$ that is not invariant under (2.198a,b). The transformation (2.198a,b) maps a point (x, u) on the curve $u = \Theta(x)$ into the point (x^*, u^*) with

$$x^* = X(x,\Theta(x);\varepsilon), \tag{2.199a}$$

$$u^* = U(x,\Theta(x);\varepsilon). \tag{2.199b}$$

For a fixed value of ε, one can eliminate x from (2.199a,b) by substitution through the inverse transformation of (2.198a), i.e., by substitution of

$$x = X(x^*,u^*;-\varepsilon)$$

into (2.199b). Then

$$u^* = U(X(x^*,u^*;-\varepsilon),\Theta(X(x^*,u^*;-\varepsilon));\varepsilon) = U(e^{-\varepsilon \mathrm{X}}x^*,\Theta(e^{-\varepsilon \mathrm{X}}x^*);\varepsilon), \tag{2.200}$$

with

$$\mathrm{X} = \sum_{i=1}^{n} \xi_i(x,u)\frac{\partial}{\partial x_i} + \eta(x,u)\frac{\partial}{\partial u} = \sum_{i=1}^{n} \xi_i(x^*,u^*)\frac{\partial}{\partial x^*_i} + \eta(x^*,u^*)\frac{\partial}{\partial u^*}$$

[cf. Exercise 2.3-5]. Replacing $(x^*,u^*,-\varepsilon)$ by (x,u,ε) in (2.200), we then have

$$u = U(e^{\varepsilon X}x, \Theta(e^{\varepsilon X}x); -\varepsilon) = U(X(x,u;\varepsilon), \Theta(X(x,u;\varepsilon)); -\varepsilon). \qquad (2.201)$$

Theorem 2.6.4-1. *Suppose* $u = \Theta(x)$ *is not an invariant surface of* (2.198a,b). *Then* (2.201) *implicitly defines a mapping of the surface* $u = \Theta(x)$ *into a one-parameter family of surfaces* $u = \phi(x;\varepsilon)$.

EXERCISES 2.6

1. For the group of transformations (2.71a,b), find invariant curves, invariant points, and invariant families of curves.
2. For the group of transformations (1.93), find invariant curves (surfaces), invariant points, and invariant families of curves (surfaces):
   (a) in (x, t)-space; and
   (b) in (x, t, u)-space.
3. Prove Theorem 2.6.1-1.
4. Geometrically interpret (2.181).
5. Prove Theorem 2.6.1-2.
6. Show that if $y = \Theta(x)$ is an invariant curve of (2.189a,b), then (2.193a,b) yields $\phi(x;\varepsilon) \equiv \Theta(x)$ for all ε.
7. Find the image $y = \phi(x;\varepsilon)$ of the curve $y = \Theta(x)$ under the rotation group

$$x^* = x\cos\varepsilon - y\sin\varepsilon,$$
$$y^* = x\sin\varepsilon + y\cos\varepsilon.$$

2.7 LOCAL TRANSFORMATIONS

For some applications to ODEs, it is essential to look at a one-parameter Lie group of point transformations (2.189a,b) acting on (x, y)-space from the point of view of transformations acting directly on the space of functions $y = y(x)$. This will lead to natural generalizations of point transformations.

2.7.1 POINT TRANSFORMATIONS

Consider again the mapping of a curve $y = \Theta(x)$ into a family of curves $y = \phi(x;\varepsilon)$ under a one-parameter Lie group of point transformations (2.189a,b). Geometrically, this transformation represents a mapping of points (x, y) into (x^*, y^*), as discussed in Section 2.6.2 [cf. Figure 2.5], giving the implicit formula (2.193b) for $y = \phi(x;\varepsilon)$. It is important to describe this mapping explicitly as a transformation of the curve $y = \Theta(x)$ to curves $y = \phi(x;\varepsilon)$. Formally, this mapping is given by

$$x^* = x,$$
$$y^* = \phi(x;\varepsilon) = (e^{\varepsilon\hat{X}}y)\Big|_{y=\Theta(x)},$$

for some infinitesimal generator $\hat{X}$ [see Figure 2.6].

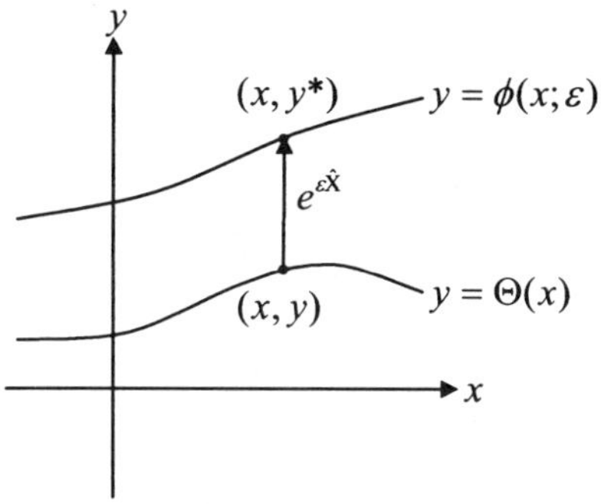

Figure 2.6. Direct mapping of a curve $y = \Theta(x)$ to curves $y = \phi(x;\varepsilon)$.

We now derive a formula for the infinitesimal generator $\hat{X}$. Under a Lie group of point transformations (2.189a,b), we have

$$x^* = x + \varepsilon\xi(x,\Theta(x)) + O(\varepsilon^2), \tag{2.202a}$$

$$y^* = \Theta(x) + \varepsilon\eta(x,\Theta(x)) + O(\varepsilon^2). \tag{2.202b}$$

The dependence of y^* on x^* defines the image $y^* = \phi(x^*;\varepsilon)$ of $y = \Theta(x)$. One eliminates x from (2.202a,b) to obtain $\phi(x^*;\varepsilon)$. Solving (2.202a) for x yields

$$x = x^* - \varepsilon\xi(x^*,\Theta(x^*)) + O(\varepsilon^2). \tag{2.203}$$

Substituting (2.203) into (2.202b) and taking the Taylor expansion in ε, we obtain

$$\phi(x^*;\varepsilon) = \Theta(x^*) + \varepsilon(\eta(x^*,\Theta(x^*)) - \xi(x^*,\Theta(x^*))\Theta'(x^*)) + O(\varepsilon^2). \tag{2.204}$$

Then, if we replace x^* by x in (2.204), the image of $y = \Theta(x)$ under the transformation (2.189a,b) is given by

$$y^* = \phi(x;\varepsilon) = \Theta(x) + \varepsilon[\eta(x,\Theta(x)) - \xi(x,\Theta(x))\Theta'(x)] + O(\varepsilon^2). \tag{2.205}$$

We now observe that the same image of $y = \Theta(x)$ can also be obtained by a transformation leaving x invariant:

$$x^* = x,$$

$$y^* = y + \varepsilon[\eta(x, y) - \xi(x, y)y'] + O(\varepsilon^2). \tag{2.206}$$

The infinitesimal generator for transformation (2.206) is given by

$$\hat{\mathrm{X}} = [\eta(x, y) - \xi(x, y)y']\frac{\partial}{\partial y}. \tag{2.207}$$

Geometrically, we have now moved from a transformation (2.189a,b) acting on (x, y)-space to a transformation (2.206) acting on the space of functions $y = y(x)$. The infinitesimal generator (2.207) is the *characteristic form* of the infinitesimal generator (2.189c).

As examples, for the scaling group (2.194a,b), we have

$$\hat{\mathrm{X}} = [2y - xy']\frac{\partial}{\partial y}, \tag{2.208}$$

and for the projection group (2.196a,b), we have

$$\hat{\mathrm{X}} = [y^2 - xyy']\frac{\partial}{\partial y}. \tag{2.209}$$

2.7.2 CONTACT AND HIGHER-ORDER TRANSFORMATIONS

We can generalize point transformations with infinitesimal generators of the characteristic form (2.207) to local transformations with infinitesimal generators of the form

$$\hat{\mathrm{X}} = \hat{\eta}(x, y, y', \ldots, y^{(k)})\frac{\partial}{\partial y} \tag{2.210}$$

involving dependence on derivatives $y^{(\ell)}$ up to some finite order $\ell = k$.

Formally, we can exponentiate (2.210) to obtain a corresponding one-parameter group of transformations acting on the space of functions $y = y(x)$:

$$x^* = x,$$
$$y^* = y + \varepsilon\hat{\eta} + O(\varepsilon^2). \tag{2.211}$$

To calculate the higher-order terms, we extend $\hat{\mathrm{X}}$ to act on $y', y'', \ldots, y^{(j)}$ by requiring that the contact conditions are preserved, i.e.,

$$(y^*)' = \frac{dy^*}{dx} = \mathrm{D}(y + \varepsilon\hat{\eta} + O(\varepsilon^2)) = y' + \varepsilon\mathrm{D}\hat{\eta} + O(\varepsilon^2),$$
$$\vdots$$
$$y^{*(j)} = \frac{dy^{*(j-1)}}{dx} = \mathrm{D}(y^{(j-1)} + \varepsilon\mathrm{D}^{j-1}\hat{\eta} + O(\varepsilon^2)) = y^{(j)} + \varepsilon\mathrm{D}^j\hat{\eta} + O(\varepsilon^2),$$

where $y^{(j)} = d^j y / dx^j$, $j \geq 1$, and D is the total derivative operator d/dx.

Consequently, the extended infinitesimal generator (*the prolongation* of $\hat{\mathrm{X}}$) is given by

$$\hat{\mathrm{X}}^{(\infty)} = \hat{\eta}\frac{\partial}{\partial y} + \hat{\eta}^{(1)}\frac{\partial}{\partial y'} + \cdots + \hat{\eta}^{(j)}\frac{\partial}{\partial y^{(j)}} + \cdots, \tag{2.212a}$$

where

$$\hat{\eta}^{(1)} = \mathrm{D}\hat{\eta} = \frac{\partial \hat{\eta}}{\partial x} + y'\frac{\partial \hat{\eta}}{\partial y} + \cdots + y^{(k+1)}\frac{\partial \hat{\eta}}{\partial y^{(k)}}, \tag{2.212b}$$

$$\hat{\eta}^{(j)} = \mathrm{D}\hat{\eta}^{(j-1)}, \quad j \geq 1. \tag{2.212c}$$

Hence, the exponentiation of the infinitesimal generator (2.210) yields the following transformation:

Definition 2.7.2-1. A one-parameter group of *local transformations* is a transformation of the form

$$x^* = x,$$

$$y^* = e^{\varepsilon \hat{\mathrm{X}}^{(\infty)}} y = y + \sum_{j=1}^{\infty} \frac{\varepsilon^j}{j!}(\hat{\mathrm{X}}^{(\infty)})^{j-1}\hat{\eta}, \tag{2.213}$$

where $\hat{\mathrm{X}}^{(\infty)}$ is given by (2.112a).

Note that one can invert (2.213) through inverse exponentiation.

A local transformation corresponds to a point transformation if and only if $\hat{\eta}$ is of the form $\hat{\eta} = \eta(x, y) - \xi(x, y)y'$ for some $\eta(x, y), \xi(x, y)$, i.e., $\hat{\eta}$ is linear in y' and has no dependence on higher derivatives of y. A local transformation (2.213) is called a *contact transformation* if $\hat{\eta}$ is of the form $\hat{\eta} = \hat{\eta}(x, y, y')$. Otherwise, a local transformation is called a *higher-order transformation.* One can show that a local transformation corresponds to an extended transformation acting on some finite-dimensional space $(x, y, y', \ldots, y^{(p)})$, $p \geq 1$, if and only if it is a contact transformation.

2.7.3 EXAMPLES OF LOCAL TRANSFORMATIONS

(1) *Scaling Group*

For the scaling group (2.194a,b), the extension of the infinitesimal generator (2.208) is given by

$$\hat{\mathrm{X}}^{(\infty)} = (2y - xy')\frac{\partial}{\partial y} + (y' - xy'')\frac{\partial}{\partial y'} - xy'''\frac{\partial}{\partial y'''} - (y''' + xy^{(4)})\frac{\partial}{\partial y'''} + \cdots$$

$$-((j-2)y^{(j)}+xy^{(j+1)})\frac{\partial}{\partial y^{(j)}}+\cdots. \tag{2.214}$$

The curve $y=\Theta(x)$ is mapped into the family of curves $y=\phi(x;\varepsilon)$ given by

$$\begin{aligned}\phi(x;\varepsilon)&=(e^{\varepsilon\hat{X}^{(\infty)}}y)\Big|_{y=\Theta(x)}\\&=[y+\varepsilon(2y-xy')+\tfrac{1}{2}\varepsilon^2(4y-2xy'-xy'+x^2y'')+O(\varepsilon^3)]\Big|_{y=\Theta(x)}\\&=\Theta(x)+\varepsilon[2\Theta(x)-x\Theta'(x)]+\tfrac{1}{2}\varepsilon^2[4\Theta(x)-3x\Theta'(x)+x^2\Theta''(x)]+O(\varepsilon^3).\end{aligned} \tag{2.215}$$

The expression (2.215) is the Taylor series of the mapping (2.195).

(2) *Projective Group*
For the projective group (2.196a,b), the extension of the infinitesimal generator (2.209) is given by

$$\hat{X}^{(\infty)}=(y^2-xyy')\frac{\partial}{\partial y}+(yy'-x(y')^2-xyy'')\frac{\partial}{\partial y'}-(3xy'y''+xyy''')\frac{\partial}{\partial y''}+\cdots. \tag{2.216}$$

Here the curve $y=\Theta(x)$ is mapped into the family of curves $y=\phi(x;\varepsilon)$ given by

$$\begin{aligned}\phi(x;\varepsilon)&=(e^{\varepsilon\hat{X}^{(\infty)}}y)\Big|_{y=\Theta(x)}\\&=[y+\varepsilon(y^2-xyy')+\tfrac{1}{2}\varepsilon^2((y^2-xyy')(2y-xy')-xy(yy'-x(y')^2-xyy''))+O(\varepsilon^3)]\Big|_{y=\Theta(x)}\\&=\Theta(x)+\varepsilon[\Theta^2(x)-x\Theta(x)\Theta'(x)]\\&\quad+\tfrac{1}{2}\varepsilon^2[2\Theta^3(x)-4x\Theta^2(x)\Theta'(x)+2x^2\Theta(x)(\Theta'(x))^2+x^2\Theta^2(x)\Theta''(x)]+O(\varepsilon^3).\end{aligned} \tag{2.217}$$

The expression (2.217) yields the *explicit* Taylor series for the mapping given by the implicit equation (2.197).

EXERCISES 2.7

1. Consider the rotation group

$$x^*=x\cos\varepsilon-y\sin\varepsilon,$$
$$y^*=x\sin\varepsilon+y\cos\varepsilon.$$

   (a) Find the extended infinitesimal generator $\hat{X}^{(\infty)}$.
   (b) Use the local transformation with the infinitesimal generator $\hat{X}^{(\infty)}$ to calculate the image of the curve $y=\Theta(x)$ to $O(\varepsilon^3)$ and compare the resulting expression with the expression obtained in Exercise 2.6-7.

2. Show that the Taylor series for (2.197) to $O(\varepsilon^3)$ agrees with (2.217).

3. Consider a point transformation

$$x^* = x + \varepsilon\xi(x,y) + O(\varepsilon^2),$$
$$y^* = y + \varepsilon\eta(x,y) + O(\varepsilon^2),$$

with extended infinitesimal generator

$$\mathrm{X}^{(\infty)} = \xi\frac{\partial}{\partial x} + \eta\frac{\partial}{\partial y} + \eta^{(1)}\frac{\partial}{\partial y'} + \cdots + \eta^{(k)}\frac{\partial}{\partial y^{(k)}} + \cdots. \tag{2.218}$$

Show that the characteristic form of the extended infinitesimal generator (2.218) is given by

$$\hat{\mathrm{X}}^{(\infty)} = \hat{\eta}\frac{\partial}{\partial y} + \hat{\eta}^{(1)}\frac{\partial}{\partial y'} + \cdots + \hat{\eta}^{(k)}\frac{\partial}{\partial y^{(k)}} + \cdots,$$

where

$$\hat{\eta} = \eta - y'\xi, \qquad \hat{\eta}^{(k)} = \eta^{(k)} - y^{(k+1)}\xi = \mathrm{D}^k\eta - \sum_{j=0}^{k}\frac{k!}{(k-j)!\,j!}y^{(k+1-j)}\mathrm{D}^j\xi, \quad k \geq 1. \tag{2.219}$$

2.8 DISCUSSION

In this chapter, we have considered one-parameter Lie groups of transformations that are completely determined by their infinitesimal transformations. Actually such groups are one-parameter *connected local Lie groups of transformations* [Gilmore (1974); Olver (1986); Ovsiannikov (1962, 1982)]. The global properties of Lie groups turn out to be unimportant for the purpose of studying the invariance of differential equations.

Using the infinitesimal generator of a one-parameter Lie group of transformations one can construct various kinds of invariants (invariant surfaces, invariant points, invariant families of surfaces). Moreover, for a one-parameter Lie group of point transformations we can determine canonical coordinates in terms of which the transformation group becomes a group of translations.

When applying Lie groups of transformations to the study of the invariance properties of a differential equation, the coordinates of the group transformations are separated into independent and dependent variables. A one-parameter Lie group of transformations acting on the space of independent and dependent variables is naturally extended (prolonged) to a one-parameter Lie group of transformations acting on an enlarged space (jet space) that includes all derivatives of the dependent variables up to a fixed finite order. This is accomplished by requiring, under the group action, preservation of derivative relations or, equivalently, the preservation of the contact conditions connecting the higher-order differentials. This requirement induces a unique extension (prolongation) of the group action to any enlarged space (higher-order jet space). Consequently, one-parameter extended (prolonged) Lie groups of transformations are characterized completely by their infinitesimals. Moreover, these extended (prolonged) infinitesimals are determined from the infinitesimals of the group action on the space of independent and dependent variables. This allows one to establish an algorithm to determine the infinitesimal transformations admitted by a given differential equation.

The study of multiparameter Lie groups of transformations reduces to the study of infinitesimal generators of one-parameter subgroups. The infinitesimal generators form a vector space called a Lie algebra that is closed under commutation. The invariance properties of a differential equation under a multiparameter Lie group of transformations can be completely characterized by its Lie algebra. The structure (commutator table) of the Lie algebra of a multiparameter group plays an essential role in applying infinitesimal transformations to differential equations.

When one considers generalizations beyond point transformations to higher-order transformations (as well as to nonlocal transformations), it turns out to be important to look at a group transformation from the point of view of mapping a given curve (surface) into another curve (surface) with the independent variable(s) fixed. This is especially necessary in studying the invariance properties of differential equations under higher-order transformations.

Our presentation of Lie groups of transformations has emphasized the essential computational and algebraic aspects of Lie group theory needed for applications to ODEs in Chapter 3 and PDEs in Chapter 4. Differential geometrical aspects of Lie group theory, although secondary in such applications, provide a complementary viewpoint and a geometrical interpretation for results covered in this chapter [Olver (1986); Warner (1983)].

The (prolonged) action of a one-parameter Lie group of transformations defines a differentiable curve in a (higher-order) jet space. Geometrically, the infinitesimal generator of the transformation represents the tangent vector of this curve at a given point. For m dependent variables and n independent variables, the jet space of order k is a differentiable manifold of dimension $(m+n)[(n+k)!/(k!\,m!)]$. A basis for the tangent space of vector fields on this manifold is defined by the coordinate vector fields which are partial derivative operators with respect to the jet space coordinates. The basis components of the tangent vector field associated to a one-parameter Lie group of transformations are thus the prolonged infinitesimals of the transformation. For an r-parameter Lie group of transformations, the associated r tangent vector fields are linearly independent and in involution, i.e., their span is an r-dimensional vector space containing all their commutators (Lie brackets). In particular, at each point in the underlying jet space, these vector fields span an r-dimensional Lie algebra. Correspondingly, the action of the transformation group defines an r-dimensional surface in the jet space.

An r-parameter Lie group of transformations on a jet space arises naturally as a representation of an underlying abstract r-dimensional connected Lie group. Such a Lie group is a differentiable manifold G on which there is a given group structure (law of composition) $\phi: G\times G \to G$ such that multiplication and inversion of group elements (points) in G are given by differentiable mappings. The mappings $L_g: G \to G$ and $R_g: G \to G$, respectively defined by left and right multiplication by a fixed group element g in G, play a central role in abstract Lie group theory. The action of these differentiable mappings on a tangent vector at the point given by the identity element e in G determines a corresponding left- or right-invariant vector field on G. The respective sets of all left- and right-invariant vector fields on an r-dimensional Lie group form r-dimensional vector spaces which possess a Lie bracket structure given by the commutator operation. In this context, Lie's three fundamental theorems have a simple geometrical meaning.

Lie's first theorem essentially states that a one-dimensional Lie subgroup of G is equivalent to an integral curve on G of a left- or right-invariant vector field determined by specifying a tangent vector at the identity element e in G. The Lie bracket of two left- or right-invariant vector fields on G measures the extent to which they are involutive, i.e., their corresponding integral curves close to form a two-dimensional differentiable submanifold (surface) in G if and only if the Lie bracket is contained in the span of these two vector fields. Lie's second and third theorems then reflect the fact that the vector spaces of left- and right-invariant vector fields on G have the structure of a Lie algebra whose commutator structure is the same at all points in G because of the invariance property of the vector fields.

Thus, associated to any abstract connected Lie group of dimension r is a unique r-dimensional Lie algebra. Conversely, corresponding to any abstract r-dimensional Lie algebra there is a unique simply-connected Lie group G. More generally, for such a Lie group, there is a one-to-one correspondence between its k-dimensional Lie subgroups and its k-dimensional Lie subalgebras. Indeed, geometrically, a k-dimensional Lie subgroup of G is a k-dimensional submanifold (surface) foliated by integral curves of left- or right-invariant vector fields on G determined by an involutive k-dimensional subspace of tangent vectors at the identity element e in G. Finally, a solvable r-dimensional Lie group is geometrically characterized by admitting an ascending chain of integral submanifolds of dimensions $1,2,\ldots,r$ generated by integral curves of involutive left- or right-invariant vector fields on the Lie group.

3

Ordinary Differential Equations (ODEs)

3.1 INTRODUCTION

Symmetries and first integrals are two fundamental structures of ordinary differential equations (ODEs). Geometrically, it is natural to view an nth-order ODE as a surface in the $(n+2)$–dimensional space whose coordinates are given by the independent variable, the dependent variable and its derivatives to order n, so that the solutions of the ODE are particular curves lying on this surface. From this point of view, a symmetry represents a motion that moves each solution curve into solution curves; a first integral represents a quantity that is conserved along each solution curve. [More precisely, a symmetry is a one-parameter group of local transformations, acting on the coordinates of the $(n+2)$–dimensional space, that maps solutions into solutions, and a first integral is a quadrature expressed by a function of the coordinates involving the independent variable, the dependent variable and its derivatives to order $n-1$, that is constant on each solution.]

In this chapter, we show how to find admitted symmetries and first integrals of an nth-order ODE. We study the integration of ODEs from these two distinct points of view.

Lie showed that if a given ODE admits a one-parameter group of point transformations (point symmetry), then the order of the ODE can be reduced by one. Moreover, the solution of the reduced ODE plus a quadrature yields the solution of the given ODE.

If an nth-order ODE admits an r-parameter solvable group of point transformations, then it can be reduced to an $(n-r)$th-order ODE plus r quadratures. When $r = n$, one can obtain the general solution of the ODE in terms of n quadratures. When $r < n$, the reduced $(n-r)$th-order ODE uses derived independent and dependent variables. In general, this ODE is not of order $n-r$ when expressed in terms of the original independent and dependent variables (typically it is still of order n).

For a first-order ODE, Lie's symmetry reduction yields the quadrature of the ODE. Lie showed that this is equivalent to finding a first integral and corresponding integrating factor of the ODE.

For an nth order-ODE, a first integral yields a quadrature reducing the order of the ODE by one. Finding a first integral of a given ODE is equivalent to obtaining an integrating factor admitted by the ODE.

If n functionally independent first integrals are known for an nth-order ODE, then one obtains the general solution of the ODE in terms of n essential constants. On the other hand, if one only knows $r < n$ functionally independent first integrals, then the ODE is reduced to an $(n-r)$th-order ODE in terms of r essential constants. In contrast to symmetry reduction, in the integrating factor approach, the reduced ODE is of order $n-r$ in terms of the original independent and dependent variables.

For a given ODE, the integrating factor method and Lie's reduction method are complementary. However, the algorithms for computing symmetries and integrating factors are similar. Symmetries are solutions of the linearization (Fréchet derivative) of the given ODE holding for *all* solutions of the given ODE. On the other hand, integrating factors are solutions of a linear system that includes the adjoint of the linearization of the given ODE holding for *all* solutions of the given ODE.

Symmetry reduction can also be applied to boundary value problems for ODEs. If a symmetry reduces the order of an ODE, the same reduction holds for any posed boundary value problem.

If an ODE admits a Lie group of transformations, then one can construct interesting special classes of solutions (*invariant solutions*) that correspond to invariant curves of the admitted Lie group of transformations. For a first-order ODE, such invariant solutions can be determined algebraically and include separatrices and singular envelope solutions. For higher-order ODEs, invariant solutions are determined either algebraically or by solving the first-order ODE for the invariant curves of the group.

3.1.1 ELEMENTARY EXAMPLES

To illustrate symmetry reduction and its connections with integrating factors for first-order ODEs, we consider two elementary examples:

(1) *Group of Translations*
The first-order ODE

$$y_1 = \frac{dy}{dx} = F(x) \tag{3.1}$$

trivially reduces to the quadrature

$$y = \int F(x)\, dx + C. \tag{3.2}$$

Obviously, the right-hand side of ODE (3.1) is characterized by no dependence on y. In particular, the one-parameter (ε) Lie group of translations

$$x^* = x, \tag{3.3a}$$

$$y^* = y + \varepsilon, \tag{3.3b}$$

is admitted by ODE (3.1) since

$$y^*_1 = \frac{dy^*}{dx^*} = \frac{dy}{dx} = y_1 \quad \text{and} \quad F(x^*) = F(x),$$

so that under the group (3.3a,b), the surface $y_1 = F(x)$ is invariant in (x, y, y_1)-space.

Moreover, it is easy to see that the ODE

$$\frac{dy}{dx} = f(x, y) \tag{3.4}$$

is invariant under the group (3.3a,b) if and only if, for any value of the parameter ε,

$$f(x^*, y^*) \equiv f(x, y+\varepsilon) \equiv f(x, y),$$

i.e., $f(x, y)$ is independent of y or, equivalently, $f(x, y) \equiv F(x)$ for some function $F(x)$. Thus the reduction of (3.1) to quadrature (3.2) is equivalent to the invariance of (3.1) under the group of translations (3.3a,b).

Under the action of the group (3.3a,b), a solution curve $y = \Theta(x)$ of ODE (3.1) maps into a curve $y^* = \Theta(x^*)$ which corresponds to the solution curve $y = \Theta(x) - \varepsilon$ of (3.1) [Figure 3.1]. Thus, from the invariance of ODE (3.1) under the group of translations (3.3a,b), we see that if $y = \Theta(x)$ is a particular solution of (3.1) then $y = \Theta(x) + C$ is the general solution of (3.1) for an arbitrary constant C.

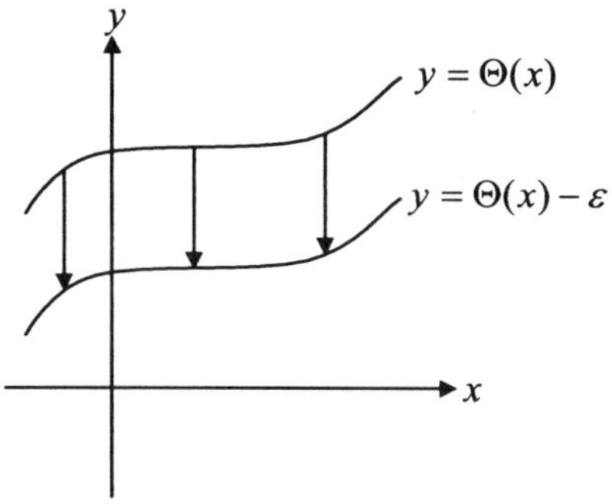

Figure 3.1

From the quadrature (3.2), we see that

$$y - \int F(x)\, dx = C \tag{3.5}$$

is a first integral of (3.1). After differentiating the first integral (3.5), we obtain

$$\frac{d}{dx}\left[y - \int F(x)\, dx\right] = y' - F(x) = 1 \cdot (y' - F(x)) = 0,$$

and hence, the function

$$\mu(x, y) \equiv 1 \tag{3.6}$$

is an integrating factor of ODE (3.1) that yields the first integral (3.5).

(2) *Group of Scalings*
The first-order ODE

$$y_1 = \frac{dy}{dx} = F\left(\frac{y}{x}\right), \tag{3.7}$$

commonly called a *homogeneous equation*, admits the one-parameter (α) group of scalings

$$x^* = \alpha x, \tag{3.8a}$$

$$y^* = \alpha x, \tag{3.8b}$$

since

$$y^*_1 = \frac{dy^*}{dx^*} = \frac{\alpha\, dy}{\alpha\, dx} = y_1 \quad \text{and} \quad F\left(\frac{y^*}{x^*}\right) = F\left(\frac{y}{x}\right).$$

Under the action of group (3.8a,b), a solution curve $y = \Theta(x)$ of ODE (3.7) maps into the curve $y^* = \Theta(x^*)$ which corresponds to the solution curve

$$y = \frac{1}{\alpha}\Theta(\alpha x) \tag{3.9}$$

of (3.7). It follows that if $y = \Theta(x)$ is a particular solution of ODE (3.7) and the curve $y - \Theta(x) = 0$ is not invariant under (3.8a,b) (i.e., $\Theta(x) \neq \lambda x$ for some fixed constant λ), then

$$y = \frac{1}{C}\Theta(Cx)$$

is the general solution of (3.7) for an arbitrary constant C.

The reduction of order of ODE (3.7) from its invariance under the group of scalings (3.8a,b) is accomplished by choosing, as new coordinates, the canonical coordinates

$$r = \frac{y}{x}, \tag{3.10a}$$

$$s = \log y. \tag{3.10b}$$

With the reparametrization $\varepsilon = \log \alpha$ $(\alpha > 0)$, the ODE (3.7) is correspondingly invariant under the one-parameter (ε) Lie group of translations

$$r^* = r, \tag{3.11a}$$

$$s^* = s + \varepsilon. \tag{3.11b}$$

Hence, from the first example it follows that, in terms of canonical coordinates (3.10a,b), the ODE (3.7) must be of the form

$$\frac{ds}{dr} = G(r), \tag{3.12}$$

for some function $G(r)$. Thus, the general solution of ODE (3.7) is given by

$$s = \int G(r)\, dr + C, \tag{3.13}$$

or, in terms of coordinates x and y,

$$y = C^* \exp\left[\int^{y/x} G(r)\, dr\right], \quad C^* = \text{const.} \tag{3.14}$$

The function $G(r)$ is determined as follows:

$$ds = \frac{1}{y}dy, \quad dr = -\frac{y}{x^2}dx + \frac{1}{x}dy,$$

and hence,

$$G(r) = \frac{ds}{dr} = \frac{y_1}{ry_1 - r^2} = \frac{F(r)}{rF(r) - r^2}, \tag{3.15}$$

where $F(r)$ is the function given in ODE (3.7).

From the quadrature (3.13), we see that

$$\log y - \int^{y/x} G(r)\,dr = C \tag{3.16}$$

is a first integral of (3.7). After differentiating (3.16) and collecting the y' terms, we obtain

$$\frac{d}{dx}\left[\log y - \int^{y/x} G(r)\,dr\right] = \frac{1}{y - xF\left(\frac{y}{x}\right)} \cdot \left(y' - F\left(\frac{y}{x}\right)\right) = 0.$$

Hence, the function

$$\mu(x, y) = \frac{1}{y - xF\left(\frac{y}{x}\right)} \tag{3.17}$$

is an integrating factor of ODE (3.7) that yields its general solution in the form (3.16).

EXERCISES 3.1

1. Consider the ODE

$$\frac{dy}{dx} = \frac{y}{x}. \tag{3.18}$$

   (a) Obtain the general solution of ODE (3.18) from its invariance under scalings:
      (i) $x^* = \alpha x, \quad y^* = \alpha y$; and
      (ii) $x^* = x, \quad y^* = \beta y$.
   (b) Find the corresponding integrating factors.
   (c) $y = \Theta(x) = x$ is a solution curve of (3.18). Find the image of this solution curve for each of the two groups in (a). Explain your answers.

2. Consider the ODE

$$\frac{dy}{dx} = A(x)B(y). \tag{3.19}$$

   (a) Find an integrating factor of (3.19).
   (b) Find a one-parameter Lie group of transformations admitted by (3.19).

3. Find the most general first order ODE $\frac{dy}{dx} = f(x, y)$ that admits the group

$$x^* = \alpha x,$$
$$y^* = \alpha^2 y.$$

4. Formulate the problem of finding one-parameter Lie groups of transformations that leave invariant the family of straight lines $y = cx$.

3.2 FIRST-ORDER ODEs

We consider applications of point symmetries to the study of a first-order ODE

$$y' = \frac{dy}{dx} = f(x, y). \tag{3.20}$$

We assume that the ODE (3.20) admits a one-parameter Lie group of point transformations, called a *point symmetry*,

$$x^* = X(x, y; \varepsilon) = x + \varepsilon\xi(x, y) + O(\varepsilon^2), \tag{3.21a}$$

$$y^* = Y(x, y; \varepsilon) = y + \varepsilon\eta(x, y) + O(\varepsilon^2), \tag{3.21b}$$

with the infinitesimal generator

$$\mathrm{X} = \xi(x, y)\frac{\partial}{\partial x} + \eta(x, y)\frac{\partial}{\partial y}. \tag{3.21c}$$

We first show how to find the general solution of ODE (3.20) from the infinitesimals $\xi(x, y)$, $\eta(x, y)$ of an admitted group (3.21a,b) from two points of view:
(i) use of canonical coordinates; and
(ii) determination of an integrating factor.

Alternatively, if a particular solution of ODE (3.20) is known, and this particular solution is not an invariant curve of point symmetry (3.21a,b), then the implicit formula (2.193a,b) yields the general solution of (3.20).

We then consider the problem of determining point symmetries (3.21a,b) admitted by a given first-order ODE (3.20). We also show how to find all first-order ODEs admitting a given point symmetry.

3.2.1 CANONICAL COORDINATES

As discussed in Section 2.3.5, given any one-parameter Lie group of point transformations (3.21a,b), there exist *canonical coordinates* $r(x,y), s(x,y)$, so that (3.21a,b) becomes the translation group

$$r^* = r, \tag{3.22a}$$

$$s^* = s + \varepsilon. \tag{3.22b}$$

These coordinates are found by solving

$$\mathrm{X}r = 0,$$
$$\mathrm{X}s = 1.$$

In terms of canonical coordinates, ODE (3.20) becomes the ODE

$$\frac{ds}{dr} = \frac{s_x + s_y y'}{r_x + r_y y'} = \frac{s_x + s_y f(x,y)}{r_x + r_y f(x,y)} = F(r,s), \tag{3.23}$$

where $F(r,s)$ is obtained by substituting x and y in terms of r and s into $[s_x + s_y f(x,y)]/[r_x + r_y f(x,y)]$. The invariance of ODE (3.20), and hence ODE (3.23), under the translation group (3.22a,b), means that $F(r, s)$ does not depend explicitly on s. Hence, ODE (3.23) must be of the form

$$\frac{ds}{dr} = G(r) = \frac{s_x + s_y f(x,y)}{r_x + r_y f(x,y)}. \tag{3.24}$$

Consequently, the general solution of ODE (3.20) is given implicitly by

$$s(x,y) = \int^{r(x,y)} G(\rho)\, d\rho + C, \quad C = \text{const.} \tag{3.25}$$

In Section 3.1.1, we solved the first-order homogeneous ODE (3.7) in terms of canonical coordinates arising from its invariance under scalings (3.8a,b). Now consider two more familiar examples in terms of canonical coordinates.

(1) *Linear Homogeneous Equation*
The first-order linear homogeneous ODE

$$y' + p(x)y = 0 \tag{3.26}$$

admits the one-parameter (α) Lie group of scaling transformations

$$x^* = x, \tag{3.27a}$$

$$y^* = \alpha y. \tag{3.27b}$$

In terms of corresponding canonical coordinates

$$r = x, \tag{3.28a}$$

$$s = \log y, \tag{3.28b}$$

ODE (3.26) becomes

$$\frac{ds}{dr} = \frac{y'}{y} = -p(r). \tag{3.29}$$

Hence, the general solution of ODE (3.26) is given by

$$s(x, y) = \log y = -\int^x p(\rho)\, d\rho + C,$$

or

$$y = \widetilde{C} \exp\left[-\int^x p(\rho)\, d\rho\right], \quad \widetilde{C} = \text{const.}$$

(2) *Linear Nonhomogeneous Equation*
The first-order linear nonhomogeneous ODE

$$y' + p(x)y = g(x) \tag{3.30}$$

admits the one-parameter (ε) Lie group of transformations

$$x^* = x, \tag{3.31a}$$

$$y^* = y + \varepsilon\phi(x), \tag{3.31b}$$

where $u = \phi(x)$ is any particular solution of the associated linear homogeneous ODE

$$u' + p(x)u = 0. \tag{3.32}$$

The infinitesimal generator corresponding to (3.31a,b) is given by

$$\mathrm{X} = \phi(x)\frac{\partial}{\partial y},$$

and hence, $\mathrm{X}s = 1$ has the solution $s = y / \phi(x)$. In terms of canonical coordinates

$$r = x, \tag{3.33a}$$

$$s = \frac{y}{\phi(x)}, \tag{3.33b}$$

the ODE (3.30) reduces to

$$\frac{ds}{dr} = \frac{g(r)}{\phi(r)},$$

which has as its general solution

$$\frac{y}{\phi(x)} = \int^x \frac{g(\rho)}{\phi(\rho)}\, d\rho + C.$$

Hence, the general solution of ODE (3.30) is given by

$$y = \phi(x) \int^{x} \frac{g(\rho)}{\phi(\rho)} d\rho + C\phi(x), \quad C = \text{const.} \tag{3.34}$$

3.2.2 INTEGRATING FACTORS

The general solution of the first-order ODE (3.20) is a family of curves

$$\omega(x, y) = \text{const.} \tag{3.35}$$

Then

$$\frac{d\omega}{dx} = \omega_x + \omega_y y' = 0, \tag{3.36}$$

and hence,

$$\omega_x + f(x, y)\omega_y = 0 \tag{3.37}$$

holds for all solutions of ODE (3.20).

We assume that ODE (3.20) admits a one-parameter Lie group of point transformations (3.21a,b). Thus, (3.21a,b) leaves invariant the family of solution curves (3.35). We further assume that, under group (3.21a,b), the solution curves (3.35) of ODE (3.20) are not invariant curves of (3.21a,b). Then, without loss of generality, the family of solution curves (3.35) satisfies [cf. Section 2.6.1]

$$\mathrm{X}\omega = \xi(x, y)\frac{\partial \omega}{\partial x} + \eta(x, y)\frac{\partial \omega}{\partial y} = 1 \tag{3.38}$$

with $\eta \neq \xi f$, where X is the infinitesimal generator (3.21c). Substituting for ω_x from (3.37) into (3.38), we obtain

$$\omega_y = \frac{1}{\eta - \xi f}, \tag{3.39a}$$

and hence,

$$\omega_x = -\frac{f}{\eta - \xi f}. \tag{3.39b}$$

Substituting (3.39a,b) into (3.36), we obtain

$$\frac{d\omega}{dx} = \frac{1}{\eta - \xi f}(y' - f).$$

Consequently,

$$\mu(x, y) = \frac{1}{\eta - \xi f} \tag{3.40}$$

is an *integrating factor* for ODE (3.20).

The integrating factor (3.40) yields the general solution of ODE (3.20) given by

$$\omega(x,y) = -\int \mu(x,y) f(x,y)\,dx + \int [\mu(x,y) + \int (\mu(x,y) f(x,y))_y\,dx]\,dy = \text{const.} \tag{3.41}$$

Conversely, one can show that if $\mu(x,y)$ is an integrating factor of a first-order ODE (3.20), then any $\xi(x,y)$ and $\eta(x,y)$ satisfying (3.40) defines an infinitesimal generator (3.21c) of a one-parameter Lie group of transformations admitted by the first-order ODE (3.20).

3.2.3 MAPPINGS OF SOLUTION CURVES

The following theorems concern one-parameter Lie groups of transformations acting on the solution curves of a first-order ODE (3.20):

Theorem 3.2.3-1. *For any function $\xi(x,y)$, the one-parameter Lie group of transformations with the infinitesimal generator*

$$\mathrm{X} = \xi(x,y)\left[\frac{\partial}{\partial x} + f(x,y)\frac{\partial}{\partial y}\right], \tag{3.42}$$

leaves invariant each solution curve of the first-order ODE $y' = f(x,y)$.

Proof. Let $y = \Theta(x)$ be a solution curve of $y' = f(x,y)$. Then

$$y' = \Theta'(x) = f(x,\Theta(x)). \tag{3.43}$$

Consider the infinitesimal generator X given by (3.42). Then

$$\mathrm{X}(y - \Theta(x)) = \xi(x,y)[f(x,y) - \Theta'(x)].$$

Hence, if $y = \Theta(x)$, then from (3.43) it follows that

$$\mathrm{X}(y - \Theta(x)) = \xi(x,\Theta(x))[f(x,\Theta(x)) - \Theta'(x)] = 0.$$

Consequently, $y = \Theta(x)$ is an invariant curve for the one-parameter Lie group of transformations with infinitesimal generator (3.42). □

Theorem 3.2.3-2. *For any function $\xi(x,y)$, the first order ODE $y' = f(x,y)$ admits a one-parameter Lie group of transformations with the infinitesimal generator $\mathrm{X} = \xi(x,y)\frac{\partial}{\partial x} + \eta(x,y)\frac{\partial}{\partial y}$ for some $\eta(x,y) \neq \xi(x,y)f(x,y)$, under which each solution curve of the ODE maps into a different solution curve of the ODE.*

Proof. Let $\omega(x,y) = \text{const}$ be the solution curves of $y' = f(x,y)$. Then

$$\omega_x + \omega_y f = 0. \tag{3.44}$$

For arbitrary $\xi(x,y)$, consider the infinitesimal generator $\mathrm{X} = \xi\frac{\partial}{\partial x} + \eta\frac{\partial}{\partial y}$, with $\eta(x,y)$ determined by the relation $\omega_y = 1/(\eta - \xi f)$. Then from $\mathrm{X}\omega = \xi\omega_x + \eta\omega_y$, after use of (3.44), we obtain $\mathrm{X}\omega = (\eta - \xi f)\omega_y = 1$. Thus the one-parameter Lie group of transformations with infinitesimal generator X maps each solution curve of the ODE into a different solution curve [cf. Section 2.6.1]. □

From Theorems 3.2.3-1 and 3.2.3-2, we see that two types of one-parameter Lie groups of transformations are admitted by any first-order ODE (3.20). Moreover, the ODE (3.20) admits infinite-parameter Lie groups of transformations of both types:

Type (i). *Trivial One-Parameter Transformation Groups*
Infinitesimal generators of the form

$$\mathrm{X} = \xi(x,y)\frac{\partial}{\partial x} + \eta(x,y)\frac{\partial}{\partial y} \quad \text{where } \frac{\eta}{\xi} \equiv f(x,y),$$

are always admitted by $y' = f(x,y)$. Here each solution curve of $y' = f(x,y)$ is an invariant curve. This type of group is useless for reducing $y' = f(x,y)$ to a quadrature since, in order to find the canonical coordinates of the group, one must first find the general solution of $y' = f(x,y)$.

Type (ii). *Nontrivial One-Parameter Transformation Groups*
For a one-parameter Lie group with an infinitesimal generator

$$\mathrm{X} = \xi(x,y)\frac{\partial}{\partial x} + \eta(x,y)\frac{\partial}{\partial y}$$

admitted by $y' = f(x,y)$ in a domain D where $\eta/\xi \neq f(x,y)$, the family of solution curves of $y' = f(x,y)$ is invariant in D, with each solution curve in D moving to a different solution curve in D. This type of group defines a *nontrivial Lie group of transformations*. It is useful for reducing $y' = f(x,y)$ to a quadrature, provided that one can solve the ODE $dy/dx = \eta/\xi$ to obtain the canonical coordinate $r(x,y)$.

The geometrical situation is illustrated in Figure 3.2. Here $\boldsymbol{\xi}^{(\mathrm{i})}(\mathbf{x}) = (\xi^{(\mathrm{i})}(x,y), \eta^{(\mathrm{i})}(x,y))$ is the infinitesimal of a one-parameter Lie group $G^{(\mathrm{i})}$ of Type (i), and $\boldsymbol{\xi}^{(\mathrm{ii})}(\mathbf{x}) = (\xi^{(\mathrm{ii})}(x,y), \eta^{(\mathrm{ii})}(x,y))$ is the infinitesimal of a one-parameter Lie group $G^{(\mathrm{ii})}$ of Type (ii), admitted by $y' = f(x,y)$; γ is any solution curve $y = \Theta(x)$ of $y' = f(x,y)$. Along γ, $\boldsymbol{\xi}^{(\mathrm{i})}(\mathbf{x})$ is tangent to γ since $\eta/\xi \equiv f$. But $\boldsymbol{\xi}^{(\mathrm{ii})}(\mathbf{x})$ is not tangent to γ since $\eta/\xi \neq f$. Consequently, $G^{(\mathrm{i})}$ leaves invariant γ whereas $G^{(\mathrm{ii})}$ maps γ into a one-parameter family of solution curves given by the implicit formula (2.193a,b) or the explicit formula (2.201) with the infinitesimal generator

$$\mathrm{X} = \xi^{(\mathrm{ii})}(x,y)\frac{\partial}{\partial x} + \eta^{(\mathrm{ii})}(x,y)\frac{\partial}{\partial y}.$$

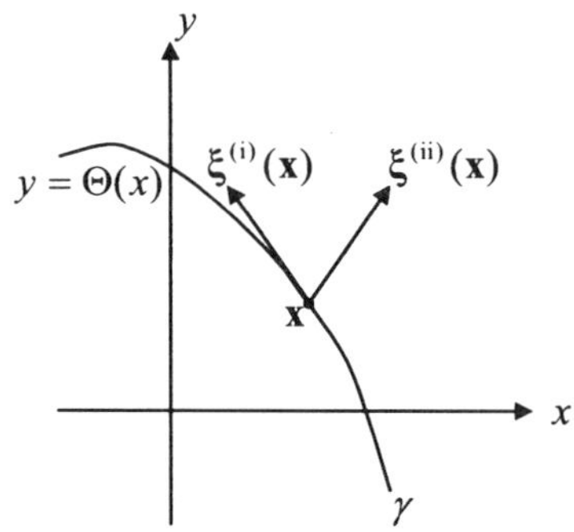

Figure 3.2. Illustration of groups of Types (i) and (ii).

3.2.4 DETERMINING EQUATION FOR SYMMETRIES OF A FIRST-ORDER ODE

A first-order ODE

$$y' = f(x,y) \tag{3.45}$$

defines a corresponding surface

$$y_1 = f(x,y) \tag{3.46}$$

in (x,y,y_1)-space with the solutions $y = \Theta(x)$ of (3.45) corresponding to points $(x,y,y_1) = (x,\Theta(x),\Theta'(x))$, i.e., $y_1 = y' = dy/dx$ when $y = \Theta(x)$ satisfies (3.45).

Consider a one-parameter Lie group of transformations

$$x^* = X(x,y;\varepsilon), \tag{3.47a}$$

$$y^* = Y(x,y;\varepsilon). \tag{3.47b}$$

Definition 3.2.4-1. The group (3.47a,b) *leaves invariant ODE* (3.45), i.e., is a *point symmetry admitted by ODE* (3.45), if and only if its first extension, defined by (2.83a–c), leaves invariant the surface (3.46).

A solution curve $y = \Theta(x)$ of (3.45) satisfies $\Theta'(x) = f(x,\Theta(x))$, and hence, lies on the surface (3.46) with $y = \Theta(x)$, $y_1 = \Theta'(x)$. Invariance of the surface (3.46) under the first extension of (3.47a,b) means that any solution curve $y = \Theta(x)$ of (3.45) maps into some solution curve $y = \phi(x;\varepsilon)$ of (3.45) under the action of the group (3.47a,b). Moreover if a transformation (3.47a,b) maps each solution curve $y = \Theta(x)$ of (3.45) into a solution curve $y=\phi(x;\varepsilon)$ of (3.45), then the surface (3.46) is invariant under (3.47a,b) with $y_1=\partial\phi(x;\varepsilon)/\partial x$. It immediately follows that *the family of solution curves of* (3.45) *is invariant under point symmetry* (3.47a,b) *if and only if* (3.45) *admits* (3.47a,b).

The following theorem results from Definition 3.2.4-1, Theorem 2.6.1-1 on the infinitesimal criterion for an invariant surface, and Theorem 2.4.2-1 on extended infinitesimals:

Theorem 3.2.4-1 (Infinitesimal Criterion for Invariance of a First-Order ODE). *Let*

$$\mathrm{X}=\xi(x,y)\frac{\partial}{\partial x}+\eta(x,y)\frac{\partial}{\partial y} \tag{3.48}$$

be the infinitesimal generator of the Lie group of transformations (3.47a,b). *Let*

$$\mathrm{X}^{(1)}=\xi(x,y)\frac{\partial}{\partial x}+\eta(x,y)\frac{\partial}{\partial y}+\eta^{(1)}(x,y,y_1)\frac{\partial}{\partial y_1} \tag{3.49}$$

be the first-extended infinitesimal generator of (3.48) *where* $\eta^{(1)}$ *is given by* (2.101) *in terms of* $\xi(x,y),\eta(x,y)$. *Then* (3.47a,b) *is admitted by a first-order ODE* (3.45) *if and only if*

$$\mathrm{X}^{(1)}(y_1-f(x,y))=\eta^{(1)}-\xi f_x-\eta f_y=0 \quad \textit{when} \quad y_1=f(x,y). \tag{3.50}$$

Proof. Left to Exercise 3.2-10. □

Explicitly, the first-extended infinitesimal of (3.48) is given by

$$\eta^{(1)}=\eta_x+[\eta_y-\xi_x]y_1-\xi_y(y_1)^2.$$

Thus, from (3.50), the first-order ODE (3.45) admits (3.48) if and only if $\xi(x,y)$ and $\eta(x,y)$ satisfy

$$\eta_x+[\eta_y-\xi_x]f-\xi_y f^2-\xi f_x-\eta f_y=0 \quad \text{for } \textit{arbitrary values} \text{ of } x \text{ and } y. \tag{3.51}$$

Equation (3.51) is the *determining equation* for the infinitesimal transformations (3.48) admitted by (3.45). The solutions of the determining equation (3.51) yield the *point symmetries* of ODE (3.45).

It is easy to check that for *any* function $\xi(x,y)$, a solution of the determining equation (3.51) is given by

$$\eta(x,y)=\xi(x,y)f(x,y). \tag{3.52}$$

This represents the trivial infinite-parameter Lie group of transformations of Type (i) that leaves each solution curve of ODE (3.45) invariant.

For any $\xi(x,y)$, it follows that

$$\eta(x,y)=\xi(x,y)f(x,y)+\chi(x,y) \tag{3.53}$$

yields the general solution of (3.51) where $\chi(x,y)$ is the general solution of the first-order linear PDE

$$\chi_x + f\chi_y - f_y\chi = 0. \tag{3.54}$$

Thus, (3.53) leads to the infinite-parameter Lie group of transformations of Type (ii) admitted by (3.45) that maps solution curves into different solution curves of (3.45). [An infinite-parameter subgroup of Type (ii) corresponds to $\eta = \chi, \xi = 0$.] Moreover, from (3.53), we see that the first-order ODE (3.45) admits

$$\mathrm{X} = \xi(x,y)\frac{\partial}{\partial x} + \eta(x,y)\frac{\partial}{\partial y}$$

if and only if it admits

$$\mathrm{Y} = [\eta(x,y) - \xi(x,y)f(x,y)]\frac{\partial}{\partial y}.$$

Consequently, the problem of finding all Lie groups of transformations (3.48), admitted by a given first-order ODE (3.45), is equivalent to finding the general solution of (3.54). But, in order to find the general solution of (3.54), we must solve the corresponding characteristic system $dx/1 = dy/f = d\chi/(f_y\chi)$, and hence, we would need to know *the general solution* of ODE (3.45). However, *any particular solution* χ of (3.54) or, equivalently, any particular solution of (3.51) with $\eta \not\equiv \xi f$, leads to a one-parameter Lie group of transformations admitted by (3.45), and hence to the general solution of (3.45) through its reduction by a quadrature. [In turn, this leads to determining the infinite-parameter Lie group of transformations admitted by (3.45).] Unfortunately, there is no general procedure to find an explicit particular solution χ of (3.54).

Next we consider the converse problem of determining all first-order ODEs that admit a given one-parameter Lie group of transformations.

3.2.5 DETERMINATION OF FIRST-ORDER ODEs INVARIANT UNDER A GIVEN GROUP

We show how to find all first-order ODEs

$$y' = f(x,y) \tag{3.55}$$

that admit a given one-parameter Lie group of transformations with the infinitesimal generator

$$\mathrm{X} = \xi(x,y)\frac{\partial}{\partial x} + \eta(x,y)\frac{\partial}{\partial y}. \tag{3.56}$$

This can be accomplished in two different ways:

(i) *Method of Canonical Coordinates.* Given the infinitesimal generator (3.56), we compute *canonical coordinates* $r(x, y)$ and $s(x, y)$ satisfying

$$\mathrm{X}r = 0, \quad \mathrm{X}s = 1, \tag{3.57}$$

so that the group corresponding to (3.56) becomes the translation group

$$r^* = r, \quad s^* = s + \varepsilon. \tag{3.58}$$

Then

$$\frac{ds}{dr} = \frac{s_x + s_y y'}{r_x + r_y y'} \tag{3.59}$$

relates y' and ds/dr. Consequently, the problem of finding all first-order ODEs (3.55) admitting (3.56) transforms to the problem of finding all first-order ODEs

$$\frac{ds}{dr} = F(r, s) \tag{3.60}$$

that admit (3.58). Clearly, $F(r, s)$ cannot depend explicitly on s. Hence, the most general first-order ODE admitting (3.56) is of the form

$$\frac{ds}{dr} = G(r), \tag{3.61}$$

where G is an arbitrary function of r. In terms of the given coordinates x and y, ODE (3.61) becomes the ODE

$$\frac{s_x(x, y) + s_y(x, y)y'}{r_x(x, y) + r_y(x, y)y'} = G(r(x, y)). \tag{3.62}$$

(ii) *Method of (Differential) Invariants.* The first-order ODE (3.55) admits (3.56) if and only if $f(x, y)$ satisfies the first-order PDE (3.51). The corresponding characteristic equations to determine $f(x, y)$ are given by

$$\frac{dx}{\xi(x, y)} = \frac{dy}{\eta(x, y)} = \frac{df}{\eta_x + (\eta_y - \xi_x)f - \xi_y f^2}. \tag{3.63}$$

The invariant

$$u(x, y) \equiv r(x, y) = \text{const} = c_1 \tag{3.64}$$

is the quadrature of the first equation of (3.63). Eliminating y through (3.64), and setting

$$f_p(x; c_1) = \frac{\eta(x, y)}{\xi(x, y)}, \tag{3.65}$$

we see that

$$f = f_p \tag{3.66}$$

is a particular solution of the second equation of (3.63):

$$\frac{df}{dx} = A + Bf + Cf^2, \tag{3.67}$$

with

$$A = A(x;c_1) = \frac{\eta_x}{\xi}, \tag{3.68a}$$

$$B = B(x;c_1) = \frac{\eta_y - \xi_x}{\xi}, \tag{3.68b}$$

$$C = C(x;c_1) = -\frac{\xi_y}{\xi}. \tag{3.68c}$$

Equation (3.67) is a first-order ODE of Riccati type, with c_1 playing the role of a parameter. Its general solution can be determined explicitly from the particular solution (3.66) through a transformation to a second-order linear ODE. Specifically, if z solves

$$z'' - \left(\frac{C'}{C} + B\right)z' + ACz = 0, \tag{3.69}$$

then

$$f = -\frac{1}{C}\frac{z'}{z} \tag{3.70}$$

solves (3.67). Hence, the general solution of (3.69) leads to the general solution of (3.67) through the well-known Riccati transformation (3.70). A particular solution of (3.69) is given by

$$z = z_p = e^{-\int Cf_p\,dx}, \tag{3.71}$$

where C is given by (3.68c), (3.64); and f_p by (3.65), (3.64). The explicit general solution of (3.69) follows from (3.71) through the method of reduction of order (to be derived from group invariance in Section 3.3.3). Consequently, one obtains the general solution of (3.67) given by

$$f = \phi(x;c_1,c_2), \tag{3.72}$$

where c_2 is a constant of integration and ϕ is a known function of its arguments. Then the general solution of (3.63) arises from the equation $c_2 = \psi(c_1)$, where ψ is an arbitrary function of c_1.

In (3.72), replacing f by y' and c_1 by $u(x, y)$, then solving for c_2, we obtain the *differential invariant* $v(x,y,y') = c_2 = \text{const}$ of the first extension of (3.56), i.e., $\mathrm{X}^{(1)}v = 0$. The general solution of (3.63) can then be expressed as $v(x,y,f) = \psi(u(x,y))$, and hence,

$$v(x,y,y') = \psi(u(x,y)) \tag{3.73}$$

is the most general first-order ODE that admits (3.56), written in terms of differential invariants.

Note that, in terms of canonical coordinates $r(x,y)$ and $s(x,y)$,

$$v(x,y,y') = \frac{ds}{dr} = \frac{s_x + s_y y'}{r_x + r_y y'}$$

satisfies $\mathrm{X}^{(1)}v = 0$ since ds/dr is invariant under (3.56), i.e. $ds^*/dr^* = ds/dr$. Consider examples for which we use both approaches to find first-order ODEs admitting specific groups:

(1) *A Scaling Group*

Suppose a first-order ODE (3.55) is invariant under the scaling group

$$x^* = e^{\varepsilon} x, \tag{3.74a}$$

$$y^* = e^{\varepsilon} y, \tag{3.74b}$$

which has the infinitesimal generator $\mathrm{X} = x\frac{\partial}{\partial x} + y\frac{\partial}{\partial y}$.

(i) Canonical coordinates for the Lie group (3.74a,b) are given by

$$r = \frac{y}{x}, \quad s = \log y.$$

Then

$$s_x = 0, \quad s_y = \frac{1}{y}, \quad r_x = -\frac{y}{x^2}, \quad r_y = \frac{1}{x}.$$

Hence, ODE (3.62) takes the form

$$\frac{y'}{ry' - r^2} = G(r). \tag{3.75}$$

Solving (3.75) for y', we find that the most general first-order ODE admitting the scaling group (3.74a,b) is given by

$$y' = H\left(\frac{y}{x}\right), \tag{3.76}$$

where H is an arbitrary function of y/x.

(ii) In terms of the method of differential invariants, the characteristic equations (3.63) become

$$\frac{dx}{x}=\frac{dy}{y}=\frac{df}{0}. \tag{3.77}$$

The quadrature of the first equation of (3.77) is given by

$$u(x,y)=\frac{y}{x}=\text{const}=c_1.$$

The second equation of (3.77) yields

$$f=\text{const}=c_2.$$

Then $c_2=\psi(c_1)$ gives $f(x,y)=\psi(y/x)$ which again yields ODE (3.76).

(2) *Rotation Group*
Suppose a first-order ODE (3.55) admits the rotation group

$$x^*=x\cos\varepsilon-y\sin\varepsilon, \tag{3.78a}$$

$$y^*=x\sin\varepsilon+y\cos\varepsilon, \tag{3.78b}$$

with the infinitesimal generator $\mathrm{X}=-y\frac{\partial}{\partial x}+x\frac{\partial}{\partial y}$.

(i) Canonical coordinates are polar coordinates

$$r=\sqrt{x^2+y^2},\quad s=\theta=\sin^{-1}\frac{y}{r}.$$

Then $r_x=x/r, r_y=y/r, s_x=-y/r^2, s_y=x/r^2$. Thus, ODE (3.62) becomes

$$\frac{-y+xy'}{x+yy'}=rG(r).$$

Consequently, the most general first-order ODE admitting (3.78a,b) is given by

$$\frac{-y+xy'}{x+yy'}=H\left(\sqrt{x^2+y^2}\right), \tag{3.79}$$

where H is an arbitrary function of $\sqrt{x^2+y^2}$.

(ii) In terms of the method of differential invariants, the characteristic equations (3.63) become

$$\frac{dx}{-y}=\frac{dy}{x}=\frac{df}{1+f^2}. \tag{3.80}$$

The quadrature of the first equation of (3.80) is given by

$$u(x,y)=\sqrt{x^2+y^2}=c_1. \tag{3.81}$$

The second characteristic equation of (3.80) now becomes

$$\frac{df}{1+f^2}=\frac{dy}{\sqrt{c_1{}^2-y^2}}. \tag{3.82}$$

Let $\alpha=\tan^{-1}f,\ \beta=\sin^{-1}(y/c_1)=\tan^{-1}(y/x)$. Then the quadrature of (3.82) yields

$$\alpha-\beta=c_2. \tag{3.83}$$

Hence, $\tan c_2=\tan(\alpha-\beta)=\psi(c_1)$ leads to

$$\frac{f-\dfrac{y}{x}}{1+\dfrac{y}{x}f}=\psi(\sqrt{x^2+y^2}). \tag{3.84}$$

Replacing f by y' in (3.84), we again get ODE (3.79).

EXERCISES 3.2

1. Let $y=\phi(x)$ be a particular solution of

$$y'+p(x)y=g(x). \tag{3.85}$$

   (a) Use this solution to find a one-parameter Lie group of transformations admitted by (3.85).
   (b) Find corresponding canonical coordinates and reduce (3.85) to a quadrature.
   (c) Illustrate for the ODE $y'+y=x$.

2. Find the integrating factor for ODE (3.30) from its invariance under (3.31a,b).

3. Show that if $\mu(x,y)$ is an integrating factor of ODE (3.20), then any $\xi(x,y),\eta(x,y)$ satisfying (3.40) defines an infinitesimal generator (3.21a) of a one-parameter Lie group of transformations admitted by (3.20).

4. (a) Characterize the infinitesimal transformation of a one-parameter Lie group of transformations that leaves invariant the family of straight lines $y=\text{const}$. Explicitly find all such groups for which $\xi(x,y)\equiv 1$.
   (b) Do the same for the family of straight lines $y/x=\text{const}$.

5. Find all first-order ODEs that admit the scalings $x^*=\alpha x,\ y^*=\alpha^k y,\ \alpha>0,$ for a fixed constant k.

6. For the first-order ODE

$$y' = f(x, y), \tag{3.86}$$

written as a differential form $M\,dx + N\,dy = 0$, introduce the associated operator

$$\mathrm{A} = N\frac{\partial}{\partial x} - M\frac{\partial}{\partial y}.$$

(a) Prove that the one-parameter Lie group with the infinitesimal generator $\mathrm{X} = \xi\frac{\partial}{\partial x} + \eta\frac{\partial}{\partial y}$ is admitted by (3.86) if and only if the commutation relation $[\mathrm{X}, \mathrm{A}] = \lambda(x, y)\mathrm{A}$ is satisfied for some function $\lambda(x, y)$.

(b) What can you say if $[\mathrm{X}, \mathrm{A}] = 0$?

(c) Illustrate for first-order ODEs that admit:

(i) $x\frac{\partial}{\partial x} + y\frac{\partial}{\partial y}$; and

(ii) $-y\frac{\partial}{\partial x} + x\frac{\partial}{\partial y}$.

7. If a first-order ODE $y' = f(x, y)$ admits two nontrivial groups with the infinitesimal generators $\mathrm{X}_i = \xi_i\frac{\partial}{\partial x} + \eta_i\frac{\partial}{\partial y}$, $i = 1, 2$, show that

$$\psi = \frac{\eta_2 - f\xi_2}{\eta_1 - f\xi_1}$$

is identically constant or a first integral of the ODE.

8. Find the most general first-order ODE that admits the one-parameter (ε) Lie group of transformations

$$x^* = x + \varepsilon,$$

$$y^* = \frac{xy}{x + \varepsilon}.$$

9. Consider the first-order ODE

$$(y - \tfrac{3}{2}x - 3)y' + y = 0.$$

(a) Find a nontrivial Lie group of transformations that is admitted by this ODE.

(b) Find the general solution of this ODE.

10. Prove Theorem 3.2.4-1.

3.3 INVARIANCE OF SECOND- AND HIGHER-ORDER ODEs UNDER POINT SYMMETRIES

Now consider the application of Lie groups of point transformations to the study of a second- or higher-order ODE

$$y^{(n)} = f(x, y, y', \dots, y^{(n-1)}), \tag{3.87}$$

or, equivalently, a surface in $(x, y, y_1, \dots, y_n)$-space,

$$y_n = f(x, y, y_1, \dots, y_{n-1}),$$

$n \geq 2$, where we use the notations

$$y^{(k)} = \frac{d^k y}{dx^k}, \quad k = 1,2,\dots,n; \quad y' = y^{(1)}, \quad y'' = y^{(2)}, \quad \text{etc.}$$

Note that the solutions $y = \Theta(x)$ of (3.87) correspond to points $(x, y, y_1, \dots, y_n) = (x, \Theta(x), \Theta'(x), \dots, \Theta^{(n)}(x))$.

Assume that ODE (3.87) admits a one-parameter Lie group of point transformations

$$x^* = X(x, y; \varepsilon) = x + \varepsilon\xi(x, y) + O(\varepsilon^2), \tag{3.88a}$$

$$y^* = Y(x, y; \varepsilon) = y + \varepsilon\eta(x, y) + O(\varepsilon^2), \tag{3.88b}$$

with infinitesimal generator

$$\mathrm{X} = \xi(x, y)\frac{\partial}{\partial x} + \eta(x, y)\frac{\partial}{\partial y}. \tag{3.88c}$$

Definition 3.3-1. A *point symmetry* of an *n*th-order ODE (3.87) is a one-parameter Lie group of point transformations (3.88a–c) admitted by (3.87).

We will show that the invariance of an *n*th-order ODE (3.87) under a point symmetry (3.88a,b) constructively leads to reducing (3.87) to an $(n - 1)$th-order ODE plus a quadrature. It will be shown that this can be done in two ways:

(i) reduction of order through canonical coordinates; and
(ii) reduction of order through differential invariants.

[In Section 3.4, we will show that reduction of order through differential invariants is a natural setting for reducing an *n*th-order ODE that admits a multiparameter Lie group of point transformations.]

We will then show how to find the point symmetries admitted by a given *n*th-order ODE. Recall that for a given first-order ODE, we saw that the admitted infinitesimals $\xi(x, y), \eta(x, y)$ satisfy a single linear PDE whose general solution could not be found without knowing the general solution of the ODE itself. Consequently, we could not systematically determine all such symmetries. But when $n \geq 2$, it turns out that the admitted infinitesimals for Lie groups of point transformations satisfy an overdetermined system of linear *determining equations* which have only a finite number of linearly independent solutions. In practice, we are usually able to compute explicitly the admitted Lie group of point transformations of an nth-order ODE if $n \geq 2$.

We will also consider the problem of finding all nth-order ODEs that admit a given one-parameter Lie group of point transformations.

3.3.1 REDUCTION OF ORDER THROUGH CANONICAL COORDINATES

The basic result is summarized in terms of the following theorem:

Theorem 3.3.1-1. *Suppose a nontrivial one-parameter Lie group of transformations* (3.88a,b), *with infinitesimal generator* (3.88c), *is admitted by an nth-order ODE* (3.87), $n \geq 2$. *Let* $r(x,y)$, $s(x,y)$ *be corresponding canonical coordinates satisfying* $\mathrm{X}r = 0$, $\mathrm{X}s = 1$. *Then the nth-order ODE* (3.87) *reduces to an* $(n-1)$*th-order ODE*

$$\frac{d^{n-1}z}{dr^{n-1}} = G\left(r, z, \frac{dz}{dr}, \ldots, \frac{d^{n-2}z}{dr^{n-2}}\right), \tag{3.89}$$

where

$$\frac{ds}{dr} = z. \tag{3.90}$$

Proof. In terms of canonical coordinates $r(x,y)$ and $s(x,y)$, we have

$$\frac{ds}{dr} = \frac{s_x + s_y y'}{r_x + r_y y'}. \tag{3.91}$$

In order for (3.91) to be nonsingular, we assume that $r_x + r_y y' \neq 0$, and hence, from $\mathrm{X}r = \xi r_x + \eta r_y = 0$, it follows that $\eta/\xi \neq y' = \Theta'(x)$ for a solution curve $y = \Theta(x)$ of (3.87). Consequently, the group (3.88a,b) acts on a domain of solution curves of ODE (3.87) where, under the action of group (3.88a,b), each solution curve of (3.87) in this domain is mapped into a different solution curve of (3.87).

Through differentiation of (3.91), we obtain

$$\frac{d^2s}{dr^2} = \left(\frac{1}{r_x + r_y y'}\right)\frac{d\left(\dfrac{s_x + s_y y'}{r_x + r_y y'}\right)}{dx} = y'' f_1\left(r, s, \frac{ds}{dr}\right) + g_1\left(r, s, \frac{ds}{dr}\right), \tag{3.92}$$

where

$$f_1\left(r, s, \frac{ds}{dr}\right) = \frac{s_y r_x - s_x r_y}{(r_x + r_y y')^3},$$

and

$$g_1\left(r,s,\frac{ds}{dr}\right)=\frac{1}{(r_x+r_y y')^3}[(y')^3(r_y s_{yy}-s_y r_{yy})+(y')^2(2r_y s_{xy}+r_x s_{yy}-2s_y r_{xy}-s_x r_{yy})$$
$$+y'(2r_x s_{xy}+r_y s_{xx}-2s_x r_{xy}-s_y r_{xx})+(r_x s_{xx}-s_x r_{xx})].$$

Next, solving (3.91) for y', we obtain

$$y'=\frac{s_x-r_x\dfrac{ds}{dr}}{r_y\dfrac{ds}{dr}-s_y}. \tag{3.93}$$

Then substituting (3.93) into (3.92) and replacing x and y in terms of r and s, we have

$$y''=\frac{d^2s}{dr^2}F_1\left(r,s,\frac{ds}{dr}\right)+G_1\left(r,s,\frac{ds}{dr}\right), \tag{3.94}$$

with $F_1=1/f_1$ and $G_1=-g_1/f_1$. Note that since r and s are canonical coordinates, it follows that $s_y r_x - s_x r_y \neq 0$ and hence, $f_1 \neq 0$. Proceeding inductively, one can show that

$$\frac{d^k s}{dr^k}=y^{(k)}f_{k-1}\left(r,s,\frac{ds}{dr}\right)+g_{k-1}\left(r,s,\frac{ds}{dr},\ldots,\frac{d^{k-1}s}{dr^{k-1}}\right) \tag{3.95}$$

for some function

$$g_{k-1}\left(r,s,\frac{ds}{dr},\ldots,\frac{d^{k-1}s}{dr^{k-1}}\right),$$

with

$$f_{k-1}\left(r,s,\frac{ds}{dr}\right)=\frac{s_y r_x - s_x r_y}{(r_x+r_y y')^{k+1}}, \quad k\geq 2.$$

This leads to

$$y^{(k)}=\frac{d^k s}{dr^k}F_{k-1}\left(r,s,\frac{ds}{dr}\right)+G_{k-1}\left(r,s,\frac{ds}{dr},\ldots,\frac{d^{k-1}s}{dr^{k-1}}\right), \tag{3.96}$$

where

$$F_{k-1}=\frac{1}{f_{k-1}} \text{ and } G_{k-1}=-\frac{g_{k-1}}{f_{k-1}}, \quad k\geq 2.$$

Note that $f_{k-1}\neq 0$ and $F_{k-1}=1/f_{k-1}\neq 0$ for $k\geq 2$.

Thus, it follows that in terms of canonical coordinates $r(x,y), s(x,y)$, the nth-order ODE (3.87) can be written as an nth-order ODE in solved form:

$$\frac{d^n s}{dr^n} = F\left(r, s, \frac{ds}{dr}, \dots, \frac{d^{n-1}s}{dr^{n-1}}\right), \tag{3.97}$$

for some function $F(r,s,ds/dr,\dots,d^{n-1}s/dr^{n-1})$. But the ODE (3.97) admits the translation group

$$r^* = r, \tag{3.98a}$$

$$s^* = s + \varepsilon. \tag{3.98b}$$

Hence, $F(r,s,ds/dr,\dots,d^{n-1}s/dr^{n-1})$ is independent of s. Consequently, the ODE (3.87) reduces to (3.89) and (3.90). □

Note that if

$$z = \phi(r; C_1, C_2, \dots, C_{n-1})$$

is the general solution of ODE (3.89), then the general solution of ODE (3.87) is given by

$$s(x,y) = \int^{r(x,y)} \phi(\rho; C_1, C_2, \dots, C_{n-1})\, d\rho + C_n,$$

where $C_1, C_2, \dots, C_n$ are n essential constants of integration. Hence, the invariance of the nth-order ODE (3.87) under a one-parameter Lie group of point transformations leads, constructively, to the reduction of (3.87) to an $(n-1)$th-order ODE plus a quadrature.

3.3.2 REDUCTION OF ORDER THROUGH DIFFERENTIAL INVARIANTS

The nth-order ODE (3.87) represented by the surface

$$F(x,y,y_1,\dots,y_n) = y_n - f(x,y,y_1,\dots,y_{n-1}) = 0 \tag{3.99}$$

admits the group (3.88a,b) if and only if the surface (3.99) is invariant, i.e.,

$$\mathrm{X}^{(n)}F = 0 \quad \text{when } F = 0,$$

where $\mathrm{X}^{(n)}$ is the nth extension of the infinitesimal generator X [cf. Section 2.4.2]. Hence, it follows that $F(x,y,y_1,\dots,y_n)$ is some function of the group's invariants

$$u(x,y), v_1(x,y,y_1), \dots, v_n(x,y,y_1,\dots,y_n), \tag{3.100}$$

where

$$\mathrm{X}u(x,y) = 0, \quad \mathrm{X}^{(k)}v_k(x,y,y_1,\dots,y_k) = 0 \quad \text{with } \frac{\partial v_k}{\partial y_k} \neq 0, \quad k = 1,2,\dots,n.$$

Clearly, for the nth extension of the group (3.88a,b), we have $u^* = u,\ v^*_k = v_k$, $k = 1, 2, \ldots, n$. Here, $v_k(x, y, y_1, \ldots, y_k)$ is a constant of integration of the characteristic equations

$$\frac{dx}{\xi(x,y)} = \frac{dy}{\eta(x,y)} = \frac{dy_1}{\eta^{(1)}(x,y,y_1)} = \cdots = \frac{dy_k}{\eta^{(k)}(x,y,y_1,\ldots,y_k)},$$

where $\eta^{(k)}$ is given by (2.100a,b), $k = 1,2,\ldots,n$.

For any set of invariants (3.100), the ODE (3.99) becomes

$$G(u, v_1, v_2, \ldots, v_n) = 0, \tag{3.101}$$

for some function $G(u, v_1, v_2, \ldots, v_n)$. We now show that one can always choose the invariants (3.100) without computing a canonical coordinate $s(x, y)$, such that (3.101) is an $(n-1)$th-order ODE. Moreover, we can do this in such a way that the nth-order ODE (3.99) will reduce to an $(n-1)$th-order ODE plus a quadrature. [Note that the same reduction holds in terms of canonical coordinates $r(x, y)$ and $s(x, y)$ with $u(x, y) = r(x, y)$ and $v_k(x, y, y_1, \ldots, y_k) = d^k s / dr^k$, $k = 1,2,\ldots,n$.]

In Section 3.2.5, we showed how to find explicit invariants $u(x, y)$ and $v_1(x, y, y_1) = v(x, y, y_1)$ of the first extension $X^{(1)}$ corresponding to the infinitesimal generator (3.88c). These invariants arose as constants of integration of the characteristic equations

$$\frac{dx}{\xi(x,y)} = \frac{dy}{\eta(x,y)} = \frac{dy_1}{\eta_x + (\eta_y - \xi_x)y_1 - \xi_y (y_1)^2}. \tag{3.102}$$

Recall that if we could determine explicitly $u(x, y)$, then the computation of $v(x, y, y_1)$ reduced to quadrature.

Since $u(x, y)$ and $v(x, y, y_1)$ are invariants under the action of the kth-extended group of (3.88a,b), it follows that dv/du is a group invariant under the action of the $(k+1)$th-extended group of (3.88a,b) since $(dv/du)^* = dv^*/du^* = dv/du$, $k \geq 1$. Continuing inductively, we see that $dv/du, d^2v/du^2, \ldots, d^{n-1}v/du^{n-1}$ are invariants of the nth-extended group of (3.88a,b). Such invariants are called *differential invariants* of the nth-extended group of (3.88a,b). Moreover, such differential invariants can be constructed for any choice of invariants $u(x, y)$ and $v(x, y, y_1)$ of the first-extended group of (3.88a,b) with $dv/\partial y_1 \neq 0$.

Then

$$\frac{dv}{du} = \frac{\dfrac{\partial v}{\partial x} + y_1 \dfrac{\partial v}{\partial y} + y_2 \dfrac{\partial v}{\partial y_1}}{\dfrac{\partial u}{\partial x} + y_1 \dfrac{\partial u}{\partial y}} = v_2(x, y, y_1, y_2) = y_2 \left[\frac{\dfrac{\partial v}{\partial y_1}}{\dfrac{\partial u}{\partial x} + y_1 \dfrac{\partial u}{\partial y}} \right] + g_1(x, y, y_1),$$

for some function $g_1(x,y,y_1)$. Inductively, one can show that

$$\frac{d^k v}{du^k} = v_{k+1}(x,y,y_1,\ldots,y_{k+1}) = y_{k+1}\left[\frac{\dfrac{\partial v}{\partial y_1}}{\left(\dfrac{\partial u}{\partial x} + y_1\dfrac{\partial u}{\partial y}\right)^k}\right] + g_k(x,y,y_1,\ldots,y_k),$$

for some function $g_k(x,y,y_1,\ldots,y_k)$, $k = 1,2,\ldots,n-1$. Consequently, the invariants $v_k(x,y,y_1,\ldots,y_k)$, $k = 2,3,\ldots,n$, are constructed as differential invariants. Moreover, it should be noted that

$$y_{k+1} = \frac{d^k v}{du^k} A_k(x,y,y_1) + B_k(x,y,y_1,\ldots,y_k),$$

where

$$A_k(x,y,y_1) = \frac{\left(\dfrac{\partial u}{\partial x} + y_1\dfrac{\partial u}{\partial y}\right)^k}{\dfrac{\partial v}{\partial y_1}},$$

$$B_k = -A_k g_k, \quad k = 1,2,\ldots,n-1.$$

Note that $A_k \neq 0$ and $1/A_k \neq 0$ for $k = 1,2,\ldots,n-1$.

Hence, it follows that *constructively*, in terms of *differential invariants*, the reduced equation (3.101) is an $(n-1)$th-order ODE

$$\frac{d^{n-1}v}{du^{n-1}} = H\left(u, v, \frac{dv}{du}, \ldots, \frac{d^{n-2}v}{du^{n-2}}\right), \tag{3.103}$$

for some function H of $u, v, dv/du, \ldots, d^{n-2}v/du^{n-2}$. Moreover, if

$$v = \phi(u; C_1, C_2, \ldots, C_{n-1})$$

is the general solution of (3.103) where $C_1, C_2, \ldots, C_{n-1}$ are arbitrary constants, then the general solution of the nth-order ODE (3.99) is found by solving the first-order ODE

$$v(x,y,y') = \phi(u(x,y); C_1, C_2, \ldots, C_{n-1}),$$

which reduces to a quadrature since it admits the group (3.88a,b).

3.3.3 EXAMPLES OF REDUCTION OF ORDER

(1) *Second-Order Linear Homogeneous Equation* (*Invariance under Scaling*)
Consider the second-order linear homogeneous ODE

$$y'' + p(x)y' + q(x)y = 0$$

or, equivalently,

$$y_2 + p(x)y_1 + q(x)y = 0. \tag{3.104}$$

The ODE (3.104) admits the one-parameter (α) Lie group of point transformations

$$x^* = x, \tag{3.105a}$$

$$y^* = \alpha y. \tag{3.105b}$$

(i) *Reduction of Order Through Canonical Coordinates.* Canonical coordinates corresponding to (3.105a,b) are given by

$$r(x,y) = x, \quad s(x,y) = \log y.$$

Then

$$z = \frac{ds}{dr} = \frac{y'}{y}, \tag{3.106}$$

so that

$$y' = yz.$$

Consequently,

$$y'' = y'z + y\frac{dz}{dr} = yz^2 + y\frac{dz}{dr}.$$

Then the second-order ODE (3.104) reduces to the following first-order ODE of Riccati type,

$$\frac{dz}{dr} + z^2 + p(r)z + q(r) = 0. \tag{3.107}$$

Note that (3.106) is the Riccati transformation relating (3.107) and (3.104).

(ii) *Reduction of Order Through Differential Invariants.* It is obvious that invariants of the first extension of (3.105a,b) are given by

$$u(x,y) = x, \quad v(x,y,y_1) = \frac{y_1}{y}.$$

The corresponding differential invariant is

$$\frac{dv}{du} = \frac{y_2}{y} - \left(\frac{y_1}{y}\right)^2 = \frac{y_2}{y} - v^2.$$

Hence,

$$y_2 = y\frac{dv}{du} + yv^2.$$

Consequently, the second-order ODE (3.104) again reduces to the Riccati equation

$$\frac{dv}{du} + v^2 + p(u)v + q(u) = 0. \tag{3.108}$$

If

$$v = \phi(u; C_1) \tag{3.109}$$

is the general solution of ODE (3.108), then from Section 3.3.2 it follows that the first-order ODE

$$v(x, y, y') = \frac{y'}{y} = \phi(x; C_1)$$

admits the group (3.105a,b). This leads to the reduction of this first-order ODE to quadrature. In particular,

$$y = C_2 e^{\int \phi(x;C_1)\,dx}.$$

(2) *Second-Order Linear Homogeneous Equation (Reduction of Order from a Given Particular Solution)*

Suppose $y = \Theta(x)$ is a particular solution of the second-order ODE (3.104). Then (3.104) admits the one-parameter (ε) Lie group of point transformations

$$x^* = x, \tag{3.110a}$$

$$y^* = y + \varepsilon\Theta(x). \tag{3.110b}$$

(i) *Reduction of Order Through Canonical Coordinates.* Canonical coordinates corresponding to (3.110a,b) are given by

$$r(x, y) = x, \quad s(x, y) = \frac{y}{\Theta(x)}.$$

Then

$$\frac{ds}{dr} = \frac{y'}{\Theta(x)} - \frac{y\Theta'(x)}{\Theta^2(x)},$$

so that

$$y' = \Theta(r)\frac{ds}{dr} + s\Theta'(r).$$

Hence,

$$y'' = 2\Theta'(r)\frac{ds}{dr} + \Theta(r)\frac{d^2 s}{dr^2} + \Theta''(r)s.$$

Let $z = ds/dr$. Then (3.104) reduces to the first-order linear ODE

$$\Theta(r)\frac{dz}{dr} + [2\Theta'(r) + p(r)\Theta(r)]z = 0.$$

(ii) *Reduction of Order Through Differential Invariants.* Clearly, $u(x,y) = x$ is an invariant of the group (3.110a,b). The invariant $v(x,y,y_1)$ of the first extension of (3.110a,b) $[\xi(x,y) = 0,\ \eta(x,y) = \Theta(x)]$ is a constant of integration of the corresponding characteristic equations (3.102), which here are given by

$$\frac{dx}{0} = \frac{dy}{\Theta(x)} = \frac{dy_1}{\Theta'(x)}.$$

Hence,

$$v(x,y,y_1) = y_1 - y\frac{\Theta'(x)}{\Theta(x)} = y_1 - y\frac{\Theta'(u)}{\Theta(u)}.$$

Then

$$\frac{dv}{du} = y_2 - y_1\frac{\Theta'(u)}{\Theta(u)} + y\left[\left(\frac{\Theta'(u)}{\Theta(u)}\right)^2 - \frac{\Theta''(u)}{\Theta(u)}\right].$$

Consequently,

$$y_1 = y\frac{\Theta'(u)}{\Theta(u)} + v, \quad y_2 = \frac{dv}{du} + v\frac{\Theta'(u)}{\Theta(u)} + y\frac{\Theta''(u)}{\Theta(u)}.$$

Hence, ODE (3.104) reduces to

$$\frac{dv}{du} + \left[\frac{\Theta'(u)}{\Theta(u)} + p(u)\right]v = 0. \tag{3.111}$$

If

$$v = \phi(u;C_1) \tag{3.112}$$

is the general solution of ODE (3.111), then the first-order ODE

$$v(x,y,y') = y' - \frac{\Theta'(x)}{\Theta(x)}y = \phi(x;C_1)$$

admits the group (3.110a,b) and, accordingly, can be reduced to a quadrature. [See example (2) of Section 3.2.1.]

(3) *Blasius Equation*

The *Blasius equation* resulting from the Prandtl–Blasius problem for a flat plate [cf. Section 1.3.1] is the ODE

$$y''' + \tfrac{1}{2} y y'' = 0,$$

or, equivalently, the surface

$$y_3 + \tfrac{1}{2} y y_2 = 0. \tag{3.113}$$

It is easy to see that the third-order ODE (3.113) admits the two-parameter (α, β) Lie group of point transformations

$$x^* = \frac{1}{\alpha} x + \beta, \tag{3.114a}$$

$$y^* = \alpha y. \tag{3.114b}$$

We treat (3.114a,b) as two one-parameter groups by considering the parameters α and β separately to reduce the Blasius equation (3.113) to two different second-order ODEs through the use of differential invariants. The reduction through canonical coordinates is left to Exercise 3.3-2. In Section 3.4.2 we will show how to reduce (3.113) directly to a first-order ODE plus two quadratures, from invariance under the group (3.114a,b).

(i) *Reduction of ODE* (3.113) *from Invariance Under Scalings.* Obvious invariants of the first extension of $x^* = (1/\alpha)x$, $y^* = \alpha y$, are given by

$$u(x, y) = xy, \quad v(x, y, y_1) = \frac{y_1}{y^2}.$$

Then

$$\frac{du}{dx} = y + x y_1 = y[1 + uv].$$

Hence,

$$y_2 = \frac{d}{dx}(y^2 v) = 2 y y_1 v + y^2 \frac{dv}{du}\frac{du}{dx} = y^3\left[2v^2 + (1 + uv)\frac{dv}{du}\right],$$

and

$$\begin{aligned} y_3 &= 3y^2 y_1\left[2v^2 + (1+uv)\frac{dv}{du}\right] + y^3\left(\frac{d}{du}\left[2v^2 + (1+uv)\frac{dv}{du}\right]\right)\frac{du}{dx} \\ &= y^4\left[(1+uv)^2\frac{d^2v}{du^2} + u(1+uv)\left(\frac{dv}{du}\right)^2 + 8v(1+uv)\frac{dv}{du} + 6v^3\right]. \end{aligned}$$

Thus, ODE (3.113) reduces to the second-order ODE

$$(1+uv)^2\frac{d^2v}{du^2}+u(1+uv)\left(\frac{dv}{du}\right)^2+(8v+\tfrac{1}{2})(1+uv)\frac{dv}{du}+6v^3+v^2=0, \tag{3.115}$$

plus a quadrature. In particular, if

$$v=\phi(u;C_1,C_2) \tag{3.116}$$

is the general solution of ODE (3.115), then the first-order ODE

$$v=\frac{y'}{y^2}=\phi(xy;C_1,C_2) \tag{3.117}$$

admits $x^*=(1/\alpha)x,\ y^*=\alpha y$. In terms of the canonical coordinates $s=\log y$ and $r=xy$, ODE (3.117) becomes the ODE

$$\frac{ds}{dr}=\frac{\phi(r;C_1,C_2)}{r\phi(r;C_1,C_2)+1}.$$

Then we obtain the general solution of the Blasius equation,

$$y=C_3\exp\left[\int^{xy}\frac{\phi(r;C_1,C_2)}{r\phi(r;C_1,C_2)+1}dr\right],$$

where C_1,C_2,C_3 are arbitrary constants.

Note that the second-order ODE (3.115) does not admit any obvious one-parameter Lie group of point transformations. In particular, the group with parameter β, i.e., $x^*=x+\beta,\ y^*=y$, *does not* induce a group of point transformations admitted by (3.115). The reason for this will be explained in Section 3.4.

(ii) *Reduction of ODE* (3.113) *from Invariance Under Translations.* Obvious invariants of the first extension of $x^*=x+\beta,\ y^*=y$, are given by

$$u(x,y)=y,\quad v(x,y,y_1)=y_1.$$

Then

$$\frac{du}{dx}=y_1=v.$$

Thus,

$$y_2=\frac{dv}{du}\frac{du}{dx}=v\frac{dv}{du},\quad y_3=v^2\frac{d^2v}{du^2}+v\left(\frac{dv}{du}\right)^2.$$

Hence, ODE (3.113) reduces to

$$v\frac{d^2v}{du^2}+\left(\frac{dv}{du}\right)^2+\tfrac{1}{2}u\frac{dv}{du}=0 \tag{3.118}$$

plus a quadrature. In particular, if

$$v = \psi(u; C_1, C_2) \tag{3.119}$$

is the general solution of ODE (3.118), then the first-order ODE

$$v = y' = \psi(y; C_1, C_2) \tag{3.120}$$

admits $x^* = x + \varepsilon$, $y^* = y$. Consequently, the general solution of the Blasius equation (3.113) is given by

$$\int^y \frac{dz}{\psi(z; C_1, C_2)} = x + C_3,$$

where C_1, C_2, C_3 are arbitrary constants.

Note that the second-order ODE (3.118) admits the obvious one-parameter (α) scaling group $u^* = \alpha u$, $v^* = \alpha^2 v$. This group is induced by the invariance of the Blasius equation (3.113) under $y^* = \alpha y$, $x^* = (1/\alpha)x$. Thus, (3.118) can be reduced to a first-order ODE. In Section 3.4, we will explain why this is possible.

3.3.4 DETERMINING EQUATIONS FOR POINT SYMMETRIES OF AN *n*th-ORDER ODE

Consider a one-parameter Lie group of point transformations

$$x^* = X(x, y; \varepsilon), \tag{3.121a}$$

$$y^* = Y(x, y; \varepsilon). \tag{3.121b}$$

Definition 3.3.4-1. The Lie group of point transformations (3.121a,b) leaves invariant an nth-order ODE (3.87), i.e., is a point symmetry admitted by ODE (3.87), if and only if its nth extension, defined by (2.89a–d) with $k = n$, leaves invariant the surface (3.99).

A solution curve $y = \Theta(x)$ of ODE (3.87) satisfies $\Theta^{(n)}(x) = f(x, \Theta(x), \Theta'(x), \ldots, \Theta^{(n-1)}(x))$ and thus lies on the surface (3.99) with $y = \Theta(x)$, $y_k = \Theta^{(k)}(x)$, $k = 1, 2, \ldots, n$. Invariance of the surface (3.99) under the nth extension of the group (3.121a,b) means that any solution curve $y = \Theta(x)$ of ODE (3.87) maps into some solution curve $y = \phi(x; \varepsilon)$ of ODE (3.87) under the action of the group (3.121a,b). Moreover, if a group of transformations (3.121a,b) maps each solution curve $y = \Theta(x)$ of ODE (3.87) into a solution curve $y = \phi(x; \varepsilon)$ of ODE (3.87), then the surface (3.99) is invariant under (3.121a,b) with $y_k = \partial^k \phi(x; \varepsilon) / \partial x^k$, $k = 1, 2, \ldots, n$. It immediately follows that the *family of all solution curves of ODE* (3.87) *is invariant under the group* (3.121a,b) *if and only if ODE* (3.87) *admits the group* (3.121a,b).

The following theorem results from Definition 3.3.4-1, Theorem 2.6.1-1 on the infinitesimal criterion for an invariant surface, and Theorem 2.4.2-1 on extended infinitesimals:

Theorem 3.3.4-1 (Infinitesimal Criterion for Invariance of an *n*th-Order ODE). *Let*

$$\mathrm{X}=\xi(x,y)\frac{\partial}{\partial x}+\eta(x,y)\frac{\partial}{\partial y} \tag{3.122}$$

be the infinitesimal generator of the one-parameter Lie group of point transformations (3.121a,b). *Let*

$$\mathrm{X}^{(n)}=\xi(x,y)\frac{\partial}{\partial x}+\eta(x,y)\frac{\partial}{\partial y}+\eta^{(1)}(x,y,y_1)\frac{\partial}{\partial y_1}+\cdots+\eta^{(n)}(x,y,y_1,\ldots,y_n)\frac{\partial}{\partial y_n} \tag{3.123}$$

be the nth-extended infinitesimal generator of (3.122), *where* $\eta^{(k)}$ *is given by* (2.100a,b) *in terms of* $\xi(x,y)$, $\eta(x,y)$ *for* $k=1,2,\ldots,n$. *Then* (3.121a,b) *is a point symmetry admitted by an nth-order ODE* (3.87) *if and only if*

$$\mathrm{X}^{(n)}(y_n-f(x,y,y_1,\ldots,y_{n-1}))=0 \quad \text{when } y_n=f,$$

or, equivalently,

$$\eta^{(n)}\Big|_{y_n=f}=\xi\frac{\partial f}{\partial x}+\eta\frac{\partial f}{\partial y}+\eta^{(1)}\frac{\partial f}{\partial y_1}+\cdots+\eta^{(n-1)}\frac{\partial f}{\partial y_{n-1}}, \tag{3.124}$$

where

$$\eta^{(k)}=\mathrm{D}^k\eta-\sum_{j=0}^{k-1}\frac{k!}{(k-j-1)!(j+1)!}y_{k-j}\mathrm{D}^{j+1}\xi$$

for $k=1,2,\ldots,n$.

Proof. Left to Exercise 3.3-9. □

More generally, an ODE $F(x,y,y_1,\ldots,y_n)=0$ admits the group (3.121a,b) with infinitesimal generator (3.122) if and only if $\mathrm{X}^{(n)}F(x,y,y_1,\ldots,y_n)=0$ when $F(x,y,y_1,\ldots,y_n)=0$.

If $f(x,y,y_1,\ldots,y_{n-1})$ is a polynomial in $y_1,y_2,\ldots,y_{n-1}$, then from Theorem 2.4.2-2 it follows that equation (3.124) (after replacing y_n by $f(x,y,y_1,\ldots,y_{n-1})$) is a polynomial in $y_1,y_2,\ldots,y_{n-1}$, whose coefficients are linear homogeneous in $\xi(x,y)$, $\eta(x,y)$, and their partial derivatives up to *n*th order. Since for any *n*th-order ODE (3.87) we can assign arbitrary values to each of $y,y_1,y_2,\ldots,y_{n-1}$ at any fixed value of *x*, it follows that the coefficient of each monomial term in (3.124) must vanish since this polynomial equation must hold for *arbitrary values* of $x,y,y_1,y_2,\ldots,y_{n-1}$. This leads to a system of *linear* *homogeneous* PDEs for $\xi(x, y)$, $\eta(x, y)$. This linear system defines the set of *determining equations* for the point symmetries admitted by the nth-order ODE (3.87). The system is *overdetermined* if $n \geq 2$ since the number of determining equations exceeds two, i.e., exceeds the number of unknowns.

In general, if $f(x, y, y_1, \ldots, y_{n-1})$ is not a polynomial in $y_1, y_2, \ldots, y_{n-1}$, then (3.124) still splits into an overdetermined system of determining equations based on the independence of the variables $x, y, y_1, y_2, \ldots, y_{n-1}$.

One can show that a second-order ODE admits at most an eight-parameter Lie group of point transformations and that an nth-order ODE ($n > 2$) admits at most an $(n+4)$ – parameter Lie group of point transformations [Lie (1893, pp. 296–298); Dickson (1924); Ovsiannikov (1982)].

We now state some theorems on the forms of infinitesimal generators that can be admitted by ODEs. These theorems cover many ODEs arising in applications and significantly simplify the many calculations involved in setting up and solving the determining equations for the infinitesimals $\xi(x, y), \eta(x, y)$. In particular, the theorems are concerned with the dependence on y of $\xi(x, y), \eta(x, y)$.

Theorem 3.3.4-2. *Consider an nth-order ODE,* $n \geq 3$, *of the form*

$$y^{(n)} = g(x, y, y')y^{(n-1)} + h(x, y, y', \ldots, y^{(n-2)}). \tag{3.125}$$

If ODE (3.125) *admits the infinitesimal generator* (3.122), *then* $\xi_y = 0$.

Theorem 3.3.4-3. *Consider an nth-order ODE,* $n \geq 3$, *of the form*

$$y^{(n)} = g(x, y)y^{(n-1)} + h(x, y, y', \ldots, y^{(n-2)}). \tag{3.126}$$

If ODE (3.126) *admits the infinitesimal generator* (3.122), *then* $\xi_y = 0, \eta_{yy} = 0$.

Theorem 3.3.4-4. *Consider a second-order ODE of the form*

$$y'' = g(x, y)y' + h(x, y). \tag{3.127}$$

If ODE (3.127) *admits the infinitesimal generator* (3.122) *with* $\xi_y = 0$, *then* $\eta_{yy} = 0$.

Theorems 3.3.4-2 to 3.3.4-4 are proved in Bluman (1990a).

Consider two examples:

(1) *The Lie Group of Transformations Acting on* R^2 *that Maps Straight Lines into Straight Lines*

This is the Lie group of point transformations admitted by the second-order ODE

$$y'' = 0. \tag{3.128}$$

The invariance criterion here is

$$\eta^{(2)} = 0 \quad \text{when } y_2 = 0,$$

where $\eta^{(2)}$ is given by (2.102), i.e., $\xi(x, y)$, $\eta(x, y)$ satisfy

$$\eta_{xx} + [2\eta_{xy} - \xi_{xx}]y_1 + [\eta_{yy} - 2\xi_y](y_1)^2 - \xi_{yy}(y_1)^3 = 0. \tag{3.129}$$

Equation (3.129) is a cubic polynomial in terms of y_1. Equating to zero the coefficients of each monomial term in (3.129), we obtain the system of *determining equations*

$$\xi_{yy} = 0, \tag{3.130a}$$

$$\eta_{xx} = 0, \tag{3.130b}$$

$$\eta_{yy} - 2\xi_{xy} = 0, \tag{3.130c}$$

$$\xi_{xx} - 2\eta_{xy} = 0. \tag{3.130d}$$

From (3.130a,b), we obtain

$$\xi = a(x)y + b(x),$$
$$\eta = c(y)x + d(y).$$

Then (3.130c,d) lead to

$$c''(y)x + d''(y) - 2a'(x) = 0, \tag{3.131a}$$

$$a''(x)y + b''(x) - 2c'(y) = 0. \tag{3.131b}$$

Taking $\partial / \partial x$ of (3.131a) and $\partial / \partial y$ of (3.131b), we see that $c''(y) = a''(x) = 0$. Thus, the solution of the determining equations (3.130a–d) is given by

$$\xi = \alpha_1 x^2 + \alpha_2 xy + \alpha_3 x + \alpha_4 y + \alpha_5,$$
$$\eta = \alpha_1 xy + \alpha_2 y^2 + \alpha_6 x + \alpha_7 y + \alpha_8,$$

where $\alpha_1, \alpha_2, ..., \alpha_8$ are arbitrary constants. This is the eight-parameter Lie group of projective transformations (2.168a,b) admitted by ODE (3.125).

(2) *Blasius Equation*
The invariance criterion for the Blasius equation

$$y''' + \tfrac{1}{2}yy'' = 0 \tag{3.132}$$

is given by

$$\eta^{(3)} + \tfrac{1}{2}y_2\eta + \tfrac{1}{2}y\eta^{(2)} = 0 \quad \text{when } y_3 = -\tfrac{1}{2}yy_2, \tag{3.133}$$

where $\eta^{(2)}, \eta^{(3)}$ are given by (2.102), (2.103). Then (3.133) becomes the polynomial equation

$$[\eta_{xxx}+\tfrac{1}{2}y\eta_{xx}]+[3\eta_{xxy}-\xi_{xxx}+y\eta_{xy}-\tfrac{1}{2}y\xi_{xx}]y_1+[3\eta_{xyy}-3\xi_{xxy}+\tfrac{1}{2}y\eta_{yy}-y\xi_{xy}](y_1)^2$$
$$-[\eta_{yyy}-3\xi_{xyy}-\tfrac{1}{2}y\xi_{yy}](y_1)^3-\xi_{yyy}(y_1)^4+[3\eta_{xy}-3\xi_{xx}+\tfrac{1}{2}y\xi_x+\tfrac{1}{2}\eta]y_2$$
$$+[3\eta_{yy}-9\xi_{xy}+\tfrac{1}{2}y\xi_y]y_1y_2-6\xi_{yy}(y_1)^2y_2-3\xi_y(y_2)^2=0. \tag{3.134}$$

The resulting determining equations for $\xi(x,y)$ and $\eta(x,y)$ are given by

$$\eta_{xxx}+\tfrac{1}{2}y\eta_{xx}=0, \tag{3.135a}$$

$$3\eta_{xxy}-\xi_{xxx}+y\eta_{xy}-\tfrac{1}{2}y\xi_{xx}=0, \tag{3.135b}$$

$$3\eta_{xyy}-3\xi_{xxy}+\tfrac{1}{2}y\eta_{yy}-y\xi_{xy}=0, \tag{3.135c}$$

$$3\xi_{xyy}+\tfrac{1}{2}y\xi_{yy}-\eta_{yyy}=0, \tag{3.135d}$$

$$\xi_{yyy}=0, \tag{3.135e}$$

$$3\eta_{xy}-3\xi_{xx}+\tfrac{1}{2}y\xi_x+\tfrac{1}{2}\eta=0, \tag{3.135f}$$

$$3\eta_{yy}-9\xi_{xy}+\tfrac{1}{2}y\xi_y=0, \tag{3.135g}$$

$$\xi_{yy}=0, \tag{3.135h}$$

$$\xi_y=0. \tag{3.135i}$$

From (3.135i), one immediately sees that $\xi=\xi(x)$, and hence, (3.135h,e) are redundant. Then (3.135g) leads to

$$\eta_{yy}=0, \tag{3.136}$$

so that (3.135c,d) are also satisfied. Taking $\partial/\partial y$ of (3.135b) and $\partial^2/\partial x\,\partial y$ of (3.135f), we are led to

$$\eta_{xy}=\xi''(x)=0. \tag{3.137}$$

Moreover, now (3.135b) is satisfied. Then

$$\xi=\alpha+\beta x, \tag{3.138a}$$

$$\eta=\gamma y+a(x). \tag{3.138b}$$

Equation (3.135f) leads to $a(x)=0, \gamma=-\beta$. No further restrictions arise from (3.135a). Consequently, the Blasius equation (3.132) only admits a two-parameter Lie group of point transformations (translations, scalings) with infinitesimals

$$\xi=\alpha+\beta x, \tag{3.139a}$$

$$\eta=-\beta y, \tag{3.139b}$$

where α and β are arbitrary constants.

If we directly use Theorem 3.3.4-3, the computations for finding $\xi(x,y)$ and $\eta(x,y)$ simplify significantly. Since ODE (3.132) is of the form (3.126), it immediately follows that $\xi_y = 0, \eta_{yy} = 0.$ Hence, the determining equations resulting from the polynomial equation (3.134) reduce to just (3.135a,b,f).

3.3.5 DETERMINATION OF *n*th-ORDER ODEs INVARIANT UNDER A GIVEN GROUP

Now we consider the problem of finding all *n*th-order ODEs that admit a given one-parameter Lie group of point transformations. This is accomplished by a simple extension of the procedure outlined in Section 3.2.4 for first-order ODEs through the use of either canonical coordinates or differential invariants.

Suppose an *n*th-order ODE (3.121) admits a given one-parameter Lie group of point transformations with infinitesimal generator (3.122). We can proceed in two obvious ways:

(i) *Method of Canonical Coordinates.* Corresponding to the infinitesimal generator (3.122), we compute canonical coordinates $r(x,y)$ and $s(x,y)$ satisfying $\mathrm{X}r = 0, \mathrm{X}s = 1,$ so that the group admitted by ODE (3.121) is now the translation group

$$r^* = r, \quad s^* = s + \varepsilon. \tag{3.140}$$

Then invariants under (3.140) are given by

$$\frac{d^k s}{dr^k}, \quad k = 1,2,\ldots,n.$$

These invariants can be expressed in terms of $x, y, y', \ldots, y^{(n)}$ through relations (3.91) and (3.95). Consequently, the most general *n*th-order ODE, written in solved form, that admits a one-parameter Lie group of point transformations with infinitesimal generator (3.122) is given by

$$\frac{d^n s}{dr^n} = G\left(r, \frac{ds}{dr}, \ldots, \frac{d^{n-1}s}{dr^{n-1}}\right), \tag{3.141}$$

where G is an arbitrary function of $r, ds/dr, \ldots, d^{n-1}s/dr^{n-1}$. Note that ODE (3.141) is an $(n-1)$th-order ODE in terms of a dependent variable $z = ds/dr$.

(ii) *Method of Differential Invariants.* Here we first find invariants $u(x,y)$ and $v(x,y,y')$ as we did for first-order ODEs. Then we compute the differential invariants

$$\frac{d^k v}{du^k}, \quad k = 1,2,\ldots,n-1,$$

which can be expressed in terms of $x, y, y', \ldots, y^{(n)}$ as indicated in Section 3.3.2. Consequently, the most general nth-order ODE (in solved form) that admits the group (3.122) can be written in the form

$$\frac{d^{n-1}v}{du^{n-1}} = H\left(u, v, \frac{dv}{du}, \ldots, \frac{d^{n-2}v}{du^{n-1}}\right), \tag{3.142}$$

where H is an arbitrary function of $u, v, dv/du, \ldots, d^{n-2}v/du^{n-2}$. Note that ODE (3.142) is an $(n-1)$th-order ODE in terms of its dependent variable v.

To find the most general nth-order ODE admitting a group of a simple form, such as a group of scalings, it is more natural to compute directly $n+1$ invariants that depend on, respectively, (x, y), (x, y, y'),..., $(x, y, y', \ldots, y^{(n)})$. The disadvantage of such a direct approach is that the reduction of order from n to $n-1$ is not automatic, as is the case when the most general nth-order ODE is obtained through either canonical coordinates or differential invariants.

As an example, we use these methods to find the most general second-order ODE that admits the scaling group (3.74a,b). The reader is referred to the calculations in Section 3.2.4, where we found the most general first-order ODE that admits (3.74a,b).

(1) *Method of Canonical Coordinates*
Canonical coordinates are given by

$$r = \frac{y}{x}, \quad s = \log y.$$

Then

$$\frac{ds}{dr} = \frac{y'}{ry' - r^2}, \quad \frac{d^2s}{dr^2} = -\frac{r^2yy''}{(ry'-r^2)^3} + \frac{2ry' - (y')^2}{(ry'-r^2)^2} = -\frac{r^2yy''}{(ry'-r^2)^3} + \frac{2ry'}{(ry'-r^2)^2} - \left(\frac{ds}{dr}\right)^2.$$

Hence, the most general second-order ODE that admits the scaling group (3.74a,b) is given by

$$y'' = 2\frac{y'(xy'-y)}{xy} + \frac{(xy'-y)^3}{x^4}G\left(\frac{y}{x}, \frac{x^2y'}{xyy'-y^2}\right), \tag{3.143}$$

where G is an arbitrary function of its arguments.

(2) *Method of Differential Invariants*
From the form of the most general first-order ODE (3.76) admitting (3.74a,b), we see that

$$u = \frac{y}{x}, \quad v = y'.$$

Then

$$\frac{dv}{du} = \frac{x^2y''}{xy'-y}.$$

Consequently, the most general second-order ODE that admits (3.74a,b) is given by

$$y'' = \frac{xy' - y}{x^2} H\left(\frac{y}{x}, y'\right), \tag{3.144}$$

where H is an arbitrary function of its arguments. Of course, (3.143) and (3.144) must yield the *same* general second-order ODE. In comparing (3.143) and (3.144), note that

$$\frac{x^2 y'}{xyy' - y^2} = \frac{y'}{\frac{y}{x} y' - \left(\frac{y}{x}\right)^2},$$

and so we see that

$$H\left(\frac{y}{x}, y'\right) = 2\left(\frac{x}{y}\right) y' + \left(y' - \frac{y}{x}\right)^2 G\left(\frac{y}{x}, \frac{y'}{\frac{y}{x}\left(y' - \frac{y}{x}\right)}\right).$$

(3) *Direct Approach*
Obvious invariants of the scaling group (3.74a,b) are given by y/x, y', yy''. Hence, the most general second-order ODE that admits (3.74a,b) is given by

$$y'' = \frac{1}{y} I\left(\frac{y}{x}, y'\right), \tag{3.145}$$

where I is an arbitrary function of its arguments. In comparing (3.144) and (3.145), note that

$$\frac{xy' - y}{x^2} = \frac{1}{y}\left(\frac{y}{x} y' - \left(\frac{y}{x}\right)^2\right),$$

and hence,

$$I\left(\frac{y}{x}, y'\right) = \left(\frac{y}{x} y' - \left(\frac{y}{x}\right)^2\right) H\left(\frac{y}{x}, y'\right).$$

EXERCISES 3.3

1. Let $y = \phi(x)$ be a particular solution of the second-order linear nonhomogeneous ODE

$$y'' + p(x)y' + q(x)y = g(x). \tag{3.146}$$

(a) Find a one-parameter Lie group of point transformations admitted by (3.146).

(b) Use canonical coordinates to reduce ODE (3.146) to a first-order ODE plus a quadrature.
(c) Use differential invariants to reduce ODE (3.146) to a first-order ODE plus a quadrature.

2. The third-order Blasius equation (3.113) admits the two-parameter (α,β) Lie group of point transformations (3.114a,b).
(a) Use canonical coordinates associated with parameter β to reduce ODE (3.113) to a second-order ODE.
(b) Find a symmetry of this second-order ODE. How is this symmetry related to the parameter α of (3.114a,b)?
(c) Find canonical coordinates for the symmetry of (b). Consequently, reduce the Blasius equation to a first-order ODE plus two quadratures.

3. Find the Lie group of point transformations admitted by the ODE

$$y'' = \alpha(y')^N,$$

where $\alpha =$ const and $N = 1,2,\ldots$ is a fixed integer. Investigate further the cases $N = 1,2,3$. Compare these cases with the Lie group admitted by $y'' = 0$. [For related ODEs, see Aguirre and Krause (1985).]

4. Find the Lie group of point transformations admitted by each of the ODEs:
(a) $y'' = y^{-3}$; and
(b) $y'' = e^{-y'}$.

5. Consider the ODE

$$2\frac{d}{dx}\left(K(y)\frac{dy}{dx}\right) + x\frac{dy}{dx} = 0 \tag{3.147}$$

that arises when seeking invariant solutions of the nonlinear heat conduction equation [Bluman and Kumei (1989b, Section 7.2.7)] from its invariance under scalings.
(a) Find a Lie group of point transformations admitted by ODE (3.147) when:
(i) $K(y) = \lambda(y+\kappa)^\nu$, where λ, κ, ν are arbitrary constants; and
(ii) $K(y) = \lambda e^{\nu y}$, where λ, ν are arbitrary constants.
(b) Find all $K(y)$ for which ODE (3.147) admits a Lie group of point transformations.

6. Consider the ODE

$$K(y')y'' + xy' - y = 0 \tag{3.148}$$

that arises when seeking invariant solutions of the nonlinear heat conduction equation from the invariance of a related potential system under scalings [Bluman and Kumei (1989a, Section 7.3.1)].
(a) Find a Lie group of point transformations admitted by ODE (3.148) when:

(i) $K(y') = \dfrac{1}{1+(y')^2}$; and

(ii) $K(y') = \dfrac{1}{1+(y')^2}\exp[\lambda \arctan y'],\ \lambda = \text{const.}$

(b) Solve ODE (3.148) when $K(y') = 1/[1+(y')^2]$.

7. Find the Lie group of point transformations admitted by the family of curves $y = p(x)$, where $p(x)$ is an arbitrary polynomial of degree n.

8. Find the most general second-order ODE that admits the scaling group $x^* = \alpha x$, $y^* = \alpha^k y$, where k is a fixed constant. Reduce the ODE to a first-order ODE plus a quadrature.

9. (a) Find the most general second-order ODE that admits the rotation group.
(b) Interpret your result geometrically.
(c) Find the general solution of the ODE

$$(x^2 + y^2)y'' + 2(y - xy')(1 + (y')^2) = 0.$$

10. Prove Theorem 3.3.4-1.

3.4 REDUCTION OF ORDER OF ODEs UNDER MULTIPARAMETER LIE GROUPS OF POINT TRANSFORMATIONS

Now we consider the invariance of an nth-order ODE under an r-parameter Lie group of point transformations, $r \geq 2$. If the corresponding r-dimensional Lie algebra is *solvable* [cf. Section 2.5.4], we show that the given nth-order ODE can be reduced to an $(n - r)$th-order ODE plus r quadratures. Theorem 2.5.2-2 shows that any two-dimensional Lie algebra is solvable. It turns out that every even-dimensional ($r = 2m$ for some integer m) Lie algebra acting on $\mathbf{R}^n$ contains a two-dimensional subalgebra. [Cohen (1911, p. 150); Dickson (1924). Both of these authors erroneously claim that this is true for *any* real Lie algebra acting on $\mathbf{R}^n$. Their proofs only hold for complex Lie algebras or real Lie algebras of even dimension. It is easy to show that the Lie algebra SO(3) corresponding to the rotation group acting on $\mathbf{R}^3$ is not solvable. One can show that there exists a three-dimensional Lie algebra acting on $\mathbf{R}^2$ that is isomorphic to SO(3) and hence not solvable.]

3.4.1 INVARIANCE OF A SECOND-ORDER ODE UNDER A TWO-PARAMETER LIE GROUP

We now show that if a second-order ODE

$$y'' = f(x, y, y') \tag{3.149}$$

admits a two-parameter Lie group of point transformations, then one can constructively reduce ODE (3.149) to two quadratures and, hence, find its general solution.

Let X_1, X_2 be basis generators of the Lie algebra of the given two-parameter Lie group of transformations, and let $X_i^{(k)}$ denote the kth-extended infinitesimal generator of X_i, $i = 1,2$. Without loss of generality, from Theorem 2.5.4-2, we can assume that

$$[X_1, X_2] = \lambda X_1 \tag{3.150}$$

for some constant λ.

Let $u(x, y), v(x, y, y')$ be invariants of $X_1^{(2)}$ such that

$$X_1 u = 0, \quad X_1^{(1)} v = 0.$$

Then the corresponding differential invariant dv / du satisfies the equation [cf. Section 3.3.2]

$$X_1^{(2)} \frac{dv}{du} = 0,$$

and hence, ODE (3.149) reduces to

$$\frac{dv}{du} = H(u, v), \tag{3.151}$$

for some function H of u, v. [Note that $\partial v / \partial y' \neq 0$.] From the commutation relation (3.150), it follows that

$$X_1 X_2 u = X_2 X_1 u + \lambda X_1 u = 0.$$

Hence,

$$X_2 u = \alpha(u) \tag{3.152a}$$

for some function α of u.

Similarly, from Theorem 2.5.2-3, it follows that

$$X_1^{(1)} X_2^{(1)} v = 0, \quad X_1^{(2)} X_2^{(2)} \frac{dv}{du} = 0.$$

Hence,

$$X_2^{(1)} v = \beta(u, v) \tag{3.152b}$$

for some function β of u, v. Since ODE (3.149) admits X_2, it follows that

$$X_2^{(2)} \left(\frac{dv}{du} - H(u, v) \right) = 0 \quad \text{when} \quad \frac{dv}{du} = H(u, v).$$

From (3.152a,b) it follows that in terms of coordinates u and v, the first extension $X_2^{(1)}$ becomes

$$\mathrm{X}_2^{(1)} = \alpha(u)\frac{\partial}{\partial u} + \beta(u,v)\frac{\partial}{\partial v},$$

which is admitted by ODE (3.151). Consider canonical coordinates $R(u,v), S(u,v)$ such that

$$\mathrm{X}_2^{(1)} R = 0, \quad \mathrm{X}_2^{(1)} S = 1.$$

Then $R(u,v), S(u,v)$ satisfy

$$\alpha(u)\frac{\partial R}{\partial u} + \beta(u,v)\frac{\partial R}{\partial v} = 0,$$
$$\alpha(u)\frac{\partial S}{\partial u} + \beta(u,v)\frac{\partial S}{\partial v} = 1.$$

Thus, the one-parameter group of translations

$$R^* = R, \tag{3.153a}$$

$$S^* = S + \varepsilon, \tag{3.153b}$$

is admitted by ODE (3.151). Hence, ODE (3.151) reduces to

$$\frac{dS}{dR} = I(R) \tag{3.154}$$

for some function I of R. Then the first-order ODE (3.154) integrates out to

$$S(u,v) = \int^{R(u,v)} I(R)\, dR + C_1, \tag{3.155}$$

for some arbitrary constant C_1. The differential equation

$$S(u(x,y), v(x,y,y')) = \int^{R(u(x,y),v(x,y,y'))} I(R)\, dR + C_1$$

admits X_1 and, hence, reduces to quadrature by the method of canonical coordinates after one determines $r(x,y)$ and $s(x,y)$ such that

$$\mathrm{X}_1 r = 0, \quad \mathrm{X}_1 s = 1.$$

Consequently, any second-order ODE that admits a two-parameter Lie group of transformations reduces completely to two quadratures.

As an example, consider the second-order linear nonhomogeneous ODE

$$y'' + p(x)y' + q(x)y = g(x). \tag{3.156}$$

Let $z = \phi_1(x)$ and $z = \phi_2(x)$ be two linearly independent solutions of the corresponding homogeneous equation

$$z'' + p(x)z' + q(x)z = 0. \tag{3.157}$$

Then ODE (3.156) admits the two-parameter $(\varepsilon_1, \varepsilon_2)$ Lie group of transformations

$$x^* = x, \tag{3.158a}$$

$$y^* = y + \varepsilon_1\phi_1(x) + \varepsilon_2\phi_2(x). \tag{3.158b}$$

The corresponding infinitesimal generators are given by

$$\mathrm{X}_1 = \phi_1(x)\frac{\partial}{\partial y}, \quad \mathrm{X}_2 = \phi_2(x)\frac{\partial}{\partial y},$$

with $[\mathrm{X}_1, \mathrm{X}_2] = 0$. Then

$$\mathrm{X}_i^{(1)} = \phi_i(x)\frac{\partial}{\partial y} + \phi_i'(x)\frac{\partial}{\partial y'}, \quad i = 1,2,$$

$$u = x, \quad v = \frac{y'}{\phi_1'} - \frac{y}{\phi_1(x)},$$

$$\mathrm{X}_2 u = \mathrm{X}_2 x = 0, \quad \mathrm{X}_2^{(1)} v = \frac{\phi_2'(x)}{\phi_1'(x)} - \frac{\phi_2(x)}{\phi_1(x)} = \frac{W(x)}{\phi_1(x)\phi_1'(x)},$$

where $W(x)$ is the Wronskian $W(x) = \phi_1\phi_2' - \phi_2\phi_1'$. Now, in terms of coordinates x and v, we have

$$\mathrm{X}_2^{(1)} = \frac{W(x)}{\phi_1(x)\phi_1'(x)}\frac{\partial}{\partial v}.$$

Canonical coordinates $R(x,v), S(x,v)$ satisfy

$$\mathrm{X}_2^{(1)} R = \frac{W}{\phi_1\phi_1'}\frac{\partial R}{\partial v} = 0, \quad \mathrm{X}_2^{(1)} S = \frac{W}{\phi_1\phi_1'}\frac{\partial S}{\partial v} = 1,$$

and hence,

$$R = x, \quad S = \frac{v\phi_1\phi_1'}{W}.$$

Consequently, by a simple calculation, we find

$$\frac{dS}{dx} = \frac{g(x)\phi_1(x)}{W(x)},$$

so that

$$S = \frac{y'\phi_1 - y\phi_1'}{W} = \int \frac{g\phi_1}{W}\,dx + C_2 \tag{3.159}$$

for some arbitrary constant C_2.

By construction, the first-order ODE (3.159) admits $X_1 = \phi_1(x)\frac{\partial}{\partial y}$. In terms of the corresponding canonical coordinates $r = x, s = y/\phi_1(x)$, ODE (3.159) reduces to

$$\frac{ds}{dx} = \frac{W}{(\phi_1)^2}\left[\int \frac{g\phi_1}{W}dx + C_2\right]. \tag{3.160}$$

But $W/(\phi_1)^2 = (\phi_2/\phi_1)'$. Hence,

$$\frac{W}{(\phi_1)^2}\int \frac{g\phi_1}{W}dx = \frac{d}{dx}\left[\frac{\phi_2}{\phi_1}\int \frac{g\phi_1}{W}dx\right] - \frac{g\phi_2}{W}.$$

Thus,

$$s = C_2\frac{\phi_2}{\phi_1} + \frac{\phi_2}{\phi_1}\int \frac{g\phi_1}{W}dx - \int \frac{g\phi_2}{W}dx + C_1,$$

for some arbitrary constant C_1, which leads to the familiar general solution

$$y = C_1\phi_1 + C_2\phi_2 + \phi_2\int \frac{g\phi_1}{W}dx - \phi_1\int \frac{g\phi_2}{W}dx \tag{3.161}$$

of ODE (3.156), derived by the variation of parameters method in standard textbooks.

3.4.2 INVARIANCE OF AN *n*th-ORDER ODE UNDER A TWO-PARAMETER LIE GROUP

Now consider an *n*th-order ODE

$$y_n = f(x, y, y_1, \ldots, y_{n-1}), \tag{3.162}$$

$n \geq 3$, assumed to be invariant under a two-parameter Lie group of point transformations with infinitesimal generators X_1, X_2 such that, without loss of generality, $[X_1, X_2] = \lambda X_1$, for some constant λ [cf. Theorem 2.5.4-2].

As in Section 3.4.1, let $u(x,y)$, $v(x,y,y_1) = v(x,y,y')$ be invariants of $X_1^{(2)}$. Then $X_1^{(2)}v_1 = 0$ where $v_1 = dv/du$, and hence, ODE (3.162) reduces to

$$\frac{d^{n-1}v}{du^{n-1}} = H\left(u, v, \frac{dv}{du}, \ldots, \frac{d^{n-2}v}{du^{n-2}}\right), \tag{3.163}$$

for some function H of $u, v, dv/du, \ldots, d^{n-2}v/du^{n-2}$.

Since $[X_1^{(k)}, X_2^{(k)}] = \lambda X_1^{(k)}$, $k = 1,2,\ldots$, it follows that

$$X_2 u = \alpha(u),$$

$$X_2^{(1)} v = \beta(u, v),$$

$$X_2^{(2)} v_1 = \gamma(u,v,v_1),$$

for some functions α, β, γ of the indicated arguments. Then

$$X_2^{(1)} = \alpha(u)\frac{\partial}{\partial u} + \beta(u,v)\frac{\partial}{\partial v},$$

with its first extension given by

$$X_2^{(2)} = \alpha(u)\frac{\partial}{\partial u} + \beta(u,v)\frac{\partial}{\partial v} + \gamma(u,v,v_1)\frac{\partial}{dv_1},$$

is admitted by ODE (3.163). Now, let $U(u,v)$, $V(u,v,v_1)$ be invariants of $X_2^{(2)}$ such that

$$X_2^{(1)} U = 0, \quad X_2^{(2)} V = 0.$$

Then

$$X_2^{(3)} \frac{dV}{dU} = 0.$$

Consequently, ODE (3.163) and, hence, ODE (3.162) reduce to

$$\frac{d^{n-2}V}{dU^{n-2}} = I\left(U, V, \frac{dV}{dU}, \ldots, \frac{d^{n-3}V}{dU^{n-3}}\right), \tag{3.164}$$

for some function I of $U, V, dV/dU, \ldots, d^{n-3}V/dU^{n-3}$. If

$$V = \phi(U; C_1, C_2, \ldots, C_{n-2})$$

is the general solution of ODE (3.164), then the first-order ODE

$$V\left(u, v, \frac{dv}{du}\right) = \phi(U(u,v); C_1, C_2, \ldots, C_{n-2}) \tag{3.165}$$

admits $X_2^{(1)} = \alpha(u)\frac{\partial}{\partial u} + \beta(u,v)\frac{\partial}{\partial v}$. Hence, ODE (3.165) reduces to a quadrature

$$v = \psi(u; C_1, C_2, \ldots, C_{n-2}, C_{n-1}).$$

But the first-order ODE

$$v(x,y,y') = \psi(u(x,y); C_1, C_2, \ldots, C_{n-2}, C_{n-1}) \tag{3.166}$$

admits X_1. Thus, ODE (3.166) reduces to a quadrature that leads to the general solution of ODE (3.162).

Hence, *if an nth-order ODE* $(n \geq 3)$ *admits a two-parameter Lie group of point transformations, then it can be reduced constructively to an* $(n-2)$*th-order ODE plus two quadratures.* Note that the order in which the operators X_1 and X_2 are used is *crucial* if $\lambda \neq 0$.

As a first example, we consider again the Blasius equation

$$y''' + \tfrac{1}{2}yy'' = 0, \tag{3.167}$$

which admits the two-parameter Lie group of point transformations [cf. (3.139a,b)] with infinitesimal generators

$$\mathrm{X}_1 = \frac{\partial}{\partial x}, \quad \mathrm{X}_2 = x\frac{\partial}{\partial x} - y\frac{\partial}{\partial y}.$$

Then

$$[\mathrm{X}_1,\ \mathrm{X}_2] = \mathrm{X}_1.$$

Invariants of $\mathrm{X}_1^{(2)}$ are given by

$$u = y, \quad v = y' = y_1, \quad v_1 = \frac{dv}{du} = \frac{y_2}{y_1}.$$

It follows that

$$\mathrm{X}_2^{(2)} = x\frac{\partial}{\partial x} - y\frac{\partial}{\partial y} - 2y_1\frac{\partial}{\partial y_1} - 3y_2\frac{\partial}{\partial y_2},$$

$$\mathrm{X}_2 u = -y = -u, \quad \mathrm{X}_2^{(1)} v = -2y_1 = -2v, \quad \mathrm{X}_2^{(2)} v_1 = -\frac{y_2}{y_1} = -v_1.$$

Then

$$\mathrm{X}_2^{(1)} U(u, v) = 0, \quad \mathrm{X}_2^{(2)} V(u, v, v_1) = 0,$$

lead to

$$U = \frac{v}{u^2}, \quad V = \frac{v_1}{u}.$$

Consequently, the third-order Blasius equation (3.167) reduces to

$$\frac{dV}{dU} = \frac{d((yy_1)^{-1}y_2)}{d(y^{-2}y_1)} = \frac{y^2y_1y_3 - y^2(y_2)^2 - y(y_1)^2y_2}{y(y_1)^2y_2 - 2(y_1)^4} = \frac{\tfrac{1}{2}y^3y_1y_2 + y^2(y_2)^2 + y(y_1)^2y_2}{2(y_1)^4 - y(y_1)^2y_2},$$

which, in terms of U and V, becomes the first-order ODE

$$\frac{dV}{dU} = \frac{V}{U}\left[\frac{\tfrac{1}{2} + V + U}{2U - V}\right]. \tag{3.168}$$

If $V = \phi(U; C_1)$ is the general solution of ODE (3.168), then the first-order ODE

$$v_1 = \frac{dv}{du} = u\phi\left(\frac{v}{u^2}; C_1\right) \tag{3.169}$$

admits $X_2^{(1)} = -u\frac{\partial}{\partial u} - 2v\frac{\partial}{\partial v}$. In terms of corresponding canonical coordinates $r = v/u^2$, $s = \log v$, ODE (3.169) becomes

$$\frac{ds}{dr} = \frac{\phi(r;C_1)}{r[\phi(r;C_1) - 2r]}. \tag{3.170}$$

This leads to the quadrature

$$v = C_2 \exp\left[\int^r \frac{\phi(\rho;C_1)}{\rho[\phi(\rho;C_1) - 2\rho]} d\rho\right], \tag{3.171}$$

where $v = y_1$, $r = y_1 / y^2$. In principle, (3.171) can be expressed in a solved form

$$y' = \psi(y;C_1,C_2)$$

that admits $X_1 = \frac{\partial}{\partial x}$ and thus reduces to a quadrature

$$\int \frac{dy}{\psi(y;C_1,C_2)} = x + C_3. \tag{3.172}$$

Equation (3.172) represents a general solution of the Blasius equation.

As a second example, we consider the third-order ODE

$$yy'\left(\frac{y}{y'}\right)'' = \pm 1, \tag{3.173}$$

that arises when one studies the group properties of the wave equation with wave speed $y(x)$ [Bluman and Kumei (1987)]. ODE (3.173) obviously admits the two-parameter Lie group of transformations

$$x^* = ax + b,$$
$$y^* = ay.$$

It is easy to see that corresponding differential invariants are given by

$$U = y', \quad V = yy''.$$

Consequently, ODE (3.173) reduces to

$$\frac{dV}{dU} = 2\frac{V}{U} \mp \frac{U}{V}. \tag{3.174}$$

Fortuitously, the first-order ODE (3.174) admits the scalings

$$U^* = \lambda U, \quad V^* = \lambda V.$$

Accordingly, one easily finds the general solution of ODE (3.174):

$$U^{-2}\left[\left(\frac{V}{U}\right)^2 \mp 1\right] = \text{const.} \tag{3.175}$$

Two cases arise depending on the sign of the constant in (3.175). We will consider the case where the constant is given by $\upsilon^2 \geq 0, \upsilon = \text{const.}$ Here it is convenient to choose first-order differential invariants, corresponding to invariance under translations in x, as new variables:

$$u = y, \quad v = y' = U.$$

Then (3.175) becomes the first-order ODE

$$\frac{dv}{du} = \frac{\sqrt{\upsilon^2 v \pm 1}}{u}. \tag{3.176}$$

The general solution of ODE (3.176) is given by

$$v = \frac{1}{2\upsilon}[(\rho u)^{\upsilon} \mp (\rho u)^{-\upsilon}], \tag{3.177}$$

where ρ is an arbitrary constant. Without loss of generality, through a uniform scaling of x and y, we can set $\rho = 1$. Hence, modulo the scalings $x^* = ax, y^* = ay$, the third-order ODE (3.173) reduces to

$$y' = \frac{1}{2\upsilon}[y^{\upsilon} \mp y^{-\upsilon}],$$

i.e., to the canonical forms

$$y' = \frac{1}{\upsilon}\sinh(\upsilon \log y) \tag{3.178a}$$

or

$$y' = \frac{1}{\upsilon}\cosh(\upsilon \log y). \tag{3.178b}$$

If the constant in (3.175) is given by $-\upsilon^2 \leq 0$, then by the same procedure one can show that finally, modulo the same scalings in x and y, the third-order ODE (3.173) has a further reduction to the canonical form

$$y' = \frac{1}{\upsilon}\sin(\upsilon \log y). \tag{3.178c}$$

The properties of the solutions to ODE (3.178c) are most interesting [Bluman and Kumei (1988)].

3.4.3 INVARIANCE OF AN nth-ORDER ODE UNDER AN r-PARAMETER LIE GROUP WITH A SOLVABLE LIE ALGEBRA

If an r-parameter Lie group of point transformations ($r \geq 3$) is admitted by an nth-order ODE ($n \geq r$), it does not always follow that one can have a reduction to an $(n-r)$th-order ODE plus r quadratures. We will show that such a reduction is always possible if the Lie algebra, $\mathcal{L}^r$, formed by the infinitesimal generators of the group, is a *solvable Lie algebra* [cf. Section 2.5-4]. Then $\mathcal{L}^r$ has a basis set of generators $\mathrm{X}_1, \mathrm{X}_2, \ldots, \mathrm{X}_r$ satisfying commutation relations of the form

$$[\mathrm{X}_i, \mathrm{X}_j] = \sum_{k=1}^{j-1} C_{ij}^k \mathrm{X}_k, \quad 1 \leq i < j, \; j = 2, \ldots, r, \tag{3.179}$$

for some real structure constants C_{ij}^k [Exercise 3.4-7]. For the same constants C_{ij}^k, the corresponding mth-extended infinitesimal generators $\mathrm{X}_j^{(m)}$ satisfy

$$[\mathrm{X}_i^{(m)}, \mathrm{X}_j^{(m)}] = \sum_{k=1}^{j-1} C_{ij}^k \mathrm{X}_k^{(m)}, \quad 1 \leq i < j, \; j = 2, \ldots, r. \tag{3.180}$$

Now consider the nth-order ODE

$$y_n = F_n(x, y, y_1, \ldots, y_{n-1}), \tag{3.181}$$

where F_n is a given function of its arguments. We assume that ODE (3.181) admits an r-parameter Lie group of point transformations ($3 \leq r \leq n$) whose infinitesimal generators X_i, $i = 1, 2, \ldots, r$, form a solvable Lie algebra and, in particular, satisfy the commutation relations (3.179) for some constants C_{ij}^k.

Let $u_{(1)}(x, y)$, $v_{(1)}(x, y, y_1)$ be invariants such that

$$\mathrm{X}_1 u_{(1)} = 0, \quad \mathrm{X}_1^{(1)} v_{(1)} = 0.$$

Then

$$\mathrm{X}_1^{(k+1)} \frac{d^k v_{(1)}}{du_{(1)}^k} = 0, \quad k = 1, 2, \ldots, n-1.$$

Let

$$v_{(1)k} = \frac{d^k v_{(1)}}{du_{(1)}^k}, \quad k = 1, 2, \ldots, n-1.$$

In terms of the invariants $u_{(1)}$, $v_{(1)}$, and the differential invariants $v_{(1)k}$, $k = 1, 2, \ldots, n-1$, of $\mathrm{X}_1^{(n)}$, ODE (3.181) reduces to an $(n-1)$th-order ODE

$$v_{(1)n-1} = F_{n-1}(u_{(1)}, v_{(1)}, v_{(1)1}, \ldots, v_{(1)n-2}), \tag{3.182}$$

for some function F_{n-1} of the indicated invariants of $X_1^{(n)}$.

From (3.179) and (3.180), it follows that

$$X_2 u_{(1)} = \alpha_1(u_{(1)}), \quad X_2^{(1)} v_{(1)} = \beta_1(u_{(1)}, v_{(1)}), \quad X_2^{(2)} v_{(1)1} = \gamma_1(u_{(1)}, v_{(1)}, v_{(1)1}),$$

for some functions $\alpha_1, \beta_1, \gamma_1$ of the indicated arguments. Hence ODE (3.182) admits the one-parameter Lie group of point transformations with infinitesimal generator

$$X_2^{(1)} = \alpha_1(u_{(1)})\frac{\partial}{\partial u_{(1)}} + \beta_1(u_{(1)}, v_{(1)})\frac{\partial}{\partial v_{(1)}},$$

whose first extension is given by

$$X_2^{(2)} = X_2^{(1)} + \gamma_1(u_{(1)}, v_{(1)}, v_{(1)1})\frac{\partial}{\partial v_{(1)1}}.$$

Let $u_{(2)}(u_{(1)}, v_{(1)})$, $v_{(2)}(u_{(1)}, v_{(1)}, v_{(1)1})$ be invariants such that

$$X_2^{(1)} u_{(2)} = 0, \quad X_2^{(2)} v_{(2)} = 0. \tag{3.183}$$

Then

$$X_2^{(2+k)} \frac{d^k v_{(2)}}{du_{(2)}^k} = 0, \quad k = 1, 2, \ldots, n-2.$$

Let

$$v_{(2)k} = \frac{d^k v_{(2)}}{du_{(2)}^k}, \quad k = 1, 2, \ldots, n-2.$$

In terms of the invariants $u_{(2)}, v_{(2)}, v_{(2)k}$, $k = 1, 2, \ldots, n-2$, of $X_2^{(n)}$ (which are also invariants of $X_1^{(n)}$), ODE (3.182) and, hence, ODE (3.181) reduce to the $(n-2)$th-order ODE

$$v_{(2)n-2} = F_{n-2}(u_{(2)}, v_{(2)}, v_{(2)1}, \ldots, v_{(2)n-3}), \tag{3.184}$$

for some function F_{n-2} of the indicated invariants of $X_2^{(n)}, X_1^{(n)}$.

From (3.179) and (3.180), it follows that

$$X_1^{(1)} X_3^{(1)} u_{(2)} = 0, \tag{3.185a}$$

$$X_2^{(1)} X_3^{(1)} u_{(2)} = 0. \tag{3.185b}$$

Then (3.185a) leads to

$$X_3^{(1)} u_{(2)} = A(u_{(1)}, v_{(1)}) \tag{3.186}$$

for some function A of $u_{(1)}, v_{(1)}$. From (3.185b), we obtain

$$X_3^{(1)}u_{(2)} = A(u_{(1)}, v_{(1)}) = \alpha_2(u_{(2)}),$$

for some function α_2 of $u_{(2)}$. Similarly,

$$X_3^{(2)}v_{(2)} = \beta_2(u_{(2)}, v_{(2)}), \quad X_3^{(3)}v_{(2)1} = \gamma_2(u_{(2)}, v_{(2)}, v_{(2)1}),$$

for some functions β_2, γ_2 of the indicated arguments. Hence, ODE (3.184) admits the point symmetry

$$X_3^{(2)} = \alpha_2(u_{(2)})\frac{\partial}{\partial u_{(2)}} + \beta_2(u_{(2)}, v_{(2)})\frac{\partial}{\partial v_{(2)}}, \tag{3.187}$$

with its first extension given by

$$X_3^{(3)} = X_3^{(2)} + \gamma_2(u_{(2)}, v_{(2)}, v_{(2)1})\frac{\partial}{\partial v_{(2)1}}.$$

Then let $u_{(3)}(u_{(2)}, v_{(2)})$, $v_{(3)}(u_{(2)}, v_{(2)}, v_{(2)1})$ be invariants such that

$$X_3^{(2)}u_{(3)} = 0, \quad X_3^{(3)}v_{(3)} = 0. \tag{3.188}$$

Consequently, we see that

$$X_3^{(3+k)}\frac{d^k v_{(3)}}{du_{(3)}^k} = 0, \quad k = 1,2,\ldots,n-3.$$

Let

$$v_{(3)k} = \frac{d^k v_{(3)}}{du_{(3)}^k}, \quad k = 1,2,\ldots,n-3.$$

In terms of the invariants $u_{(3)}, v_{(3)}, v_{(3)k}$, $k = 1,2,\ldots,n-3$, of $X_3^{(n)}$ (which are also invariants of $X_1^{(n)}, X_2^{(n)}$), ODE (3.184) and, hence, ODE (3.181) reduce to the $(n-3)$th-order ODE

$$v_{(3)n-3} = F_{n-3}(u_{(3)}, v_{(3)}, v_{(3)1}, \ldots, v_{(3)n-4}), \tag{3.189}$$

for some function F_{n-3} of the indicated invariants.

We continue inductively and suppose that for $q = 3,\ldots,m$, $m < r$,

$$u_{(q)}(u_{(q-1)}, v_{(q-1)}), \quad v_{(q)}(u_{(q-1)}, v_{(q-1)}, v_{(q-1)1})$$

are invariants such that

$$X_p^{(q-1)}u_{(q)} = 0, \quad X_p^{(q)}v_{(q)} = 0, \quad p = 1,2,\ldots,q,$$

$$X_p^{(q+k)}\frac{d^k v_{(q)}}{du_{(q)}^k} = 0, \quad k = 1,2,\ldots,n-q \text{ for } 1 \le p \le q.$$

Let $v_{(q)k} = d^k v_{(q)} / du_{(q)}{}^k$, $k = 1,2,\ldots,n-q$. Then the nth-order ODE (3.181) reduces to the $(n-m)$th-order ODE

$$v_{(m)n-m} = F_{n-m}(u_{(m)}, v_{(m)}, v_{(m)1}, \ldots, v_{(m)n-m-1}), \tag{3.190}$$

for some function F_{n-m} of the invariants of $\mathrm{X}_m^{(n)}, \mathrm{X}_{m-1}^{(n)}, \ldots, \mathrm{X}_2^{(n)}, \mathrm{X}_1^{(n)}$.

To go from step m to step $m+1$, we proceed as follows: From (3.180), it follows that

$$\mathrm{X}_j^{(m-1)} \mathrm{X}_{m+1}^{(m-1)} u_{(m)} = 0, \quad j = 1,2,\ldots,m.$$

Then $\mathrm{X}_1^{(m-1)} \mathrm{X}_{m+1}^{(m-1)} u_{(m)} = 0$ leads to

$$\mathrm{X}_{m+1}^{(m-1)} u_{(m)} = A_1(u_{(1)}, v_{(1)}, v_{(1)1}, \ldots, v_{(1)m-2}),$$

for some function A_1 of the invariants of $\mathrm{X}_1^{(m-1)}$. Similarly, $\mathrm{X}_2^{(m-1)} \mathrm{X}_{m+1}^{(m-1)} u_{(m)} = 0$ leads to

$$A_1 = A_2(u_{(2)}, v_{(2)}, v_{(2)1}, \ldots, v_{(2)m-3}),$$

for some function A_2 of the invariants of $\mathrm{X}_2^{(m-1)}, \mathrm{X}_1^{(m-1)}$. Then $\mathrm{X}_l^{(m-1)} \mathrm{X}_{m+1}^{(m-1)} u_{(m)} = 0$ leads to

$$A_1 = A_l(u_{(l)}, v_{(l)}, v_{(l)1}, \ldots, v_{(l)m-l-1}),$$

for some function A_l of the invariants of $\mathrm{X}_l^{(m-1)}, \mathrm{X}_{l-1}^{(m-1)}, \ldots, \mathrm{X}_1^{(m-1)}, 1 \le l \le m-2$. Now $\mathrm{X}_{m-1}^{(m-1)} \mathrm{X}_{m+1}^{(m-1)} u_{(m)} = 0$ leads to

$$A_1 = A_{m-1}(u_{(m-1)}, v_{(m-1)}),$$

for some function A_{m-1} of the invariants $u_{(m-1)}, v_{(m-1)}$ of $\mathrm{X}_{m-1}^{(m-1)}, \mathrm{X}_{m-2}^{(m-1)}, \ldots, \mathrm{X}_1^{(m-1)}$. Finally, $\mathrm{X}_m^{(m-1)} \mathrm{X}_{m+1}^{(m-1)} u_{(m)} = 0$ leads to

$$\mathrm{X}_{m+1}^{(m-1)} u_{(m)} = A_1 = \alpha_m(u_{(m)}),$$

for some function α_m of $u_{(m)}$.

Similarly, one can show that

$$\mathrm{X}_{m+1}^{(m)} v_{(m)} = \beta_m(u_{(m)}, v_{(m)}), \quad \mathrm{X}_{m+1}^{(m+1)} v_{(m)1} = \gamma_m(u_{(m)}, v_{(m)}, v_{(m)1}),$$

for some functions β_m, γ_m of the indicated arguments. Hence,

$$\mathrm{X}_{m+1}^{(m)} = \alpha_m(u_{(m)}) \frac{\partial}{\partial u_{(m)}} + \beta_m(u_{(m)}, v_{(m)}) \frac{\partial}{\partial v_{(m)}},$$

with its first extension given by

$$X_{m+1}^{(m+1)} = X_{m+1}^{(m)} + \gamma_m(u_{(m)}, v_{(m)}, v_{(m)1})\frac{\partial}{\partial v_{(m)1}},$$

is admitted by ODE (3.190) since ODE (3.181) admits X_{m+1}. Now let $u_{(m+1)}(u_{(m)}, v_{(m)})$, $v_{(m+1)}(u_{(m)}, v_{(m)}, v_{(m)1})$ be invariants such that

$$X_{m+1}^{(m)} u_{(m+1)} = 0, \quad X_{m+1}^{(m+1)} v_{(m+1)} = 0.$$

Then

$$X_{m+1}^{(m+1+k)} \frac{d^k v_{(m+1)}}{du_{(m+1)}^k} = 0, \quad k = 1,2,\ldots,n-m-1.$$

Let

$$v_{(m+1)k} = \frac{d^k v_{(m+1)}}{du_{(m+1)}^k}, \quad k = 1,2,\ldots,n-m-1.$$

In terms of the invariants $u_{(m+1)}, v_{(m+1)}, v_{(m+1)k}$, $k = 1,2,\ldots,n-m-1$, of $X_{m+1}^{(n)}$ (which are also invariants of $X_1^{(n)}, X_2^{(n)},\ldots, X_m^{(n)}$), ODE (3.190) and, hence, ODE (3.181) reduce to the $(n-m-1)$th-order ODE

$$v_{(m+1)n-m-1} = F_{n-m-1}(u_{(m+1)}, v_{(m+1)}, v_{(m+1)1},\ldots, v_{(m+1)n-m-2}), \tag{3.191}$$

for some function F_{n-m-1} of the indicated invariants of X_{m+1}^{n}.

Finally, two cases are distinguished:

Case I. $3 \le r < n$.
Here ODE (3.181) reduces to an $(n-r)$th-order ODE

$$v_{(r)n-r} = F_{n-r}(u_{(r)}, v_{(r)}, v_{(r)1},\ldots, v_{(r)n-r-1}), \tag{3.192}$$

for some function F_{n-r} of the invariants of $X_r^{(n)}$, plus r quadratures. The quadratures arise as follows:

Suppose

$$v_{(r)} = \phi_r(u_{(r)}; C_1, C_2,\ldots, C_{n-r})$$

is the general solution of ODE (3.192). Then the first-order ODE

$$v_{(r)}(u_{(r-1)}, v_{(r-1)}, v_{(r-1)1}) = \phi_r(u_{(r)}(u_{(r-1)}, v_{(r-1)}); C_1, C_2,\ldots, C_{n-r})$$

admits

$$\mathrm{X}_r^{(r-1)} = \alpha_{r-1}(u_{(r-1)})\frac{\partial}{\partial u_{(r-1)}} + \beta_{r-1}(u_{(r-1)}, v_{(r-1)})\frac{\partial}{\partial v_{(r-1)}},$$

which leads to a quadrature

$$v_{(r-1)} = \phi_{r-1}(u_{(r-1)}; C_1, C_2, \ldots, C_{n-r+1})$$

for some function ϕ_{r-1} of the indicated arguments. Continuing inductively, assume we have obtained

$$v_{(k)} = \phi_k(u_{(k)}; C_1, C_2, \ldots, C_{n-k}).$$

Then the first-order ODE

$$v_{(k)}(u_{(k-1)}, v_{(k-1)}, v_{(k-1)1}) = \phi_k(u_{(k)}(u_{(k-1)}, v_{(k-1)}); C_1, C_2, \ldots, C_{n-k})$$

admits

$$\mathrm{X}_k^{(k-1)} = \alpha_{k-1}(u_{(k-1)})\frac{\partial}{\partial u_{(k-1)}} + \beta_{k-1}(u_{(k-1)}, v_{(k-1)})\frac{\partial}{\partial v_{(k-1)}}, \quad k = r, r-1, \ldots, 1 \quad [v_{(0)} = y],$$

which leads to the quadratures

$$v_{(k-1)} = \phi_{k-1}(u_{(k-1)}; C_1, C_2, \ldots, C_{n-k+1}), \quad k = r, r-1, \ldots, 1 \quad [v_{(0)} = y]$$

for some functions ϕ_{k-1} of the indicated arguments.

Case II. $3 \le r = n$.
Here ODE (3.181) reduces to a first-order ODE

$$v_{(n-1)1} = F_1(u_{(n-1)}, v_{(n-1)}) \tag{3.193}$$

for some function F_1 of the invariants $u_{(n-1)}, v_{(n-1)}$ of $\mathrm{X}_{n-1}^{(n)}$, plus $n-1$ quadratures that are obtained as demonstrated for Case I. The first-order ODE (3.193) reduces to a quadrature since ODE (3.193) admits

$$\mathrm{X}_n^{(n-1)} = \alpha_{n-1}(u_{(n-1)})\frac{\partial}{\partial u_{(n-1)}} + \beta_{n-1}(u_{(n-1)}, v_{(n-1)})\frac{\partial}{\partial v_{(n-1)}}.$$

Thus, the solution of ODE (3.175) is reduced to n quadratures.

Note that in reducing an nth-order ODE to an $(n-r)$th-order ODE plus r quadratures, from its invariance under an r-parameter Lie group of point transformations whose infinitesimal generators form an r-dimensional solvable Lie algebra, *one does not need to determine the intermediate ODEs of orders* $n-1, n-2, \ldots, n-r+2$. *Moreover, in Case* I, *one does not need to determine the intermediate ODE of order* $n-r+1$.

As an example, consider the fourth-order ODE

$$\left[yy'\left(\frac{y}{y'}\right)''\right]' = 0, \tag{3.194}$$

that arises when studying the invariance properties of the wave equation in an inhomogeneous medium [Bluman and Kumei (1987, 1988)]. ODE (3.194) obviously admits the three-parameter (a, b, c) Lie group of point transformations

$$x^* = ax + b, \tag{3.195a}$$

$$y^* = cy. \tag{3.195b}$$

The corresponding infinitesimal generators $\mathrm{X}_1 = \frac{\partial}{\partial x}$ (parameter b), $\mathrm{X}_2 = x\frac{\partial}{\partial x}$ (parameter a), and $\mathrm{X}_3 = y\frac{\partial}{\partial y}$ (parameter c) satisfy

$$[\mathrm{X}_1, \mathrm{X}_2] = \mathrm{X}_1, \quad [\mathrm{X}_2, \mathrm{X}_3] = 0, \quad [\mathrm{X}_1, \mathrm{X}_3] = 0,$$

which are commutation relations of the form (3.179). To carry out the reduction algorithm, we first need the following extended infinitesimal generators:

$$\mathrm{X}_1^{(1)} = \frac{\partial}{\partial x}, \quad \mathrm{X}_2^{(1)} = x\frac{\partial}{\partial x} - y_1\frac{\partial}{\partial y_1}, \quad \mathrm{X}_2^{(2)} = x\frac{\partial}{\partial x} - y_1\frac{\partial}{\partial y_1} - 2y_2\frac{\partial}{\partial y_2},$$

$$\mathrm{X}_3^{(1)} = y\frac{\partial}{\partial y} + y_1\frac{\partial}{\partial y_1}, \quad \mathrm{X}_3^{(2)} = y\frac{\partial}{\partial y} + y_1\frac{\partial}{\partial y_1} + y_2\frac{\partial}{\partial y_2},$$

$$\mathrm{X}_3^{(3)} = y\frac{\partial}{\partial y} + y_1\frac{\partial}{\partial y_1} + y_2\frac{\partial}{\partial y_2} + y_3\frac{\partial}{\partial y_3}.$$

From

$$\mathrm{X}_1 u_{(1)} = 0, \quad \mathrm{X}_1^{(1)} v_{(1)} = 0, \quad v_{(1)1} = \frac{dv_{(1)}}{du_{(1)}},$$

we get

$$u_{(1)} = y, \quad v_{(1)} = y_1, \quad v_{(1)1} = \frac{y_2}{y_1}.$$

Then

$$\alpha_1(u_{(1)}) = \mathrm{X}_2 u_{(1)} = 0, \quad \beta_1(u_{(1)}, v_{(1)}) = \mathrm{X}_2^{(1)} v_{(1)} = -v_{(1)},$$

$$\gamma_1(u_{(1)}, v_{(1)}, v_{(1)1}) = \mathrm{X}_2^{(2)} v_{(1)1} = -\frac{y_2}{y_1} = -v_{(1)1}.$$

Thus, in terms of $u_{(1)}, v_{(1)}, v_{(1)1}$, we have

$$\mathrm{X}_2^{(1)} = -u_{(1)}\frac{\partial}{\partial u_{(1)}}, \quad \mathrm{X}_2^{(2)} = -u_{(1)}\frac{\partial}{\partial u_{(1)}} - u_{(1)1}\frac{\partial}{\partial u_{(1)1}}.$$

Now, from

$$\mathrm{X}_2^{(1)}u_{(2)} = 0, \quad \mathrm{X}_2^{(2)}v_{(2)} = 0, \quad v_{(2)1} = \frac{dv_{(2)}}{du_{(2)}},$$

we find that

$$u_{(2)} = u_{(1)} = y, \quad v_{(2)} = \frac{v_{(1)1}}{v_{(1)}} = \frac{y_2}{(y_1)^2}, \quad v_{(2)1} = \frac{y_1y_3 - 2(y_2)^2}{(y_1)^4}.$$

Then

$$\alpha_2(u_{(2)}) = \mathrm{X}_3^{(1)}u_{(2)} = y = u_{(2)}, \quad \beta_2(u_{(2)}, v_{(2)}) = \mathrm{X}_3^{(2)}v_{(2)} = -\frac{y_2}{(y_1)^2} = -v_{(2)},$$

$$\gamma_2(u_{(2)}, v_{(2)}, v_{(2)1}) = \mathrm{X}_3^{(3)}v_{(2)1} = \frac{4(y_2)^2 - 2y_1y_3}{(y_1)^4} = -2v_{(2)1}.$$

Thus, in terms of $u_{(2)}, v_{(2)}, v_{(2)1}$, we have

$$\mathrm{X}_3^{(2)} = u_{(2)}\frac{\partial}{\partial u_{(2)}} - v_{(2)}\frac{\partial}{\partial v_{(2)}}, \quad \mathrm{X}_3^{(3)} = u_{(2)}\frac{\partial}{\partial u_{(2)}} - v_{(2)}\frac{\partial}{\partial v_{(2)}} - 2v_{(2)1}\frac{\partial}{\partial v_{(2)1}}.$$

Now, from $\mathrm{X}_3^{(2)}u_{(3)} = 0$, $\mathrm{X}_3^{(3)}v_{(3)} = 0$, we get

$$u_{(3)} = u_{(2)}v_{(2)} = \frac{yy_2}{(y_1)^2}, \quad v_{(3)} = (u_{(2)})^2 v_{(2)1} = \frac{y^2[y_1y_3 - 2(y_2)^2]}{(y_1)^4}. \tag{3.196}$$

It must now follow that ODE (3.194) reduces to

$$\frac{dv_{(3)}}{du_{(3)}} = F(u_{(3)}, v_{(3)}),$$

for some function F of $u_{(3)}, v_{(3)}$. We now find $F(u_{(3)}, v_{(3)})$. We have

$$\frac{dv_{(3)}}{du_{(3)}} = \frac{y}{(y_1)^2}\left[\frac{y(y_1)^2y_4 - 7yy_1y_2y_3 + 2(y_1)^3y_3 - 4(y_1)^2(y_2)^2 + 8y(y_2)^3}{(y_1)^2y_2 + yy_1y_3 - 2y(y_2)^2}\right]. \tag{3.197}$$

ODE (3.194) can be expressed as

$$y^2(y_1)^2y_4 = 4y(y_1)^2(y_2)^2 + 5y^2y_1y_2y_3 - (y_1)^4y_2 - 3y(y_1)^3y_3 - 4y^2(y_2)^3. \tag{3.198}$$

Replacing $y^2(y_1)^2y_4$ in ODE (3.197) through substitution of (3.198), we obtain

$$\frac{dv_{(3)}}{du_{(3)}} = -\frac{(y_1)^2 + 2yy_2}{(y_1)^2} = -(1 + 2u_{(3)}), \tag{3.199}$$

which *fortunately* reduces to the quadrature

$$v_{(3)} = -[u_{(3)} + (u_{(3)})^2 + c_1]. \tag{3.200}$$

After (3.196) is substituted for $v_{(3)}$ and $u_{(3)}$, the quadrature (3.200) becomes

$$(u_{(2)})^2 v_{(2)1} = -[u_{(2)}v_{(2)} + (u_{(2)}v_{(2)})^2 + c_1]. \tag{3.201}$$

ODE (3.201) admits

$$X_3^{(2)} = u_{(2)}\frac{\partial}{\partial u_{(2)}} - v_{(2)}\frac{\partial}{\partial v_{(2)}},$$

with corresponding canonical variables given by

$$R = u_{(2)}v_{(2)}, \quad S = \log v_{(2)}.$$

Then (3.201) transforms to

$$\frac{dS}{dR} = \frac{1}{R} + \frac{1}{R^2 - c_1}. \tag{3.202}$$

Consider the case when the constant $c_1 > 0$, and let $c_1 = (C_1)^2$. Then

$$S = \log R + \log\left(\frac{R - C_1}{R + C_1}\right)^{1/(2C_1)} + c_2,$$

with arbitrary constant c_2. Consequently,

$$v_{(2)} = \phi(u_{(2)}; C_1, C_2) = \frac{C_1}{u_{(2)}}\left(\frac{1 + A(u_{(2)})}{1 - A(u_{(2)})}\right), \tag{3.203}$$

with $A(u_{(2)}) = (C_2 / u_{(2)})^{2C_1}$ in terms of arbitrary constants C_1, C_2. Then the first-order ODE resulting from (3.203), namely,

$$\frac{v_{(1)1}}{v_{(1)}} = \phi(u_{(1)}; C_1, C_2), \tag{3.204}$$

admits $X_2^{(1)} = -v_{(1)}\dfrac{\partial}{\partial v_{(1)}}$. Hence, ODE (3.204) reduces to

$$\frac{dv_{(1)}}{v_{(1)}} = \phi(u_{(1)}; C_1, C_2)\, du_{(1)},$$

which integrates out to

$$v_{(1)} = \psi(y; C_1, C_2, C_3) = C_3 \exp\left[\int^y \phi(u_{(1)}; C_1, C_2)\, du_{(1)}\right]. \tag{3.205}$$

Finally, the first-order ODE

$$y_1 = \frac{dy}{dx} = \psi(y; C_1, C_2, C_3)$$

admits $X_1 = \frac{\partial}{\partial x}$ and thus reduces to the quadrature

$$\int^y \frac{dy}{\psi(y; C_1, C_2, C_3)} = x + C_4,$$

yielding a general solution of ODE (3.194). The case $c_1 = -(C_1)^2$, substituted into (3.202), would yield another general solution of ODE (3.194) in solved form.

In using the reduction algorithm to reduce an nth-order ODE to an $(n-r)$th-order ODE plus r quadratures, from invariance of the nth-order ODE under an r-parameter Lie group of point transformations whose infinitesimal generators form an r-dimensional solvable Lie algebra, we determine, iteratively, invariants $u_{(i)}, v_{(i)}, v_{(i)1}$ and coefficients $\alpha_i(u_{(i)}), \beta_i(u_{(i)}, v_{(i)}), \gamma_i(u_{(i)}, v_{(i)}, v_{(i)1})$ such that

$$\begin{aligned}(u_{(1)}, v_{(1)}, v_{(1)1}) &\to (\alpha_1, \beta_1, \gamma_1) \to (u_{(2)}, v_{(2)}, v_{(2)1}) \to (\alpha_2, \beta_2, \gamma_2) \to \cdots \\ &\to (u_{(r-1)}, v_{(r-1)}, v_{(r-1)1}) \to (\alpha_{r-1}, \beta_{r-1}, \gamma_{r-1}) \to (u_{(r)}, v_{(r)}).\end{aligned}$$

The nth-order ODE reduces to an $(n-r)$th-order ODE in terms of the variables $u_{(r)}, v_{(r)}$. The quadratures follow from reversing the arrows of this iterative procedure.

3.4.4 INVARIANCE OF AN OVERDETERMINED SYSTEM OF ODEs UNDER AN r-PARAMETER LIE GROUP WITH A SOLVABLE LIE ALGEBRA

Now consider an overdetermined system of two ODEs of orders m and n given by

$$f(x, y, y', \ldots, y^{(m)}) = 0, \tag{3.206a}$$

$$g(x, y, y', \ldots, y^{(n)}) = 0, \tag{3.206b}$$

$m \le n$. We assume that each of the ODEs (3.206a,b) admits the same r-parameter Lie group with a solvable Lie algebra $\mathcal{L}^r$, with a basis set of r infinitesimal generators satisfying commutation relations of the form (3.179). Then, from Section 3.4.3, there exist invariants $u_{(r)} = u(x, y, y', \ldots, y^{(r-1)})$, $v_{(r)} = v(x, y, y', \ldots, y^{(r)})$ of $\mathcal{L}^r$, such that ODEs (3.206a and b), respectively, reduce to the equivalent overdetermined system of equations

$$F\left(u, v, \frac{dv}{du}, \ldots, \frac{d^{m-r}v}{du^{m-r}}\right) = 0, \tag{3.207a}$$

$$G\left(u, v, \frac{dv}{du}, \ldots, \frac{d^{n-r}v}{du^{n-r}}\right) = 0, \tag{3.207b}$$

for some functions F and G of the indicated invariants.

Now suppose $v = \Phi(u)$ solves the system of equations (3.207a,b). Then any solution of the ODE

$$v(x, y, y', \ldots, y^{(r)}) = \Phi(u(x, y, y', \ldots, y^{(r-1)})) \tag{3.208}$$

solves the given system of ODEs (3.206a,b). If $\Phi(u)$ depends on $m-r$ essential constants, then we obtain a general solution of the system of ODEs (3.206a,b). Since u and v are invariants of $\mathcal{L}^r$, it follows that ODE (3.208) is also invariant under $\mathcal{L}^r$. Hence, in principle, ODE (3.208) can be reduced constructively to r quadratures, with r essential constants $c_1, \ldots, c_r$. Thus, we obtain a function $\Psi(x, y; c_1, \ldots, c_r)$ for which the equation

$$\Psi(x, y; c_1, \ldots, c_r) = 0 \tag{3.209}$$

yields an explicit solution of the overdetermined system of ODEs (3.206a,b).

The previous considerations have assumed that solutions of the system of ODEs (3.206a,b) satisfy

$$\frac{dv}{du} = \frac{(v_x + v_y y' + \cdots + v_{y^{(r-1)}} y^{(r)})}{(u_x + u_y y' + \cdots + u_{y^{(r-1)}} y^{(r)})} \neq 0.$$

Alternatively, solutions of the system of ODEs (3.206a,b) can be obtained by considering

$$u = A, \tag{3.210a}$$

$$v = B, \tag{3.210b}$$

for some constants A and B. From (3.210a,b), a solution of the system of ODEs (3.206a,b) satisfies the system of ODEs

$$u(x, y, y', \ldots, y^{(r-1)}) = A, \tag{3.211a}$$

$$v(x, y, y', \ldots, y^{(r)}) = B. \tag{3.211b}$$

Since the system of ODEs (3.211a,b) is invariant under the infinitesimal generators of $\mathcal{L}^r$, it follows that the solution of (3.211a,b) reduces constructively to quadratures. Thus, one can find all solutions of the system of ODEs (3.211a,b).

Most important, the considerations outlined above allow one to generate solutions of the system of ODEs (3.206a,b) without first determining the general solution of either of these ODEs.

As an example, consider the following system of ODEs that arises when studying the invariance properties of the wave equation in an inhomogeneous medium [Bluman and Kumei (1987, 1988)] with wave speed $y(x)$:

$$[yy'(y(y')^{-1})'']' = 0, \tag{3.212a}$$

$$\left\{y^2\left[\frac{(y^{-1}y')'''}{2(y^{-1}y')'+(y^{-1}y')^2}+3\frac{[2(y^{-1}y')^3-2(y^{-1}y')(y^{-1}y')'(y^{-1}y')''-((y^{-1}y')'')^2]}{[2(y^{-1}y')'+(y^{-1}y')^2]^2}\right]\right\}'=0. \tag{3.212b}$$

Recall that the fourth-order ODE (3.212a) [cf. (3.194)] admits the three-parameter Lie group of transformations (3.195a,b) with infinitesimal generators $X_1 = \frac{\partial}{\partial x}$, $X_2 = x\frac{\partial}{\partial x}$, and $X_3 = y\frac{\partial}{\partial y}$. The fifth-order ODE (3.212b) also admits this same group. Consequently, the Lie group of point transformations (3.195a,b) is a solvable three-parameter Lie group of point symmetries admitted by the overdetermined system of ODEs (3.212a,b). A convenient choice of differential invariants of this three-parameter group is given by

$$u = \frac{yy''}{(y')^2}, \tag{3.213a}$$

$$v = \frac{y^2y'''}{(y')^3}. \tag{3.213b}$$

Consequently, the ODEs (3.212a,b), respectively, reduce to the ODEs

$$(2u^2-u-v)\left(\frac{dv}{du}-2u+1\right)=0, \tag{3.214a}$$

and

$$2u\Gamma+\left(\Gamma_u+\Gamma_v\frac{dv}{du}+\Gamma_{v_1}\frac{d^2v}{du^2}\right)(u+v-2u^2)=0, \tag{3.214b}$$

with

$$\Gamma(u,v,v_1)=(1-u)+\frac{(u+v-2u^2)v_1+(2u^2-3uv-9u+6v+4)}{2u-1}+3\frac{(v-2u+1)^2}{(2u-1)^2},$$

where $v_1 = dv/du$.

From (3.214a), the following two cases arise naturally:

$$v = 2u^2 - u, \tag{3.215}$$

$$v = u^2 - u + \delta, \quad \delta = \text{const.} \tag{3.216}$$

We now determine separately the compatibility between (3.215) and (3.216) with (3.214b).

Case I. $v - 2u^2 - u.$
In this case, it is easy to check that (3.214b) is satisfied identically so that we have a solution of the system of ODEs (3.212a,b) defined by $v = \Phi_1(u) = 2u^2 - u$ which, in terms of the differential invariants (3.213a,b), yields the third-order ODE

$$\frac{y^2 y'''}{(y')^3} = 2\frac{y^2 (y'')^2}{(y')^4} - \frac{yy''}{(y')^2}. \tag{3.217}$$

The ODE (3.217) must admit the solvable group (3.195a,b). It is easy to show that ODE (3.217) can be expressed in the form $du/dx = 0$ and, hence, we have

$$u = \frac{yy''}{(y')^2} = \text{const} = \lambda.$$

From this equation, it is easy to show that the general solution of ODE (3.217) is given by

$$y = (c_1 + c_2 x)^{c_3}, \tag{3.218}$$

yielding a three-parameter family of solutions of the system of ODEs (3.212a,b).

Case II. $v = u^2 - u + \delta.$
In this case, the substitution of (3.216) into ODE (3.214b) leads to the compatibility equation

$$(u^2 - \delta)(1 - 4\delta) = 0. \tag{3.219}$$

We now set to zero each factor in (3.219). Clearly, the first factor again yields the solution (3.218). The second factor yields $\delta = 1/4$, so that now we have all other solutions of the system of ODEs (3.214a,b) as given by $v = \Phi_2(u) = u^2 - u + \frac{1}{4}$. In terms of the differential invariants (3.213a,b), this corresponds to all solutions of the ODE

$$\frac{y^2 y'''}{(y')^3} = \frac{y^2 (y'')^2}{(y')^4} - \frac{yy''}{(y')^2} + \frac{1}{4}. \tag{3.220}$$

It follows that the third-order ODE (3.220) must admit the solvable three-parameter Lie group (3.195a,b) and, hence, can be reduced completely to quadratures. As differential invariants of this three-parameter group, it is convenient to choose

$$U = y', \quad V = yy''. \tag{3.221}$$

Then ODE (3.220) reduces to

$$\frac{dV}{dU} = \frac{V}{U} + \frac{U^3}{4V}$$

whose general solution, from scaling invariance, is given by

$$V^2 = \tfrac{1}{4}U^4 + \alpha U^2, \qquad \alpha = \text{const.} \tag{3.222}$$

From the differential invariants (3.221), we see that $V = yU\,dU/dy$, and thus, (3.222) becomes the first-order ODE

$$y^2\left(\frac{dU}{dy}\right)^2 = \tfrac{1}{4}U^2 + \alpha, \tag{3.223}$$

which is invariant under scalings in y. In terms of $y' = U$, the integration of (3.223) yields two families of ODEs in terms of arbitrary constants β and $\gamma > 0$:

$$y' = \beta[(\gamma y)^{1/2} \pm (\gamma y)^{-1/2}]. \tag{3.224}$$

The solution of ODE (3.224) yields two further three-parameter families of solutions of the given system of ODEs (3.216a,b):

$$\sqrt{y} - \arctan(c_1\sqrt{y}) = c_2(x + c_3), \tag{3.225}$$

and

$$\sqrt{y} + \log\left|\frac{\sqrt{y} - c_1}{\sqrt{y} + c_1}\right|^{1/2} = c_2(x + c_3). \tag{3.226}$$

Hence, all solutions of the system of ODEs (3.214a,b) are given by (3.218), (3.225), and (3.226). It is left to Exercise 3.4-6 to find all solutions of the system of ODEs (3.214a,b) that satisfy the system of ODEs given by (3.217) and (3.220), i.e., to find all functions which lie in the intersection of the family of functions satisfying (3.218) and either (3.225) or (3.226).

Further details of the work in this section appear in Bluman and Kumei (1989a).

EXERCISES 3.4

1. For each of the following second-order ODEs, find an admitted two-parameter Lie group of point transformations and use appropriate differential invariants to find the general solution of the ODEs:
   (a) $y'' + Ay' + By + C = 0$ where A, B, C are constants;
   (b) $y'' + \dfrac{A}{x}y' + \dfrac{B}{x^2}y = 0$ where A, B are constants;

(c) $(x^2+y^2)y''+2(y-xy')(1+(y')^2)=0$;

(d) $yy''+(y')^2=1$; and

(e) $y''+\dfrac{y'}{y^2}-\dfrac{1}{xy}=0$.

2. For each of the following two parameter Lie groups of transformations whose Lie algebras are spanned by the set $\{\mathrm{X}_1, \mathrm{X}_2\}$, find all admitted second- and third-order ODEs:

   (a) $\mathrm{X}_1=\dfrac{\partial}{\partial x}, \mathrm{X}_2=\dfrac{\partial}{\partial y}$;

   (b) $\mathrm{X}_1=x\dfrac{\partial}{\partial x}, \mathrm{X}_2=y\dfrac{\partial}{\partial y}$;

   (c) $\mathrm{X}_1=x\dfrac{\partial}{\partial x}-y\dfrac{\partial}{\partial y}, \mathrm{X}_2=x\dfrac{\partial}{\partial x}+y\dfrac{\partial}{\partial y}$;

   (d) $\mathrm{X}_1=\dfrac{\partial}{\partial x}, \mathrm{X}_2=x\dfrac{\partial}{\partial x}$; and

   (e) $\mathrm{X}_1=\dfrac{\partial}{\partial x}, \mathrm{X}_2=x\dfrac{\partial}{\partial x}+y\dfrac{\partial}{\partial y}$.

3. Consider the fourth-order ODE (3.194), invariant under the three-parameter Lie group of point transformations (3.195a,b).

   (a) Show that if $\mathrm{X}_1=y\dfrac{\partial}{\partial y}, \mathrm{X}_2=\dfrac{\partial}{\partial x}, \mathrm{X}_3=x\dfrac{\partial}{\partial x}$, then commutation relations of the form (3.179) are satisfied. Accordingly, solve ODE (3.194) by the reduction algorithm.

   (b) What happens when trying the reduction algorithm with $\mathrm{X}_1=x\dfrac{\partial}{\partial x}, \mathrm{X}_2=y\dfrac{\partial}{\partial y}$, $\mathrm{X}_3=\dfrac{\partial}{\partial x}$?

4. Let $\phi_1(x), \phi_2(x), \ldots, \phi_n(x)$ be n linearly independent solutions of the nth-order linear homogeneous ODE

$$y^{(n)}+p_1(x)y^{(n-1)}+\cdots+p_{n-1}(x)y'+p_n(x)y=0.$$

   (a) Find an n-parameter Lie group of point transformations that is admitted by the nth-order linear nonhomogeneous ODE

$$y^{(n)}+p_1(x)y^{(n-1)}+\cdots+p_{n-1}(x)y'+p_n(x)y=g(x). \tag{3.227}$$

   Show that the corresponding Lie algebra is solvable.

   (b) Use group invariance to obtain the general solution of ODE (3.227). In particular, find the well-known formula obtained by the variation of parameters method.

5. The overdetermined system of second-order ODEs given by

$$x^2y'' + xy' - y = 0, \tag{3.228a}$$

$$yy'' - 2(y')^2 = 0, \tag{3.228b}$$

admits $X_1 = x\frac{\partial}{\partial x}, X_2 = y\frac{\partial}{\partial y}$. Find the general solution of the system of ODEs (3.228a,b) in three ways by using differential invariants corresponding to:
(a) X_1;
(b) X_2; and
(c) the set $\{X_1, X_2\}$.

6. Find the general solution of the overdetermined system consisting of the ODEs (3.217) and (3.220).

7. Show that if an r-dimensional Lie algebra is solvable, then one can find a spanning (basis) set of infinitesimal generators $\{X_1, X_2, \ldots, X_r\}$ so that commutation relations of the form (3.179) hold.

8. (a) Let a second-order ODE admit a two-parameter Lie group of transformations with infinitesimal generators X_1, X_2 such that $[X_1, X_2] = 0$, i.e., the Lie algebra is Abelian. Suppose "canonical coordinates" $R(x, y), S(x, y)$ can be found such that

$$X_1R = 1, \quad X_2R = 0, \quad X_1S = 0, \quad X_2S = 1. \tag{3.229}$$

Transform the given ODE to (R, S) coordinates and reduce it to two quadratures.
(b) Show that if the Lie algebra of a two-parameter Lie group of transformations acting on R^2 is Abelian, then it is does not necessarily follow that one can find "canonical coordinates" $R(x, y), S(x, y)$ satisfying relations (3.229). Explain this geometrically and give a specific example.

9. Find the most general second- and third-order ODEs admitting the three-parameter group with infinitesimal generators

$$X_1 = (1+x^2)\frac{\partial}{\partial x} + xy\frac{\partial}{\partial y}, \quad X_2 = xy\frac{\partial}{\partial x} + (1+y^2)\frac{\partial}{\partial y}, \quad X_3 = y\frac{\partial}{\partial x} - x\frac{\partial}{\partial y}. \tag{3.230}$$

Note that the corresponding Lie algebra is not solvable [cf. Exercise 2.5-13].

3.5 CONTACT SYMMETRIES AND HIGHER-ORDER SYMMETRIES

Now we consider the invariance of an nth-order ODE under contact transformations when $n \geq 2$, and under higher-order local transformations when $n \geq 3$ [cf. Section 2.7.2].

Definition 3.5-1. A second- or higher-order ODE

$$y^{(n)} = f(x, y, y', \ldots, y^{(n-1)}), \tag{3.231}$$

$n \ge 2$, admits a one-parameter group of local transformations

$$\begin{aligned} x^* &= x, \\ y^* &= y + \varepsilon\hat{\eta}(x, y, y', \ldots, y^{(\ell)}) + O(\varepsilon^2) \end{aligned} \tag{3.232}$$

with infinitesimal generator

$$\hat{\mathrm{X}} = \hat{\eta}(x, y, y', \ldots, y^{(\ell)})\frac{\partial}{\partial y} \tag{3.233}$$

if and only if (3.232) maps any solution $y = \Theta(x)$ of ODE (3.231) into solutions $y = \Theta^*(x) = (e^{\varepsilon\hat{\mathrm{X}}^{(\infty)}}y)\Big|_{y=\Theta(x)}$ of ODE (3.231), where $\hat{\mathrm{X}}^{(\infty)}$ is the extended infinitesimal generator given by (2.212).

In particular, the group (3.232) leaves invariant ODE (3.231) if and only if the curve $y = \Theta^*(x) = (e^{\varepsilon\hat{\mathrm{X}}^{(\infty)}}y)\Big|_{y=\Theta(x)}$ satisfies ODE (3.231) whenever a curve $y = \Theta(x)$ does. The highest order ℓ of the derivatives of y appearing in $\hat{\eta}$ is called the *order* of the local transformation (3.232).

Definition 3.5-2. A one-parameter group of local transformations (3.232) of order $0 \le \ell \le n-1$ admitted by ODE (3.231) is a *symmetry of order* ℓ of (3.231).

When $\ell = 1$, a local transformation (3.232) is a *point symmetry* of ODE (3.231) if $\hat{\eta}(x, y, y')$ is linear in y'. In particular,

$$\hat{\mathrm{X}} = [\eta(x, y) - \xi(x, y)y']\frac{\partial}{\partial y} \quad \text{and} \quad \mathrm{X} = \xi(x, y)\frac{\partial}{\partial x} + \eta(x, y)\frac{\partial}{\partial y}$$

have the same action on solutions $y = \Theta(x)$ of ODE (3.231). More generally, if $\hat{\eta}(x, y, y')$ is not linear in y', then a local transformation (3.232) for $\ell = 1$ is a *contact symmetry* of ODE (3.231). When $\ell \ge 2$, a local transformation (3.232) is a *higher-order symmetry* [cf. Section 2.7.2] of ODE (3.231). The infinitesimal $\hat{\eta}(x, y, y', \ldots, y^{(\ell)})$ of (3.232) is called the *symmetry characteristic*.

We will show how to find the contact symmetries and higher-order symmetries admitted by a given nth-order ODE, $n \ge 2$. As is the case for point symmetries of a first-order ODE, when $\ell = n-1$, the infinitesimal $\hat{\eta}(x, y, y', \ldots, y^{(n-1)})$ for a symmetry of order $n-1$ satisfies a linear homogeneous PDE whose general solution cannot be found unless we know the general solution of the ODE itself. However, when $\ell < n-1$, the determining equation for the infinitesimal $\hat{\eta}(x, y, y', \ldots, y^{(\ell)})$, in general, splits into an overdetermined system of linear homogeneous PDEs which has only a finite number of linearly independent solutions.

Most important, we will give a reformulation of the differential invariant method for reduction of order [cf. Section 3.3.2] that allows an extension of this method to include contact and higher-order symmetries.

3.5.1 DETERMINING EQUATIONS FOR CONTACT SYMMETRIES AND HIGHER-ORDER SYMMETRIES

We start by giving the infinitesimal criterion for the invariance of an nth-order ODE (3.231) under a one-parameter group of local transformations (3.232) with infinitesimal generator (3.233) of order $0 \le \ell \le n-1$.

Geometrically, an infinitesimal generator (3.233) acting on the space of solutions of a given ODE (3.231) corresponds to a vector field $\hat{\mathbf{X}}^{(n)}$ tangent to the surface defined by (3.231) in $(x, y, y_1, \ldots, y_n)$ – space,

$$y_n = f(x, y, y_1, \ldots, y_{n-1}), \tag{3.234}$$

given by

$$\begin{aligned}\hat{\mathbf{X}}^{(n)} &= \hat{\eta}\frac{\partial}{\partial y} + \mathrm{D}\hat{\eta}\frac{\partial}{\partial y_1} + \cdots + \mathrm{D}^n\hat{\eta}\frac{\partial}{\partial y_n} \quad \text{on } y_n = f \\ &= \hat{\eta}\frac{\partial}{\partial y} + \mathbf{D}\hat{\eta}\frac{\partial}{\partial y_1} + \cdots + \mathbf{D}^n\hat{\eta}\frac{\partial}{\partial y_n},\end{aligned} \tag{3.235}$$

where

$$\hat{\eta} = \hat{\eta}(x, y, y_1, \ldots, y_\ell), \quad \mathrm{D} = \frac{\partial}{\partial x} + \sum_{i=1}^{\infty} y_i \frac{\partial}{\partial y_{i-1}} \quad \text{with } y_0 = y,$$

and

$$\mathbf{D} = \mathrm{D}\,|_{y_n = f} = \frac{\partial}{\partial x} + y_1\frac{\partial}{\partial y} + \cdots + y_{n-1}\frac{\partial}{\partial y_{n-2}} + f(x, y, y_1, \ldots, y_{n-1})\frac{\partial}{\partial y_{n-1}} \tag{3.236}$$

is the total derivative operator associated with the surface (3.234). Since (3.235) is a tangent vector field, it is an infinitesimal generator of a one-parameter Lie group of local transformations on the surface (3.234) given by

$$x^* = x, \tag{3.237a}$$

$$y^* = e^{\varepsilon\hat{\mathbf{X}}^{(n)}} y, \tag{3.237b}$$

$$y^*_j = e^{\varepsilon\hat{\mathbf{X}}^{(n)}} y_j, \quad j = 1, 2, \ldots, n-1, \tag{3.237c}$$

$$y^*_n = e^{\varepsilon\hat{\mathbf{X}}^{(n)}} y_n = f(x, y^*, y^*_1, \ldots, y^*_{n-1}). \tag{3.237d}$$

Solutions $y=\Theta(x)$ of ODE (3.231) satisfy $\Theta^{(n)}(x)=f(x,\Theta(x),\Theta'(x),\ldots,\Theta^{(n-1)}(x))$ and thus represent curves lying on the surface (3.234) with $y=\Theta(x)$, $y_j=\Theta^{(j)}(x)$, $j=1,2,\ldots,n$. If a solution curve $y=\Theta(x)$ of ODE (3.231) is not itself invariant under the transformation group (3.237a–d), then it is mapped into another solution curve $y=\phi(x;\varepsilon)=(e^{\varepsilon\hat{X}^{(n)}}y)\big|_{y=\Theta(x)}$ of ODE (3.231) for any value of ε.

Definition 3.5.1-1. An nth-order ODE (3.231) is invariant under a one-parameter group of local transformations (3.232) if and only if the corresponding surface (3.234) in $(x,y,y_1,\ldots,y_n)$-space admits the one-parameter Lie group of transformations (3.237a–d).

The following theorem results from Definitions 3.5-2 and 3.5.1-1 and Theorem 2.6.7-1 on the infinitesimal criterion for an invariant surface:

Theorem 3.5.1-1 (Infinitesimal Criterion for Invariance of an ODE). *A vector field* (3.235) *is the infinitesimal generator of a symmetry of order* $\ell<n$ *admitted by the nth-order ODE* (3.231) *if and only if*

$$\hat{\mathbf{X}}^{(n)}(y_n-f)=0, \tag{3.238}$$

or, equivalently,

$$\mathbf{D}^n\hat{\eta}=f_y\hat{\eta}+f_{y_1}\mathbf{D}\hat{\eta}+\cdots+f_{y_{n-1}}\mathbf{D}^{n-1}\hat{\eta}. \tag{3.239}$$

Equation (3.239) is called the *symmetry determining equation* for ODE (3.231), and its solutions $\hat{\eta}(x,y,y_1,\ldots,y_\ell)$ are the symmetry characteristics up to order $0\le\ell\le n-1$ admitted by the nth-order ODE (3.231) or, equivalently, the surface (3.234).

Since for any nth-order ODE, we can assign arbitrary values to each of $y,y_1,\ldots,y_{n-1}$ at any value of x, it follows that the symmetry determining equation (3.239) is a linear homogeneous PDE for $\hat{\eta}$ with independent variables $x,y,y_1,\ldots,y_{n-1}$. If $\ell=n-1$, then the general solution of (3.239) cannot be found unless one knows the general solution of the given ODE (3.231). However, if $\ell<n-1$, then the symmetry determining equation (3.239) reduces to an overdetermined system of linear homogeneous PDEs which has a finite number of linearly independent solutions for $\hat{\eta}$. There exist efficient computational algorithms to solve such systems [cf. Head (1992); Hereman (1996); Wolf (2002a)] and, typically, the solutions here can be found explicitly although the computational complexity grows rapidly as the order ℓ increases.

If a given nth-order ODE (3.231) admits a point symmetry, then one can make a simplifying ansatz to solve the symmetry determining equation (3.239) for higher-order symmetries. In particular, if ODE (3.231) admits a scaling symmetry $x\to\alpha^q x$, $y\to\alpha^p y$, then the symmetry determining equation (3.239) admits the scaling symmetry $\hat{\eta}\to\alpha^r\hat{\eta}$, $x\to\alpha^q x$, $y\to\alpha^p y$, $y_1\to\alpha^{p-q}y_1,\ldots,y_{n-1}\to\alpha^{p-(n-1)q}y_{n-1}$, for arbitrary r = const. As a consequence, one can seek invariant solutions of the symmetry determining equation (3.239) of the form

$$\hat{\eta} = x^r g\left(\frac{y^q}{x^p}, \frac{(y_1)^q}{x^{p-q}}, \ldots, \frac{(y_{n-1})^q}{x^{p-(n-1)q}}\right).$$

Hence, (3.239) reduces to an overdetermined linear system of PDEs in terms of the scaling invariant variables $y^q/x^p, (y_1)^q/x^{p-1}, \ldots, (y_{n-1})^q/x^{p-(n-1)q}$. Similarly, if ODE (3.231) admits a translation symmetry $x \to x+\varepsilon, y \to y$, or $x \to x, y \to y+\varepsilon$, then the symmetry determining equation (3.239) admits the translation symmetry $x \to x+\varepsilon$, $y \to y$, or $x \to x, y \to y+\varepsilon$, $y_1 \to y_1, \ldots, y_{n-1} \to y_{n-1}$ together with the scaling symmetry $\hat{\eta} \to \alpha^r \hat{\eta}$ for arbitrary r = const. Consequently, here one can seek invariant solutions of the symmetry determining equation (3.239) of the form

$$\hat{\eta} = e^{rx} g(y, y_1, \ldots, y_{n-1}) \quad \text{or} \quad \hat{\eta} = e^{ry} g(x, y_1, \ldots, y_{n-1}),$$

leading to a reduction of (3.239) to an overdetermined linear system of PDEs in terms of the translation invariant variables $y, y_1, \ldots, y_{n-1}$, or $x, y_1, \ldots, y_{n-1}$.

More generally, if one can find a point symmetry $\widetilde{\mathrm{X}}$ of the *symmetry determining equation* (3.239) (which need not necessarily be a point symmetry admitted by the given ODE (3.231)), then one can consider a corresponding ansatz $\widetilde{\mathrm{X}}\hat{\eta} = r\hat{\eta}$, for arbitrary r = const, to seek invariant solutions $\hat{\eta}$ in terms of the corresponding n invariants $u_1, \ldots, u_n$ determined by solving $\widetilde{\mathrm{X}}u_i(x, y, y_1, \ldots, y_{n-1}) = 0$ through use of the admitted point symmetry

$$\widetilde{\mathrm{X}} = \tilde{\xi}(x, y, y_1, \ldots, y_{n-1})\frac{\partial}{\partial x} + \tilde{\eta}(x, y, y_1, \ldots, y_{n-1})\frac{\partial}{\partial y} + \tilde{\eta}_1(x, y, y_1, \ldots, y_{n-1})\frac{\partial}{\partial y_1}$$
$$+ \cdots + \tilde{\eta}_{n-1}(x, y, y_1, \ldots, y_{n-1})\frac{\partial}{\partial y_{n-1}}.$$

Through use of canonical coordinates [cf. Section 3.2.5], one can show that the point symmetries admitted by the symmetry determining equation (3.239) include all point symmetries admitted by the given ODE (3.231). The use of the ansatz $\widetilde{\mathrm{X}}\hat{\eta} = r\hat{\eta}$ is illustrated in the examples in Section 3.5.2.

3.5.2 EXAMPLES OF CONTACT SYMMETRIES AND HIGHER-ORDER SYMMETRIES

(1) *Contact Symmetries*

As a first example, consider the elementary third-order linear ODE

$$y''' = 0, \tag{3.240}$$

corresponding to the surface $y_3 = 0$ in (x, y, y_1, y_2, y_3) – space. The symmetry determining equation (3.239) for contact symmetries $\hat{\eta}(x, y, y_1)$ admitted by ODE (3.240) is given by

$$\begin{aligned}\mathbf{D}^3\hat{\eta} = \hat{\eta}_{xxx} + 3y_1\hat{\eta}_{xxy} + 3(y_1)^2\hat{\eta}_{xyy} + (y_1)^3\hat{\eta}_{yyy} \\ + 3[\hat{\eta}_{xxy_1} + 2y_1\hat{\eta}_{xyy_1} + (y_1)^2\hat{\eta}_{yyy_1} + \hat{\eta}_{xy} + y_1\hat{\eta}_{yy}]y_2 \\ + 3[\hat{\eta}_{xy_1y_1} + y_1\hat{\eta}_{yy_1y_1} + \hat{\eta}_{yy_1}](y_2)^2 + \hat{\eta}_{y_1y_1y_1}(y_2)^3 = 0,\end{aligned} \tag{3.241}$$

where

$$\mathbf{D} = \frac{\partial}{\partial x} + y_1\frac{\partial}{\partial y} + y_2\frac{\partial}{\partial y_1}.$$

With respect to the coefficients of like powers of y_2, PDE (3.241) splits into the overdetermined linear system

$$\hat{\eta}_{xxx} + 3y_1\hat{\eta}_{xxy} + 3(y_1)^2\hat{\eta}_{xyy} + (y_1)^3\hat{\eta}_{yyy} = 0, \tag{3.242a}$$

$$\hat{\eta}_{xxy_1} + 2y_1\hat{\eta}_{xyy_1} + (y_1)^2\hat{\eta}_{yyy_1} + \hat{\eta}_{xy} + y_1\hat{\eta}_{yy} = 0, \tag{3.242b}$$

$$\hat{\eta}_{xy_1y_1} + y_1\hat{\eta}_{yy_1y_1} + \hat{\eta}_{yy_1} = 0, \tag{3.242c}$$

$$\hat{\eta}_{y_1y_1y_1} = 0. \tag{3.242d}$$

It is straightforward to solve (3.242a–d) to obtain the three admitted contact symmetries

$$\hat{\eta}_1 = (y_1)^2, \quad \hat{\eta}_2 = (2y - xy_1)y_1, \quad \hat{\eta}_3 = (2y - xy_1)^2, \tag{3.243}$$

in addition to the seven admitted point symmetries

$$\hat{\eta}_4 = 1, \quad \hat{\eta}_5 = x, \quad \hat{\eta}_6 = x^2, \quad \hat{\eta}_7 = y, \quad \hat{\eta}_8 = y_1, \quad \hat{\eta}_9 = xy_1, \quad \hat{\eta}_{10} = 2xy - x^2y_1. \tag{3.244}$$

As a second example, consider the third-order nonlinear ODE

$$y''' = 6x\frac{(y'')^3}{(y')^2} + 6\frac{(y'')^2}{y'}, \tag{3.245}$$

or, equivalently, the surface $y_3 = 6x(y_2)^3(y_1)^{-2} + 6(y_2)^2(y_1)^{-1}$ in (x, y, y_1, y_2, y_3) – space, which admits the scaling symmetries $x \to \alpha x,\ y \to \beta y$, and the translation symmetry $x \to x,\ y \to y + \varepsilon$. For ODE (3.245), the contact symmetry determining equation (3.239) for $\hat{\eta}(x, y, y_1)$ becomes

$$\mathbf{D}^3\hat{\eta} - 18x(y_2)^2(y_1)^{-2}\mathbf{D}^2\hat{\eta} + 12x(y_2)^3(y_1)^{-3}\mathbf{D}\hat{\eta} - 12y_2(y_1)^{-1}\mathbf{D}^2\hat{\eta} + 6(y_2)^2(y_1)^{-2}\mathbf{D}\hat{\eta} = 0 \tag{3.246}$$

with $\hat{\eta}_{y_1y_1} \neq 0$, where

$$\mathbf{D}=\frac{\partial}{\partial x}+y_1\frac{\partial}{\partial y}+y_2\frac{\partial}{\partial y_1}+6(xy_2+y_1)(y_2)^2(y_1)^{-2}\frac{\partial}{\partial y_2}.$$

One can see that (3.246) is a cubic polynomial in y_2 and hence it splits into a linear system of four third-order PDEs. Rather than look for the general solution directly, we seek solutions by using ansatzes based on the invariants of the point symmetries admitted by the given ODE (3.245).

First, since (3.245) admits translations in y, we seek solutions of the contact symmetry determining equation (3.246) of the form

$$\hat{\eta}=e^{py}g(x,y_1), \qquad p=\text{const}, \tag{3.247}$$

with $g_{y_1y_1}\neq 0$. For simplicity, we initially choose $p=0$. Substitution of (3.247) into (3.246) leads to the linear system of PDEs

$$g_{xxx}=0, \tag{3.248a}$$

$$y_1 g_{xxy_1}-4g_{xx}=0, \tag{3.248b}$$

$$2g_x-6xg_{xx}-2y_1g_{xy_1}+(y_1)^2g_{xy_1y_1}=0, \tag{3.248c}$$

$$12xg_x-18xy_1g_{xy_1}+6y_1g_{y_1}+6(y_1)^2g_{y_1y_1}+(y_1)^3g_{y_1y_1y_1}=0. \tag{3.248d}$$

One can readily solve (3.248a–d) to obtain

$$g=c_1+c_2(y_1)^{-1}+c_3(y_1)^{-2}+c_4xy_1+c_5x(y_1)^2+c_6x^2(y_1)^4$$

with arbitrary constants c_i. It is not hard to show that the contact symmetry determining equation (3.246) has no solutions of the form (3.247) when $p\neq 0$, from the compatibility of the resulting linear system of PDEs for $g(x,y_1)$. Hence, the ansatz (3.247) yields the four admitted contact symmetries

$$\hat{\eta}_1=(y_1)^{-1}, \quad \hat{\eta}_2=(y_1)^{-2}, \quad \hat{\eta}_3=x(y_1)^2, \quad \hat{\eta}_4=x^2(y_1)^4. \tag{3.249}$$

Next, using the x and y scalings admitted by the given ODE (3.245), we seek solutions of (3.246) of the form

$$\hat{\eta}=x^q y^p g(z), \tag{3.250}$$

where $z=xy^{-1}y_1$, $p,q=\text{const}$, and $g''(z)\neq 0$. Substitution of (3.250) into the contact symmetry determining equation (3.246) yields an overdetermined linear system of four third-order ODEs for $g(z)$. The compatibility conditions for this overdetermined system yield the solutions

$$g=9z^2-12z+4 \ \ (q=0,\, p=2), \tag{3.251a}$$

$$g=3-2z^{-1} \ \ (q=1,\, p=0), \tag{3.251b}$$

$$g=3z^3-2z^2 \ \ (q=-1,\, p=3). \tag{3.251c}$$

Hence, the ansatz (3.250) yields three more contact symmetries admitted by ODE (3.245), given by

$$\hat{\eta}_5 = 9x^2(y_1)^2 - 12xyy_1 + 4y^2, \quad \hat{\eta}_6 = 3x - 2y(y_1)^{-1}, \quad \hat{\eta}_7 = 3x^2(y_1)^3 - 2xy(y_1)^2. \tag{3.252}$$

More generally, through analysis of the overdetermined linear system of PDEs arising from the contact symmetry determining equation (3.246) for ODE (3.245), one can show that *all* solutions $\hat{\eta}(x, y, y_1)$ of (3.246) are yielded by the ansatzes (3.247) and (3.250). Hence, the contact symmetries admitted by ODE (3.245) consist of (3.249) and (3.252).

As a final example, we again consider the Blasius equation

$$y''' + \tfrac{1}{2}yy'' = 0, \tag{3.253}$$

with the corresponding surface given by $y_3 = -\tfrac{1}{2}yy_2$ in $(x, y, y_1, y_2, y_3)-$space. The symmetry determining equation (3.239) for contact symmetries $\hat{\eta}(x, y, y_1)$ admitted by ODE (3.253) takes the form

$$\mathbf{D}^3\hat{\eta} + \tfrac{1}{2}y_2\hat{\eta} + y\mathbf{D}^2\hat{\eta} = 0, \tag{3.254}$$

where

$$\mathbf{D} = \frac{\partial}{\partial x} + y_1\frac{\partial}{\partial y} + y_2\frac{\partial}{\partial y_1} - \tfrac{1}{2}yy_2\frac{\partial}{\partial y_2}.$$

Hence, (3.254) is a cubic polynomial in y_2 and, thus, splits into an overdetermined linear system of four PDEs. The coefficient of $(y_2)^3$ in (3.254) yields

$$\hat{\eta}_{y_1y_1y_1} = 0, \tag{3.255}$$

and thus,

$$\hat{\eta} = \alpha(x, y) + \beta(x, y)y_1 + \gamma(x, y)(y_1)^2 \tag{3.256}$$

for some functions of α, β, γ of x, y. Consequently, (3.254) becomes a polynomial in y_1. Then the coefficients of $y_1(y_2)^2$, $(y_2)^2$, and $(y_1)^2 y_2$ in (3.254), respectively, yield

$$\gamma_y = 0, \quad \beta_y = \tfrac{2}{3}\gamma - 2\gamma_x, \quad \gamma = 12\beta_{yy}.$$

From these equations, it is easy to show that $\gamma = 0$, and hence $\hat{\eta}$ is at most linear in y_1. Thus, the Blasius equation (3.253) admits no contact symmetries.

(2) *Higher-Order Symmetries*

As an example, we consider the fourth-order ODE

$$y^{(4)} = \frac{4}{3}\frac{(y''')^2}{y''}, \tag{3.257}$$

or, equivalently, the surface $y_4 = \frac{4}{3}(y_3)^2(y_2)^{-1}$ in $(x, y, y_1, y_2, y_3, y_4)-$space, which admits the scaling symmetries $x \to \alpha x, y \to \beta y$ and the translation symmetries $x \to x + \varepsilon_1$, $y \to y + \varepsilon_2$. The symmetry determining equation (3.239) for second-order symmetries $\hat{\eta}(x, y, y_1, y_2)$ admitted by ODE (3.257) is given by

$$\mathbf{D}^4\hat{\eta} - \frac{8}{3}y_3(y_2)^{-1}\mathbf{D}^3\hat{\eta} + \frac{4}{3}(y_3)^2(y_2)^{-2}\mathbf{D}^2\hat{\eta} = 0, \tag{3.258}$$

where

$$\mathbf{D} = \frac{\partial}{\partial x} + y_1\frac{\partial}{\partial y} + y_2\frac{\partial}{\partial y_1} + y_3\frac{\partial}{\partial y_2} + \frac{4}{3}(y_3)^2(y_2)^{-1}\frac{\partial}{\partial y_3}.$$

PDE (3.258) is a fourth-degree polynomial in y_3. The coefficient of $(y_3)^4$ yields

$$(y_2)^3\hat{\eta}_{y_2y_2y_2y_2} + \frac{16}{3}(y_2)^2\hat{\eta}_{y_2y_2y_2} + \frac{44}{9}y_2\hat{\eta}_{y_2y_2} + \frac{8}{27}\hat{\eta}_{y_2} = 0, \tag{3.259}$$

which has the general solution

$$\hat{\eta} = (y_2)^{-1/3}\alpha(x, y, y_1) + (y_2)^{2/3}\beta(x, y, y_1) + (y_2)^{1/3}\gamma(x, y, y_1) + \kappa(x, y, y_1) \tag{3.260}$$

for some functions $\alpha, \beta, \gamma, \kappa$ of x, y, y_1. Then if we divide the symmetry determining equation (3.258) by a power of y_2, it splits with respect to powers of $(y_2)^{1/3}$ into separate equations. The coefficients of $(y_3)^3$ in these equations yield

$$y_1\alpha_y + \alpha_x = 0, \quad \beta_{y_1} = 0, \quad \gamma_{y_1} = 0, \tag{3.261}$$

and thus, $\alpha = \alpha(z, y_1)$, $\beta = \beta(x, y)$, $\gamma = \gamma(x, y)$, with $z = xy_1 - y$.

From the coefficients of the remaining powers of y_3, we find that

$$\alpha = a_0 + a_1y_1 + a_2z, \tag{3.262a}$$

$$\beta = b_0 + b_1x + b_2x^2 + b_3x^3, \tag{3.262b}$$

$$\gamma = c_0 + c_1x + c_2y + c_3xy + c_4x^2, \tag{3.262c}$$

$$\kappa = k_0 + k_1y + k_2x + k_3y_1 + k_4xy_1, \tag{3.262d}$$

for arbitrary constants a_i, b_i, c_i, k_i. Consequently, the fourth-order ODE (3.257) admits the 12 second-order symmetries given by

$$\hat{\eta}_1 = (y_2)^{-1/3}, \quad \hat{\eta}_2 = y_1(y_2)^{-1/3}, \quad \hat{\eta}_3 = (xy_1 - y)(y_2)^{-1/3}, \quad \hat{\eta}_4 = (y_2)^{2/3}, \quad \hat{\eta}_5 = x(y_2)^{2/3},$$
$$\hat{\eta}_6 = x^2(y_2)^{2/3}, \quad \hat{\eta}_7 = x^3(y_2)^{2/3}, \quad \hat{\eta}_8 = (y_2)^{1/3}, \quad \hat{\eta}_9 = x(y_2)^{1/3}, \quad \hat{\eta}_{10} = x^2(y_2)^{1/3},$$

$$\hat{\eta}_{11} = xy(y_2)^{1/3}, \quad \hat{\eta}_{12} = y(y_2)^{1/3}. \tag{3.263}$$

We note that ODE (3.257) also admits five point symmetries, corresponding to (3.362d), but no contact symmetries.

As a second example, we consider the fourth-order ODE

$$(yy'(y(y')^{-1})'')' = 0, \tag{3.264}$$

that arises in the study of the invariance properties of the wave equation with a wave speed $y(x)$. The surface represented by ODE (3.264) in $(x, y, y_1, y_2, y_3, y_4)$ – space is given by

$$y_4 = f(y, y_1, y_2, y_3) = 5(y_1)^{-1} y_2 y_3 - 3y^{-1} y_1 y_3 - y^{-2}(y_1)^2 y_2 + 4y^{-1}(y_2)^2 - 4(y_1)^{-2}(y_2)^3. \tag{3.265}$$

The symmetry determining equation for second-order symmetries $\hat{\eta}(x, y, y_1, y_2)$ admitted by ODE (3.264) is given by

$$\mathbf{D}((y_1\hat{\eta} + y\mathbf{D}\hat{\eta})(2y(y_1)^{-3}(y_2)^2 - (y_1)^{-1}y_2 - y(y_1)^{-2}y_3) + yy_1\mathbf{D}^2((y_1)^{-1}\hat{\eta} - y(y_1)^{-2}\mathbf{D}\hat{\eta})) = 0, \tag{3.266}$$

where

$$\mathbf{D} = \frac{\partial}{\partial x} + y_1\frac{\partial}{\partial y} + y_2\frac{\partial}{\partial y_1} + y_3\frac{\partial}{\partial y_2} + f(y, y_1, y_2, y_3)\frac{\partial}{\partial y_3}.$$

It is not hard to see that (3.266) is a fourth-degree polynomial in y_3. The coefficient of $(y_3)^4$ yields

$$\hat{\eta} = \alpha(x, y, y_1) + \beta(x, y, y_1)y_2 + \gamma(x, y, y_1)(y_2)^2 + \kappa(x, y, y_1)(y_2)^3 \tag{3.267}$$

for some functions $\alpha, \beta, \gamma, \kappa$ of x, y, y_1. Then the coefficients of $y_2(y_3)^3$ and $(y_2)^3(y_3)^2$ in (3.266) yield

$$y_1\kappa_{y_1} + 4\kappa = 0, \tag{3.268}$$

$$25(y_1)^2\kappa_{y_1y_1} + 135y_1\kappa_{y_1} + 56\kappa = 0. \tag{3.269}$$

The compatibility between (3.268) and (3.269) then leads to

$$\kappa = 0.$$

Similarly, the coefficients of $(y_3)^3$, $(y_2)^2(y_3)^2, y_2(y_3)^2, (y_2)^3 y_3$ in (3.266) lead to

$$\gamma = \beta = 0.$$

Hence, ODE (3.264) admits no second-order symmetries. Finally, for $\hat{\eta} = \alpha(x, y, y_1)$, it is straightforward to show that the coefficients of $(y_3)^3, (y_3)^2, y_3$ in (3.266) lead to

$$\alpha = a_0 + a_1 y + a_2 y_1 + a_3 x y_1$$

for arbitrary constants a_i. Therefore, ODE (3.264) admits four point symmetries consisting of translations in x and y, and scalings in x and y, but no contact symmetries.

3.5.3 REDUCTION OF ORDER USING POINT SYMMETRIES IN CHARACTERISTIC FORM

Consider a second-order ODE given by a surface

$$F(x,y,y_1,y_2) = y_2 - f(x,y,y_1) = 0 \tag{3.270}$$

that is assumed to admit a one-parameter Lie group of point transformations with the once-extended infinitesimal generator

$$\mathrm{X}^{(1)} = \xi(x,y)\frac{\partial}{\partial x} + \eta(x,y)\frac{\partial}{\partial y} + \eta^{(1)}(x,y,y_1)\frac{\partial}{\partial y_1}, \tag{3.271}$$

where $\eta^{(1)} = \mathrm{D}\eta - y_1 \mathrm{D}\xi$ and

$$\mathrm{D} = \frac{\partial}{\partial x} + y_1\frac{\partial}{\partial y} + y_2\frac{\partial}{\partial y_1}.$$

In Section 3.3.2, we showed how ODE (3.270) can be reduced, constructively, to a first-order ODE of the form $dv/du = H(u,v)$ in terms of invariants $u(x,y)$, $v(x,y,y_1)$ of the infinitesimal generator (3.271). In particular, u and v are given by the constants of integration that arise in solving the characteristic equations

$$\frac{dx}{\xi} = \frac{dy}{\eta} = \frac{dy_1}{\mathrm{D}\eta - y_1\mathrm{D}\xi}, \tag{3.272}$$

which result from

$$\mathrm{X}^{(1)}u(x,y) = \xi u_x + \eta u_y = 0, \tag{3.273a}$$

$$\mathrm{X}^{(1)}v(x,y,y_1) = \xi v_x + \eta v_y + (\mathrm{D}\eta - y_1\mathrm{D}\xi)v_{y_1} = 0 \quad [v_{y_1} \not\equiv 0]. \tag{3.273b}$$

We now demonstrate an analogous reduction of order of ODE (3.270) in terms of the invariants that arise from expressing the infinitesimal generator (3.271) in its characteristic form

$$\hat{\mathbf{X}}^{(1)} = \hat{\eta}\frac{\partial}{\partial y} + \hat{\eta}^{(1)}\frac{\partial}{\partial y_1} \quad \text{on } F = 0, \tag{3.274}$$

where

$$\hat{\eta} = \eta - y_1\xi, \quad \hat{\eta}^{(1)} = \mathbf{D}\hat{\eta} = \mathrm{D}\eta - y_1\mathrm{D}\xi - f\xi \quad \text{on } F = 0 \tag{3.275}$$

with $\mathbf{D} = \frac{\partial}{\partial x} + y_1 \frac{\partial}{\partial y} + f \frac{\partial}{\partial y_1}$.

Since (3.274) involves no motion on x, it immediately follows that x is an invariant, $\hat{\mathbf{X}}^{(1)}x = 0$. A second invariant $w(x, y, y_1)$, satisfying

$$\hat{\mathbf{X}}^{(1)}w(x, y, y_1) = \hat{\eta} w_y + \hat{\eta}^{(1)} w_{y_1} = 0 \quad [w_{y_1} \not\equiv 0], \tag{3.276}$$

is obtained as a constant of integration in solving the characteristic equation

$$\frac{dy}{\hat{\eta}} = \frac{dy_1}{\hat{\eta}^{(1)}} \quad [x = \text{const}]. \tag{3.277}$$

Then

$$w^{(1)}(x, y, y_1, y_2) = \frac{dw}{dx} = \frac{\partial w}{\partial x} + y_1 \frac{\partial w}{\partial y} + y_2 \frac{\partial w}{\partial y_1} \quad [(w^{(1)})_{y_2} \not\equiv 0]$$

is obviously a differential invariant. Hence, in terms of the invariants $x, w, w^{(1)}$, it follows that the surface (3.270) is given by

$$\frac{\partial w}{\partial y_1} F(x, y, y_1, y_2) = w^{(1)} - \hat{H}(x, w) \equiv \hat{F}(x, w, w^{(1)}) = 0$$

for some function $\hat{H}(x, w) = \mathbf{D}w$. Thus, the second-order ODE $y'' = f(x, y, y')$ reduces, constructively, to the first-order ODE

$$\frac{dw}{dx} = \hat{H}(x, w). \tag{3.278}$$

Most important, we note that in contrast to the standard reduction in terms of invariants $u(x, y)$, $v(x, y, y_1)$ of $\mathrm{X}^{(1)}$, the reduction (3.278) is given directly in terms of the original independent variable x of the given second-order ODE (3.270). The two reduction of order methods are the same if $\hat{\mathbf{X}}^{(1)} = \mathrm{X}^{(1)}$ on $F = 0$, which happens if the given point symmetry is of the form $\xi \equiv 0, \eta \not\equiv 0$.

We now give examples of this direct reduction of order and compare it to the standard reduction method.

(1) *Translation Symmetry*
Consider the second-order nonlinear harmonic oscillator equation

$$y'' + \omega^2 y + g(y) = 0, \quad \omega = \text{const}, \tag{3.279}$$

where $g(y) = G'(y)$ for some nonlinear potential $G(y)$. The corresponding surface in $(x, y, y_1, y_2)-$space, given by

$$F(x, y, y_1, y_2) = y_2 + \omega^2 y + g(y) = 0, \tag{3.280}$$

admits the translation symmetry $\xi = 1, \eta = 0$ [i.e., $\hat{\eta} = -y_1$] with the extended infinitesimal generator in characteristic form

$$\hat{\mathbf{X}}^{(1)} = -y_1 \frac{\partial}{\partial y} + (\omega^2 y + g(y)) \frac{\partial}{\partial y_1}. \tag{3.281}$$

An invariant satisfying $\hat{\mathbf{X}}^{(1)} w(x, y, y_1) = 0$ $[w_{y_1} \not\equiv 0]$ is found by solving the characteristic equation (3.277), which here becomes the separable first-order ODE

$$\frac{dy_1}{dy} = -\frac{\omega^2 y + g(y)}{y_1}. \tag{3.282}$$

From the constant of integration of (3.282), we obtain the invariant

$$w = \tfrac{1}{2}(y_1)^2 + \tfrac{1}{2}\omega^2 y^2 + G(y). \tag{3.283}$$

Then

$$\mathbf{D}w = y_1 w_y - [\omega^2 y + g(y)] w_{y_1} = \omega^2 y y_1 + g(y) y_1 - [\omega^2 y + g(y)] y_1 = 0,$$

and hence, the harmonic oscillator equation (3.279) is reduced to the trivial first-order ODE

$$\frac{dw}{dx} = 0. \tag{3.284}$$

Since the solution of ODE (3.284) is $w = \text{const} = c$, the reduction (3.284) leads to the quadrature of ODE (3.279) given by

$$\tfrac{1}{2}(y_1)^2 + \tfrac{1}{2}\omega^2 y^2 + G(y) = c, \tag{3.285}$$

which is simply the harmonic oscillator energy.

In comparison, standard invariants satisfying $\mathrm{X}^{(1)} u = u_x = 0$, $\mathrm{X}^{(1)} v = v_x = 0$, are given by $u = y$, $v = y_1$. Thus, ODE (3.279) reduces to the first-order ODE

$$\frac{dv}{du} = \frac{y_2}{y_1} = -\frac{\omega^2 y + g(y)}{y_1} = -\frac{\omega^2 u + g(u)}{v}. \tag{3.286}$$

We note that ODEs (3.286) and (3.282) are the same first-order ODEs with the roles of the reduced ODE and the equation for the first-order invariant interchanged in the two reduction procedures.

(2) *Scaling Symmetry*
Consider the second-order Euler equation

$$x^2 y'' + 4xy' + 2y = 0, \tag{3.287}$$

with the corresponding surface

$$F(x, y, y_1, y_2) = y_2 + 4x^{-1} y_1 + 2x^{-2} y = 0 \tag{3.288}$$

in (x, y, y_1, y_2) – space. This surface admits the scaling symmetry $\xi = x, \eta = 0$ [i.e., $\hat{\eta} = -xy_1$], with the extended infinitesimal generator in characteristic form

$$\hat{\mathbf{X}}^{(1)} = -xy_1 \frac{\partial}{\partial y} + (3y_1 + 2x^{-1}y)\frac{\partial}{\partial y_1}. \tag{3.289}$$

The characteristic equation (3.277) for finding an invariant $\hat{\mathbf{X}}^{(1)}w(x, y, y_1) = 0$ reduces to solving the homogeneous ODE

$$\frac{dy_1}{dy} = -3x^{-1} - 2x^{-2}\frac{y}{y_1} \quad [x = \text{const}]. \tag{3.290}$$

The constant of integration arising from the solution of ODE (3.290) yields

$$w = (y_1y^{-1} + 2x^{-1})^2(y_1y^{-1} + x^{-1})^{-1}y = (y_1 + 2yx^{-1})^2(y_1 + yx^{-1})^{-1}. \tag{3.291}$$

Then one finds that

$$\mathbf{D}w = w_x + y_1w_y - [4x^{-1}y_1 + 2x^{-2}y]w_{y_1} = -x^{-1}(y_1 + 2yx^{-1})^2(y_1 + yx^{-1})^{-1} = -x^{-1}w. \tag{3.292}$$

Hence, the Euler equation (3.287) reduces to the first-order linear ODE

$$\frac{dw}{dx} = -x^{-1}w. \tag{3.293}$$

Thus, $xw = \text{const} = c$ and, consequently, (3.291) leads to a quadrature of (3.287) given by the surface

$$(xy_1 + 2y)^2(xy_1 + y)^{-1} = c. \tag{3.294}$$

Finally, note that this surface inherits the scaling symmetry $x \to \alpha x, y \to y$ admitted by ODE (3.287). Hence, one can solve ODE (3.294) to obtain the complete quadrature of the Euler equation (3.287).

In comparison, $\mathrm{X}^{(1)} = x\frac{\partial}{\partial x} - y_1\frac{\partial}{\partial y_1}$ has the invariants $u = y, v = xy_1$, satisfying $\mathrm{X}^{(1)}u = xu_x = 0, \mathrm{X}^{(1)}v = xv_x - y_1v_{y_1} = 0$. Then, in terms of these invariants, the Euler equation (3.287) has the standard reduction

$$\frac{dv}{du} = \frac{y_1 + xy_2}{y_1} = -3 - 2\frac{u}{v}. \tag{3.295}$$

Note that ODE (3.295) is the same as ODE (3.290).

We briefly consider the situation for an nth-order ODE $y^{(n)} = f(x, y, y', \ldots, y^{(n-1)})$ represented as a surface

$$F(x, y, y_1, \ldots, y_n) = y_n - f(x, y, y_1, \ldots, y_{n-1}) = 0. \tag{3.296}$$

If ODE (3.296) admits a point symmetry $\mathrm{X} = \xi(x,y)\frac{\partial}{\partial x} + \eta(x,y)\frac{\partial}{\partial y}$ with its extended infinitesimal generator in characteristic form given by

$$\hat{\mathbf{X}}^{(n-1)} = \hat{\eta}\frac{\partial}{\partial y} + \hat{\eta}^{(1)}\frac{\partial}{\partial y_1} + \cdots + \hat{\eta}^{(n-1)}\frac{\partial}{\partial y_{n-1}} \quad \text{on } F = 0, \tag{3.297a}$$

$$\hat{\eta} = \eta(x,y) - y_1\xi(x,y), \quad \hat{\eta}^{(j)} = \mathbf{D}^j\hat{\eta}, \quad j = 1,2,\dots,n-1, \tag{3.297b}$$

where

$$\mathbf{D} = \frac{\partial}{\partial x} + y_1\frac{\partial}{\partial y} + \cdots + y_{n-1}\frac{\partial}{\partial y_{n-2}} + f\frac{\partial}{\partial y_{n-1}},$$

then one is able to find n functionally independent invariants x, w_i, $i = 1,2,\dots,n-1$, satisfying $\hat{\mathbf{X}}^{(n)}w_i = 0$, which are determined as constants of integration in solving the characteristic equations

$$\frac{dy}{\hat{\eta}} = \frac{dy_1}{\hat{\eta}^{(1)}} = \cdots = \frac{dy_{n-1}}{\hat{\eta}^{(n-1)}} \quad [x = \text{const}]. \tag{3.298}$$

The differential invariants $w_i^{(1)} = dw_i/dx, i = 1,2,\dots,n-1$, can be shown to lead to a direct reduction of the given nth-order ODE (3.296) to a system of first-order ODEs with x as the independent variable.

3.5.4 REDUCTION OF ORDER USING CONTACT SYMMETRIES AND HIGHER-ORDER SYMMETRIES

The direct reduction of order method presented in Section 3.5.3, using point symmetries in characteristic form (i.e., as first-order symmetries) admitted by an nth-order ODE, generalizes naturally to using admitted contact symmetries and higher-order symmetries. We show this generalization by means of the two examples (3.245) and (3.257) considered in Section 3.5.2.

(1) *Contact Symmetries*
The third-order ODE (3.245), represented by the surface

$$F = y_3 - 6x(y_1)^{-2}(y_2)^3 - 6(y_1)^{-1}(y_2)^2 = 0 \tag{3.299}$$

in (x, y, y_1, y_2, y_3) – space, admits seven contact symmetries given by (3.249) and (3.252). Here, we carry out the direct reduction of order method through use of the contact symmetry $\hat{\eta} = (y_1)^{-1}$ corresponding to the infinitesimal generator

$$\hat{\mathbf{X}}^{(2)} = \hat{\eta}\frac{\partial}{\partial y} + \hat{\eta}^{(1)}\frac{\partial}{\partial y_1} + \hat{\eta}^{(2)}\frac{\partial}{\partial y_2} \quad \text{on } F = 0, \tag{3.300}$$

where

$$\hat{\eta}=(y_1)^{-1},\quad \hat{\eta}^{(1)}=-(y_1)^{-2}y_2,\quad \hat{\eta}^{(2)}=-[6x(y_1)^{-4}(y_2)^3+4(y_1)^{-3}(y_2)^2]. \tag{3.301}$$

First, we determine the invariants satisfying

$$\hat{\mathbf{X}}^{(2)}w(x,y,y_1,y_2)=(y_1)^{-1}w_y-(y_1)^{-2}y_2w_{y_1}-[6x(y_1)^{-4}(y_2)^3+4(y_1)^{-3}(y_2)^2]w_{y_2}=0. \tag{3.302}$$

Clearly, one invariant is $w=x$. Two more invariants arise as constants of integration in solving the characteristic equations $[x=\text{const}]$

$$\frac{dy}{(y_1)^{-1}}=-\frac{dy_1}{(y_1)^{-2}y_2}=-\frac{dy_2}{6x(y_1)^{-4}(y_2)^3+4(y_1)^{-3}(y_2)^2}. \tag{3.303}$$

Note that we have some freedom of choice of independent variable in solving (3.303). If we choose y_1 as the independent variable, then (3.303) becomes the system of ODEs

$$\frac{dy_2}{dy_1}=6x(y_1)^{-2}(y_2)^2+4(y_1)^{-1}y_2, \tag{3.304a}$$

$$\frac{dy}{dy_1}=-y_1(y_2)^{-1}. \tag{3.304b}$$

Clearly, ODE (3.304a) admits the scaling symmetry $y_1\to\alpha y_1,\, y_2\to\alpha y_2$ and, hence, is easily solved to yield the invariant

$$w_1=2x(y_1)^3+(y_1)^4(y_2)^{-1}. \tag{3.305}$$

Then we solve (3.305) for $(y_2)^{-1}$ and substitute it into (3.304b), which becomes a linear homogeneous ODE for y. The solution of ODE (3.304b) thereby yields a second invariant

$$w_2=y-\tfrac{1}{2}w_1(y_1)^{-2}-2xy_1. \tag{3.306}$$

Then we find that

$$\mathbf{D}w_1=(w_1)_x+(w_1)_{y_1}y_2+6(w_1)_{y_2}(y_1)^{-2}(y_2)^3(x+y_1(y_2)^{-1})=0, \tag{3.307}$$

and

$$\mathbf{D}w_2=(w_2)_x+(w_2)_y y_1+(w_2)_{y_1}y_2-\tfrac{1}{2}(y_1)^{-2}\mathbf{D}w_1=0, \tag{3.308}$$

after using (3.307) and (3.305), where

$$\mathbf{D}=\frac{\partial}{\partial x}+y_1\frac{\partial}{\partial y}+y_2\frac{\partial}{\partial y_1}+6(y_1)^{-2}(y_2)^3(x+y_1(y_2)^{-1})\frac{\partial}{\partial y_2}.$$

Hence, the differential invariants dw_1/dx and dw_2/dx, determined by (3.307) and (3.308), yield a reduction of ODE (3.299) to the system of trivial first-order ODEs

$$\frac{dw_1}{dx} = 0, \tag{3.309a}$$

$$\frac{dw_2}{dx} = 0. \tag{3.309b}$$

Then the solution $w_1 = \text{const} = c_1$, $w_2 = \text{const} = c_2$ of (3.309a,b) yields two quadratures of ODE (3.299) given by

$$2x(y_1)^3 + (y_1)^4(y_2)^{-1} = c_1, \tag{3.310}$$

$$y - \tfrac{1}{2}c_1(y_1)^{-2} - 2xy_1 = c_2. \tag{3.311}$$

Thus, we obtain a reduction of the third-order ODE (3.299) to a first-order ODE given by the surface

$$2x(y_1)^3 + (c_2 - y)(y_1)^2 + \tfrac{1}{2}c_1 = 0. \tag{3.312}$$

Finally, note that this surface admits the symmetry $x \to \alpha x$, $y \to \alpha^{2/3}(y - c_2) + c_2$, which is inherited from the scaling symmetries and translation symmetry of (3.299). Consequently, one can solve ODE (3.312) to obtain the complete quadrature of the given ODE (3.299).

(2) *Second-Order Symmetries*

The fourth-order ODE (3.257), represented by the surface

$$F = y_4 - \tfrac{4}{3}(y_2)^{-1}(y_3)^2 = 0 \tag{3.313}$$

in $(x, y, y_1, y_2, y_3, y_4)$ – space, admits the 12 second-order symmetries given by (3.263). We now apply the direct reduction of order method, using one of the admitted second-order symmetries: $\hat{\eta} = y_1(y_2)^{-1/3}$. The corresponding extended infinitesimal generator is given by

$$\hat{\mathbf{X}}^{(3)} = \hat{\eta}\frac{\partial}{\partial y} + \hat{\eta}^{(1)}\frac{\partial}{\partial y_1} + \hat{\eta}^{(2)}\frac{\partial}{\partial y_2} + \hat{\eta}^{(3)}\frac{\partial}{\partial y_3} \quad \text{on } F = 0, \tag{3.314a}$$

where

$$\hat{\eta}^{(1)} = (y_2)^{2/3} - \tfrac{1}{3}y_1(y_2)^{-4/3}y_3, \quad \hat{\eta}^{(2)} = \tfrac{1}{3}(y_2)^{-1/3}y_3, \quad \hat{\eta}^{(3)} = \tfrac{1}{3}(y_2)^{-4/3}(y_3)^2. \tag{3.314b}$$

First, we determine invariants satisfying

$$\hat{\mathbf{X}}^{(3)}w(x, y, y_1, y_2, y_3)$$
$$= y_1(y_2)^{-1/3}w_y + (y_2)^{-4/3}[(y_2)^2 - \tfrac{1}{3}y_1y_3]w_{y_1} + \tfrac{1}{3}(y_2)^{-1/3}y_3w_{y_2} + \tfrac{1}{3}(y_2)^{-4/3}(y_3)^2w_{y_3} = 0. \tag{3.315}$$

An obvious invariant is $w = x$. Three additional invariants w_1, w_2, w_3 arise as constants of integration in solving the characteristic equations $[x = \text{const}]$

$$\frac{dy}{y_1(y_2)^{-1/3}} = \frac{dy_1}{(y_2)^{-4/3}[(y_2)^2 - \frac{1}{3}y_1y_3]} = \frac{3dy_2}{(y_2)^{-1/3}y_3} = \frac{3dy_3}{(y_2)^{-4/3}(y_3)^2}. \tag{3.316}$$

If we choose y_3 as the independent variable, then (3.316) becomes the system of ODEs

$$\frac{dy_2}{dy_3} = (y_3)^{-1}y_2, \tag{3.317a}$$

$$\frac{dy_1}{dy_3} = 3(y_3)^{-2}(y_2)^2 - (y_3)^{-1}y_1, \tag{3.317b}$$

$$\frac{dy}{dy_3} = 3(y_3)^{-2}y_1y_2. \tag{3.317c}$$

We see that (3.317a) is a linear homogeneous ODE in terms of the dependent variable y_2. If the solution of (3.317a) is substituted into (3.317b), we obtain a linear inhomogeneous ODE in terms of the dependent variable y_1. In turn, after the substitution of the solution of (3.317a,b) into (3.317c), we see that (3.317c) becomes a separable ODE in terms of the dependent variable y. Hence, the constants of integration of the system of ODEs (3.317a–c) yield the three invariants

$$w_1 = (y_3)^{-1}y_2, \tag{3.318a}$$

$$w_2 = y_1y_3 - \tfrac{3}{2}(y_2)^2, \tag{3.318b}$$

$$w_3 = y - 9(y_3)^{-2}(y_2)^3 + 3(y_3)^{-1}y_1y_2. \tag{3.318c}$$

Then we find that

$$\mathbf{D}w_1 = -\tfrac{1}{3}, \tag{3.319a}$$

$$\mathbf{D}w_2 = -2y_2y_3 + \tfrac{4}{3}(y_2)^{-1}y_1(y_3)^2 = \tfrac{4}{3}(w_1)^{-1}w_2, \tag{3.319b}$$

$$\mathbf{D}w_3 = 0, \tag{3.319c}$$

where

$$\mathbf{D} = \frac{\partial}{\partial x} + y_1\frac{\partial}{\partial y} + y_2\frac{\partial}{\partial y_1} + y_3\frac{\partial}{\partial y_2} + \frac{4}{3}(y_3)^2(y_2)^{-1}\frac{\partial}{\partial y_3}.$$

Hence, the differential invariants dw_1/dx, dw_2/dx, dw_3/dx, determined by (3.319a–c), reduce the fourth-order ODE (3.313) to the system of first-order ODEs

$$\frac{dw_1}{dx} = -\frac{1}{3}, \tag{3.320a}$$

$$\frac{dw_2}{dx} = \frac{4}{3}(w_1)^{-1}w_2, \tag{3.320b}$$

$$\frac{dw_3}{dx} = 0. \tag{3.320c}$$

Consequently, after solving (3.320a–c), we obtain

$$w_1 = -\tfrac{1}{3}x + c_1, \quad w_2 = c_2(-\tfrac{1}{3}x + c_1)^{-4}, \quad w_3 = c_3, \tag{3.321}$$

yielding the three quadratures

$$(y_3)^{-1}y_2 + \tfrac{1}{3}x = \text{const} = c_1, \tag{3.322a}$$

$$[y_1y_3 - \tfrac{3}{2}(y_2)^2](-\tfrac{1}{3}x + c_1)^4 = \text{const} = c_2, \tag{3.322b}$$

$$y - 9(y_3)^{-2}(y_2)^3 + 3(y_3)^{-1}y_1y_2 = \text{const} = c_3. \tag{3.322c}$$

Substitution of y_2 and y_3 from (3.322a,b) into (3.322c) then reduces the given fourth-order ODE (3.313) to a first-order ODE,

$$y_1 = \pm\frac{1}{c_1 - \frac{1}{3}x}\sqrt{\tfrac{1}{9}(c_3 - y)^2 + 6c_2}, \tag{3.323}$$

which is separable. Hence, the solution of (3.323) yields the complete quadrature of ODE (3.313), i.e., the general solution,

$$(c_1 - \tfrac{1}{3}x)^{\pm 1}(y - c_3 + \sqrt{(c_3 - y)^2 + 54c_2}) = \text{const} = c_4$$

or, equivalently,

$$y = \tilde{c}_3 + \tilde{c}_4(\tilde{c}_1 - x) + \tilde{c}_2(\tilde{c}_1 - x)^{-1} \tag{3.324}$$

for some renamed constants $\tilde{c}_i$.

EXERCISES 3.5

1. Consider a one-parameter Lie group of point transformations

$$x^* = x + \varepsilon\xi(x,y) + O(\varepsilon^2),$$
$$y^* = y + \varepsilon\eta(x,y) + O(\varepsilon^2),$$

with infinitesimal generator $\mathrm{X} = \xi\frac{\partial}{\partial x} + \eta\frac{\partial}{\partial y}$, admitted by an nth-order ODE (3.231).

Show that the corresponding vector field $\hat{\mathbf{X}}^{(n)}$, tangent to the surface (3.234) representing ODE (3.231), is given by

$$\hat{\mathbf{X}}^{(n)} = \hat{\eta}\frac{\partial}{\partial y} + \hat{\eta}^{(1)}\frac{\partial}{\partial y_1} + \cdots + \hat{\eta}^{(n)}\frac{\partial}{\partial y_n} \quad \text{on } F = 0,$$

where

$$\hat{\eta} = \eta - y_1\xi, \quad \hat{\eta}^{(k)} = \mathrm{D}^k\eta - \sum_{j=0}^{k} \frac{k!}{(k-j)!\,j!} y_{k+1-j}\mathrm{D}^j\xi, \quad k = 1,2,\ldots,n-1,$$

with

$$y_n = f(x, y, y_1, \ldots, y_n) \text{ and } \hat{\eta}^{(n)} = f_y\hat{\eta} + f_{y_1}\hat{\eta}^{(1)} + \cdots + f_{y_{n-1}}\hat{\eta}^{(n-1)}.$$

2. Find the point symmetries admitted by the nonlinear Duffing equation $y'' + ay' + by + y^3 = 0, \quad a, b = \text{const.}$

3. Find the point symmetries admitted by the ODE $y'' = 2(y')^2 \cot y + \sin y \cos y$ [Stephani (1989)]. Show that these symmetries form an SO(3) Lie algebra.

4. Find the contact symmetries admitted by the ODEs:
   (a) $y''' = 0$ [verify (3.243)];
   (b) $y''' = x(x-1)(y'')^3 - 2x(y'')^2 + y''$; and
   (c) $y''' = y\left(\dfrac{y''}{y'}\right)^3$.

5. Find the contact symmetries admitted by the third-order ODE $y''' + yy' = 0$. This ODE arises when describing traveling wave solutions of the Korteweg–de Vries (KdV) equation [see Exercise 4.1-2].

6. Find the second-order symmetries admitted by the fourth-order ODE $y^{(4)} = (y)^{-1}y'y'''$.

7. Find the symmetries up to second-order of the fourth-order ODE [Sheftel (1997)] $y^{(4)} = y^{-5/3}$. Show that the admitted point symmetries form a three-dimensional Lie algebra with commutators given by $[\hat{\mathrm{X}}_1, \hat{\mathrm{X}}_2] = 2\hat{\mathrm{X}}_1$, $[\hat{\mathrm{X}}_2, \hat{\mathrm{X}}_3] = 2\hat{\mathrm{X}}_3$, $[\hat{\mathrm{X}}_1, \hat{\mathrm{X}}_3] = -\hat{\mathrm{X}}_2$.

8. Show that the Blasius equation (3.253) admits no second-order symmetries based on the ansatz $\widetilde{\mathrm{X}}\hat{\eta} = r\hat{\eta}$ for either the translation symmetry or for the scaling symmetry admitted by (3.253).

9. Find all third-order ODEs admitting the contact symmetry $\hat{\eta} = (y_1)^{-1}$.

10. Find all fourth-order ODEs admitting the second-order symmetry $\hat{\eta} = y_2$.

11. Use the direct reduction of order method to reduce:
    (a) the second-order linear ODE (3.104) from its invariance under the scaling symmetry (3.105a,b);

(b) the Blasius equation (3.253) from its invariance under translations in x;
(c) the third-order ODE (3.245) from its invariance under
   (i) scalings in x;
   (ii) scalings in y;
   (iii) the contact symmetry $\hat{\eta}_2$ of (3.249); and
   (iv) the contact symmetry $\hat{\eta}_3$ of (3.249);
(d) the fourth-order ODE (3.257) from its invariance under
   (i) translations in x;
   (ii) translations in y; and
   (iii) each of the second-order symmetries $\hat{\eta}_1, \hat{\eta}_4, \hat{\eta}_5, \ldots, \hat{\eta}_{12}$ given by (3.263).

12. Consider the Thomas–Fermi equation

$$y'' = x^{-1/2} y^{3/2}. \tag{3.325}$$

(a) Show that ODE (3.325) admits the scaling symmetry $\hat{\mathrm{X}} = (3y + xy_1)\frac{\partial}{\partial y}$. Use direct reduction of order to reduce (3.325) from its invariance under $\hat{\mathrm{X}}$.
(b) Show that the equation $\hat{\mathbf{X}}^{(1)} w(x, y, y_1) = 0$ for the invariant $w(x, y, y_1)$ is equivalent to the reduced first-order ODE obtained by the standard reduction method.
(c) Show that although the invariant $w(x, y, y_1)$ cannot be found explicitly, the corresponding reduced ODE $dw/dx = \hat{H}(x, w)$ is given implictly by

$$\theta_w \frac{dw}{dx} + \theta_x = \frac{3x^{1/2} y^{5/2} - 4\theta^2}{3y + x\theta}, \tag{3.326}$$

where $y_1 = \theta(y, w; x)$ is the solution of $\hat{\mathbf{X}}^{(1)} w(x, y, y_1) = 0$. In particular, show that there is no essential y-dependence in (3.326) and, consequently, (3.326) is a first-order ODE in terms of variables x and w.

3.6 FIRST INTEGRALS AND REDUCTION OF ORDER THROUGH INTEGRATING FACTORS

We generalize the classical treatment, presented in Section 3.2.2, of first integrals and integrating factors for first-order ODEs to second- and higher-order ODEs.

Definition 3.6-1. A *first integral* of an nth-order ODE

$$y^{(n)} = f(x, y, y', \ldots, y^{(n-1)}) \tag{3.327}$$

is a function $\psi(x, y, y', \ldots, y^{(n-1)})$ with an essential dependence on $y^{(n-1)}$ satisfying

$$\frac{d\psi}{dx} = 0 \quad \text{when } y^{(n)} = f, \tag{3.328}$$

i.e., $\psi(x, y, y', ..., y^{(n-1)})$ is constant for every solution $y = \Theta(x)$ of ODE (3.327).

Since a first integral $\psi(x, y, y', ..., y^{(n-1)})$ of ODE (3.327) satisfies $\psi(x, y, y', \ldots, y^{(n-1)}) = \text{const} = c$ for each solution $y = \Theta(x)$ of (3.327), it represents a conserved quantity for any solution $y = \Theta(x)$. Moreover, a first integral provides a quadrature which reduces (3.327) to an $(n-1)$th-order ODE in terms of the original variables $x, y, y', ..., y^{(n-1)}$. If one knows r first integrals of (3.327) which are functionally independent, i.e., none of them is a function combination of the others, then the nth-order ODE (3.327) is reduced to an $(n-r)$th-order ODE in terms of r essential constants and $n-r+1$ variables $x, y, y', ..., y^{(n-r-1)}$. In particular, any n functionally independent first integrals yield a general solution of ODE (3.327) involving n essential constants. These constants represent n independent conserved quantities for the solutions $y = \Theta(x)$ of the ODE.

As we showed in Section 3.2.2 in the case of a first order ODE, it is well-known that finding a first integral is equivalent to finding an integrating factor. The same is true for second- and higher-order ODEs.

Definition 3.6-2. An *integrating factor* of an nth-order ODE (3.327) is a function $\Lambda(x, y, y', ..., y^{(\ell)}) \not\equiv 0$, $0 \le \ell \le n-1$, that satisfies

$$\Lambda(x, y, y', ..., y^{(\ell)})(y^{(n)} - f(x, y, y', ..., y^{(n-1)})) = \frac{d}{dx}\psi(x, y, y', ..., y^{(n-1)}) \tag{3.329}$$

for some function $\psi(x, y, y', ..., y^{(n-1)})$ which has an essential dependence on $y^{(n-1)}$. The highest-order ℓ of the derivatives of y in $\Lambda(x, y, y', ..., y^{(\ell)})$ is called the order of the integrating factor.

From (3.329) it follows that $d\psi / dx = 0$ when $y^{(n)} = f(x, y, y', \ldots, y^{(n-1)})$ and, hence, $\psi(x, y, y', \ldots, y^{(n-1)}) = \text{const} = c$ on solutions $y = \Theta(x)$ of ODE (3.327) for which $\Lambda(x, y, y', \ldots, y^{(\ell)})$ is nonsingular. In particular, if $\Lambda(x, y, y', \ldots, y^{(\ell)})$ is nonsingular for an arbitrary function $y(x)$, then $\psi(x, y, y', \ldots, y^{(n-1)}) = \text{const} = c$ holds on every solution of (3.327) and thus determines a first integral of ODE (3.327). Conversely, if $\psi(x, y, y', \ldots, y^{(n-1)})$ is a first integral of ODE (3.327), then one can easily show that (3.329) holds with $\Lambda = \partial\psi / \partial y^{(n-1)}$ defining the corresponding integrating factor of (3.327). Hence, all first integrals of ODE (3.327) arise from integrating factors satisfying the linear relation (3.329) for arbitrary functions $y(x)$. Equation (3.329) is called the *characteristic equation* for first integrals and integrating factors. Note that, as a consequence of the linearity of the characteristic equation (3.329), the set of all integrating factors and the set of all first integrals of a given ODE (3.327), respectively, form vector spaces.

We will first derive a necessary and sufficient system of linear determining equations whose solutions yield the integrating factors of any given nth-order ODE (3.327). We will also derive a line integral formula that yields a first integral corresponding to an integrating factor through the characteristic equation (3.329). Note that the local existence theory for solutions of an ODE [Coddington (1961)] guarantees that n functionally independent first integrals $\psi_1(x,y,y',\ldots,y^{(n-1)}),\ldots,$ $\psi_n(x,y,y',\ldots,y^{(n-1)})$ exist for any nth-order ODE (3.327). Since any function of $\psi_1(x,y,y',\ldots,y^{(n-1)}),\ldots,\psi_n(x,y,y',\ldots,y^{(n-1)})$ is also a first integral, one sees that a given ODE (3.327) admits an infinite number of integrating factors with an essential dependence on $y^{(n-1)}$. Thus, the determining system for integrating factors of order $n-1$ of ODE (3.327) always has infinitely many solutions. However, for integrating factors of order $\ell < n-1$, the determining system reduces to an overdetermined system of linear PDEs that has at most a finite number of linearly independent solutions. This situation is analogous to that of the determining equation for symmetries of ODE (3.327) [cf. Section 3.5].

Definition 3.6-3. An integrating factor $\Lambda(x,y,y',\ldots,y^{(\ell)})$ of order $\ell \le n-1$ of an nth-order ODE (3.327) is of *point-form* if $\ell = 1$ and Λ is linear in y', i.e.,

$$\Lambda(x,y,y') = \alpha(x,y) + \beta(x,y)y'.$$

Otherwise, for $\ell = 1$, an integrating factor is called *first-order*, and for $\ell \ge 2$, *higher-order*.

We will show how to solve the determining system to obtain *all* integrating factors of orders $0 \le \ell < n-1$ of an nth-order ODE (3.327). We will also present effective ansatzes for directly finding particular solutions of the integrating factor determining system. Moreover, the use of such ansatzes is essential for obtaining integrating factors of order $\ell = n-1$. These ansatzes lead to a splitting of the integrating factor determining system into an overdetermined linear system of PDEs. Most important, in all cases one can solve these equations by a simple algorithmic procedure, analogous to that for solving the determining equation for symmetries of orders $\ell \le n-1$ of ODE (3.327). This will be illustrated through many examples.

3.6.1 FIRST-ORDER ODEs

Consider a first-order ODE (3.327) or, equivalently, a surface

$$y_1 = f(x,y). \tag{3.330}$$

A first integral of ODE (3.330) is any function $\psi(x,y) = \text{const} = c$ on the surface (3.330) such that $\psi_y \not\equiv 0$, and thus satisfies

$$(\mathrm{D}\psi)\big|_{y_1=f} = \mathbf{D}\psi = \psi_x + f\psi_y = 0, \tag{3.331}$$

where

$$\mathrm{D} = \frac{\partial}{\partial x} + y_1 \frac{\partial}{\partial y} \tag{3.332a}$$

and

$$\mathbf{D} = \mathrm{D}\big|_{y_1=f} = \frac{\partial}{\partial x} + f\frac{\partial}{\partial y} \tag{3.332b}$$

(i.e., y_1 is eliminated through ODE (3.330)). Hence, for an arbitrary function $y(x)$, one has

$$\mathrm{D}\psi = (y_1 - f)\Lambda, \tag{3.333}$$

where, from (3.331),

$$\Lambda(x,y) = \psi_y = -\frac{1}{f}\psi_x, \tag{3.334}$$

which is the integrating factor corresponding to the first integral $\psi(x,y)$. Conversely, any function $\Lambda(x,y) \not\equiv 0$ satisfying (3.333) for arbitrary values of x, y, y_1, for some function $\psi(x,y)$, is an integrating factor of ODE (3.330) with the corresponding first integral $\psi(x,y)$. We can obtain necessary and sufficient defining conditions for integrating factors and first integrals of ODE (3.330) by the elimination of $\Lambda(x,y)$ and $\psi(x,y)$, respectively, in (3.334).

Theorem 3.6.1-1. *The integrating factors of ODE* (3.330) *are the solutions* $\Lambda(x,y) \not\equiv 0$ *of the determining equation*

$$\Lambda_x + (f\Lambda)_y = 0. \tag{3.335}$$

For a given integrating factor, the corresponding first integral $\psi(x,y)$ *of ODE* (3.330) *is given by the line integral*

$$\psi = \int_C [-\Lambda f\, dx + \Lambda\, dy], \tag{3.336}$$

where C *is a path curve from any point* $(\tilde{x},\tilde{y})$ *to the point* (x,y) *in* $(x,y)-$*space. A change in* $(\tilde{x},\tilde{y})$ *changes* (3.336) *by the addition of a constant. If* $f(x,y)$ *and* $\Lambda(x,y)$ *are nonsingular, then the line integral* (3.336) *is independent of the path curve* C.

Proof. Suppose $\Lambda(x,y)$ is an integrating factor of ODE (3.330). Then, from (3.334), we have the pair of equations

$$-\Lambda f = \psi_x, \quad \Lambda = \psi_y. \tag{3.337}$$

By cross-differentiation and commutativity of partial derivatives, we see that $\Lambda(x,y)$ satisfies (3.335), which is just the integrability condition for solving (3.337).

Conversely, suppose $\Lambda(x,y)$ is a solution of the integrating factor determining equation (3.335). Since the integrability condition holds for solving (3.337), it follows that there exists a $\psi(x,y)$ satisfying (3.334). The fundamental theorem of calculus for line integrals of gradients then yields (3.336). Moreover, from the integrating factor determining equation (3.335), it follows that (3.336) is independent of the path curve C when $f(x,y)$ and $\Lambda(x,y)$ are nonsingular. □

There is an alternative formulation of Theorem 3.6.1-1 that is more useful when one considers its generalization to higher-order ODEs. First, writing out the integrating factor determining equation (3.335) and using (3.332a), we have

$$0=\Lambda_x+f\Lambda_y+f_y\Lambda=\mathrm{D}\Lambda+(f-y_1)\Lambda_y+f_y\Lambda=-\mathrm{E}_1(\theta), \tag{3.338a}$$

with

$$\theta(x,y,y_1)=(y_1-f(x,y))\Lambda(x,y), \tag{3.338b}$$

where

$$\mathrm{E}_1=\frac{\partial}{\partial y}-\mathrm{D}\frac{\partial}{\partial y_1} \tag{3.339}$$

denotes a truncated Euler operator. One can easily show that the operator (3.339) annihilates total derivatives of arbitrary differentiable functions of x and y. In particular, for any function $\psi(x,y)$, if we let $\theta(x,y,y_1)=\mathrm{D}\psi(x,y)$ and define $\Psi_1=\theta_{y_1}$, $\Psi_0=\theta_y-\mathrm{D}\Psi_1$, then

$$(\Psi_1)_{y_1}=0, \tag{3.340}$$

$$\Psi_0=0, \tag{3.341}$$

which hold by the identities $\Psi_1=\psi_y$, $\Psi_0=(\mathrm{D}\psi)_y-\mathrm{D}\psi_y$. Conversely, if (3.340) and (3.341) hold for some function $\theta(x,y,y_1)$, where $\Psi_1=\theta_{y_1}$, $\Psi_0=\theta_y-\mathrm{D}\Psi_1$, then (3.340) yields $\theta_{y_1y_1}=0$ and so we have

$$\theta=Ay_1+B \tag{3.342}$$

for some functions $A(x,y)$, $B(x,y)$. Then (3.341) yields

$$A_x-B_y=0,$$

which is just the integrability condition for there to exist a function $\psi(x,y)$ such that $A=\psi_y$, $B=\psi_x$. Hence, from (3.342), we obtain $\theta=\psi_x+\psi_y y_1=\mathrm{D}\psi$. Moreover, since then $\theta_{y_1}=\psi_y$ and $\theta-y_1\theta_{y_1}=\psi_x$ are functions depending only on x and y, we have

$$\psi(x,y) = \int_C [\theta(x,y,0)\,dx + \theta_{y_1}(x,y,0)\,dy], \tag{3.343}$$

which gives a line integral formula for $\psi(x,y)$ in terms of $\theta(x,y,y_1)$, to within an arbitrary constant, where C is any path curve from a point $(\tilde{x},\tilde{y})$ to the point (x,y) in (x,y) – space. Then substitution of (3.338b) into (3.343) yields the line integral (3.336).

The determining equation (3.335) for integrating factors is a first-order linear homogeneous PDE which has an infinite number of solutions. In particular, if $\Lambda(x,y)$ is an integrating factor of ODE (3.330) with corresponding first integral $\psi(x,y)$, then since any function $F(\psi)$ is automatically a first integral of ODE (3.330), it follows that $F'(\psi)\Lambda$ is also an integrating factor as obtained from (3.334). Moreover, this represents the general solution of the integrating factor determining equation (3.335). From (3.336), we see that any particular solution $\Lambda = \Lambda_1(x,y)$ of (3.335) yields a corresponding first integral $\psi = \psi_1(x,y)$ and thus reduces ODE (3.330) to the quadrature $\psi_1(x,y) =$ const $= c_1$. However, in general, one cannot obtain any solution of the integrating factor determining equation (3.335) without knowing the general solution of ODE (3.330).

Consequently, for a given ODE (3.330), one often seeks to determine if it admits an integrating factor of a specific form. Two simple ansatzes are based on elimination of variables.

If we seek $\Lambda = \alpha(x)$, then from the integrating factor determining equation (3.335) we find that $\alpha' + \alpha f_y = 0$ and, hence, $f(x,y)$ must satisfy $f_y = -\alpha'/\alpha$, leading to $f(x,y) = -(\alpha'/\alpha)y + \beta$, for some function $\beta = \beta(x)$. Thus, ODE (3.330) admits an integrating factor depending only on x if and only if $f(x,y)$ is linear in y. Then $\Lambda = e^{-\int A(x)dx}$ is the integrating factor, where $y_1 = f(x,y) = A(x)y + B(x)$.

Alternatively, if we seek $\Lambda = \alpha(y)$, then the integrating factor determining equation (3.335) yields $\alpha' f + \alpha f_y = 0$, and so $f(x,y)$ must satisfy $f_y / f = -\alpha'/\alpha$. This leads to $f = \beta/\alpha$, for some function $\beta = \beta(x)$. Thus, ODE (3.330) admits an integrating factor depending only on y if and only if $f(x,y)$ is separable in x and y. Then $\Lambda = 1/B(y)$ is the integrating factor, where $y_1 = f(x,y) = A(x)B(y)$.

A more effective general ansatz is based on separation of variables. Consider $\Lambda = \alpha(x)\beta(y)$. Then, from the integrating factor determining equation (3.335), we find that $(\alpha'/\alpha) + (\beta'/\beta)f + f_y = 0$ and, hence, by integration with respect to y, we obtain $f\beta = -(\alpha'/\alpha)\int \beta\,dy + \gamma$, for some function $\gamma = \gamma(x)$. Thus, ODE (3.330) admits an integrating factor of separable form if and only if

$$y_1 = f(x,y) = \frac{A(x)C(y) + B(x)}{C'(y)}, \tag{3.344}$$

for some functions $A(x)$, $B(x)$, $C(y)$. The integrating factor is then given by

$$\Lambda(x,y) = e^{-\int A(x)dx} C'(y), \tag{3.345}$$

with the corresponding first integral given by

$$\psi(x,y) = e^{-\int A(x)dx} C(y) - \int e^{-\int A(x)dx} B(x)\,dx. \tag{3.346}$$

The class of ODEs (3.344) includes all linear ODEs, i.e., corresponding to $C(y) = y$; all separable ODEs, i.e., corresponding to $A(x) \equiv 0$ or $B(x) \equiv 0$; and all Bernoulli ODEs, i.e., corresponding to $C(y) = y^r$, $r = \text{const}$.

More generally, for any given integrating factor ansatz $\Lambda = \alpha(x,y)$, one can solve the integrating factor determining equation (3.335) for $f(x,y)$ to obtain

$$y_1 = f(x,y) = -\frac{1}{\alpha(x,y)} \int \alpha_x(x,y)\,dy + \frac{\beta(x)}{\alpha(x,y)}, \tag{3.347}$$

which yields the most general ODE admitting the given integrating factor. Conversely, if one can match the form of (3.347) to a given ODE (3.330) for some $\alpha(x,y)$ and $\beta(x)$, then one immediately obtains an integrating factor $\Lambda = \alpha(x,y)$.

3.6.2 DETERMINING EQUATIONS FOR INTEGRATING FACTORS OF SECOND-ORDER ODEs

We now consider a second-order ODE

$$y'' = f(x,y,y') \tag{3.348}$$

or, equivalently, a surface given by

$$y_2 = f(x,y,y_1). \tag{3.349}$$

Theorem 3.6.2-1. *A function $\psi(x,y,y')$ with an essential dependence on y' is a first integral of ODE* (3.348), *satisfying $\psi(x,y,y') =$ const $= c$ on the surface* (3.349), *if and only if*

$$(\mathrm{D}\psi)\big|_{y_2=f} = \mathbf{D}\psi = \psi_x + y_1\psi_y + f\psi_{y_1} = 0, \tag{3.350}$$

where

$$\mathrm{D} = \frac{\partial}{\partial x} + y_1\frac{\partial}{\partial y} + y_2\frac{\partial}{\partial y_1} \tag{3.351a}$$

and

$$\mathbf{D} = \mathrm{D}\big|_{y_2=f} = \frac{\partial}{\partial x} + y_1\frac{\partial}{\partial y} + f\frac{\partial}{\partial y_1} = \mathrm{D} - (y_2 - f)\frac{\partial}{\partial y_1}. \tag{3.351b}$$

From (3.351a,b), one sees that (3.350) is equivalent to the characteristic equation

$$\mathrm{D}\psi = \Lambda(y_2 - f) \tag{3.352a}$$

holding for arbitrary values of x, y, y_1, y_2, with

$$\Lambda(x, y, y_1) = \psi_{y_1}, \tag{3.352b}$$

which is the integrating factor corresponding to the first integral. Conversely, if a function $\Lambda(x, y, y_1) \not\equiv 0$ satisfies (3.352a,b) for some $\psi(x, y, y_1)$, then $\Lambda(x, y, y')$ is an integrating factor of ODE (3.348) and $\psi(x, y, y')$ is the corresponding first integral. We now derive a determining system that yields all integrating factors $\Lambda(x, y, y')$ of ODE (3.348).

Consider the truncated Euler operator

$$\mathrm{E}_2 = \frac{\partial}{\partial y} - \mathrm{D}\frac{\partial}{\partial y_1} + \mathrm{D}^2\frac{\partial}{\partial y_2}. \tag{3.353}$$

One can easily show that operator (3.353) is connected with annihilating total derivatives of any differentiable function of x, y, y_1. Let

$$\theta(x, y, y_1, y_2) = \mathrm{D}\psi(x, y, y_1), \tag{3.354}$$

and introduce the notations

$$\Psi_2 = \theta_{y_2}, \quad \Psi_1 = \theta_{y_1} - \mathrm{D}\Psi_2, \quad \Psi_0 = \theta_y - \mathrm{D}\Psi_1 = \mathrm{E}_2(\theta). \tag{3.355}$$

Theorem 3.6.2-2. *A function $\theta(x, y, y_1, y_2)$ is a total derivative* (3.354) *if and only if it satisfies*

$$(\Psi_2)_{y_2} = (\Psi_1)_{y_2} = 0, \tag{3.356a}$$

$$\Psi_0 = 0, \tag{3.356b}$$

on the entire $(x, y, y_1, y_2)-$space. *In particular, if $\theta(x, y, y_1, y_2)$ satisfies* (3.356a,b), *then* (3.354) *holds with*

$$\psi(x, y, y_1) = \int_C [(\theta(x, y, y_1, 0) - y_1\Psi_1(x, y, y_1))\,dx + \Psi_1(x, y, y_1)\,dy + \Psi_2(x, y, y_1)\,dy_1], \tag{3.357}$$

where C is any path curve from a point $(\tilde{x}, \tilde{y}, \tilde{y}_1)$ to (x, y, y_1).

Proof. Suppose a function $\theta(x, y, y_1, y_2)$ satisfies (3.354) for some $\psi(x, y, y_1)$. Then, using the identities

$$(\mathrm{D}\psi)_y = \mathrm{D}\psi_y, \quad (\mathrm{D}\psi)_{y_1} = \mathrm{D}\psi_{y_1} + \psi_y, (\mathrm{D}\psi)_{y_2} = \psi_{y_1}, \tag{3.358}$$

one can easily verify that E_2 annihilates $D\psi$. Hence, $\theta = D\psi$ satisfies $\Psi_0 = 0$. Furthermore, from (3.358), one sees that $(\Psi_2)_{y_2} = (\Psi_1)_{y_2} = 0$ are identities, with

$$\Psi_2 = \psi_{y_1}, \quad \Psi_1 = \psi_y. \tag{3.359}$$

Thus, (3.356a,b) holds. This establishes the "only if" part of the theorem.

Let

$$\Phi = \theta - y_1\Psi_1 - y_2\Psi_2. \tag{3.360}$$

Then, using (3.354) and (3.359), it follows that

$$\Phi = \psi_x. \tag{3.361}$$

Hence, from (3.359) and (3.361), we see that the integrability conditions for the existence of a function $\psi(x, y, y_1)$ satisfying (3.354) are given by

$$(\Psi_2)_{y_2} = (\Psi_1)_{y_2} = \Phi_{y_2} = 0, \tag{3.362a}$$

$$(\Psi_2)_y = (\Psi_1)_{y_1}, \tag{3.362b}$$

$$(\Psi_2)_x = \Phi_{y_1}, \quad (\Psi_1)_x = \Phi_y. \tag{3.362c}$$

Now suppose a function $\theta(x, y, y_1, y_2)$ satisfies (3.356a,b). Using (3.355), we have

$$\begin{aligned}(\Psi_2)_y &= \theta_{yy_2} = (\Psi_0 + D\Psi_1)_{y_2}\\ &= (\Psi_1)_{y_1} + (\Psi_0)_{y_2} + D((\Psi_1)_{y_2}).\end{aligned}$$

Thus, we obtain (3.362b) from (3.356a,b). Next, using (3.355) and (3.360), we find that

$$\begin{aligned}\Phi_{y_2} &= \theta_{y_2} - \Psi_2 - y_1(\Psi_1)_{y_2} - y_2(\Psi_2)_{y_2} = -y_1(\Psi_1)_{y_2} - y_2(\Psi_2)_{y_2},\\ \Phi_{y_1} &= \theta_{y_1} - \Psi_1 - y_1(\Psi_1)_{y_1} - y_2(\Psi_2)_{y_1}\\ &= (\Psi_2)_x + y_1((\Psi_2)_y - (\Psi_1)_{y_1}),\\ \Phi_y &= \theta_y - y_1(\Psi_1)_y - y_2(\Psi_2)_y\\ &= (\Psi_1)_x + \Psi_0 + y_2((\Psi_1)_{y_1} - (\Psi_2)_y),\end{aligned}$$

which yield (3.362a,c) from (3.356a,b) and (3.362b). Hence, the integrability conditions (3.362a–c) hold as a consequence of (3.356a,b). This establishes the "if" part of the theorem.

Finally, from the relations (3.359), (3.360), and (3.361), by the fundamental theorem of calculus for gradients, we obtain

$$\begin{aligned}\psi &= \int_C [\psi_x\,dx + \psi_y\,dy + \psi_{y_1}\,dy_1]\\ &= \int_C [\Phi\,dx + \Psi_1\,dy + \Psi_2\,dy_1]\end{aligned} \tag{3.363}$$

to within an arbitrary constant, where C is any path curve from a point $(\tilde{x},\tilde{y},\tilde{y}_1)$ to (x,y,y_1). Then integral (3.363), combined with (3.360) and (3.362a), yields the line integral formula (3.357). □

Now, by applying Theorem 3.6.2-2 to the characteristic equation (3.352a), we obtain a necessary and sufficient determining system for all integrating factors $\Lambda(x,y,y_1)$ of ODE (3.349). Let

$$\theta(x,y,y_1,y_2) = (y_2 - f(x,y,y_1))\Lambda(x,y,y_1). \tag{3.364}$$

Then (3.356a) reduces to an identity and (3.356b) is linear in y_2. Consequently, (3.356b) splits into two equations

$$A(x,y,y_1) = B(x,y,y_1) = 0,$$

where the functions A and B are given by

$$\Psi_0 = A(x,y,y_1)y_2 + B(x,y,y_1) = \mathrm{E}_2(\theta).$$

Explicitly, this yields

$$2\Lambda_y + \Lambda_{xy_1} + y_1\Lambda_{yy_1} + (f\Lambda)_{y_1y_1} = 0, \tag{3.365a}$$

$$-(f\Lambda)_y + (f\Lambda)_{xy_1} + y_1(f\Lambda)_{yy_1} + \Lambda_{xx} + (y_1)^2\Lambda_{yy} + 2y_1\Lambda_{xy} = 0. \tag{3.365b}$$

For any $\Lambda(x,y,y_1)$ satisfying the integrating factor determining system (3.365a,b), it follows from (3.359–3.361) and (3.364) that we have

$$\Psi_2 = \psi_{y_1} = \Lambda, \quad \Psi_1 = \psi_y = -\Lambda_x - y_1\Lambda_y - (f\Lambda)_{y_1}, \quad \Phi = \psi_x = -y_1\Psi_1 - f\Psi_2.$$

Hence, the first integral formula (3.357) reduces to

$$\psi(x,y,y_1) = \int_C [(y_1\Lambda_x + (y_1)^2\Lambda_y + y_1(f\Lambda)_{y_1} - f\Lambda)\,dx - (\Lambda_x + (f\Lambda)_{y_1} + y_1\Lambda_y)\,dy + \Lambda\,dy_1], \tag{3.366}$$

where C is a path curve from a point $(\tilde{x},\tilde{y},\tilde{y}_1)$ to (x,y,y_1). If $f(x,y,y_1)$ and $\Lambda(x,y,y_1)$ are nonsingular, then the curve C can be chosen arbitrarily, and a change in $(\tilde{x},\tilde{y},\tilde{y}_1)$ just changes (3.366) by the addition of a constant. Most important, if $f(x,y,y_1)$ or $\Lambda(x,y,y_1)$ are singular, then one can choose a path curve C so that the line integral (3.366) is nonsingular. This will be illustrated in examples in Section 3.6.3. Thus, the following theorem has been proved:

Theorem 3.6.2-3. *The integrating factors of ODE* (3.349) *are the solutions* $\Lambda(x,y,y_1) \not\equiv 0$ *of the integrating factor determining system* (3.365a,b). *For a given integrating factor, the corresponding first integral of ODE* (3.349) *is given by the line integral* (3.366). *Conversely, every first integral arises from a corresponding integrating factor through the characteristic equation* (3.352a,b).

The integrating factor determining system (3.365a,b) is a linear homogeneous system of second-order PDEs for $\Lambda(x,y,y_1)$. If one knows two functionally independent first integrals $\psi_1(x,y,y_1)$ and $\psi_2(x,y,y_1)$ of ODE (3.349), i.e., ψ_1 is not equal to a function of ψ_2, then the general solution of (3.365a,b) is given by $\Lambda(x,y,y_1) = F_{\psi_1}\Lambda_1 + F_{\psi_2}\Lambda_2$ where $F = F(\psi_1,\psi_2)$ is an arbitrary function of ψ_1 and ψ_2, with $\Lambda_1 = (\psi_1)_{y_1}$, $\Lambda_2 = (\psi_2)_{y_1}$. Hence, finding all solutions of the integrating factor determining system (3.365a,b) is equivalent to solving the original ODE (3.349). However, any particular solution of (3.365a,b) yields a first integral (3.366) which reduces ODE (3.349) to a first-order ODE given by the surface $\psi(x,y,y_1) = \text{const} = c$. If two solutions of the integrating factor determining system (3.365a,b) yield functionally independent first integrals $\psi_1(x,y,y_1)$ and $\psi_2(x,y,y_1)$, then ODE (3.349) is reduced to quadrature by the elimination of y_1 in the two equations $\psi_1(x,y,y_1) = \text{const} = c_1$, $\psi_2(x,y,y_1) = \text{const} = c_2$. This represents, implicitly, the general solution of ODE (3.349) in terms of two essential constants c_1 and c_2.

Note that if two first integrals are related by $\psi_2 = F(\psi_1)$ for some function F, then $\Lambda_2 = F'(\psi_1)\Lambda_1$ holds for the corresponding integrating factors. This gives a criterion for checking when two solutions of the integrating factor determining system yield functionally independent first integrals.

Lemma 3.6.2-1 (Criterion for Integrating Factors to Yield Two Functionally Independent First Integrals). *Two integrating factors $\Lambda_1(x,y,y_1)$ and $\Lambda_2(x,y,y_1)$ of ODE* (3.349) *determine functionally independent first integrals $\psi_1(x,y,y_1)$ and $\psi_2(x,y,y_1)$, $\psi_2 \neq F(\psi_1)$, if and only if for all functions $F(\psi_1)$, $\Lambda_2 \neq F'(\psi_1)\Lambda_1$ or, equivalently, $\Lambda_1 \neq F'(\psi_2)\Lambda_2$ where ψ_1 and ψ_2 are given, respectively, in terms of Λ_1 and Λ_2 by the line integral formula* (3.366).

As a corollary of Lemma 3.6.2-1, we have the following useful sufficient criterion for the functional independence of first integrals:

Theorem 3.6.2-4. *If two integrating factors $\Lambda_1(x,y,y_1)$ and $\Lambda_2(x,y,y_1)$ of ODE* (3.349) *satisfy*

$$\Lambda_2/\Lambda_1 \not\equiv \text{const}, \quad (\Lambda_2/\Lambda_1)_{y_1} = 0, \tag{3.367}$$

then the corresponding first integrals $\psi_1(x,y,y_1)$ and $\psi_2(x,y,y_1)$, given by the line integral formula (3.366), *are functionally independent.*

Proof. From (3.367), we have $\Lambda_2/\Lambda_1 = c(x,y) \neq \text{const}$ for some function $c(x,y)$. Then from Lemma 3.6.2-1 we see that $\Lambda_1(x,y,y_1)$ and $\Lambda_2(x,y,y_1)$ determine the functionally dependent first integrals $\psi_1(x,y,y_1)$ and $\psi_2(x,y,y_1)$ if and only if the relation $c(x,y) = F'(\psi_1)$ holds for some function $F(\psi_1)$. Taking $\partial/\partial y_1$ of this relation, we obtain $F''(\psi_1)\Lambda_1 = 0$, and thus, $F'(\psi_1) = \text{const}$. But, since $c(x,y) \neq \text{const}$, it follows that $\psi_1(x,y,y_1)$ and $\psi_2(x,y,y_1)$ must be functionally independent. □

3.6.3 FIRST INTEGRALS OF SECOND-ORDER ODEs

We now consider several effective algorithmic methods for finding solutions of the integrating factor determining system (3.365a,b) for a second-order ODE (3.349). The situation is analogous to solving the symmetry determining equation [cf. Section 3.5.1] where, for a given second-order ODE (3.349), one can find all of its finite number of admitted point symmetries (if any exist) but, in general, one needs specific ansatzes in order to find admitted first-order symmetries (contact symmetries).

We also show how to construct a first integral of a second-order ODE (3.349) from an admitted integrating factor through the line integral formula (3.366) and obtain a corresponding reduction of order of the ODE.

(1) *Point-Form Ansatzes*
If one considers a point-form integrating factor

$$\Lambda = \alpha(x,y) + \beta(x,y)y_1, \tag{3.368}$$

then the integrating factor determining system (3.365a,b) reduces to an overdetermined linear system of PDEs for $\alpha(x,y)$ and $\beta(x,y)$ given by

$$2\alpha_y + \beta_x + 3y_1\beta_y + 2\beta f_{y_1} + \alpha f_{y_1y_1} + y_1\beta f_{y_1y_1} = 0, \tag{3.369a}$$

$$\begin{aligned}&\alpha_{xx} + y_1\beta_{xx} + (y_1)^2\alpha_{yy} + (y_1)^3\beta_{yy} + 2y_1\alpha_{xy} + 2(y_1)^2\beta_{xy}\\ &-(\alpha f)_y + (\beta f)_x + (\alpha f_{y_1})_x + y_1(\beta f_{y_1})_x + y_1(\alpha f_{y_1})_y + (y_1)^2(\beta f_{y_1})_y = 0.\end{aligned} \tag{3.369b}$$

The system (3.369a,b) has at most a finite number of linearly independent solutions $\alpha(x,y), \beta(x,y)$. For a given solution $\alpha(x,y)$ and $\beta(x,y)$, the corresponding first integral (3.366) of ODE (3.349) is given by

$$\begin{aligned}\psi(x,y,y_1) = \int_C [(-f\alpha + y_1(\alpha_x + \alpha f_{y_1}) + (y_1)^2(\beta_x + \alpha_y + \beta f_{y_1}) + (y_1)^3\beta_y)dx\\ -(\alpha_x + \alpha f_{y_1} + f\beta + y_1(\beta_x + \alpha_y + \beta f_{y_1}) + (y_1)^2\beta_y)dy + (\alpha + \beta y_1)dy_1]\end{aligned} \tag{3.370}$$

for any path curve C from $(\tilde{x},\tilde{y},\tilde{y}_1)$ to (x,y,y_1).

We now establish a useful classification result:

Lemma 3.6.3-1. *A second-order ODE* (3.349) *admits an integrating factor of point-form* (3.368) *if and only if* (3.349) *is of the form*

$$\begin{aligned}y_2 = -\tfrac{1}{2}(\log\beta)_y(y_1)^2 - [\tfrac{1}{2}(\log\beta)_x + \beta^{-1/2}(\alpha\beta^{-1/2})_y]y_1\\ -\beta^{-1/2}(\alpha\beta^{-1/2})_x - (\alpha+\beta y_1)^{-1}(\gamma_x + \gamma_y y_1)\end{aligned} \tag{3.371}$$

for some functions $\alpha(x,y)$, $\beta(x,y)$, $\gamma(x,y)$. *The corresponding first integral of ODE* (3.349) *is given by*

$$\psi(x,y,y_1)=\tfrac{1}{2}\beta^{-1}(\alpha+\beta y_1)^2+\gamma. \tag{3.372}$$

Proof. From (3.352b), we have

$$\psi_{y_1}=\Lambda=\alpha+\beta y_1. \tag{3.373}$$

After integrating (3.373) with respect to y_1, we find that

$$\begin{aligned}\psi&=\tfrac{1}{2}\beta(y_1)^2+\alpha y_1+\tilde{\gamma}\\ &=\tfrac{1}{2}\beta^{-1}\Lambda^2+\gamma,\ \gamma=\tilde{\gamma}-\tfrac{1}{2}\beta^{-1}\alpha^2,\end{aligned} \tag{3.374}$$

for some function $\tilde{\gamma}(x,y)$, which thus yields the first integral (3.372). Then substitution of (3.374) into the characteristic equation (3.352a) gives

$$\begin{aligned}\mathrm{D}\psi&=\Lambda\beta^{-1}\mathrm{D}\Lambda-\tfrac{1}{2}\Lambda^2\beta^{-2}\mathrm{D}\beta+D\gamma=\Lambda(\beta^{-1}\mathrm{D}\alpha+\tfrac{1}{2}(-\beta^{-2}\alpha+\beta^{-1}y_1)\mathrm{D}\beta+\Lambda^{-1}\mathrm{D}\gamma+y_2)\\ &=\Lambda(y_2-f).\end{aligned}$$

Hence, we obtain

$$f=-\tfrac{1}{2}\beta^{-1}D\beta-\beta^{-1}D\alpha+\tfrac{1}{2}\beta^{-2}\alpha D\beta-(\alpha+\beta y_1)^{-1}D\gamma,$$

which yields ODE (3.371). □

Note that if one can match a given ODE (3.349) to the form (3.371) for some $\alpha(x,y)$, $\beta(x,y)$, $\gamma(x,y)$, then one obtains an integrating factor (3.368) with first integral (3.372). Most important, the classification given by Lemma 3.6.3-1 leads to a stronger counterpart of Lemma 3.6.2-1 and Theorem 3.6.2-4 for the functional independence of first integrals arising from integrating factors of point-form.

Theorem 3.6.3-1. *Suppose ODE* (3.349) *admits two integrating factors of point-form,*

$$\Lambda_1=\alpha_1(x,y)+\beta_1(x,y)y_1,\quad \Lambda_2=\alpha_2(x,y)+\beta_2(x,y)y_1\quad [\Lambda_1\neq\Lambda_2]. \tag{3.375}$$

Then the integrating factors (3.375) *determine functionally independent first integrals of ODE* (3.349) *if, when* $f_{y_1y_1y_1}\not\equiv 0$, Λ_1 *and* Λ_2 *are linearly independent or, when* $f_{y_1y_1y_1}\equiv 0$, $\Lambda_1+c_1\sqrt{\beta_1}$ *and* $\Lambda_2+c_2\sqrt{\beta_2}$ *are linearly independent for all constants* c_1, c_2.

Proof. Suppose $\psi_2=F(\psi_1)$ for some function F, where $\psi_1(x,y,y_1)$ and $\psi_2(x,y,y_1)$ are functionally dependent first integrals corresponding to the integrating factors $\Lambda_1=\alpha_1+\beta_1y_1=(\psi_1)_{y_1}$ and $\Lambda_2=\alpha_2+\beta_2y_1=(\psi_2)_{y_1}$. The case $\beta_1=\beta_2=0$ is covered by Theorem 3.6.2-4. So, henceforth, we assume that $\beta_1\not\equiv 0$. To proceed, we see that since $\Lambda_2=F'(\psi_1)\Lambda_1$ is linear in y_1, it must satisfy

$$(F'\Lambda_1)_{y_1y_1} = 3\beta_1 F''\Lambda_1 + F'''(\Lambda_1)^3 = 0. \tag{3.376}$$

There are now two cases to consider: $F'''(\psi_1) \equiv 0$ or $F'''(\psi_1) \not\equiv 0$. In the first case, (3.376) gives $F''(\psi_1) = 0$ and so we immediately have $F(\psi_1) = c_1\psi_1 + c_2$ for some constants c_1 and c_2. Then $\Lambda_2 = F'(\psi_1)\Lambda_1 = c_1\Lambda_1$ yields

$$\beta_2 = c_1\beta_1, \quad \alpha_2 = c_1\alpha_1.$$

Next, in the case $F'''(\psi_1) \not\equiv 0$, (3.376) yields $3F''/F''' = -(\Lambda_1)^2/\beta_1$. It then follows from (3.372) that

$$-\frac{3}{2}\frac{F''}{F'''} = \frac{1}{2}\frac{(\Lambda_1)^2}{\beta_1} = \psi_1 - \gamma_1. \tag{3.377}$$

But, since F depends just on ψ_1, we must have $\gamma_1 = \text{const}$ and, consequently,

$$\frac{F'''}{F''} = -\frac{3}{2}\frac{1}{\psi_1 - \gamma_1}.$$

Thus, $F(\psi_1) = c_0 + c_1\psi_1 + c_2\sqrt{\psi_1 - \gamma_1}$ for some constants c_0, c_1, and $c_2 \neq 0$. Substitution of the right-hand side of (3.377) into $\Lambda_2 = F'(\psi_1)\Lambda_1 = [c_1 + \tfrac{1}{2}c_2(\psi_1 - \gamma_1)^{-1/2}]\Lambda_1$ then yields

$$\beta_2 = c_1\beta_1, \quad \alpha_2 = c_1\alpha_1 + c_2\tfrac{1}{\sqrt{2}}\sqrt{\beta_1}. \tag{3.378}$$

Finally, from (3.377) combined with (3.371) and (3.372), we have that in this case $f(x, y, y_1)$ is quadratic in y_1. Hence, $f_{y_1y_1y_1} = 0$. □

As an example to illustrate how to find point-form integrating factors, we consider the nonlinear Duffing equation

$$y'' + ay' + by + y^3 = 0, \quad a = \text{const} \geq 0, \; b = \text{const} \geq 0. \tag{3.379}$$

This describes a nonlinear damped oscillator, where a is the damping constant and $\sqrt{b}/2\pi$ is the frequency for undamped linearized oscillations. For point-form integrating factors (3.368) of ODE (3.379), the integrating factor determining system (3.369a,b) is given by

$$2\alpha_y + \beta_x - 2a\beta + 3\beta_y y_1 = 0, \tag{3.380a}$$

$$\begin{aligned}(\beta_{yy})(y_1)^3 + (2\beta_{xy} - a\beta_y + \alpha_{yy})(y_1)^2 + (-2a\beta_x + \beta_{xx} + 2\alpha_{xy})y_1 \\ + b\alpha + 3y^2\alpha - a\alpha_x + \alpha_{xx} - by\beta_x - y^3\beta_x + by\alpha_y + y^3\alpha_y = 0,\end{aligned} \tag{3.380b}$$

which are polynomial equations in terms of y_1. Hence, the coefficients of like powers of y_1 yield the determining equations

$$\beta_y = 0, \quad \alpha_{yy} = 0, \tag{3.381a}$$

$$2\alpha_y + \beta_x - 2a\beta = 0, \tag{3.381b}$$

$$(b+3y^2)\alpha - a\alpha_x + \alpha_{xx} - (by+y^3)\beta_x + (by+y^3)\alpha_y = 0. \tag{3.381c}$$

From (3.381a), we see that $\beta = \beta(x)$ and

$$\alpha = \alpha_0(x) + y\alpha_1(x) \tag{3.382}$$

for some functions $\alpha_0(x), \alpha_1(x)$. Then (3.381b) yields

$$\alpha_1 = a\beta - \tfrac{1}{2}\beta', \tag{3.383}$$

and (3.381c) becomes a polynomial equation in terms of y, which splits into the equations

$$\alpha_0 = 0, \tag{3.384a}$$

$$\alpha_1 = \tfrac{1}{4}\beta', \tag{3.384b}$$

$$\alpha_1'' - a\alpha_1' + 2b\alpha_1 = b\beta'. \tag{3.384c}$$

By combining (3.383) and (3.384b), we obtain

$$\beta = \beta_0 e^{(4a/3)x}, \quad \beta_0 = \text{const}, \tag{3.385a}$$

$$\alpha_1 = \tfrac{1}{3}a\beta_0 e^{(4a/3)x}. \tag{3.385b}$$

Finally, (3.384c) yields $\tfrac{2}{9}a^3 = ba$, and so

$$a = 0 \quad \text{or} \quad b = \tfrac{2}{9}a^2. \tag{3.386}$$

Hence, ODE (3.379) admits a single point-form integrating factor given by

$$\Lambda = (\tfrac{1}{3}ay + y_1)e^{(4a/3)x}, \tag{3.387}$$

where the constants a and b are restricted to satisfy (3.386). A corresponding first integral is obtained from the line integral (3.370). If we choose the path curve C to be a piecewise straight line, parallel to the coordinate axes, from $(0,0,0)$ to (x,y,y_1), then (3.370) yields

$$\begin{aligned}\psi &= \int_0^x 0\,dx + \int_0^y (\tfrac{1}{9}a^2y + y^3)e^{(4a/3)x}dy + \int_0^{y_1} (\tfrac{1}{3}ay + y_1)e^{(4a/3)x}dy_1 \\ &= (\tfrac{1}{18}a^2y^2 + \tfrac{1}{4}y^4 + \tfrac{1}{3}ayy_1 + \tfrac{1}{2}(y_1)^2)e^{(4a/3)x}.\end{aligned} \tag{3.388}$$

We thus obtain a reduction of ODE (3.379), with $a = 0$ or $b = \frac{2}{9}a^2$, to the first-order ODE given by $\psi(x, y, y_1) = \text{const} = c$. [In Section 3.7.3, we will show how to obtain a second first integral leading to the complete quadrature of ODE (3.379).]

(2) *Symmetry-Type Ansatzes*
If a second-order ODE (3.349) admits a point symmetry

$$\mathrm{X} = \xi(x, y)\frac{\partial}{\partial x} + \eta(x, y)\frac{\partial}{\partial y}, \tag{3.389}$$

then the first-extended generator

$$\mathrm{X}^{(1)} = \xi\frac{\partial}{\partial x} + \eta\frac{\partial}{\partial y} + \eta^{(1)}\frac{\partial}{\partial y_1}, \quad \eta^{(1)} = \mathrm{D}\eta - y_1\mathrm{D}\xi \tag{3.390}$$

maps first integrals of ODE (3.349) into first integrals since, geometrically, a first integral is constant on every solution curve on the surface (3.349) in (x, y, y_1, y_2)–space on which the second-extended generator $\mathrm{X}^{(2)}$ is a tangent vector field [cf. Section 3.5.1]. Thus, $\mathrm{X}^{(1)}$ describes a geometrical motion within the vector space of first integrals of ODE (3.349).

Theorem 3.6.3-2. *Suppose* $\psi(x, y, y_1)$ *is a first integral of ODE* (3.349) *with integrating factor* $\Lambda(x, y, y_1)$. *Then, under any point symmetry* (3.389) *admitted by ODE* (3.349),

$$\widetilde{\psi} = \mathrm{X}^{(1)}\psi + \widetilde{c}, \quad \widetilde{c} = \text{const}, \tag{3.391}$$

yields a first integral with integrating factor

$$\widetilde{\Lambda} = \mathrm{X}^{(1)}\Lambda + R\Lambda, \tag{3.392}$$

where

$$R \equiv \frac{\partial \eta^{(1)}}{\partial y_1} = \eta_y - \xi_x - 2y_1\xi_y. \tag{3.393}$$

Proof. We apply $\mathrm{X}^{(2)}$ to the characteristic equation (3.352a), which yields

$$(\mathrm{X}^{(1)}\Lambda)(y_2 - f) + \Lambda(\mathrm{X}^{(2)}(y_2 - f)) = \mathrm{X}^{(2)}\mathrm{D}\psi. \tag{3.394}$$

Then, from (2.100a,b) and (3.124), we obtain

$$\begin{aligned} \mathrm{X}^{(2)}(y_2 - f) &= \eta^{(2)} - f_x\xi - f_y\eta - f_{y_1}\eta^{(1)} \\ &= \mathrm{D}^2\eta - y_1\mathrm{D}^2\xi - 2y_2\mathrm{D}\xi - f_{y_1}(\mathrm{D}\eta - y_1\mathrm{D}\xi) - f_x\xi - f_y\eta \\ &= (\eta_y - y_1\xi_y - 2\mathrm{D}\xi)(y_2 - f). \end{aligned} \tag{3.395}$$

Hence, the left-hand side of (3.394) becomes

$$(\mathrm{X}^{(1)}\Lambda + (R - \mathrm{D}\xi)\Lambda)(y_2 - f).$$

We next evaluate the right-hand side of (3.394) by using the operator identity

$$\mathrm{D}\mathrm{X}^{(2)} - \mathrm{X}^{(2)}\mathrm{D} = (\mathrm{D}\xi)\mathrm{D}$$

that holds on differentiable functions of x, y, y_1 [Exercise 3.6-17]. This yields

$$\mathrm{D}(\mathrm{X}^{(1)}\psi) = (\mathrm{X}^{(1)}\Lambda + R\Lambda)(y_2 - f), \tag{3.396}$$

and hence, (3.392) is an integrating factor corresponding to the first integral (3.391). □

From Theorem 3.6.3-2, we see that every point symmetry admitted by ODE (3.349) induces a point symmetry on the vector space of its integrating factors. The explicit generator in (x, y, y_1, Λ) – space is given by

$$\widetilde{\mathrm{X}} = \xi\frac{\partial}{\partial x} + \eta\frac{\partial}{\partial y} + \eta^{(1)}\frac{\partial}{\partial y_1} + R\Lambda\frac{\partial}{\partial \Lambda}, \tag{3.397}$$

which maps integrating factors into integrating factors of ODE (3.349). Consequently, one can consider an ansatz to seek integrating factors invariant under (3.397) to within a scaling. In particular, for such an ansatz, one has

$$\widetilde{\mathrm{X}}\Lambda = r\Lambda, \quad r = \text{const}, \tag{3.398}$$

or, equivalently,

$$\xi\Lambda_x + \eta\Lambda_y + (\eta_x + y_1\eta_y - y_1\xi_x - (y_1)^2\xi)\Lambda_{y_1} + (\eta_y - \xi_x - 2y_1\xi_y - r)\Lambda = 0. \tag{3.399}$$

We can solve the first-order linear PDE (3.399) by the method of characteristics in terms of invariants $u(x,y)$ and $v(x,y,y_1)$ $[v_{y_1} \neq 0]$ of $\mathrm{X}^{(1)}$ [cf. Section 3.3.2] given by solving (3.102). This leads to

$$\Lambda = \exp\left(\int\frac{r-R}{\eta^{(1)}}dy_1\right)w(u,v), \tag{3.400a}$$

in terms of an arbitrary function $w(u,v)$, where $\eta^{(1)} = \eta^{(1)}(u,v,y_1)$, $R = R(u,v,y_1)$, with x and y eliminated in terms of u and v.

Alternatively, if $\xi \not\equiv 0$, then one can have

$$\Lambda = \exp\left(\int\frac{r-R}{\xi}dx\right)w(u,v), \tag{3.400b}$$

in terms of an arbitrary function $w(u,v)$, where $\xi = \xi(x,u)$, $R = R(x,u,v)$, with y and y_1 eliminated in terms of u and v.

If $\xi = 0$, then one can have

$$\Lambda = \exp\left(\int\frac{r-R}{\eta}dy\right)w(u,v), \tag{3.400c}$$

in terms of an arbitrary function $w(u,v)$, where $\eta=\eta(y,u)$, $R=R(y,u,v)$, with x and y_1 eliminated in terms of u and v.

Note that the ansatz (3.398) reduces to the scaling invariance of Λ corresponding to

$$X^{(1)}\Lambda = s\Lambda, \qquad s=\text{const}, \tag{3.401}$$

if and only if $R=\text{const}$, with $s=r-R$. Moreover, from Theorem 3.6.3-2, this condition on R also corresponds to the scaling invariance of $\psi(x,y,y_1)$ under $X^{(1)}$, i.e.,

$$X^{(1)}\psi = r\psi+\widetilde{c}, \qquad r=\text{const}=s+R, \tag{3.402}$$

to within some constant $\widetilde{c}$. All point symmetries satisfying $R=\text{const}$ are easily classified.

Lemma 3.6.3-2. *A point symmetry* (3.389) *of a second-order ODE* (3.349) *has* $R\equiv\partial\eta^{(1)}/\partial y_1=\text{const}=c$ *if and only if*

$$\eta=\alpha(x)+(\beta'(x)+c)y, \qquad \xi=\beta(x), \tag{3.403}$$

for some functions $\alpha(x), \beta(x)$. *In particular, all translations* $[\xi=\text{const}=a,$ $\eta=\text{const}=b]$ *and all scalings* $[\xi=ax, \eta=by,\ a=\text{const},\ b=\text{const}]$ *satisfy* $R=\text{const}$.

Proof. Left to Exercise 3.6-19. □

Thus, from Lemma 3.6.3-2 and Theorem 3.6.3-2, it follows that any translation and scaling symmetries admitted by a given ODE (3.349) are automatically inherited by its integrating factor determining system (3.365a,b). Hence, if ODE (3.349) is invariant under such symmetries, then one can consider a simple ansatz

$$\Lambda = w(u,v) \tag{3.404}$$

in terms of corresponding invariants $u(x,y), v(x,y,y_1)$.

We now illustrate the use of ansatzes (3.400a–c) and (3.404) through several examples.

As an elementary first example, consider a general second-order linear homogeneous ODE [cf. Section 3.3.3]

$$y_2+p(x)y_1+q(x)y=0, \tag{3.405}$$

which admits the scaling $y\to\lambda y, x\to x$ and the shifts $x\to x,\ y\to y+\varepsilon\phi(x)$ where $\phi(x)$ is any solution of ODE (3.405), i.e., $\phi''+p(x)\phi'+q(x)\phi=0$. Respective symmetry invariants are given by $u_{(1)}=x, v_{(1)}=y_1/y$, and $u_{(2)}=x, v_{(2)}=y_1-y\phi'/\phi$. Using the ansatz (3.404) with the unique joint invariant $u=x$, we seek solutions of the integrating factor determining system (3.369a,b) of the form

$$\Lambda = w(x). \tag{3.406}$$

Substitution of (3.406) into (3.369a,b) yields a single equation

$$w'' - (p(x)w)' + q(x)w = 0, \tag{3.407}$$

which is the adjoint of ODE (3.405). ODE (3.407) is well-known as the determining equation for a classical integrating factor of (3.405). If one knows a solution $w(x)$ of ODE (3.407), then a first integral of ODE (3.405) is obtained from (3.370) through

$$\begin{aligned}\psi(x,y,y_1) &= \int_C [(qyw + y_1 w')\,dx - (w' - pw)\,dy + w\,dy_1] \\ &= \int_0^x 0\,dx - (w' - pw)\int_0^y dy + w\int_0^{y_1} dy_1 \\ &= -(w' - pw)y + wy_1,\end{aligned}$$

where C is a piecewise straight line, parallel to the coordinate axes, from $(0,0,0)$ to (x,y,y_1).

Now consider the second-order linear ODE (3.405) with, for example, $p(x) = 4/x$, $q(x) = 2/x^2$, which yields the Euler equation

$$y_2 + \frac{4}{x}y_1 + \frac{2}{x^2}y = 0 \tag{3.408}$$

admitting the additional scaling symmetry $x \to \lambda x$, $y \to y$. In Section 3.5.3, we derived a first integral (3.294) of ODE (3.408) using symmetry reduction. Here we seek integrating factors and corresponding first integrals using the ansatz (3.400b,c) based on invariance under scalings in x and y and shifts in y.

For the x scaling symmetry [$\eta = 0$ and $\xi = x$], we have $R = -1$ from (3.393), and thus ansatz (3.400b) yields $\Lambda = x^{r+1}w(y, xy_1)$ in terms of the scaling invariants x and xy_1. Similarly, for the y scaling symmetry [$\eta = y$ and $\xi = 0$], ansatz (3.400c) yields $\Lambda = y^{r-1}w(x, y_1/y)$ in terms of the scaling invariants x and y_1/y. Finally, for the shift symmetry [$\eta = \phi(x)$ and $\xi = 0$], we have $R = 0$, and hence, $\Lambda = e^{ry/\phi}w(x, y_1 - y\phi'/\phi)$. Consequently, the common invariant form for Λ is given by

$$\Lambda = x^s, \quad s = \text{const.} \tag{3.409}$$

Substitution of (3.409) into the integrating factor determining equation (3.407) easily leads to the solutions

$$\Lambda_1 = x^2, \quad \Lambda_2 = x^3. \tag{3.410}$$

From Theorem 3.6.2-4, we see that the solutions (3.410) determine two functionally independent first integrals given by the line integral formula (3.370). Since ODE (3.408) is singular at $x = 0$, we choose the path curve C to be a piecewise straight line from $(\tilde{x},0,0)$ to (x,y,y_1), parallel to the coordinate axes, with $\tilde{x} \neq 0$. Then (3.370) yields the first integrals

$$\psi_1 = \int_{\tilde{x}}^{x} 0\,dx + \int_0^y 2x\,dy + \int_0^{y_1} x^2\,dy_1 = 2xy + x^2 y_1, \tag{3.411a}$$

$$\psi_2 = \int_{\tilde{x}}^{x} 0dx + \int_0^y x^2 dy + \int_0^{y_1} x^3 dy_1 = x^2 y + x^3 y_1. \tag{3.411b}$$

Hence, $\psi_1 = \text{const} = c_1$ and $\psi_2 = \text{const} = c_2$ yield two quadratures giving the complete reduction of ODE (3.408). Explicitly, by eliminating y_1 in the equations $\psi_1 = c_1$ and $\psi_2 = c_2$, we have

$$y = c_1 x^{-1} - c_2 x^{-2}. \tag{3.412}$$

For a final example, consider the ODE

$$y_2 = 2\frac{(xy_1 - y)(1 + (y_1)^2)}{x^2 + y^2}, \tag{3.413}$$

which admits the rotation symmetry [cf. Exercise 3.3-9]

$$\mathrm{X} = y\frac{\partial}{\partial x} - x\frac{\partial}{\partial y}, \quad \text{i.e., } \xi = y,\ \eta = -x. \tag{3.414}$$

From (3.393), we see that $R = -2y_1$. Since $R \neq \text{const}$, we seek integrating factors of (3.413) given by the general ansatz (3.400a) in terms of the rotation invariants

$$u = x^2 + y^2, \quad v = \frac{(y - xy_1)^2}{1 + (y_1)^2}, \tag{3.415}$$

satisfying $\mathrm{X}u = 0$, $\mathrm{X}^{(1)}v = 0$. Since $\eta^{(1)} = -(1 + (y_1)^2)$, the resulting ansatz is given by

$$\Lambda = e^{-r \arctan y_1}(1 + (y_1)^2)^{-1} w(u, v), \qquad r = \text{const}.$$

For simplicity, we take $r = 0$. From (3.398), this corresponds to invariance of Λ under $\widetilde{\mathrm{X}} = \mathrm{X}^{(1)} + R\Lambda\frac{\partial}{\partial \Lambda}$. Hence, for the ansatz

$$\Lambda = (1 + (y_1)^2)^{-1} w(u, v), \tag{3.416}$$

the integrating factor determining system (3.365a,b) becomes

$$2uw_u + (9v - 6u)w_v + 2u(v - u)w_{vu} + 4v(v - u)w_{vv} = 0, \tag{3.417a}$$

$$3u^2 w_u + 4v^2 w_v + 4uv(u - v)w_{vu} + 2u^2(u - v)w_{uu} = 0. \tag{3.417b}$$

It is easy to see that $w = \text{const}$ satisfies (3.417a,b). Hence we obtain a single integrating factor

$$\Lambda_1 = (1 + (y_1)^2)^{-1}. \tag{3.418}$$

The corresponding first integral of ODE (3.413) is given by the line integral formula (3.366), which here becomes

$$\psi_1 = \int_C \left[\frac{2y}{x^2+y^2}dx - \frac{2x}{x^2+y^2}dy + \frac{1}{1+(y_1)^2}dy_1 \right]. \tag{3.419}$$

Since the integrand in (3.419) is singular at $x = y = 0$, we take C to be a piecewise straight line from $(\widetilde{x}, \widetilde{y}, 0)$ to (x, y, y_1), parallel to the coordinate axes, with $\widetilde{x} \neq 0$ and $\widetilde{y} \neq 0$. This yields

$$\begin{aligned} \psi_1 &= \int_{\widetilde{x}}^{x} \frac{2\widetilde{y}}{x^2+\widetilde{y}^2}dx + \int_{\widetilde{y}}^{y} \frac{-2x}{x^2+y^2}dy + \int_0^{y_1} \frac{1}{1+(y_1)^2}dy_1 \\ &= -2\arctan\frac{y}{x} + \arctan y_1 - 2\arctan\frac{\widetilde{y}}{\widetilde{x}} + \pi. \end{aligned} \tag{3.420}$$

Setting $\widetilde{y} = 0$ and letting $\widetilde{\psi}_1 = \tan\psi_1$, we obtain the simplified first integral

$$\widetilde{\psi}_1 = \frac{y_1(y^2-x^2)+2xy}{y^2-x^2-2xyy_1}. \tag{3.421}$$

We next obtain another integrating factor from the determining system (3.417a,b) by exploiting the scaling symmetry $x \to \lambda x, y \to \lambda y$ admitted by ODE (3.413). If we further restrict the ansatz (3.416) to be scaling invariant, then $w(u,v) = w(z)$, with $z = v/u$, and hence,

$$\Lambda = (1+(y_1)^2)^{-1}w(z). \tag{3.422}$$

This simplifies the integrating factor determining system (3.417a,b) to a single ODE

$$u(4v-3u)w' + 2v(v-u)w'' = 0. \tag{3.423}$$

The general solution of ODE (3.423) is a linear combination of $w_1 = 1$ and

$$w_2 = (uv^{-1}-1)^{1/2}. \tag{3.424}$$

Thus, from (3.424), we obtain a second integrating factor

$$\Lambda_2 = (1+(y_1)^2)^{-1}\left(\frac{(x^2+y^2)(1+(y_1)^2)}{(y-xy_1)^2} - 1\right)^{1/2} = \frac{x+yy_1}{(y-xy_1)(1+(y_1)^2)}. \tag{3.425}$$

We now check that the integrating factor (3.425) satisfies the criterion of Lemma 3.6.2-1 to yield a first integral that is functionally independent of (3.420). First observe from (3.419) that since $\mathrm{X}^{(1)}\psi_1 = y(\psi_1)_x - x(\psi_1)_y - (1+(y_1)^2)(\psi_1)_{y_1} = 1$ is nonzero, it follows that ψ_1 is not a function only of $z = v/u$. It then follows from ansatz (3.422) and (3.424) that $w_2(z) = \Lambda_2/\Lambda_1 \neq F'(\psi_1)$ for any function $F(\psi_1)$. Thus, the first integrals corresponding to Λ_1 and Λ_2 are functionally independent. From the line integral formula (3.366), we obtain

$$\psi_2 = \int_C \left[\frac{y_1(y^2 - x^2) - 2xy}{(y - xy_1)(x^2 + y^2)} dx - \frac{x^2 - y^2 + 2xyy_1}{(y - xy_1)(x^2 + y^2)} dy + \frac{x + y_1 y}{(y - xy_1)(1 + (y_1)^2)} dy_1 \right].$$

Since this integrand is singular at $x = y = 0$, we take C from $(\tilde{x}, \tilde{y}, 0)$ to (x, y, y_1), with $\tilde{x} \neq 0, \tilde{y} \neq 0$, using a piecewise straight line parallel to the axes. This yields the first integral

$$\begin{aligned}\psi_2 &= \int_{\tilde{x}}^{x} \frac{2x}{(x^2 + \tilde{y}^2)} dx + \int_{\tilde{y}}^{y} \frac{y^2 - x^2}{y(x^2 + y^2)} dy + \int_0^{y_1} \frac{x + y_1 y}{(y - xy_1)(1 + (y_1)^2)} dy_1 \\ &= -\log(y - xy_1) + \tfrac{1}{2}\log(1 + (y_1)^2) + \log(x^2 + y^2) - \log(\tilde{x}^2 + \tilde{y}^2) + \log \tilde{y}.\end{aligned} \tag{3.426}$$

By exponentiation of (3.426), we obtain the simplified first integral

$$\tilde{\psi}_2 = \frac{(1 + (y_1)^2)^{1/2}(x^2 + y^2)}{y - xy_1} \tag{3.427}$$

to within a multiplicative constant. The first integrals (3.427) and (3.421) lead to the complete reduction (quadrature) of ODE (3.413) given by $\tilde{\psi}_1 = \text{const} = c_1$, $\tilde{\psi}_2 = \text{const} = c_2$. Explicitly, by eliminating y_1, we obtain

$$c_2(y + c_1 x) = \sqrt{1 + (c_1)^2}\,(x^2 + y^2), \tag{3.428}$$

which is the general solution of ODE (3.413).

(3) *Elimination of Variables Ansatzes*

We now consider ansatzes involving elimination of one of the variables x or y. Note that the elimination of y_1 is a special case of the point-form ansatz (3.368).

First of all, suppose

$$\Lambda = \mu(x, y_1), \tag{3.429}$$

i.e., Λ has no dependence on y. From the characteristic equation (3.352a,b), it is straightforward to classify all second-order ODEs (3.349) admitting integrating factors of the form (3.429). We integrate $\psi_{y_1} = \mu(x, y_1)$ with respect to y_1, which gives

$$\psi = a(x, y_1) + b(x, y), \quad a_{y_1} = \mu. \tag{3.430}$$

Substitution of (3.430) into the characteristic equation (3.352a) yields

$$-a_{y_1} f = a_x + b_x + y_1 b_y. \tag{3.431}$$

Thus, from (3.430) and (3.431), we obtain the following classification result:

Theorem 3.6.3-3. *A second-order ODE* (3.349) *admits an integrating factor of the form* (3.429) *if and only if ODE* (3.349) *is of the form*

$$y_2 = f(x,y,y_1) = \frac{y_1 v_y(x,y) + v_x(x,y) - \int \mu_x(x,y_1)\,dy_1}{\mu(x,y_1)} \tag{3.432}$$

for some functions $v(x,y)$, $\mu(x,y_1)$. *The corresponding first integral is given by*

$$\psi(x,y,y_1) = \int \mu(x,y_1)\,dy_1 - v(x,y). \tag{3.433}$$

If one can match a given ODE (3.349) to the form (3.432) for some $v(x,y)$, $\mu(x,y_1)$, then one immediately obtains a first integral (3.433).

Now consider integrating factors of the form

$$\Lambda = \mu(y,y_1). \tag{3.434}$$

By the same steps as in proving Theorem 3.6.3-3, it is straightforward to classify all second-order ODEs (3.349) admitting integrating factors of the form (3.434). This leads to the following result:

Theorem 3.6.3-4. *A second-order ODE* (3.349) *admits an integrating factor of the form* (3.434) *if and only if ODE* (3.349) *is of the form*

$$y_2 = f(x,y,y_1) = \frac{y_1\left(v_y(x,y) - \int \mu_y(y,y_1)\,dy_1\right) + v_x(x,y)}{\mu(y,y_1)} \tag{3.435}$$

for some functions $v(x,y)$, $\mu(y,y_1)$. *The corresponding first integral is given by*

$$\psi(x,y,y_1) = \int \mu(y,y_1)\,dy_1 - v(x,y). \tag{3.436}$$

If one can match a given ODE (3.349) to the form (3.435) for some $v(x,y)$, $\mu(y,y_1)$, then one immediately has a first integral (3.436).

The matching of a given ODE (3.349) to (3.432) or (3.435) can be carried out algorithmically [Cheb-Terrab and Roche (1999)]. In particular, from the integrating factor determining system (3.365a,b), one can easily derive necessary and sufficient conditions for a function $f(x,y,y_1)$ to satisfy (3.432) or (3.435).

Theorems 3.6.3-3 and 3.6.3-4 can be combined to obtain a result for integrating factors depending only on y_1.

Corollary 3.6.3-1. *A second-order ODE* (3.349) *admits an integrating factor of the form* $\Lambda = \mu(y_1)$ *if and only if ODE* (3.349) *is of the form*

$$y_2 = f(x,y,y_1) = \frac{v_x(x,y) + y_1 v_y(x,y)}{\mu(y_1)} \tag{3.437}$$

for some functions $\nu(x,y)$, $\mu(y_1)$. *The corresponding first integral is given by*

$$\psi(x,y,y_1) = \int \mu(y_1)\,dy_1 - \nu(x,y). \tag{3.438}$$

We note that a special case of an ODE of the form (3.437) is given by the *separable* second-order ODE

$$y_2 = \frac{a(y)}{b(y_1)}, \tag{3.439}$$

for any functions $a(y)$, $b(y_1)$.

An ODE (3.439) reduces to quadrature since the first integral (3.438) yields

$$\psi = -\int a(y)\,dy + \int y_1 b(y_1)\,dy_1 = \text{const} = c_1,$$

which represents a first-order separable ODE [cf. Section 3.6.1] given by an algebraic expression in y, y_1:

$$\int y_1 b(y_1)\,dy_1 = c_1 + \int a(y)\,dy. \tag{3.440}$$

Hence, if we denote the solved form of (3.440) by $y_1 = g(y;c_1)$, then the quadrature

$$\int \frac{1}{g(y;c_1)}\,dy - x = \text{const} = c_2 \tag{3.441}$$

yields the general solution of ODE (3.439).

3.6.4 DETERMINING EQUATIONS FOR INTEGRATING FACTORS OF THIRD- AND HIGHER-ORDER ODEs

We now consider a third- or higher-order ODE

$$y^{(n)} = f(x,y,y',\ldots,y^{(n-1)}), \quad n \geq 3, \tag{3.442}$$

or, equivalently, the surface given by

$$y_n - f(x,y,y_1,\ldots,y_{n-1}) = 0. \tag{3.443}$$

Theorem 3.6.4-1. *A function* $\psi(x,y,y',\ldots,y^{(n-1)})$ *with an essential dependence on* $y^{(n-1)}$ *is a first integral of ODE* (3.442), *satisfying* $\psi(x,y,y',\ldots,y^{(n-1)}) = \text{const} = c$ *on the surface* (3.443), *if and only if*

$$\mathrm{D}\psi\,|_{y_n=f} = \mathbf{D}\psi = \psi_x + y_1\psi_y + \cdots + y_{n-1}\psi_{y_{n-2}} + f\psi_{y_{n-1}} = 0, \tag{3.444}$$

where

$$\mathrm{D} = \frac{\partial}{\partial x} + y_1 \frac{\partial}{\partial y} + \cdots + y_n \frac{\partial}{\partial y_{n-1}} \tag{3.445a}$$

and

$$\mathbf{D} = \mathrm{D}\,|_{y_n=f} = \mathrm{D} - (y_n - f)\frac{\partial}{\partial y_{n-1}}. \tag{3.445b}$$

A corresponding integrating factor of order ℓ *of ODE* (3.442) *is a function* $\Lambda(x, y, y', ..., y^{(\ell)}) \not\equiv 0,\ 0 \le \ell \le n-1,$ *given by the characteristic equation*

$$\mathrm{D}\psi = (y_n - f)\Lambda \tag{3.446a}$$

with

$$\Lambda(x, y, y', ..., y^{(\ell)}) = \psi_{y_{n-1}}, \tag{3.446b}$$

which is equivalent to (3.444). *All first integrals of ODE* (3.442) *arise through integrating factors satisfying* (3.446a,b).

We now derive a determining system that yields the integrating factors of all orders, $0 \le \ell \le n-1$, for ODE (3.442). Consider the truncated Euler operator

$$\mathrm{E}_n = \sum_{i=0}^{n} (-\mathbf{D})^i \frac{\partial}{\partial y_i} \quad \text{with } y_0 = y. \tag{3.447}$$

Let

$$\theta(x, y, y_1, \ldots, y_n) = \mathrm{D}\psi(x, y, y_1, \ldots, y_{n-1}), \tag{3.448}$$

and define, inductively,

$$\Psi_n = \theta_{y_n}, \tag{3.449a}$$

$$\Psi_{n-k} = \theta_{y_{n-k}} - \mathrm{D}\Psi_{n-k+1}, \quad k = 1,2,\ldots,n, \tag{3.449b}$$

where

$$\Psi_0 = \mathrm{E}_n(\theta). \tag{3.449c}$$

The following result shows that the truncated Euler operator (3.447) is connected with annihilating total derivatives of differentiable functions of $x, y, y_1, \ldots, y_{n-1}$:

Theorem 3.6.4-2. *A function* $\theta(x, y, y_1, \ldots, y_n)$ *is a total derivative* (3.448) *if and only if it satisfies*

$$\frac{\partial \Psi_{n-k}}{\partial y_n} = 0, \quad k = 0, 1, \ldots, n-1, \tag{3.450a}$$

$$\Psi_0 = 0, \tag{3.450b}$$

on the entire $(x, y, y_1, \ldots, y_n)-$space. *Then* (3.448) *holds with*

$$\psi(x, y, y_1, \ldots, y_{n-1}) = \int_C \left[\theta(x, y, y_1, \ldots, y_n)\, dx + \sum_{i=1}^{n} \Psi_i(x, y, y_1, \ldots, y_{n-1})(dy_{i-1} - y_i\, dx) \right]_{y_n=0}, \tag{3.451}$$

where C *is any path curve from a point* $(\tilde{x}, \tilde{y}, \tilde{y}_1, \ldots, \tilde{y}_{n-1})$ *to* $(x, y, y_1, \ldots, y_{n-1})$.

Proof. We break up the proof into "if" and "only if" parts. First of all, suppose a function $\theta(x, y, y_1, \ldots, y_n)$ satisfies (3.448) for some $\psi(x, y, y_1, \ldots, y_{n-1})$. Then from the identities

$$(\mathrm{D}\psi)_y = \mathrm{D}\psi_y, \tag{3.452a}$$

$$(\mathrm{D}\psi)_{y_i} = \mathrm{D}\psi_{y_i} + \psi_{y_{i-1}}, \quad i = 1, 2, \ldots, n \text{ with } y_0 = y, \tag{3.452b}$$

it follows that

$$\mathrm{E}_n(\mathrm{D}\psi) = \sum_{i=0}^{n} (-\mathrm{D})^i (\mathrm{D}\psi)_{y_i} = -\sum_{i=0}^{n} (-\mathrm{D})^{i+1} \psi_{y_i} + \sum_{i=1}^{n} (-\mathrm{D}^i)\psi_{y_{i-1}} = (-1)^n \mathrm{D}^n \psi_{y_n} = 0$$

since $\psi_{y_n} = 0$. Thus, (3.450b) holds. In addition, substitution of (3.448) into (3.449a,b) yields

$$\Psi_i = \psi_{y_{i-1}}, \quad i = 1, 2, \ldots, n, \tag{3.453}$$

and hence, we obtain (3.450a) since $(\Psi_i)_{y_n} = \psi_{y_n y_{i-1}} = 0$ from $\psi_{y_n} = 0$.

Let

$$\Phi = \theta - \sum_{i=1}^{n} y_i \Psi_i, \tag{3.454}$$

which satisfies

$$\Phi = \psi_x \tag{3.455}$$

from (3.448) and (3.453). Then we see that the integrability conditions for the existence of $\psi(x, y, y_1, \ldots, y_{n-1})$ are given by

$$\frac{\partial \Phi}{\partial y_n} = \frac{\partial \Psi_i}{\partial y_n} = 0, \quad i = 1,2,\ldots,n, \tag{3.456a}$$

$$\frac{\partial \Psi_i}{\partial y_j} = \frac{\partial \Psi_{j+1}}{\partial y_{i-1}}, \quad j = 0,1,\ldots,n-1, \ i = 1,2,\ldots,n, \tag{3.456b}$$

$$\frac{\partial \Phi}{\partial y_j} = \frac{\partial \Psi_{j+1}}{\partial x}, \quad j = 0,1,\ldots,n-1. \tag{3.456c}$$

Conversely, suppose a function $\theta(x, y, y_1, \ldots, y_n)$ satisfies (3.450a,b). We now proceed to show that the integrability conditions (3.456a–c) are satisfied. Let

$$\Psi_{i,j} = \frac{\partial \Psi_i}{\partial y_{j-1}} - \frac{\partial \Psi_j}{\partial y_{i-1}}, \quad i = 1,2,\ldots,n, \ j = i,\ldots,n, \tag{3.457a}$$

$$\Psi_{i,n+1} = \frac{\partial \Psi_i}{\partial y_n}, \quad i = 0,1,\ldots,n. \tag{3.457b}$$

We begin by establishing the following useful recursion identities

$$\Psi_{j,n+1} = -\Psi_{j+1,n} - \mathrm{D}\Psi_{j+1,n+1}, \quad j = 0,1,\ldots,n-1, \tag{3.458a}$$

$$\Psi_{j,m} = -\Psi_{j+1,m-1} - \mathrm{D}\Psi_{j+1,m}, \quad j = 1,\ldots,m-1, \ m = 1,2,\ldots,n, \tag{3.458b}$$

$$\Psi_{m,m} = 0, \quad m = 1,2,\ldots,n,$$

holding independently of (3.450a,b). To obtain (3.458b), one first applies $\partial/\partial y_m$ to (3.449b) with $n-k=j$ and subtracts $\partial/\partial y_j$ of (3.449b) with $n-k=m$. Then one combines terms by using (3.457a). Similarly, one obtains (3.458a).

Now consider $\Psi_{m,k}$ for $k > m \geq 1$. If $k - m = 2\ell + 1$, then using (3.458b) iteratively, we have that $\Psi_{m,m+2\ell+1}$ is a linear combination of $\mathrm{D}^{2i}\Psi_{j,j+1}$, $j = m + \ell + i$, $i = 0,1,\ldots,\ell$. Similarly, if $k - m = 2\ell$, then $\Psi_{m,m+2\ell}$ is a linear combination of $\mathrm{D}^{2i+1}\Psi_{j,j+1}$, $j = m + \ell + i$, $i = 0,1,\ldots,\ell-1$. To proceed, we consider (3.458a) with $j = n-2$, which yields

$$-\Psi_{n-1,n} = \Psi_{n-2,n+1} + \mathrm{D}\Psi_{n-1,n+1} = \frac{\partial \Psi_{n-2}}{\partial y_n} + \mathrm{D}\frac{\partial \Psi_{n-1}}{\partial y_n} = 0$$

from (3.450a). Likewise, for $j = n-4, n-6, \ldots, 1$ or 0 (for n odd or even, respectively), by using (3.450a) and (3.458a) iteratively, we obtain

$$\Psi_{m,m+1} = 0, \quad m = n-\ell,\ldots,n-1, \tag{3.459}$$

where $\ell = n/2$ if n is even, or $\ell = (n-1)/2$ if n is odd. Hence, after combining the above results, we have

$$\Psi_{m,k} = 0 \tag{3.460}$$

for all $1 \le m \le k \le n$, which yields (3.456b). Then from (3.454) and (3.457a,b), we obtain

$$\frac{\partial \Phi}{\partial y_k} = \theta_{y_k} - \Psi_k - \sum_{i=1}^{n} y_i \frac{\partial \Psi_i}{\partial y_k} = \frac{\partial \Psi_{k+1}}{\partial x} + \sum_{i=1}^{n} y_i \Psi_{k+1,i}, \quad k = 0,1,\ldots,n-1,$$

through use of (3.449b) as well as (3.450b) in the case $k = 0$. Hence, this yields (3.456c) from (3.460). Finally, we obtain (3.456a) immediately from (3.450a). Thus, all of the integrability conditions (3.456a–c) hold as a consequence of (3.450a,b), and so there exists a $\psi(x, y, y_1, \ldots, y_{n-1})$ satisfying (3.448).

Finally, from the relations (3.453) and (3.455) for the partial derivatives of $\psi(x, y, y_1, \ldots, y_{n-1})$ in terms of $\theta(x, y, y_1, \ldots, y_n)$, the fundamental theorem of calculus for gradients yields

$$\psi = \int_C \left[\psi_x \, dx + \sum_{i=0}^{n-1} \psi_{y_i} \, dy_i \right] = \int_C \left[\Phi \, dx + \sum_{i=1}^{n} \Psi_i \, dy_{i-1} \right] \tag{3.461}$$

to within a constant, where C is any path curve from a point $(\tilde{x}, \tilde{y}, \tilde{y}_1, \ldots, \tilde{y}_{n-1})$ to $(x, y, y_1, \ldots, y_{n-1})$. Thus, (3.461), (3.454), (3.456a) yield the line integral (3.451). □

Remarkably, through the identities (3.458a,b), the equations in the system (3.450a,b) of Theorem 3.6.4-2 can be reduced to a *simpler system of half as many equations* as follows. We use the notation $[q]$ to denote the greatest integer less than or equal to a given rational number q.

Lemma 3.6.4-1. *Equations* (3.450a) *for* $k = 2m$, $m = 0,1,\ldots,[n/2]$, *together with* (3.450b), *are equivalent to the system of* $2 + [n/2]$ *independent equations given by*

$$\theta_{y_n y_n} = 0, \tag{3.462a}$$

$$\theta_{y_k y_k}\Big|_{y_n=0} = -\sum_{i=1}^{n-k} \sum_{j=0}^{i} (-1)^i (i+j) \frac{(i-1)!}{(i-j)!\, j!} (\mathrm{D}^{i-j} \theta_{y_{k+i} y_{k-j}})\Big|_{y_n=0}, \quad k = \left[\frac{n+1}{2}\right], \ldots, n-1, \tag{3.462b}$$

$$\sum_{i=0}^{n} (-1)^i (\mathrm{D}^i \theta_{y_i})\Big|_{y_n=0} = 0. \tag{3.462c}$$

The remaining equations (3.450a) *for* $k = 2m+1$, $m = 0,1,\ldots,[(n-1)/2]$, *are (differential) linear combinations of* (3.462a,b).

Proof. We use (3.458a,b) recursively. For $j = n-1$, we obtain

$$\frac{\partial \Psi_{n-1}}{\partial y_n} = -\mathrm{D}\frac{\partial \Psi_n}{\partial y_n}.$$

Then, for $j = n-2$ and $j = n-3$, we find that

$$\frac{\partial \Psi_{n-2}}{\partial y_n} = \mathrm{D}^2 \frac{\partial \Psi_n}{\partial y_n} - \Psi_{n-1,n}, \quad \frac{\partial \Psi_{n-3}}{\partial y_n} = -\mathrm{D}^3 \frac{\partial \Psi_n}{\partial y_n} + 2\mathrm{D}\Psi_{n-1,n}.$$

Continuing, we find that $\partial \Psi_{n-2\ell-1} / \partial y_n$, $\ell = 0,1,\ldots,[(n-1)/2]$, is a differential linear combination of $\partial \Psi_n / \partial y_n$ and (if $\ell \geq 1$) $\Psi_{j,j+1}$, $j = n-\ell,\ldots,n-1$. In addition, we see that $\Psi_{j,j+1}$, $j = [(n+1)/2],\ldots, n-1$, is a differential linear combination of $\partial \Psi_{n-2i} / \partial y_n$, $i = 0,1,\ldots,n-j$. Thus, $\partial \Psi_{n-2\ell-1} / \partial y_n$ can be expressed as a differential linear combination of $\partial \Psi_{n-2i} / \partial y_n$, $i = 0,1,\ldots,\ell$, which establishes the second part of the lemma. It now follows from (3.458a,b) that the equations $\partial \Psi_{n-2\ell} / \partial y_n = 0$, $\ell = 0,1,\ldots,[n/2]$, are equivalent to the system

$$\frac{\partial \Psi_n}{\partial y_n} = 0, \tag{3.463}$$

$$\Psi_{j,j+1} = 0, \quad j = \left[\frac{n+1}{2}\right],\ldots,n-1. \tag{3.464}$$

Finally, we note that (3.463) and (3.449a,b) show that Ψ_{n-k} equals $\Psi_{n-k}\big|_{y_n=0}$ plus (if $k \geq 1$) a polynomial of degree k in y_n with coefficients given by differential linear combinations of $\partial \Psi_{n-i} / \partial y_n$ for $i = 0,1,\ldots,k-1$, $k = 1,2,\ldots,n$. Consequently, we conclude that (3.450a,b) is equivalent to the system

$$\frac{\partial \Psi_n}{\partial y_n} = 0, \quad \Psi_0\big|_{y_n=0} = 0, \quad \Psi_{j,j+1}\big|_{y_n=0} = 0, \quad j = [(n+1)/2],\ldots,n-1,$$

which are, respectively, given by (3.462a–c) through (3.457a,b) and (3.449a,b). Furthermore, since the terms $\theta_{y_k y_k}$, $k = [(n+1)/2],\ldots,n$, in (3.462a,b) are linearly independent, we see that the equations (3.462a–c) are independent. This establishes the first part of the lemma.

Now, from the characteristic equation (3.446a), let

$$\theta(x, y, y_1, \ldots, y_n) = (y_n - f(x, y, y_1, \ldots, y_{n-1}))\Lambda(x, y, y_1, \ldots, y_\ell), \quad 0 \leq \ell \leq n-1. \tag{3.465}$$

By taking $\partial / \partial y_{n-k}$ of (3.465), $k = 0,1,\ldots,n$, we obtain

$$\theta_{y_n} = \Lambda, \tag{3.466a}$$

$$\theta_{y_{n-k}}\Big|_{y_n=0} = -(f\Lambda)_{y_{n-k}}, \quad k = 1,2,\ldots,n. \tag{3.466b}$$

Then, using (3.449a,b), we have

$$\Psi_n = \theta_{y_n} = \Lambda \tag{3.467a}$$

and

$$\begin{aligned}\Psi_{n-k}\Big|_{y_n=0} &= \sum_{j=0}^{k}((-\mathrm{D})^j\theta_{y_{n-k+j}})\Big|_{y_n=0}\\ &= \sum_{j=0}^{k-1}-(-\mathrm{D}_{n-1})^j(f\Lambda)_{y_{n-k+j}} + (-\mathrm{D}_{n-1})^k\Lambda, \quad k = 1,2,\ldots,n,\end{aligned} \tag{3.467b}$$

where D_{n-1} is the truncated total derivative operator defined by

$$\mathrm{D}_k = \frac{\partial}{\partial x} + \sum_{i=1}^{k} y_i \frac{\partial}{\partial y_{i-1}}, \quad k \geq 1. \tag{3.468}$$

Finally, from (3.454) and (3.468), we obtain

$$\begin{aligned}\Phi\Big|_{y_n=0} &= \theta\Big|_{y_n=0} - \sum_{i=1}^{n-1} y_i\Psi_i\Big|_{y_n=0}\\ &= -f\Lambda + \sum_{i=1}^{n-i} y_i\left(\sum_{j=0}^{n-i-1}(-\mathrm{D}_{n-1})^j(f\Lambda)_{y_{i+j}} - (-\mathrm{D}_{n-1})^{n-i}\Lambda\right).\end{aligned} \tag{3.469}$$

Hence, for all integrating factors of ODE (3.443), Theorem 3.6.4-2 and Lemma 3.6.4-1 give a necessary and sufficient determining system consisting of the $1+[n/2]$ equations (3.462b,c). The total system can be written out explicitly as follows, using (3.466a,b) and (3.467a,b):

$$\begin{aligned}&(f\Lambda)_{y_{n-m}y_{n-m}} + \sum_{i=1}^{m-1}\sum_{j=0}^{i}\frac{(i+j)(i-1)!}{(i-j)!\,j!}(-1)^i(\mathrm{D}_{n-1})^{i-j}(f\Lambda)_{y_{n-m+i}y_{n-m-j}}\\ &\quad + (-1)^{m+1}\sum_{i=0}^{m}\frac{(i+m)(m-1)!}{(m-i)!\,i!}(\mathrm{D}_{n-1})^{m-i}\Lambda_{y_{n-m-i}} = 0, \quad m = 1,\ldots,\left[\frac{n}{2}\right],\end{aligned} \tag{3.470a}$$

$$\sum_{i=0}^{n-1}(-1)^i(\mathrm{D}_{n-1})^i(f\Lambda)_{y_i} + (-1)^{n-1}(\mathrm{D}_{n-1})^n\Lambda = 0. \tag{3.470b}$$

Furthermore, through (3.467a,b) and (3.469) combined with (3.451), we obtain the first integral $\psi(x,y,y_1,\ldots,y_{n-1})$ corresponding to the integrating factor $\Lambda(x,y,y_1,\ldots,y_\ell)$ from the explicit line integral formula

$$\psi = \int_C\left[\Lambda(dy_{n-1} - f\,dx) + \sum_{i=1}^{n-1}\Big(\sum_{j=0}^{n-i-1}-(-\mathrm{D}_{n-1})^j(f\Lambda)_{y_{i+j}} + (-\mathrm{D}_{n-1})^{n-i}\Lambda\Big)(dy_{i-1} - y_i\,dx)\right], \tag{3.471}$$

where C is a path curve from any point $(\tilde{x},\tilde{y},\tilde{y}_1,\ldots,\tilde{y}_{n-1})$ to the point $(x,y,y_1,\ldots,y_{n-1})$ in $(x,y,y_1,\ldots,y_{n-1})$ – space. If $f(x,y,y_1,\ldots,y_{n-1})$ and $\Lambda(x,y,y_1,\ldots,y_\ell)$ are nonsingular, then C can be chosen arbitrarily and, hence, in any convenient way to simplify the integral. Most important, if $f(x,y,y_1,\ldots,y_{n-1})$ or $\Lambda(x,y,y_1,\ldots,y_\ell)$ are singular, then some path curve C can be chosen so that the line integral (3.471) is nonsingular. □

Theorem 3.6.4-3. *The integrating factors of order* $0\le\ell\le n-1$ *of ODE* (3.443) *are the solutions* $\Lambda(x,y,y_1,\ldots,y_\ell)\not\equiv 0$ *of the determining system consisting of the* $1+[n/2]$ *equations* (3.470a,b). *For a given integrating factor, the corresponding first integral of ODE* (3.443) *is given by the line integral formula* (3.471).

The integrating factor determining system (3.470a,b) is a system of $1+[n/2]$ linear homogeneous PDEs of order n for $\Lambda(x,y,y_1,\ldots,y_\ell)$. Any solution of (3.470a,b) yields a first integral (3.471) which provides a reduction of the order of ODE (3.443) by one, yielding an $(n-1)$th-order ODE represented by the surface $\psi(x,y,y_1,\ldots,y_{n-1})=\text{const}=c$. If one knows $1<k\le n$ solutions of the integrating factor determining system (3.470a,b), such that the resulting k first integrals $\psi_1,\ldots,\psi_k$ are functionally independent, i.e., none is a function of the others, then ODE (3.443) is reduced to an $(n-k)$th-order ODE given by the elimination of $y_{n-1},\ldots,y_{n-k+1}$ in the equations $\psi_i(x,y,y_1,\ldots,y_{n-1})=\text{const}=c_i$, $i=1,2,\ldots,k$. Hence, if $k=n$, this yields a complete reduction of ODE (3.443) to quadrature.

It is straightforward to see that a set of first integrals ψ_i, $i=1,2,\ldots,k$, is functionally dependent if and only if $F(\psi_1,\ldots,\psi_k)=0$ holds for some nonconstant function $F(\psi_1,\ldots,\psi_k)$. Consequently, from the characteristic equation (3.446a,b), it follows that a set of integrating factors Λ_i, $i=1,2,\ldots,k\le n$, of ODE (3.443) determines k functionally independent first integrals ψ_i, $i=1,2,\ldots,k$, if and only if $\sum_{i=1}^{k}F_{\psi_i}\Lambda_i\ne 0$ holds for all functions $F(\psi_1,\ldots,\psi_k)\not\equiv\text{const}$, where each ψ_i and Λ_i are related through (3.446a,b) and (3.471). For first integrals with integrating factors of order $\ell<n-1$, there is a stronger criterion for functional independence.

Theorem 3.6.4-4. *A set of integrating factors* Λ_i, $i=1,2,\ldots,k$, *of order* $0\le\ell\le n-k$ *of ODE* (3.443) *determines* k *functionally independent first integrals* ψ_i, $i=1,2,\ldots,k$, *if and only if the integrating factors are linearly independent, i.e.,*

$$\sum_{i=1}^{k} c_i\Lambda_i \ne 0 \tag{3.472}$$

holds for all constants $c_i\ne 0$.

Proof. Suppose

$$\sum_{i=1}^{k} c_i \Lambda_i = 0 \tag{3.473}$$

holds where $c_i = \text{const}$. Let $\widetilde{F} = \sum_{i=1}^{k} c_i \psi_i$ where $\{\psi_i\}$ is a set of k first integrals corresponding to the set of k integrating factors $\{\Lambda_i\}$. Then, from the characteristic equation (3.446a,b) and (3.473), it immediately follows that

$$\widetilde{F}_{y_{n-1}} = \sum_{i=1}^{k} c_i (\psi_i)_{y_{n-1}} = \sum_{i=1}^{k} c_i \Lambda_i = 0. \tag{3.474}$$

Hence, using (3.474) and (3.445b), we have $D\widetilde{F} = \mathbf{D}\widetilde{F} = \sum_{i=1}^{k} c_i \mathbf{D}\psi_i = 0$. Consequently, $\widetilde{F} = \text{const} = c$, so that the set of first integrals $\{\psi_i\}$ satisfies

$$F(\psi_1, \ldots, \psi_k) = 0, \tag{3.475}$$

where $F = \widetilde{F} - c$. Hence, the set $\{\psi_i\}$ is functionally dependent.

Conversely, suppose (3.475) holds where $\{\psi_i\}$ is a set of k first integrals with the corresponding set of k integrating factors $\{\Lambda_i\}$ being of order $\ell \le n-k$. By taking $\partial / \partial y_{n-1}$ of (3.475) and using (3.446b), we obtain (3.473) with $c_i = F_{\psi_i}$. We now show that $c_i = \text{const}$. Since the set $\{\psi_i\}$ is functionally dependent, we suppose that at most $1 \le r < k$ of these first integrals are functionally independent and denote them as $\psi_1, \ldots, \psi_r$. Hence, through reduction of order of ODE (3.443), the variables $y_{n-1}, \ldots, y_{n-r}$ in each expression Λ_i can be eliminated in terms of $\psi_1, \ldots, \psi_r$, and $x, y, y_1, \ldots, y_{n-r-1}$. Then $\partial \Lambda_i / \partial \psi_{r+1} = \cdots = \partial \Lambda_i / \partial \psi_k = 0$, and, furthermore, since each Λ_i is assumed to be of order $\ell \le n-k < n-r$, it follows that $\partial \Lambda_i / \partial y_m = \sum_{j=1}^{r} (\partial \Lambda_i / \partial \psi_j)(\partial \psi_j / \partial y_m) = 0$, $m = n-r, \ldots, n-1$. Thus, since the functional independence of $\psi_1, \ldots, \psi_r$, implies that the $r \times r$ Jacobian matrix of partial derivatives $\partial \psi_j / \partial y_m$ is invertible, we obtain

$$\frac{\partial \Lambda_i}{\partial \psi_j} = 0 \quad \text{for all } i, j = 1, \ldots, k.$$

Now we take $\partial / \partial \psi_j$ of (3.473) with $c_i = F_{\psi_i}$, and use the commutativity of partial derivatives $\partial c_i / \partial \psi_j = F_{\psi_i \psi_j} = \partial c_j / \partial \psi_i$, to obtain $\sum_{i=1}^{k} \Lambda_i (\partial c_j / \partial \psi_i) = 0$. Then we use $\Lambda_i = \partial \psi_i / \partial y_{n-1}$ and the chain rule to obtain

$$\frac{\partial c_j}{\partial y_{n-1}} = 0. \tag{3.476}$$

Since $c_j = F_{\psi_j}$ also satisfies $\mathbf{D}c_j = \mathbf{D}F_{\psi_j} = \sum_{i=1}^{k} F_{\psi_j\psi_i}\mathbf{D}\psi_i = 0$, it then follows from (3.476) and (3.445b) that $\mathrm{D}c_j = 0$. Hence, for $j = 1,\ldots,k$, we have $c_j = \text{const}$ in (3.473), which completes the proof. □

A sufficient criterion, generalizing Theorem 3.6.2-4 for functional independence of first integrals, can be given with less restriction on the order of corresponding integrating factors than required in Theorem 3.6.4-4.

Theorem 3.6.4-5. *If* $2 \le k \le n$ *integrating factors* $\Lambda_i(x, y, y_1, \ldots, y_\ell)$, $i = 1,2,\ldots,k$, *of any order* $0 \le \ell \le n-1$ *of ODE* (3.443) *are linearly independent and satisfy* $(\Lambda_i / \Lambda_j)_{y_{n-r}} = 0$, $r = 1,\ldots,k-1$, $i \le j = 1,2,\ldots,k$, *then the corresponding first integrals* ψ_i, $i = 1,2,\ldots,k$, *given by* (3.471), *are functionally independent.*

Proof. Left to Exercise 3.6-23. □

The general solution of the integrating factor determining system (3.470a,b) is given by

$$\Lambda = \sum_{i=1}^{n} F_{\psi_i} \Lambda_i, \tag{3.477}$$

where $F(\psi_1,\ldots,\psi_n)$ is an arbitrary function of n functionally independent first integrals $\psi_i(x, y, y_1, \ldots, y_{n-1})$ with corresponding integrating factors $\Lambda_i(x, y, y_1, \ldots, y_\ell)$, $i = 1,2,\ldots,n$, arising from the characteristic equation (3.446a,b). Hence, finding all solutions of (3.470a,b) is equivalent to solving the original nth-order ODE (3.443). However, it suffices to find just $1 \le k \le n$ solutions that yield functionally independent first integrals in order to reduce the order of ODE (3.443) by k.

For integrating factors of order $\ell < n-1$, the determining system (3.470a,b) splits into an overdetermined linear system of PDEs and thus has a finite number of linearly independent solutions. In practice, it is useful to consider ansatzes for further simplifying and reducing this overdetermined system. Moreover, for finding integrating factors of order $\ell = n-1$, there is no inherent splitting of the determining system (3.470a,b) and, hence, it is necessary to use ansatzes so as to obtain an overdetermined system from (3.470a,b) with at most a finite number of linearly independent solutions. Most important, in all these situations the resulting systems of determining equations for integrating factors can be solved by the same algorithmic procedure used in solving the analogous determining equations for symmetries of order $\ell \le n-1$ [cf. Section 3.5].

We now summarize several effective ansatzes, generalizing those given in Section 3.6.3 to third- and higher-order ODEs, for obtaining solutions of the integrating factor determining system (3.470a,b). Examples illustrating the use of these ansatzes will be given in Section 3.6.5.

(1) *Point-Form Ansatzes*
If one considers integrating factors of ODE (3.443) of point-form

$$\Lambda = \alpha(x,y) + \beta(x,y)y_1, \tag{3.478}$$

then the integrating factor determining system (3.470a,b) becomes an overdetermined system of $1+[n/2]$ linear homogeneous PDEs for $\alpha(x,y)$ and $\beta(x,y)$.

Theorem 3.6.4-6. *An nth-order ODE* (3.443), $n \geq 3$, *admits an integrating factor of point-form* (3.478) *if and only if*

$$y_n = f(x,y,y_1,\dots,y_{n-1}) = \frac{h_x + \sum_{i=0}^{n-2} y_{i+1}h_{y_i} - y_{n-1}(\alpha_x + y_1(\alpha_y + \beta_x) + (y_1)^2\beta_y + y_2\beta)}{\alpha + \beta y_1} \tag{3.479}$$

for some function $h(x,y,y_1,\dots,y_{n-2})$. *In particular, if* $n>3$, *it is necessary that* $f(x,y,y_1,\dots,y_{n-1})$ *be at most linear in* y_{n-1}; *if* $n=3$, *it is necessary that* $f(x,y,y_1,\dots,y_{n-1})$ *be at most quadratic in* y_{n-1}.

Proof. We start from the identity

$$(\alpha + \beta y_1)y_n = \mathrm{D}(\alpha y_{n-1} + \beta y_1 y_{n-1}) - y_{n-1}(\mathrm{D}\alpha + y_1\mathrm{D}\beta + y_2\beta).$$

Then, from the characteristic equation (3.446a), it follows that (3.478) is an integrating factor of $y_n = f(x,y,y_1,\dots,y_{n-1})$ if and only if for some function $\psi(x,y,y_1,\dots,y_{n-1})$, we have

$$\mathrm{D}(\psi - (\alpha + \beta y_1)y_{n-1}) = -(\alpha + \beta y_1)f - y_{n-1}(\mathrm{D}\alpha + y_1\mathrm{D}\beta + y_2\beta).$$

Thus, we obtain the relation

$$f = \frac{\mathrm{D}h - y_{n-1}(\mathrm{D}\alpha + y_1\mathrm{D}\beta + y_2\beta)}{\alpha + \beta y_1}, \tag{3.480}$$

where

$$h = (\alpha + \beta y_1)y_{n-1} - \psi. \tag{3.481}$$

Then $\partial/\partial y_n$ of (3.480) immediately yields $h_{y_{n-1}} = 0$ if $n \geq 3$. Hence, we obtain ODE (3.479).

Conversely, for any function $h(x,y,y_1,\dots,y_{n-2})$, from (3.480) and (3.481) we see that the characteristic equation (3.446a) for ODE (3.479) is satisfied for the integrating factor (3.478) if $n \geq 3$, where $\psi = (\alpha + \beta y_1)y_{n-1} - h$. □

In practice, the simplest way to determine if a given third- or higher-order ODE (3.443) admits an integrating factor of point-form is to first verify that the necessary conditions given in Theorem 3.6.4-6 are satisfied. Then one solves the resulting integrating factor determining system (3.470a,b). Alternatively, note that if one can match the form of (3.479) to a given ODE (3.443), then one immediately obtains the integrating factor (3.478), with the corresponding first integral given by (3.481).

(2) *Elimination of Variables Ansatz*
More generally, for an nth-order ODE (3.443), $n \geq 3$, if one considers integrating factors $\Lambda(x, y, y_1, \ldots, y_\ell)$ of order $1 \leq \ell < n-1$, i.e., depending on variables y_i up to some order strictly *less* than the order of the highest derivative appearing in $f(x, y, y_1, \ldots, y_{n-1})$, then the integrating factor determining system (3.470a,b) again reduces to an overdetermined system of $1+[n/2]$ linear homogeneous PDEs, which has at most a finite number of linearly independent solutions $\Lambda(x, y, y_1, \ldots, y_\ell)$. There exist efficient computational algorithms to solve such systems [Wolf (2002a,b)] and, hence, it is straightforward to find all integrating factors of order less than $n-1$ for a given nth-order ODE (3.443). However, the computational complexity grows quickly as n increases.

(3) *Symmetry-Type Ansatzes*
If an nth-order ODE (3.442) admits a point symmetry

$$\mathrm{X} = \xi(x, y)\frac{\partial}{\partial x} + \eta(x, y)\frac{\partial}{\partial y}, \tag{3.482}$$

then since the corresponding surface (3.443) is invariant under the nth-extended generator

$$\mathrm{X}^{(n)} = \xi\frac{\partial}{\partial x} + \eta\frac{\partial}{\partial y} + \sum_{i=1}^{n} \eta^{(i)}\frac{\partial}{\partial y_i}$$

with $\eta^{(i)}$ given by (2.100a,b), it follows that the $(n-1)$th-extended generator $\mathrm{X}^{(n-1)}$ maps first integrals of ODE (3.442) into first integrals since any first integral of ODE (3.442) is constant for every solution curve on the surface (3.443).

Theorem 3.6.4-7. *Suppose $\psi(x, y, y_1, \ldots, y_{n-1})$ is a first integral with integrating factor $\Lambda(x, y, y_1, \ldots, y_\ell)$ for ODE* (3.443). *Then, under any point symmetry* (3.482) *admitted by ODE* (3.443),

$$\widetilde{\psi} = \mathrm{X}^{(n-1)}\psi + \widetilde{c}, \quad \widetilde{c} = \text{const}, \tag{3.483}$$

yields a first integral with integrating factor given by

$$\widetilde{\Lambda} = \mathrm{X}^{(n-1)}\Lambda + R_{n-1}\Lambda, \tag{3.484}$$

where

$$R_{n-1} \equiv \frac{\partial \eta^{(n-1)}}{\partial y_{n-1}} = \eta_y - n\xi_y y_1 - (n-1)\xi_x. \tag{3.485}$$

Proof. Left to Exercise 3.6-27. □

Consequently, every point symmetry admitted by ODE (3.443) induces a point symmetry on the vector space of its integrating factors. The explicit generator in $(x, y, y_1, \ldots, y_{n-1}, \Lambda)$ – space is given by

$$\widetilde{\mathrm{X}} = \xi \frac{\partial}{\partial x} + \eta \frac{\partial}{\partial y} + \sum_{i=1}^{n-1} \eta^{(i)} \frac{\partial}{\partial y_i} + R_{n-1} \Lambda \frac{\partial}{\partial \Lambda}, \tag{3.486}$$

which maps integrating factors into integrating factors of ODE (3.443). Hence, as an ansatz, one can consider integrating factors invariant under (3.486),

$$\widetilde{\mathrm{X}}\Lambda = r\Lambda, \quad r = \text{const}, \tag{3.487}$$

or, equivalently,

$$\xi \Lambda_x + \eta \Lambda_y + \sum_{i=1}^{n-1} \eta^{(i)} \Lambda_{y_i} + (\eta_y - n\xi_y y_1 - (n-1)\xi_x - r)\Lambda = 0. \tag{3.488}$$

If we now obtain invariants $u(x, y)$ and $v(x, y, y_1)$ of (3.443) by solving $\mathrm{X}u = 0$, $\mathrm{X}^{(1)}v = 0$ $[v_{y_1} \neq 0]$, and introduce the differential invariants $v_i(x, y, y_1, \ldots, y_{i+1}) = d^i v / du^i$, satisfying $\mathrm{X}^{(i+1)} v_i = 0$, for $i = 1, \ldots, n-2$, then we have the general solution of (3.448) given by

$$\Lambda = \exp\left(\int \frac{r - R_{n-1}(u, v, y_1)}{\eta^{(1)}(u, v, y_1)} dy_1 \right) w(u, v, v_1, \ldots, v_{n-2}), \tag{3.489}$$

for an arbitrary function $w(u, v, v_1, \ldots, v_{n-2})$, with x and y eliminated in terms of u and v. Alternatively, if $\xi \neq 0$ or $\eta \neq 0$, then we can have

$$\Lambda = \exp\left(\int \frac{r - R_{n-1}(x, u, v)}{\xi(x, u)} dx \right) w(u, v, v_1, \ldots, v_{n-2}) \tag{3.490a}$$

or

$$\Lambda = \exp\left(\int \frac{r - R_{n-1}(y, u, v)}{\eta(y, u)} dy \right) w(u, v, v_1, \ldots, v_{n-2}), \tag{3.490b}$$

in terms of an arbitrary function $w(u, v, v_1, \ldots, v_{n-2})$.

These ansatzes correspond to the scaling invariance of $\Lambda(x, y, y_1, \ldots, y_\ell)$ and $\psi(x, y, y_1, \ldots, y_{n-1})$ under $\mathrm{X}^{(n-1)}$, given by

$$\mathrm{X}^{(n-1)}\Lambda = s\Lambda, \quad s = \text{const}, \tag{3.491}$$

$$\mathrm{X}^{(n-1)}\psi = r\psi + \widetilde{c}, \quad r = \text{const}, \; \widetilde{c} = \text{const}, \tag{3.492}$$

if and only if $R_{n-1} = r - s = \text{const}$, which holds for the following class of point symmetries:

Lemma 3.6.4-2. *A point symmetry* (3.482) *of the nth-order ODE* (3.443) *has* $R_{n-1} \equiv \partial\eta^{(n-1)} / \partial y_{n-1} = \text{const} = c$ *if and only if*

$$\eta = \alpha(x) + ((n-1)\beta'(x) + c)y, \quad \xi = \beta(x), \tag{3.493}$$

for some functions $\alpha(x)$, $\beta(x)$. *In particular, all translations and scalings satisfy* $R_{n-1} = \text{const}$.

Proof. Left to Exercise 3.6-19. □

Hence, note that any translations and scalings admitted by a given ODE (3.443) are automatically inherited by its integrating factor determining system (3.470a,b), so that one can consider the simple ansatz

$$\Lambda = w(u, v, v_1, \ldots, v_{n-2}), \tag{3.494}$$

i.e., $r = R_{n-1}$ in (3.489) since $R_{n-1} = \text{const}$.

3.6.5 EXAMPLES OF FIRST INTEGRALS OF THIRD- AND HIGHER-ORDER ODEs

Consider a third-order ODE

$$y''' = f(x, y, y', y'') \tag{3.495}$$

represented as a surface

$$y_3 = f(x, y, y_1, y_2). \tag{3.496}$$

From Theorem 3.6.4-3, we see that the integrating factors $\Lambda(x, y, y_1, y_2)$ of ODE (3.496) are given by the explicit determining system

$$2\Lambda_{y_1} + \Lambda_{y_2 x} + y_1\Lambda_{y_2 y} + y_2\Lambda_{y_2 y_1} + (f\Lambda)_{y_2 y_2} = 0, \tag{3.497a}$$

$$\begin{aligned}
&3y_2\Lambda_{xy} + 3y_1y_2\Lambda_{yy} + 3(y_2)^2\Lambda_{yy_1} + \Lambda_{xxx} + (y_1)^3\Lambda_{yyy} + (y_2)^3\Lambda_{y_1y_1y_1} + 3y_1\Lambda_{xxy} \\
&\quad + 3y_2\Lambda_{xxy_1} + 3(y_1)^2\Lambda_{xyy} + 3(y_2)^2\Lambda_{xy_1y_1} + 3(y_1)^2y_2\Lambda_{yyy_1} + 3y_1(y_2)^2\Lambda_{yy_1y_1} + 6y_1y_2\Lambda_{xyy_1} \\
&\quad + (f\Lambda)_y - (f\Lambda)_{xy_1} - y_1(f\Lambda)_{yy_1} - y_2(f\Lambda)_{y_1y_1} + y_2(f\Lambda)_{yy_2} + (f\Lambda)_{xxy_2} + (y_1)^2(f\Lambda)_{yyy_2} \\
&\quad + (y_2)^2(f\Lambda)_{y_1y_1y_2} + 2y_1(f\Lambda)_{xyy_2} + 2y_2(f\Lambda)_{xy_1y_2} + 2y_1y_2(f\Lambda)_{yy_1y_2} = 0,
\end{aligned} \tag{3.497b}$$

with the corresponding first integrals $\psi(x, y, y_1, y_2)$ of ODE (3.496) given by the explicit line integral formula

$$\begin{aligned}
\psi = \int_C [(&-(f\Lambda)_{y_1} + (f\Lambda)_{xy_2} + y_1(f\Lambda)_{yy_2} + y_2(f\Lambda)_{y_1y_2} + y_2\Lambda_y \\
&+ \Lambda_{xx} + (y_1)^2\Lambda_{yy} + (y_2)^2\Lambda_{y_1y_1} + 2y_1\Lambda_{xy} + 2y_2\Lambda_{xy_1} + 2y_1y_2\Lambda_{yy_1})(dy - y_1\,dx) \\
&- ((f\Lambda)_{y_2} + \Lambda_x + y_1\Lambda_y + y_2\Lambda_{y_1})(dy_1 - y_2\,dx) + \Lambda(dy_2 - f\,dx)].
\end{aligned} \tag{3.498}$$

We now illustrate how to find solutions of the integrating factor determining system (3.497a,b) through the algorithmic methods summarized in Section 3.6.4. We also show the calculation of corresponding first integrals through the line integral formula (3.498) and illustrate reduction of order from first integrals for third-order ODEs.

As a first example, consider the ODE

$$y_3 = -yy_1, \tag{3.499}$$

arising from seeking traveling wave solutions of the Korteweg–de Vries (KdV) equation [Exercise 4.1-2]. The point symmetries admitted by ODE (3.499) consist of the translation symmetry $x \to x+\varepsilon, y \to y$ and the scaling symmetry $x \to \lambda x,\ y \to \lambda^{-2} y.$ To begin, we observe that since ODE (3.499) does not involve y_2, it satisfies the necessary condition of Theorem 3.6.4-6 to admit point-form integrating factors

$$\Lambda = \alpha(x,y) + \beta(x,y)y_1. \tag{3.500}$$

Substitution of (3.500) into the integrating factor determining system (3.497a,b) yields, respectively,

$$\beta = 0 \tag{3.501}$$

and, after (3.501) is used,

$$3\alpha_{xy}y_2 + 3\alpha_{yy}y_1y_2 + \alpha_{yyy}(y_1)^3 + 3\alpha_{xyy}(y_1)^2 + 3\alpha_{xxy}y_1 + \alpha_{xxx} - y\alpha_x = 0. \tag{3.502}$$

From the splitting of (3.502) with respect to y_1 and y_2, we obtain $\alpha_{xy} = \alpha_{yy} = 0$ and $y\alpha_x = \alpha_{xxx}$. This immediately gives

$$\alpha = \alpha_0 + \alpha_1 y, \quad \alpha_0 = \text{const}, \ \alpha_1 = \text{const}.$$

Hence, ODE (3.499) admits two integrating factors of point-form, given by

$$\Lambda_1 = 1, \quad \Lambda_2 = y. \tag{3.503}$$

Using the line integral formula (3.498) with C chosen to be a piecewise straight line from $(0,0,0,0)$ to (x,y,y_1,y_2) parallel to the coordinate axes, we obtain the corresponding first integrals

$$\psi_1 = \int_C [y\,dy + dy_2] = \tfrac{1}{2}y^2 + y_2 \tag{3.504a}$$

and

$$\psi_2 = \int_C [(y^2 + y_2)\,dy - y_1\,dy_1 + y\,dy_2] = \tfrac{1}{3}y^3 - \tfrac{1}{2}(y_1)^2 + yy_2. \tag{3.504b}$$

From Theorem 3.6.4-4, we see that the first integrals (3.504a,b) are functionally independent. Hence, we have two quadratures $\psi_1 = \text{const} = c_1$, $\psi_2 = \text{const} = c_2$, which lead to a reduction of the third-order ODE (3.499) to a first-order ODE

$$y_1 = \pm\sqrt{2c_1 y - \tfrac{1}{3}y^3 - 2c_2}.$$

This reduced ODE is separable, which results from the invariance of the integrating factors (3.503) under the translation symmetry $x \to x+\varepsilon, y \to y$ admitted by ODE (3.499) (but not under the admitted scaling symmetry). Thus, we obtain an additional first integral [cf. (3.441)]

$$\psi_3 = \pm\int \frac{dy}{\sqrt{2c_1 y - \frac{1}{3}y^3 - 2c_2}} - x, \tag{3.505}$$

which obviously is functionally independent of ψ_1 and ψ_2. Thus, $\psi_3 = \text{const} = c_3$ yields the complete quadrature (i.e., the general solution) of ODE (3.499).

For a second example, consider the third-order ODE

$$y''' = 6x\frac{(y'')^3}{(y')^2} + 6\frac{(y'')^2}{y'} \tag{3.506}$$

or, equivalently, the surface

$$y_3 - 6x(y_2)^3(y_1)^{-2} - 6(y_2)^2(y_1)^{-1} = 0, \tag{3.507}$$

which admits contact symmetries as shown in Section 3.5.2. First we observe that (3.507) is cubic in y_2 and, hence, from Theorem 3.6.4-6, it does not admit any integrating factor of point-form $\Lambda = \alpha(x,y) + \beta(x,y)y_1$. Consequently, we instead make use of the symmetry-type ansatz (3.490a,b) to seek integrating factors. From Section 3.5.2 [cf. Exercise 3.5-5], we see that the point symmetries of (3.507) consist of translations in y, scalings in x, and scalings in y. For the y translation symmetry [$\eta = 1$, $\xi = 0$], the ansatz (3.490b) yields $\Lambda = e^{ry}w(x, y_1, y_2)$ in terms of invariants x, y_1, y_2. For the y scaling symmetry [$\eta = y$, $\xi = 0$], the ansatz (3.490b) yields $\Lambda = y^{r-1}w(x, y_1/y, y_2/y)$ in terms of invariants $x, y_1/y, y_2/y$. Similarly, the x scaling symmetry [$\eta = 0, \xi = x$] leads to $\Lambda = x^{r+2}w(y, xy_1, x^2y_2)$. Hence the common joint invariant form for $\Lambda(x, y, y_1, y_2)$ is given by

$$\Lambda = x^r (y_1)^s w(u), \quad u = \frac{xy_2}{y_1}, \; r = \text{const}, \; s = \text{const}. \tag{3.508}$$

Substitution of (3.508) into the integrating factor determining system (3.497a,b) yields the equations

$$u(2u+1)(3u+1)w'' + (36u^2 + (21+s)u + r + 1)w' + (36u + 12 + 2s)w = 0, \quad (3.509a)$$

$$\begin{aligned}
&(2u+1)(3u+1)u^3(u-1)^2 w''' \\
&\quad - 3u^2(u-1)((-20+4s)u^3 + (2+4r+3s)u^2 + (6+3r+s)u + r)w'' \\
&\quad + 3u(2(s-3)(s-8)u^4 + (-46+7s+s^2-22r+4rs)u^3 \\
&\quad + (-2+6r+7s+s^2+2rs)u^2 + r(9+r+2s)u + r(r-1))w' \\
&\quad - (-12(s-3)(s-2)u^4 - (s-2)(24+30r+5s+s^2)u^3 \\
&\qquad - 3r(6r+5s+s^2)u^2 - 3r(r-1)(4+s)u - r(r-1)(r-2))w = 0. \quad (3.509b)
\end{aligned}$$

We take d/du of (3.509a), eliminate w''' through (3.509b), and then eliminate w'' through (3.509a). This leads to $Aw' + Bw = 0$, where

$$A = ((s+1)u + r - 1)(su + r - 2),$$

$$B = \frac{(su + r - 2)(12(1+s)u^3 + (18(r-1) + (1+s)(6+s))u^2 + 2(r-1)(s+6)u + r(r-1))}{u(2u+1)(3u+1)}.$$

Hence, if $(s,r) \neq (0,2)$ or $(-1,1)$, so that $(A,B) \not\equiv (0,0)$, then we obtain a separable first-order ODE

$$\frac{w'}{w} = -\frac{12(1+s)u^3 + (18(r-1) + (1+s)(6+s))u^2 + 2(r-1)(s+6)u + r(r-1)}{u(2u+1)(3u+1)((s+1)u + r - 1)}. \quad (3.510)$$

This has the solution

$$w = (3u+1)^p (2u+1)^q ((s+1)u + r - 1)u^{-r}, \quad p = 3r - s - 5, \ q = -2r + s + 2, \quad (3.511)$$

which can be readily checked to satisfy both equations (3.509a,b) including the cases $(s,r) = (0,2), (-1,1)$. Thus, we have the family of integrating factors

$$\Lambda = (y_1)^{s+r}(y_2)^{-r}\left(\frac{3xy_2}{y_1} + 1\right)^{3r-s-5}\left(\frac{2xy_2}{y_1} + 1\right)^{s-2r+2}\left((s+1)\frac{xy_2}{y_1} + r - 1\right) \quad (3.512)$$

depending on two free parameters (r,s). One expects that (3.512) yields at most two functionally independent first integrals.

To obtain simple expressions, we choose $r = 2$ and $s = 1, 2$ in (3.512), which yields two integrating factors

$$\Lambda_1 = \frac{(y_1)^3}{(y_2)^2}, \quad \Lambda_2 = \frac{(y_1)^4}{(y_2)^2}. \quad (3.513)$$

Since $\Lambda_1/\Lambda_2=(y_1)^{-1}\not\equiv$ const does not involve y_2, from Theorem 3.6.4-5 we see that Λ_1 and Λ_2 yield functionally independent first integrals ψ_1 and ψ_2, given by the line integral formula (3.498). Hence, we obtain the first integrals

$$\psi_1=\int_C\left[(-3(y_1)^2)\,dx-(6xy_1+\frac{3(y_1)^2}{(y_2)^2})dy_1+\frac{(y_1)^3}{(y_2)^2}dy_2\right]=-3x(y_1)^2-\frac{(y_1)^3}{y_2} \tag{3.514a}$$

and

$$\psi_2=\int_C\left[(-2(y_1)^3)\,dx-(6x(y_1)^2+\frac{4(y_1)^3}{y_2})dy_1+\frac{(y_1)^4}{(y_2)^2}dy_2\right]=-2x(y_1)^3-\frac{(y_1)^4}{y_2} \tag{3.514b}$$

with C chosen to be a piecewise straight line from $(0,0,0,\tilde{y}_2)$ to (x,y,y_1,y_2), parallel to the coordinate axes, with $\tilde{y}_2\neq 0$. Note that it is now straightforward to verify that the family of integrating factors (3.512) reduces to

$$\begin{aligned}\Lambda&=(-\psi_1)^{p+1}(-\psi_2)^q(q+1)\Lambda_2+(-\psi_1)^p(-\psi_2)^{q+1}(p+1)\Lambda_1\\&=F_{\psi_1}\Lambda_1+F_{\psi_2}\Lambda_2,\quad F=(-1)^{1+p+q}(\psi_1)^{p+1}(\psi_2)^{q+1},\quad p=3r-s-5,\,q=s-2r+2,\end{aligned}$$

and hence, the corresponding family of first integrals is a function combination of (3.514a,b).

We now obtain a third functionally independent first integral by considering (3.509a,b) in the special case when $(s,r)=(0,2)$. Since (3.511) with $s=0,\ r=2$ yields a solution of (3.509a,b), given by

$$w=\frac{u+1}{u^2}(3u+1)(2u+1)^{-2},$$

we can use reduction of order to find the general solution of (3.509a,b) in this case. This yields a second solution

$$w=u^{-2},$$

which leads to the integrating factor

$$\Lambda_3=\frac{(y_1)^2}{(y_2)^2}. \tag{3.515}$$

Note that Λ_3 is linearly independent of Λ_1 and Λ_2. A corresponding first integral is obtained from the line integral formula (3.498) with the same path curve C as before. This yields

$$\psi_3 = \int_C \left[(-6y_1)\,dx + 2\,dy - (6x + \frac{2y_1}{y_2})\,dy_1 + \frac{(y_1)^2}{(y_2)^2}\,dy_2 \right] = 2y - 6xy_1 - \frac{(y_1)^2}{y_2}, \tag{3.516}$$

which is functionally independent of (3.514a,b) since $(\psi_1)_y = (\psi_2)_y = 0$ but $(\psi_3)_y \neq 0$.

The first integrals (3.514a,b) and (3.516) yield three quadratures $\psi_1 = \text{const} = c_1$, $\psi_2 = \text{const} = c_2$, $\psi_3 = \text{const} = c_3$, giving the complete reduction of (3.507) to an algebraic equation through the elimination of y_2 and y_1. Explicitly, solving (3.514a) for y_2 and substituting into (3.514b) and (3.516), we have

$$x(y_1)^3 + c_1 y_1 - c_2 = 0,$$
$$3x(y_1)^2 + (c_3 - 2y)y_1 - c_1 = 0.$$

After algebraically combining these equations to eliminate y_1, we obtain

$$\begin{aligned} &3x(c_1(c_3 - 2y) - 9xc_2)^2 - c_1((c_3 - 2y)^2 + 12xc_1)^2 \\ &\quad + (c_3 - 2y)(c_1(c_3 - 2y) - 9xc_2)((c_3 - 2y)^2 + 12xc_1) = 0, \end{aligned} \tag{3.517}$$

which is the general solution of ODE (3.507).

Finally, consider a fourth-order ODE

$$y^{(4)} = f(x, y, y', y'', y''') \tag{3.518}$$

represented as a surface

$$y_4 - f(x, y, y_1, y_2, y_3) = 0. \tag{3.519}$$

From (3.470a,b), we see that the determining system for the integrating factors $\Lambda(x, y, y_1, y_2, y_3)$ of ODE (3.519) is given by

$$2\Lambda_{y_2} + D_3\Lambda_{y_3} + (f\Lambda)_{y_3y_3} = 0, \tag{3.520a}$$

$$2\Lambda_y + 3D_3\Lambda_{y_1} + (D_3)^2\Lambda_{y_2} + D_3(f\Lambda)_{y_2y_3} + 2(f\Lambda)_{y_1y_3} - (f\Lambda)_{y_2y_2} = 0, \tag{3.520b}$$

$$(f\Lambda)_y - D_3(f\Lambda)_{y_1} + (D_3)^2(f\Lambda)_{y_2} - (D_3)^3(f\Lambda)_{y_3} - (D_3)^4\Lambda = 0, \tag{3.520c}$$

with $D_3 = \frac{\partial}{\partial x} + y_1\frac{\partial}{\partial y} + y_2\frac{\partial}{\partial y_1} + y_3\frac{\partial}{\partial y_2}$. The line integral formula for the corresponding first integrals is given by

$$\begin{aligned} \psi = \int_C [&(-(f\Lambda)_{y_1} + D_3(f\Lambda)_{y_2} - (D_3)^2(f\Lambda)_{y_3} - (D_3)^3\Lambda)(dy - y_1\,dx) \\ &+ (-(f\Lambda)_{y_2} + D_3(f\Lambda)_{y_3} + (D_3)^2\Lambda)(dy_1 - y_2\,dx) \\ &+ (-(f\Lambda)_{y_3} - D_3\Lambda)(dy_2 - y_3\,dx) + \Lambda(dy_3 - f\,dx)]. \end{aligned} \tag{3.521}$$

Through the algorithmic methods presented in Section 3.6.4, we now give an example to illustrate how to solve the integrating factor determining system (3.520a–c) and calculate first integrals from (3.521) to obtain a reduction of order.

Consider the fourth-order ODE

$$(yy'(y/y')'')'=0$$

that arises in the study of the wave equation with wave speed $y(x)$. This ODE is equivalent to the surface

$$y_4=-\frac{(y_1)^2y_2}{y^2}+\frac{4(y_2)^2}{y}-\frac{4(y_2)^3}{(y_1)^2}-\frac{3y_1y_3}{y}+\frac{5y_2y_3}{y_1} \tag{3.522}$$

in $(x,y,y_1,y_2,y_3,y_4)-$space. We now find all integrating factors $\Lambda(x,y,y_1,y_2)$ of ODE (3.522) up to second-order. From the integrating factor determining system (3.520a–c), we see that the first equation (3.520a) yields $\Lambda_{y_2}=0$ and so Λ is at most a first-order integrating factor. The second equation (3.520b) then becomes linear in y_2. The coefficient of y_2 yields

$$3\Lambda+5y_1\Lambda_{y_1}+(y_1)^2\Lambda_{y_1y_1}=0,$$

which is an Euler equation [cf. Section 3.6.3] with the general solution $\Lambda=\alpha(x,y)(y_1)^{-1}+\beta(x,y)(y_1)^{-3}$. The remaining terms in (3.520b) now split with respect to powers of y_1, giving

$$y\beta_y-2\beta=0,\quad y\alpha_y-2\alpha=0,\quad \beta_x=\alpha_x=0.$$

This yields $\alpha=\alpha_0y^2$, $\beta=\beta_0y^2$, with $\alpha_0=\text{const}$, $\beta_0=\text{const}$. We then find that $\Lambda=\alpha_0y^2(y_1)^{-1}+\beta_0y^2(y_1)^3$ satisfies the third equation (3.520c). Hence, we obtain two integrating factors given by

$$\Lambda_1=y^2(y_1)^{-1},\quad \Lambda_2=y^2(y_1)^{-3}. \tag{3.523}$$

Since these are first-order integrating factors, from Theorem 3.6.4-4 it follows that the corresponding first integrals are functionally independent. The line integral formula (3.521) gives

$$\begin{aligned}\psi_1&=\int_C\left[\left(y_2+\frac{2yy_3}{y_1}-\frac{4y(y_2)^2}{(y_1)^2}\right)dy+\left(\frac{4y^2(y_2)^2}{(y_1)^3}-\frac{y_3y^2}{(y_1)^2}\right)dy_1+\left(y-\frac{4y_2y^2}{(y_1)^2}\right)dy_2+\frac{y^2}{y_1}dy_3\right]\\&=yy_2-\frac{2y^2(y_2)^2}{(y_1)^2}+\frac{y^2y_3}{y_1}\end{aligned} \tag{3.524a}$$

and

$$\psi_2 = \int_C \left[\left(\frac{y_2}{(y_1)^2} + \frac{2yy_3}{(y_1)^3} - \frac{2y(y_2)^2}{(y_1)^4} \right) dy + \left(-\frac{2yy_2}{(y_1)^3} + \frac{4y^2(y_2)^2}{(y_1)^5} - \frac{3y^2y_3}{(y_1)^4} \right) dy_1 \right.$$
$$\left. + \left(\frac{y}{(y_1)^2} - \frac{2y_2y^2}{(y_1)^4} \right) dy_2 + \frac{y^2}{(y_1)^3} dy_3 \right]$$
$$= \frac{yy_2}{(y_1)^2} - \frac{y^2(y_2)^2}{(y_1)^4} + \frac{y^2y_3}{(y_1)^3}, \tag{3.524b}$$

where C is chosen to be a piecewise straight line from $(0,0,\tilde{y}_1,0,0)$ to (x,y,y_1,y_2,y_3), parallel to the coordinate axes, with $\tilde{y}_1 \neq 0$. The first integrals (3.524a,b) yield two quadratures $\psi_1 = \text{const} = c_1$, $\psi_2 = \text{const} = c_2$, leading to a reduction of the fourth-order ODE (3.522) to a second-order ODE

$$y_2 = \pm \frac{y_1}{y} \sqrt{c_2(y_1)^2 - c_1}. \tag{3.525}$$

The ODE (3.525) is separable [cf. (3.439)], and hence, it immediately admits two quadratures, yielding

$$\int \frac{1}{\cosh((c_2)^{1/2}(\log y + c_3))} dy = \left(\frac{c_1}{c_2} \right)^{1/2} x + c_4, \tag{3.526}$$

which is a general solution of ODE (3.522).

EXERCISES 3.6

1. Consider the general second-order linear ODE
$$y'' + p(x)y' + q(x)y = g(x).$$
   (a) Find its integrating factors of the form $\Lambda(x,y)$ and corresponding first integrals when $g(x) \equiv 0$ and $g(x) \not\equiv 0$.
   (b) In the case $p(x) = \text{const}$, $q(x) = \text{const}$, find its point-form integrating factors $\Lambda = \alpha(x,y) + \beta(x,y)y_1$ and corresponding first integrals.

2. Consider the nonlinear van der Pol oscillator
$$y'' - c(1 - ay^2)y' + by^p = 0, \quad a,b,c,p = \text{const}. \tag{3.527}$$
   Find its point-form integrating factors $\Lambda = \alpha(x,y) + \beta(x,y)y_1$ and corresponding first integrals. Show that two functionally independent first integrals are obtained if

and only if $p = 3$ and $ac = -3b$, and give the complete quadrature of (3.527) in this case.

3. Consider the fully nonlinear Duffing oscillator

$$y'' + ay^3 = 0,\ a = \text{const.} \tag{3.528}$$

   (a) Show that the only point-form integrating factor admitted by (3.528) is given by $\Lambda = y_1$.
   (b) Show that the reduced first-order ODE $\psi = \text{const} = c$ given by the first integral ψ corresponding to $\Lambda = y_1$ is separable. Hence, obtain the complete quadrature of (3.528).
   (c) Find the integrating factor of (3.528) corresponding to the quadrature of this separable reduced ODE.
   (d) Consider the integrating factor ansatz (3.400b,c) for (3.528) using the joint invariants of the translation symmetry $x \to x + \varepsilon, y \to y$ and the scaling symmetry $x \to \lambda x, y \to \lambda^{2/(1-p)} y$ of (3.528). Show that this ansatz yields a first-order integrating factor of (3.528) that is not of point-form and gives a first integral functionally independent of the one arising from the integrating factor $\Lambda = y_1$.

4. Consider the variable frequency oscillator [Mimura and Nôno (1994)]

$$y'' + (y')^2 y = 0. \tag{3.529}$$

   (a) Find the point-form integrating factors $\Lambda = \alpha(x, y) + \beta(x, y)y_1$ and corresponding first integrals of ODE (3.529).
   (b) Find the integrating factors given by the ansatz (3.400b,c) using the joint invariants of the scaling symmetry $x \to \lambda x,\ y \to y$ and translation symmetry $x \to x + \varepsilon,\ y \to y$ of ODE (3.529). Find the corresponding first integrals.
   (c) Obtain the complete quadrature of ODE (3.529).

5. The Thomas–Fermi equation is given by

$$y'' = x^{-1/2} y^{3/2}. \tag{3.530}$$

   (a) Show that ODE (3.530) admits no point-form integrating factors.
   (b) Find the first-order integrating factors of ODE (3.530) given by the ansatz (3.400b,c), using the invariants of the scaling symmetry $x \to \lambda x,\ y \to \lambda^{-3} y$ admitted by (3.530).

6. Find all first-order integrating factors $\Lambda(x, y, y_1)$ and corresponding of the third-order ODE $y''' = 0$. By using the general solution of this ODE, find all second-order integrating factors $\Lambda(x, y, y_1, y_2)$ and corresponding first integrals.

7. Consider the Blasius equation

$$y''' + \tfrac{1}{2} y y'' = 0. \tag{3.531}$$

   (a) Show that the third-order ODE (3.531) admits no first-order integrating factors.

(b) Find the second-order integrating factors of (3.531) given by the ansatz (3.490a,b) using the joint invariants of the translation symmetry $x \to x+\varepsilon,\ y \to y$ and scaling symmetry $x \to \lambda x,\ y \to \lambda^{-1} y$. Show that ODE (3.531) admits no integrating factors of the form $\Lambda = y^r y_1^s \alpha(u),\ u = y_2 / y^3\ [r, s = \text{const}]$.

8. Consider the KdV traveling wave ODE (3.499).
   (a) The two admitted point-form integrating factors (3.503) of ODE (3.499) lead to a reduced first-order ODE. Find the integrating factor of ODE (3.499) corresponding to the quadrature of this reduced ODE.
   (b) Find the first-order integrating factors of ODE (3.499).
   (c) Find the second-order integrating factors of ODE (3.499) given by the ansatz $\Lambda = y^r y_1^s \alpha(u)\ [r, s = \text{const}]$, using the scaling symmetry invariant $u = y_2 / y^2$.

9. Consider the fourth-order ODE

$$y^{(4)} = \tfrac{4}{3}(y''')^2 / y''. \tag{3.532}$$

   (a) Find the point-form integrating factors of ODE (3.532).
   (b) Find the first- and second-order integrating factors of ODE (3.532).
   (c) Find the third-order integrating factors of ODE (3.532) given by the ansatz (3.490a,b), using the joint invariants of the x and y translation symmetries and the x and y scaling symmetries of (3.532).

10. Classify all second-order ODEs $y'' = f(x, y, y')$ admitting an integrating factor of the form $\Lambda = \mu(x, y)$.

11. Classify all second-order ODEs $y'' = f(x, y, y')$ admitting integrating factors of the form:
    (a) $\Lambda = 1 / y_1$;
    (b) $\Lambda = (y_1)^2$; and
    (c) $\Lambda = e^{y_1}$.

12. Find the necessary and sufficient conditions on a function $f(x, y, y_1)$ such that the second-order ODE $y'' = f(x, y, y')$ admits an integrating factor of the form:
    (a) $\Lambda = \mu(x, y_1)$; and
    (b) $\Lambda = \mu(y, y_1)$.

13. Classify all third- and higher-order ODEs admitting an integrating factor of the form $\Lambda = \mu(x, y, y_1)$.

14. Consider the truncated Euler operator $\mathrm{E}_1 = \dfrac{\partial}{\partial y} - \mathrm{D}\dfrac{\partial}{\partial y_1}$ on (x, y, y_1)-space, where

$$\mathrm{D} = \frac{\partial}{\partial x} + y_1 \frac{\partial}{\partial y}.$$

(a) Show that the equation

$$\mathrm{E}_1(\theta(x,y,y_1)) = 0 \tag{3.533a}$$

on (x,y,y_1)-space can be explicitly solved by the following steps to obtain

$$\theta(x,y,y_1) = \mathrm{D}\psi(x,y),$$
$$\psi(x,y) = \int_0^1 x\theta(\lambda x,0,0) + y\theta_{y_1}(x,\lambda y,\lambda y_1)\,d\lambda. \tag{3.533b}$$

In (3.533a), replace y by $Y(\lambda) = \lambda y$ and y_1 by $Y_1(\lambda) = \mathrm{D}Y(\lambda) = \lambda y_1$, and then multiply by $dY/d\lambda = y$ to get

$$\begin{aligned} 0 &= y\left(\frac{\partial}{\partial Y}\theta(x,Y,Y_1) - \mathrm{D}\frac{\partial}{\partial Y_1}\theta(x,Y,Y_1)\right) \\ &= -\mathrm{D}\left(-y\frac{\partial}{\partial Y_1}\theta(x,Y,Y_1)\right) + \frac{\partial}{\partial \lambda}\theta(x,Y,Y_1). \end{aligned} \tag{3.534}$$

Finally, integrate (3.534) from $\lambda = 0$ to $\lambda = 1$, and use the identity

$$\theta(x,0,0) = \int_0^1 x\theta(\lambda x,0,0)\,d\lambda,$$

which leads to (3.533b).

(b) Show that for an appropriate path curve C, the line integral formula (3.336) reduces to (3.533b) to within a constant.

15. For a general first-order ODE $y' = f(x,y)$, solve the integrating factor equations $dx/1 = dy/f = d\Lambda/(-f_y\Lambda)$ to obtain the general solution $\Lambda = F'(\psi_1)\Lambda_1$, where $\Lambda_1 = (\psi_1)_y$ is *any* particular solution of (3.335) and F is an arbitrary function of ψ_1.

16. Show that the truncated Euler operator $\mathrm{E}_2 = \dfrac{\partial}{\partial y} - \mathrm{D}\dfrac{\partial}{\partial y_1} + \mathrm{D}^2\dfrac{\partial}{\partial y_2}$, with $\mathrm{D} = \dfrac{\partial}{\partial x} + y_1\dfrac{\partial}{\partial y} + y_2\dfrac{\partial}{\partial y_1}$, on (x,y,y_1,y_2)-space can be inverted similarly to (3.533a,b). Show that the line integral formula (3.366) reduces to this inversion for an appropriate path curve C.

17. Prove the identity

$$[\mathrm{D},\mathrm{X}^{(n)}] = (\mathrm{D}\xi)\mathrm{D} + (\mathrm{D}\eta^{(n)})\frac{\partial}{\partial y_n},$$

where $X^{(n)}$ is the nth extension (2.99f) of the operator $X = \xi(x,y)\frac{\partial}{\partial x} + \eta(x,y)\frac{\partial}{\partial y}$, $\eta^{(n)}$ is the nth extension (2.100a,b) of η, and D is the total derivative operator (3.445a) on $(x, y, y_1, \ldots, y_n)$ - space.

18. State and prove the converse of Theorem 3.6.3-1.

19. Prove Lemmas 3.6.3-2 and 3.6.4-2.

20. Prove Theorem 3.6.3-4 and Corollary 3.6.3-1.

21. Show that Ψ_{n-k} equals $\Psi_{n-k}\big|_{y_n=0}$ plus a polynomial of kth degree in y_n with coefficients given by differential linear combinations of $\partial\Psi_{n-i}/\partial y_n$, $0 \le i \le k$, $k = 1,2,\ldots,n$.

22. Show that the system of equations $\Psi_{j,j+1}\big|_{y_n=0} = 0$, $j = [(n+1)/2],\ldots,n-1$, and $\Psi_0\big|_{y_n=0} = 0$ explicitly yield (3.462b,c).

23. Prove Theorem 3.6.4-5.

24. Prove Theorem 3.6.4-7.

3.7 FUNDAMENTAL CONNECTIONS BETWEEN INTEGRATING FACTORS AND SYMMETRIES

We now discuss important connections between the determining systems derived in Sections 3.5 and 3.6, respectively, for symmetries and integrating factors of second- and higher-order ODEs.

For an ODE with a variational principle, all first integrals can be shown to arise from invariance of the action functional under one-parameter groups of local transformations [cf. Section 2.7.2] through the classical Noether's Theorem [Noether (1918); Boyer (1967); Olver (1986)]. In particular, Noether's Theorem states that a one-parameter local transformation group leaves invariant the given action functional if and only if the infinitesimal of the transformations, in characteristic form [cf. Section 3.5.1], is an integrating factor of the given ODE. Clearly, since every such one-parameter local transformation group leaves invariant the extremals of the action functional, it gives rise to a corresponding symmetry of the ODE. However, all symmetries of the ODE need not necessarily arise from local transformation groups of the action functional. For example, often an action functional is *not* invariant under scalings admitted by the corresponding ODE. If a symmetry of an ODE corresponds to a one-parameter local transformation group of an action functional for the ODE, then it is called a variational symmetry. Consequently, from Noether's Theorem, it follows that integrating factors are the same as variational symmetries when an ODE possesses a variational principle.

Existence of a variational principle for an ODE can be expressed as a condition on the linear operator associated with the linearization of the ODE (i.e., its Fréchet derivative). In particular, an ODE admits a variational principle if and only if its linearization operator is self-adjoint. [Moreover, then an action functional can be constructed from the independent and dependent variables of the ODE through an explicit formula [Olver (1986)].]

Whether or not an ODE admits a variational principle, the linearization operator of the ODE is directly connected with the determining equation for symmetries of the ODE. From Theorem 3.5.1-1, symmetry characteristics are the solutions of the linearization of the ODE holding on the *entire* solution space of the ODE. As a result, if the linearization operator of an ODE is self-adjoint, then the integrating factors of the ODE are the same as the variational symmetries of the ODE and, thus, in this situation, the integrating factor determining system is equivalent to the symmetry determining equation augmented by conditions for a symmetry to be variational. If the linearization operator of an ODE is not self-adjoint, then the integrating factors are no longer symmetries but instead turn out to be directly connected with the solutions of adjoint linearization of the ODE, as is familiar in the classical case of second-order linear ODEs [cf. Section 3.6.3].

We show that for an nth-order ODE, whether or not its linearization operator is self-adjoint, the integrating factor determining system of $1+[n/2]$ equations, arising from Theorem 3.6.4-3, is equivalent to the *adjoint* equation of the symmetry determining equation, augmented by $[n/2]$ extra equations. These extra determining equations are called the *adjoint invariance conditions*, while the solutions of the adjoint equation of the symmetry determining equation are called *adjoint-symmetries*. Thus, the integrating factors of a nth-order ODE are those adjoint-symmetries that satisfy the adjoint invariance conditions.

In the case when an nth-order ODE admits a variational principle, the symmetry determining equation is the same as its adjoint equation, so here adjoint-symmetries are symmetries. The adjoint invariance conditions are then equivalent to the condition for a symmetry to be variational. We explicitly identify the variational symmetry condition as $[n/2]$ determining equations obtained by splitting up the integrating factor determining system into the symmetry determining equation augmented by $[n/2]$ extra determining equations.

Finally, we compare the calculations for integrating factors, symmetries, and adjoint-symmetries. In particular, we show that the class of nth-order ODEs admitting integrating factors of a given form is of a cardinality similar to that of the class of nth-order ODEs admitting symmetries of the same form.

3.7.1 ADJOINT-SYMMETRIES

Consider an nth-order ODE

$$y^{(n)}=f(x,y,y',\ldots,y^{(n-1)}) \tag{3.535}$$

represented by the surface

$$F(x,y,y_1,\ldots,y_n)=y_n-f(x,y,y_1,\ldots,y_{n-1})=0. \tag{3.536}$$

The linearization operator of ODE (3.536) is given by

$$\mathrm{L}_F = \mathrm{D}^n - \sum_{i=0}^{n-1} f_{y_i} \mathrm{D}^i, \tag{3.537}$$

where

$$\mathrm{D} = \frac{\partial}{\partial x} + \sum_{i=1}^{\infty} y_i \frac{\partial}{\partial y_{i-1}} \quad \text{with} \quad y_0 = y.$$

The adjoint linearization operator is given by (through formal integration by parts)

$$\mathrm{L}*_F = (-1)^n \mathrm{D}^n - \sum_{i=0}^{n-1} \sum_{j=0}^{i} (-1)^i \frac{i!}{(i-j)!\,j!} (\mathrm{D}^{i-j} f_{y_i}) \mathrm{D}^j. \tag{3.538}$$

The operator (3.538) can also be introduced by the following relation:

Lemma 3.7.1-1. *The operators* L_F *and* $\mathrm{L}*_F$ *satisfy the identity*

$$W\mathrm{L}_F V - V\mathrm{L}*_F W = \mathrm{D}S[W,V;F], \tag{3.539a}$$

where S *is the trilinear function defined by*

$$S[W,V;U] = \sum_{i=0}^{n-1} \sum_{j=0}^{i} (-1)^j (\mathrm{D}^{i-j} V) \mathrm{D}^j (W U_{y_{i+1}}) \tag{3.539b}$$

for arbitrary functions $U(x,y,y_1,\ldots,y_n), V(x,y,y_1,\ldots,y_n), W(x,y,y_1,\ldots,y_n)$.

Proof. The identity (3.539a,b) is verified by a direct expansion of both sides of (3.539a) through use of the definitions (3.537) and (3.538) for L_F and $\mathrm{L}*_F$. □

Definition 3.7.1-1. An nth-order ODE (3.535) is *self-adjoint* if and only if $\mathrm{L}_F = \mathrm{L}*_F$. In particular, self-adjointness is equivalent to the $n+1$ conditions

$$(-1)^n = 1, \quad \text{i.e., } n \text{ is even}, \tag{3.540a}$$

$$f_{y_i} = \sum_{j=0}^{n-i-1} (-1)^{i+j} \frac{(i+j)!}{i!\,j!} \mathrm{D}^j f_{y_{i+j}}, \quad i = 0,1,\ldots,n-1. \tag{3.540b}$$

Let

$$\mathbf{L}_F = \mathrm{L}_F\big|_{F=0} = \mathbf{D}^n - \sum_{i=0}^{n-1} f_{y_i} \mathbf{D}^i, \tag{3.541a}$$

$$\mathbf{L}*_F = \mathrm{L}*_F\big|_{F=0} = (-1)\mathbf{D}^n - \sum_{i=0}^{n-1} \sum_{j=0}^{i} (-1)^i \frac{i!}{(i-j)!\,j!} (\mathbf{D}^{i-j} f_{y_i}) \mathbf{D}^j, \tag{3.541b}$$

where

$$\mathbf{D} = \mathrm{D}\big|_{F=0} = \frac{\partial}{\partial x} + \sum_{i=1}^{n-1} y_i \frac{\partial}{\partial y_{i-1}} + f(x, y, y_1, \ldots, y_{n-1}) \frac{\partial}{\partial y_{n-1}}.$$

The operators (3.541a,b) are the restrictions of operators (3.537) and (3.538) to the surface (3.536). From Theorem 3.5.1-1, we see that the symmetries in characteristic form of order $0 \le \ell \le n-1$ of ODE (3.535) are the solutions $\hat{\eta}(x, y, y_1, \ldots, y_\ell)$ of the symmetry determining equation

$$\mathbf{L}_F \hat{\eta} = \mathbf{D}^n \hat{\eta} - \sum_{i=0}^{n-1} f_{y_i} \mathbf{D}^i \hat{\eta} = 0. \tag{3.542}$$

Definition 3.7.1-2. The *adjoint-symmetries* of order $0 \le \ell \le n-1$ of ODE (3.535) are the solutions $\omega(x, y, y_1, \ldots, y_\ell)$ of the adjoint-symmetry determining equation

$$\mathbf{L}^*{}_F\, \omega = (-1)^n \mathbf{D}^n \omega - \sum_{i=0}^{n-1} (-1)^i \mathbf{D}^i (f_{y_i} \omega) = 0. \tag{3.543}$$

Geometrically, symmetries of ODE (3.535) describe motions on the surface (3.536) [cf. Section 3.5.1]. When ODE (3.535) is self-adjoint, its adjoint-symmetries are the same as its symmetries in characteristic form. However, if ODE (3.535) is not self-adjoint, then, in general, its adjoint-symmetries are not symmetries (i.e., the only common solution of the determining equations (3.542) and (3.543) is $\hat{\eta} = \omega = 0$), and there is no obvious geometrical invariance or motion related to the solutions of the adjoint-symmetry determining equation (3.543).

We note that the conditions for self-adjointness of ODE (3.535) can be formulated equivalently in terms of $\mathbf{L}_F$ and $\mathbf{L}^*{}_F$.

Lemma 3.7.1-2. *An nth-order ODE* (3.535) *is self-adjoint if and only if* $f(x, y, y_1, \ldots, y_{n-1})$ *satisfies* $\mathbf{L}_F = \mathbf{L}^*{}_F$, *which is equivalent to the* $n+1$ *conditions*

$$(-1)^n = 1,$$

$$f_{y_j} = \sum_{i=0}^{n-j-1} (-1)^{i+j} \frac{(i+j)!}{i!\, j!} \mathbf{D}^i f_{y_{i+j}}, \quad j = 0, 1, \ldots, n-1.$$

In particular, it is necessary (*but not sufficient unless* $n = 2$) *that* $f_{y_{n-1}} = 0$.

Proof. Left to Exercise 3.7-5. □

A similar discussion applies to an nth-order ODE

$$F(x, y, y_1, \ldots, y_n) = 0, \quad F_{y_n} \ne 0 \tag{3.544}$$

that is not in a solved form (3.536) in terms of y_n. Symmetries in characteristic form $\hat{\eta}(x,y,y_1,\ldots,y_\ell)$ of an ODE (3.544) are solutions of the determining equation $\mathbf{L}_F\hat{\eta}=0$, and adjoint-symmetries $\omega(x,y,y_1,\ldots,y_\ell)$ of (3.544) are solutions of the adjoint equation $\mathbf{L}^*_F\,\omega=0$, where $\mathbf{L}_F$ and $\mathbf{L}^*_F$ are the restrictions of the operators

$$\mathrm{L}_F=\sum_{i=0}^{n} F_{y_i}\mathrm{D}^i \quad\text{and}\quad \mathrm{L}^*_F=\sum_{i=0}^{n}\sum_{j=0}^{i}(-1)^i\frac{i!}{(i-j)!\,j!}(\mathrm{D}^{i-j}F_{y_i})\mathrm{D}^j$$

to the surface $F(x,y,y_1,\ldots,y_n)=0$. The criterion for self-adjointness, $\mathrm{L}_F=\mathrm{L}^*_F$, of ODE (3.544) is equivalent to the $n+2$ conditions

$$(-1)^n=1,\quad \text{i.e.,}\quad n \text{ is even,} \tag{3.545a}$$

$$F_{y_j}=\sum_{i=0}^{n-j}(-1)^{i+j}\frac{(i+j)!}{i!\,j!}\mathrm{D}^iF_{y_{i+j}},\quad j=0,1,\ldots,n. \tag{3.545b}$$

In particular, it is necessary (but not sufficient unless $n=2$) that ODE (3.544) take the form $F=Ay_n+B$ with $A_{y_n}=B_{y_n}=0$, $2B_{y_{n-1}}=nD_{n-1}A$ and, if $n\geq 3$, $A_{y_{n-1}}=0$, where D_{n-1} is the truncated total derivative operator (3.468).

3.7.2 ADJOINT INVARIANCE CONDITIONS AND INTEGRATING FACTORS

From Theorem 3.6.4-3, recall that the integrating factors of order $0\leq\ell\leq n-1$ of an nth-order ODE (3.535) are the solutions of the determining system of $1+[n/2]$ equations (3.470a,b). In particular, (3.470b) is the Euler operator equation [cf. (3.447)] $\mathrm{E}_n(F\Lambda)\big|_{y_n=0}=0$ which follows from Lemma 3.6.4-1. We now show that (3.470b) plus a certain differential linear combination of (3.470a) yields the adjoint-symmetry determining equation (3.543).

Lemma 3.7.2-1. *Let* E_n *be the truncated Euler operator* (3.447) *in* $(x,y,y_1,\ldots,y_n)$*-space. For any function* $\omega(x,y,y_1,\ldots,y_\ell)$, $0\leq\ell\leq n-1$, *define inductively*

$$\Psi_n=\omega,\quad \Psi_{n-k}=\frac{\partial(F\omega)}{\partial y_{n-k}}-\mathrm{D}\Psi_{n-k+1},\quad k=1,\ldots,n, \tag{3.546a}$$

$$\Omega_n=\frac{\partial\Psi_n}{\partial y_n}=0,\quad \Omega_{n-k}=\frac{\partial\Psi_{n-k}}{\partial y_n}+\mathrm{D}\frac{\partial\Psi_{n-k+1}}{\partial y_n},\qquad k=1,\ldots,n, \tag{3.546b}$$

where $F(x,y,y_1,\ldots,y_n)$ *is given by the surface* (3.536). *Then* $\mathrm{E}_n(F\omega)\big|_{y_n=0}=0$ *holds if and only if*

$$\mathbf{L}^*_F\,\omega-\sum_{i=0}^{n-1}(-1)^i\mathbf{D}^i(f\,\Omega_i\big|_{y_n=0})=0. \tag{3.547}$$

Proof. From (3.447) we note that $\mathrm{E}_n(F\omega) = \Psi_0$. Now taking $\partial / \partial y_n$ of (3.546a) yields the relation

$$\Omega_{n-k} = \frac{\partial \omega}{\partial y_{n-k}} - \frac{\partial \Psi_{n-k+1}}{\partial y_{n-1}}, \quad k = 1,2,\dots,n, \tag{3.548}$$

which allows (3.546a) to be expressed as

$$\Psi_{n-k} = F\Omega_{n-k} - f_{y_{n-k}}\omega - \mathbf{D}\Psi_{n-k+1}, \quad k = 1,2,\dots,n. \tag{3.549}$$

Then, by finite induction on the index k in (3.549), we obtain

$$\begin{aligned}\Psi_0 &= F\Omega_0 - f_y\omega - \mathbf{D}(F\Omega_1 - f_{y_1}\omega - \mathbf{D}(\cdots - \mathbf{D}(F\Omega_{n-1})))\\ &= \mathbf{L}^*{}_F\,\omega + \sum_{i=0}^{n-1} (-1)^i \mathbf{D}^i(F\Omega_i)\\ &= \mathrm{E}_n(F\omega).\end{aligned}$$

Hence, this establishes (3.547). □

The remaining $[n/2]$ determining equations (3.470a) for integrating factors in Theorem 3.6.4-3 are equivalent to the system of equations $\partial\Psi_i / \partial y_n = 0,\ i = 0,1,\dots,n-1$. Through relation (3.546b), this system can be written as $\Omega_i = 0,\ i = 0,1,\dots,n-1$. Then, from Lemma 3.6.4-1, we see that these equations in turn are equivalent to the simpler system of *half as many equations*

$$\Omega_{n-2m}\big|_{y_n=0} = 0, \quad m = 1,\dots,[n/2]. \tag{3.550}$$

We refer to each equation

$$\Omega_{n-k}\big|_{y_n=0} = \omega_{y_{n-k}} + \sum_{j=1}^{k-1} (-1)^{j-1}(\mathbf{D}^{j-1}(f_{y_{n-k+j}}\omega))_{y_{n-1}} + (-1)^k(\mathbf{D}^{k-1}\omega)_{y_{n-1}} = 0, \quad k = 1,2,\dots,n, \tag{3.551}$$

obtained inductively from (3.548) and (3.549), as an *adjoint invariance equation*, and we call the system of equations (3.550) the *adjoint invariance conditions* on $\omega(x, y, y_1, \dots, y_\ell)$. Hence, Lemma 3.7.2-1 and Theorem 3.6.4-3 lead immediately to the following main result:

Theorem 3.7.2-1. *The integrating factors of order* $0 \le \ell \le n-1$ *of ODE* (3.535) *are those adjoint-symmetries of order* ℓ *of* (3.535) *that satisfy the* $[n/2]$ *adjoint invariance conditions* (3.550). *Explicitly,* $\Lambda(x, y, y_1, \dots, y_\ell)$ *is an integrating factor of ODE* (3.535) *if and only if*

$$(-1)^n \mathbf{D}^n \Lambda - \sum_{i=0}^{n-1} (-1)^i \mathbf{D}^i (f_{y_i} \Lambda) = 0, \tag{3.552a}$$

$$\Lambda_{y_{n-2m}} + \sum_{i=1}^{2m-1} (-1)^{i-1} (\mathbf{D}^{i-1}(f_{y_{n-2m+i}} \Lambda))_{y_{n-1}} + (\mathbf{D}^{2m-1}\Lambda)_{y_{n-1}} = 0, \quad m = 1, \ldots, [n/2]. \tag{3.552b}$$

The version (3.552a,b) of the integrating factor determining system leads to a useful ansatz for finding integrating factors as follows. Clearly, every integrating factor of ODE (3.535) is an adjoint-symmetry, but not conversely if $n \geq 2$, since the adjoint invariance conditions need to be satisfied for an adjoint-symmetry of ODE (3.535) to be an integrating factor. Note that if $\omega(x, y, y_1, \ldots y_\ell)$ satisfies the adjoint-symmetry determining equation (3.543), then so does $\omega\psi$, where $\psi(x, y, y_1, \ldots y_{n-1})$ is *any* function satisfying $\mathbf{D}\psi = 0$, i.e., $\psi(x, y, y_1, \ldots y_{n-1})$ is a first integral of ODE (3.535). Hence, if one already knows $k \geq 1$ first integrals $\psi_1(x, y, y_1, \ldots y_{n-1}), \ldots, \psi_k(x, y, y_1, \ldots y_{n-1})$ that are functionally independent, and if $\omega(x, y, y_1, \ldots y_\ell)$ is an adjoint-symmetry that is *not* an integrating factor of ODE (3.535), then one can seek a multiplier function $\psi = \psi(\psi_1, \ldots, \psi_k)$ so that

$$\Lambda = \omega\psi \tag{3.553}$$

satisfies the adjoint invariance conditions (3.552b). In particular, (3.552b) reduces to a system of $[n/2]$ *first-order* linear homogeneous PDEs for $\psi(\psi_1, \ldots, \psi_k)$. Thus, the ansatz (3.553) allows one to seek integrating factors through the use of any known adjoint-symmetries and first integrals. Examples illustrating this ansatz will be given in Section 3.7.3.

3.7.3 EXAMPLES OF FINDING ADJOINT-SYMMETRIES AND INTEGRATING FACTORS

The determining equation (3.543) for adjoint-symmetries $\omega(x, y, y_1, \ldots, y_\ell)$ of order ℓ of a given nth-order ODE (3.535) is an nth-order linear homogeneous PDE in terms of $n+1$ independent variables $x, y, y_1, \ldots, y_{n-1}$. Hence, for $\ell = n-1$, the adjoint-symmetry determining equation (3.543) has infinitely many solutions. However, for $0 \leq \ell < n-1$, equation (3.543) in general splits into an overdetermined linear system of PDEs with at most a finite number of linearly independent solutions. These solutions can be calculated by the same algorithmic procedure as that used for finding symmetries of order $0 \leq \ell < n-1$ [cf. Section 3.5.1].

If a second- or higher-order ODE (3.535) admits a point symmetry $\hat{\eta} = \eta(x, y) - \xi(x, y) y_1$, so that the surface (3.536) is invariant [i.e., $\hat{\mathbf{X}}^{(n)} F = 0$ under the generator $\hat{\mathbf{X}}^{(n)}$ defined by (3.235)], then through the use of canonical coordinates

[Exercise 3.7-7] one can show that the adjoint-symmetry determining equation (3.543) admits the corresponding point symmetry

$$\mathrm{X}^{(n)} = \xi\frac{\partial}{\partial x} + \eta\frac{\partial}{\partial y} + \sum_{i=1}^{n-1} \eta^{(i)}\frac{\partial}{\partial y_i}.$$

This allows one to make a simplifying ansatz to seek solutions of (3.543). In particular, if ODE (3.535) admits a scaling $\xi = qx$, $\eta = py$, i.e., $x \to \alpha^q x$, $y \to \alpha^p y$ $[p, q = \text{const}]$, then the adjoint-symmetry determining equation (3.543) admits the scalings

$$\omega \to \alpha^r \omega, \quad x \to \alpha^q x, \quad y_i \to \alpha^{p-iq} y_i, \quad i = 0,1,\ldots,n-1,$$

for arbitrary $r = \text{const}$, and so one can seek invariant solutions of (3.543) [see Section 4.2.1] of the form

$$\omega = x^r \rho(y^q x^{-p}, x^{q-p}(y_1)^q, \ldots, x^{(n-1)q-p}(y_{n-1})^q) \quad \text{if } q \neq 0, \tag{3.554a}$$

$$\omega = y^r \rho(x, y^{-1}y_1, \ldots, y^{-1}y_{n-1}) \quad \text{if } p \neq 0. \tag{3.554b}$$

The ansatzes (3.554a,b) respectively reduce the adjoint-symmetry determining equation (3.543) to an overdetermined linear system of PDEs in terms of invariant variables $x^{iq-p}(y_i)^q$, $i = 0,1,\ldots,n-1$, or $x, y^{-1}y_i, i = 1,\ldots,n-1$. Similarly, if ODE (3.535) admits a translation $\xi = 1$, $\eta = 0$, i.e., $x \to x+\varepsilon, y \to y$, or $\xi = 0$, $\eta = 1$, i.e., $x \to x, y \to y+\varepsilon$, then the adjoint-symmetry determining equation (3.543) admits the translation

$$x \to x+\varepsilon,\ y \to y, \quad \text{or} \quad x \to x,\ y \to y+\varepsilon,$$

$$y_i \to y_i, \quad i = 1,\ldots,n-1,$$

together with the scaling

$$\omega \to e^{r\varepsilon}\omega$$

for arbitrary $r = \text{const}$. Consequently, one can seek invariant solutions of (3.543) of the form

$$\omega = e^{rx}\rho(y, y_1,\ldots,y_{n-1}) \quad \text{or} \quad \omega = e^{ry}\rho(x, y_1,\ldots,y_{n-1}). \tag{3.555}$$

Each ansatz (3.555) reduces the adjoint-symmetry determining equation (3.535) to an overdetermined linear system of PDEs in terms of invariant variables y, y_i or x, y_i, $i = 1,\ldots,n-1$.

The previous ansatzes are obvious counterparts of the ansatzes presented in Section 3.5.1 for solving the symmetry determining equation for second- and higher-order ODEs. We now give examples that illustrate the calculation of adjoint-symmetries.

As a first example, consider the nonlinear Duffing equation

$$y'' + ay' + by + y^3 = 0, \quad b = \tfrac{2}{9}a^2, \quad a = \text{const}, \tag{3.556}$$

for which an integrating factor of point-form was obtained in Section 3.6.3 [cf. (3.387)]. Here we will compare the point symmetries and point-form adjoint-symmetries of ODE (3.556), and then obtain a second integrating factor through the ansatz (3.553).

The linearization operator of (3.556) is given by $\mathrm{L}_F = \mathrm{D}^2 + a\mathrm{D} + \frac{2}{9}a^2 + 3y^2$, which is not a self-adjoint operator since $\mathrm{L}^*_F = \mathrm{D}^2 - a\mathrm{D} + \frac{2}{9}a^2 + 3y^2 \neq \mathrm{L}_F$ if $a \neq 0$. Hence, in this case, ODE (3.556) is not self-adjoint, and so its adjoint-symmetries are not symmetries. The determining equation (3.543) for the adjoint-symmetries $\omega(x, y, y_1)$ of ODE (3.556) is given by

$$\mathbf{D}^2\omega - a\mathbf{D}\omega + (\tfrac{2}{9}a^2 + 3y^2)\omega = 0, \tag{3.557}$$

where

$$\mathbf{D} = \frac{\partial}{\partial x} + y_1\frac{\partial}{\partial y} - (ay_1 + \tfrac{2}{9}a^2 y + y^3)\frac{\partial}{\partial y_1}.$$

If we consider point-form adjoint-symmetries $\omega = \alpha(x, y) + \beta(x, y)y_1$, then the adjoint-symmetry determining (3.557) reduces to a system of four linear PDEs arising from the coefficients of like powers of y_1:

$$\beta_{yy} = 0, \tag{3.558a}$$

$$-4a\beta_y + \alpha_{yy} + 2\beta_{xy} = 0, \tag{3.558b}$$

$$2a^2\beta - (\tfrac{2}{3}a^2 y + 3y^3)\beta_y - 3a\beta_x - 2a\alpha_y + 2\alpha_{xy} + \beta_{xx} = 0, \tag{3.558c}$$

$$(y^3 + \tfrac{2}{9}a^2 y)(2a\beta\alpha - 2\beta_x - \alpha_y) + (3y^2 + \tfrac{2}{9}a^2)\alpha - a\alpha_x + \alpha_{xx} = 0. \tag{3.558d}$$

After integrating (3.558a,b) with respect to y, we obtain $\beta = \beta_0(x) + \beta_1(x)y$ and $\alpha = (2a\beta_1(x) - \beta_1'(x))y^2 + \alpha_0(x) + \alpha_1(x)y$ for some functions $\alpha_0(x), \alpha_1(x), \beta_0(x), \beta_1(x)$. Then (3.558cd) leads to $\alpha_0 = 0$, $\beta_1 = 0$, $\alpha_1 = \beta_0' - a\beta_0$, and $3\beta_0'' - 7a\beta_0' + 4a^2\beta_0 = 0$. Hence, we obtain

$$\beta = c_1 e^{ax} + c_2 e^{(4a/3)x}, \quad \alpha = \tfrac{1}{3}c_2 a e^{(4a/3)x} y.$$

Thus, ODE (3.556) admits two point-form adjoint-symmetries given by

$$\omega_1 = e^{(4a/3)x}(\tfrac{1}{3}ay + y_1), \tag{3.559a}$$

$$\omega_2 = e^{ax}y_1. \tag{3.559b}$$

By a similar calculation, the point symmetries $\hat{\eta} = \alpha(x, y) + \beta(x, y)y_1$ of ODE (3.556) are given by

$$\hat{\eta}_1 = e^{(a/3)x}(\tfrac{1}{3}ay + y_1), \quad \hat{\eta}_2 = y_1.$$

Thus, the adjoint-symmetries (3.559a,b) are not symmetries of ODE (3.556) except when $a = 0$.

From Section 3.6.3 [cf. (3.387)], we see that the adjoint-symmetry (3.559a) is an integrating factor of (3.556), while (3.559b) is an adjoint-symmetry of ODE (3.556) which is *not* an integrating factor of (3.556). In particular, the adjoint-symmetry (3.559b) fails to satisfy the adjoint invariance condition of Theorem 3.7.2-1,

$$\Lambda_y - a\Lambda_{y_1} + (\mathbf{D}\Lambda)_{y_1} = 0, \tag{3.560}$$

which is necessary and sufficient for an adjoint-symmetry ω to be an integrating factor $\Lambda = \omega$ of ODE (3.556). We now use the adjoint-symmetry (3.559b) to seek an integrating factor

$$\Lambda = \psi(\psi_1)e^{ax}y_1, \tag{3.561}$$

depending on a function $\psi(\psi_1)$, where

$$\psi_1 = e^{(4a/3)x}(\tfrac{1}{4}y^4 + \tfrac{1}{2}(\tfrac{1}{3}ay + y_1)^2) \tag{3.562}$$

is the first integral corresponding to the integrating factor (3.559a). Note that (3.561) satisfies the adjoint-symmetry determining equation (3.557) for arbitrary $\psi(\psi_1)$ since $\mathbf{D}\psi_1 = 0$. Substitution of (3.561) into the adjoint invariance condition (3.560) yields

$$-a\psi + (y_1(\psi_1)_y - (ay_1 + \tfrac{2}{9}a^2y + y^3)(\psi_1)_{y_1})\psi' = -a(\psi + \tfrac{4}{3}\psi_1\psi') = 0,$$

which reduces to a first-order separable ODE $\psi'/\psi = -\tfrac{3}{4}\psi_1$. Hence, $\psi(\psi_1) = c(\psi_1)^{-3/4}$, $c = \text{const}$, and thus, we obtain

$$\Lambda = y_1e^{ax}(\psi_1)^{-3/4} = y_1(\tfrac{1}{4}y^4 + \tfrac{1}{2}(\tfrac{1}{3}ay + y_1)^2)^{-3/4}, \tag{3.563}$$

giving an integrating factor of ODE (3.556). Since $\omega_1/\Lambda = (1 + \tfrac{1}{3}a(y/y_1))e^{(a/3)x}(\psi_1)^{3/4}$ is clearly not a function of only ψ_1, from Lemma 3.6.2-1 it follows that Λ yields a first integral ψ_2 that is functionally independent of ψ_1. Using the line integral formula (3.366) to calculate ψ_2, we have

$$\psi_2 = \int_C [(\tfrac{2}{9}a^2y + y^3 + ay_1)z^{-3/4}\,dy + y_1z^{-3/4}\,dy_1], \tag{3.564}$$

with

$$z(y, y_1) = \tfrac{1}{4}y^4 + \tfrac{1}{2}(\tfrac{1}{3}ay + y_1)^2. \tag{3.565}$$

Here we choose C to be a path curve such that $z = \text{const}$ in the (y, y_1)-plane, which is conveniently parametrized by $y = Y(\lambda), y_1 = Y_1(\lambda)$, satisfying $Y(0) = \tilde{y}$, $Y_1(0) = \tilde{y}_1$, $Y(1) = y$, $Y_1(1) = y_1$, as follows:

$$\frac{dY}{d\lambda} = z_{y_1}(Y, Y_1) = \tfrac{1}{3}aY + Y_1,$$

$$\frac{dY_1}{d\lambda} = -z_y(Y, Y_1) = -Y^3 - \tfrac{1}{9}a^2Y - \tfrac{1}{3}aY_1.$$

This system is readily integrated to yield

$$z = \tfrac{1}{4}Y^4 + \tfrac{1}{2}(\tfrac{1}{3}aY + Y_1)^2 = \text{const}, \tag{3.566a}$$

$$\lambda = \int_{\tilde{y}}^{Y} \frac{d\tau}{(2z - \frac{1}{2}\tau^4)^{1/2}}. \tag{3.566b}$$

Then, after we combine terms and simplify using (3.566a), the first integral (3.564) becomes

$$\psi_2 = \int_0^1 \tfrac{4}{3} a z^{1/4}\, d\lambda = \tfrac{4}{3} a z^{1/4} \int_{\tilde{y}}^{y} \frac{d\tau}{(2z - \frac{1}{2}\tau^4)^{1/2}} \tag{3.567}$$

with z given by (3.565).

The first integrals (3.563) and (3.567) together yield the quadrature of ODE (3.556) given by $\psi_1 = \text{const} = c_1$, $\psi_2 = \text{const} = c_2$. Explicitly, we have

$$c_2 = \tfrac{4}{3} a (c_1)^{1/4} e^{-(1/3)ax} \int_{\tilde{y}}^{y} \frac{d\tau}{(2c_1 e^{-(4a/3)x} - \frac{1}{2}\tau^4)^{1/2}}. \tag{3.568}$$

As a second example, we return to the third-order ODE

$$y''' = 6x(y')^{-2}(y'')^3 + 6(y')^{-1}(y'')^2, \tag{3.569}$$

which admits contact symmetries as shown in Section 3.5.2. In Section 3.6.5, we obtained three second-order integrating factors of ODE (3.569) that led to its complete quadrature [cf. (3.517)]. Here we seek the first-order adjoint-symmetries admitted by ODE (3.569). Since ODE (3.569) is of third-order, it is not self-adjoint, and hence, its adjoint-symmetries are not symmetries. The linearization operator of (3.569) is given by

$$\mathrm{L}_F = \mathbf{D}^3 - (18x(y_2)^2(y_1)^{-2} + 12y_2(y_1)^{-1})\mathbf{D}^2 + (12x(y_2)^3(y_1)^{-3} + 6(y_2)^2(y_1)^{-2})\mathbf{D}.$$

From the adjoint operator $\mathbf{L}^*_F$, the determining equation (3.543) for adjoint-symmetries $\omega(x, y, y_1)$ of ODE (3.569) becomes

$$\mathbf{D}^3\omega + \mathbf{D}^2(18x(y_2)^2(y_1)^{-2}\omega + 12y_2(y_1)^{-1}\omega) + \mathbf{D}(12x(y_2)^3(y_1)^{-3}\omega + 6(y_2)^2(y_1)^{-2}\omega) = 0, \tag{3.570}$$

where

$$\mathbf{D} = \frac{\partial}{\partial x} + y_1\frac{\partial}{\partial y} + y_2\frac{\partial}{\partial y_1} + f\frac{\partial}{\partial y_2}, \quad f = 6x(y_2)^3(y_1)^{-2} + 6(y_2)^2(y_1)^{-1}.$$

It is not hard to show that the adjoint-symmetry determining equation (3.570) is a polynomial equation of sixth degree in terms of y_2 and, thus, reduces to a linear system of seven equations arising from the coefficients of like powers of y_2. The equation given by the coefficient of $(y_2)^6$ immediately yields $\omega = 0$. Hence we find that ODE (3.569) admits no first-order adjoint-symmetries, in contrast to its seven admitted contact symmetries and three admitted point symmetries [cf. Section 3.5.2].

As a final example, consider the fourth-order ODE

$$(yy'(y(y')^{-1})'')' = 0 \tag{3.571}$$

that arises in the study of the wave equation with wave speed $y(x)$. In Section 3.5.2, we showed that the admitted symmetries of ODE (3.571) up to second-order consist only of point symmetries given by x translation and x, y scalings. In Section 3.6.5, we obtained two first-order integrating factors of ODE (3.571), which led to a second-order separable ODE [cf. (3.525)] and thereby yielded the quadrature of ODE (3.571). Here, we find the adjoint-symmetries of ODE (3.571) up to second-order and then apply ansatz (3.553) to obtain additional integrating factors of (3.571).

We first show that ODE (3.571) is not self-adjoint. Expressing (3.571) in solved form, we obtain

$$y_4 = f(y, y_1, y_2, y_3) = -\frac{(y_1)^2 y_2}{y^2} + 4\frac{(y_2)^2}{y} - 4\frac{(y_2)^3}{(y_1)^2} - 3\frac{y_1 y_3}{y} + 5\frac{y_2 y_3}{y_1} = 0. \tag{3.572}$$

Since $f_{y_3} = -3y^{-1}y_1 + 5(y_1)^{-1}y_2 \neq 0$, from Lemma 3.7.1-2 it immediately follows that ODE (3.572) fails to be self-adjoint. The same conclusion holds for ODE (3.571) in its original form, as seen from (3.545b). Hence, adjoint-symmetries of ODE (3.572) are not symmetries of (3.572).

The determining equation (3.543) for adjoint-symmetries $\omega(x, y, y_1, y_2)$ of ODE (3.572) is most easily derived by starting from the linearization of ODE (3.571) rather than that of (3.572). This leads to

$$\begin{aligned} &y\mathbf{D}((y_2(y_1)^{-1} - 2y(y_2)^2(y_1)^{-3} + yy_3(y_1)^{-2})\mathbf{D}\widetilde{\omega}) \\ &\qquad + (y_1)^{-1}\mathbf{D}^2(yy_1\mathbf{D}\widetilde{\omega}) + \mathbf{D}(y(y_1)^{-2}\mathbf{D}^2(yy_1\mathbf{D}\widetilde{\omega})) = 0 \end{aligned} \tag{3.573}$$

with $\widetilde{\omega} = y_1 y^{-2}\omega$, where

$$\mathbf{D} = \frac{\partial}{\partial x} + y_1\frac{\partial}{\partial y} + y_2\frac{\partial}{\partial y_1} + y_3\frac{\partial}{\partial y_2} + f\frac{\partial}{\partial y_3}.$$

We find that the adjoint-symmetry determining equation (3.573) is a quartic polynomial in terms of y_3 and, thus, reduces to an overdetermined linear system of five PDEs for $\omega(x, y, y_1, y_2)$. To avoid the complexity of solving this system directly, we instead exploit the x translation symmetry and the x, y scaling symmetries of ODE (3.572) to look for solutions $\omega(x, y, y_1, y_2)$ based on ansatzes (3.554a,b) and (3.555) using the common joint invariant

$$u = y(y_1)^{-2}y_2.$$

Thus, we consider

$$\omega = y^r (y_1)^s \alpha(u), \quad r = \text{const}, \quad s = \text{const}. \tag{3.574}$$

Then the coefficient of $(y_3)^4$ in the adjoint-symmetry determining equation (3.573) yields $\alpha^{(4)} = 0$, and hence,

$$\alpha = \alpha_0 + \alpha_1 u + \alpha_2 u^2 + \alpha_3 u^3, \quad \alpha_i = \text{const}. \tag{3.575}$$

The coefficients of the remaining powers of y_3 in (3.573) now become a system of polynomial equations in u which reduce to algebraic equations for $r, s, \alpha_0, \alpha_1, \alpha_2, \alpha_3$. First, from the coefficients of $(y_3)^3 u^7$ and $(y_3)^2 u^7$, we find that $\alpha_3 = 0$. Next, from the coefficients of $(y_3)^3 u^6, (y_3)^2 u^6, y_3 u^6$, we find that $(s+1)\alpha_2 = 0$. This leads to two cases: $s = -1$ or $\alpha_2 = 0$. If $s = -1$, we find that the coefficients of $(y_3)^2 u^5$ and $y_3 u^5$ yield $\alpha_1 = r - 2 = 0$. Hence, we obtain

$$\alpha = \alpha_0 + \alpha_2 u^2, \quad s = -1, \ r = 2, \tag{3.576}$$

which is readily checked to satisfy the adjoint-symmetry determining equation (3.573). Finally, if $\alpha_2 = 0$, the coefficients of $(y_3)^2 u^5$ and $y_3 u^5$ just yield $\alpha_1 = 0$. Then, from the coefficients of $(y_3)^2 u^4$ and $y_3 u^4$, we obtain $s = -3$. The remaining coefficients yield $r = 2$. Hence, we have

$$\alpha = \alpha_0, \quad s = -3, \ r = 2. \tag{3.577}$$

Therefore, (3.576) and (3.577) yield three adjoint-symmetries given by

$$\omega_1 = y^2 (y_1)^{-1}, \quad \omega_2 = y^2 (y_1)^{-3}, \quad \omega_3 = y^4 (y_2)^2 (y_1)^{-5}. \tag{3.578}$$

From results in Section 3.6.5 [cf. (3.523)], it follows that the two first-order adjoint-symmetries ω_1 and ω_2 are integrating factors of ODE (3.572), while the second-order adjoint-symmetry ω_3 is not an integrating factor of ODE (3.572). We now use ω_3 to obtain a higher-order integrating factor of ODE (3.572) by means of the ansatz (3.553), which here becomes

$$\Lambda = y^4 (y_2)^2 (y_1)^{-5} \psi(\psi_1, \psi_2) \tag{3.579}$$

depending on a function $\psi(\psi_1, \psi_2)$, where

$$\psi_1 = y y_2 - 2y^2 (y_1)^{-2} (y_2)^2 + y^2 (y_1)^{-1} y_3, \tag{3.580a}$$

$$\psi_2 = y (y_1)^{-2} y_2 - y^2 (y_1)^{-4} (y_2)^2 + y^2 (y_1)^{-3} y_3, \tag{3.580b}$$

are the first integrals corresponding to the integrating factors $\Lambda_1 = \omega_1$, $\Lambda_2 = \omega_2$. Since $\mathbf{D}\psi_1 = \mathbf{D}\psi_2 = 0$, it follows that ansatz (3.579) satisfies the adjoint-symmetry determining equation (3.573) for an arbitrary function $\psi(\psi_1, \psi_2)$. Then, from Theorem 3.7.2-1, we see that ansatz (3.579) yields an integrating factor of ODE (3.572) if and only if it satisfies the adjoint invariance conditions

$$\Lambda_{y_2} + (-3y^{-1}y_1 + 5y_2(y_1)^{-1})\Lambda_{y_3} + (\mathbf{D}\Lambda)_{y_3} = 0, \tag{3.581a}$$

$$\begin{aligned}\Lambda_y &+ (-2y_1y_2y^{-2} + 8(y_2)^3(y_1)^{-3} - 3y^{-1}y_3 - 5y_2(y_1)^{-2}y_3)\Lambda_{y_3} - (3y^{-1} + 5y_2(y_1)^{-1})\Lambda \\ &- (\mathbf{D}((-(y_1)^2y^{-2} + 8y_2(y_1)^{-1} - 12(y_2)^2(y_1)^{-2} + 5y_3(y_1)^{-1})\Lambda))_{y_3} \\ &+ (\mathbf{D}^2((-3y_1y^{-1} + 5y_2(y_1)^{-1})\Lambda) + \mathbf{D}^3\Lambda)_{y_3} = 0.\end{aligned} \tag{3.581b}$$

After some simplifications, we find that the adjoint invariance conditions (3.581a,b) reduce to the single PDE

$$\psi_2\psi_{\psi_2} + \psi_1\psi_{\psi_1} + 2\psi = 0. \tag{3.582}$$

This has the general solution

$$\psi = (\psi_2)^{-2}\tilde{\psi}\left(\frac{\psi_1}{\psi_2}\right),$$

where $\tilde{\psi}$ is an arbitrary function of its argument. Hence, by the relation (3.477) between first integrals and integrating factors, we obtain an integrating factor

$$\Lambda_3 = y^4(y_2)^2(y_1)^{-5}(\psi_2)^{-2} = y^4(y_2)^2(y_1)^{-1}(yy_2 - y^2(y_1)^{-2}(y_2)^2 + y^2(y_1)^{-1}y_3)^{-2} \tag{3.583}$$

and a corresponding functionally dependent first integral

$$\psi_3 = \frac{\psi_1}{\psi_2} = (y_1)^2 - y^2(y_2)^2(yy_2 - y^2(y_2)^2(y_1)^{-2} + y^2(y_1)^{-1}y_3)^{-1}. \tag{3.584}$$

3.7.4 NOETHER'S THEOREM, VARIATIONAL SYMMETRIES, AND INTEGRATING FACTORS

An nth-order ODE

$$F(x, y, y', \ldots, y^{(n)}) = 0, \quad \frac{\partial F}{\partial y^{(n)}} \not\equiv 0, \tag{3.585}$$

has a variational formulation if its solutions $y = \Theta(x)$ on a domain $x \in [a,b]$ correspond to the extremals of an action functional

$$S[y(x)] = \int_a^b \mathcal{L}(x, y, y', \ldots, y^{(n)})\, dx. \tag{3.586}$$

In particular, on the space of functions $y(x)$, consider a one-parameter group of local transformations $x^* = x$, $y^* = y + \varepsilon V(x) + O(\varepsilon^2)$, with the infinitesimal generator

$$\hat{\mathrm{X}}^{(\infty)} = V(x)\frac{\partial}{\partial y} + V'(x)\frac{\partial}{\partial y'} + \cdots \tag{3.587}$$

in terms of an arbitrary function $V(x)$. The corresponding variation of the action (3.586) is given by

$$\hat{\mathrm{X}}^{(\infty)}S[y(x)] = \int_a^b \hat{\mathrm{X}}^{(\infty)}\mathcal{L}(x,y,y',\ldots,y^{(n)})\,dx$$
$$= \int_a^b V(x)\hat{\mathrm{E}}_n(\mathcal{L}(x,y,y',\ldots,y^{(n)}))\,dx + A(V;x,y,y',\ldots,y^{(2n-1)})\Big|_{x=a}^{x=b}, \tag{3.588}$$

where

$$\hat{\mathrm{E}}_n = \frac{\partial}{\partial y} - \frac{d}{dx}\frac{\partial}{\partial y'} + \cdots + (-1)^n \frac{d^n}{dx^n}\frac{\partial}{\partial y^{(n)}} \tag{3.589}$$

is the standard Euler operator in the calculus of variations [Olver (1986)], and where, through integration by parts,

$$A(V;x,y,y',\ldots,y^{(2n-1)}) = V\frac{\partial\mathcal{L}}{\partial y'} + \left(V'\frac{\partial\mathcal{L}}{\partial y''} - V\frac{d}{dx}\frac{\partial\mathcal{L}}{\partial y''}\right) + \cdots$$
$$+ \sum_{j=1}^{n}(-1)^{j-1}V^{(n-j)}\frac{d^{j-1}}{dx^{j-1}}\frac{\partial\mathcal{L}}{\partial y^{(n)}} \quad [V^{(0)} = V]. \tag{3.590}$$

Now suppose $V(x)$ and its derivatives $V'(x), V''(x),$ etc. vanish at the end-points $x = a$, $x = b$. Consequently, the end-point values of $y(x), y'(x), y''(x),$ etc. are fixed under the transformation generated by (3.587). Then the equation

$$\hat{\mathrm{X}}^{(\infty)}S[y(x)] = \int_a^b V(x)\hat{\mathrm{E}}_n(\mathcal{L}(x,y,y',\ldots,y^{(n)}))\,dx = 0 \tag{3.591}$$

is a necessary condition for $y(x)$ to be an extremal of the action (3.586). Since $V(x)$ is arbitrary within the domain $x \in (a,b)$, it follows that the extremals must satisfy the Euler–Lagrange ODE

$$\frac{\partial}{\partial y}\mathcal{L}(x,y,y',\ldots,y^{(n)}) - \frac{d}{dx}\frac{\partial}{\partial y'}\mathcal{L}(x,y,y',\ldots,y_{(n)}) + \cdots$$
$$+ (-1)^n\frac{d^n}{dx^n}\frac{\partial}{\partial y^{(n)}}\mathcal{L}(x,y,y',\ldots,y^{(n)}) = 0. \tag{3.592a}$$

Hence, a given ODE (3.585) corresponds to the extremals of an action (3.586) if and only if

$$F(x,y,y',\ldots,y^{(n)}) = \hat{\mathrm{E}}_n(\mathcal{L}(x,y,y',\ldots,y^{(n)})) \tag{3.592b}$$

holds for some function $\mathcal{L}(x,y,y',\ldots,y^{(n)})$, for all functions $y(x)$. Any two functions $\mathcal{L}$ that differ by a total derivative dW/dx of an arbitrary differentiable function $W(x,y,y',\ldots,y^{(n-1)})$ yield the same Euler-Lagrange equation (3.592b), since one can show that $\hat{\mathrm{E}}_n$ annihilates a function if and only if it is of the form dW/dx. We will see that in (3.592b) there is no loss of generality in allowing the highest order derivative of y in $\mathcal{L}$ to be the same as the highest order derivative of y in F.

The variation (3.588) of the action is equivalent to the identity

$$\hat{X}^{(n)}\mathcal{L}(x,y,y_1,\dots,y_n) = V\mathrm{E}_k(\mathcal{L}(x,y,y_1,\dots,y_n)) + \mathrm{D}_k A(V;x,y,y_1,\dots,y_{k-1}), \quad k=2n, \tag{3.593}$$

holding in $(x,y,y_1,\dots,y_k)$-space, where $\hat{X}^{(n)}$ denotes the nth-extension of the generator $\hat{X} = V\frac{\partial}{\partial y}$, $V = V(x,y,y_1,\dots,y_\ell)$ is an arbitrary function (with $\ell \le n$), D_k is the truncated total derivative operator (3.468), and

$$\mathrm{E}_k = \sum_{i=0}^{k} (-\mathrm{D}_k)^i \frac{\partial}{\partial y_i}, \quad k \ge 0, \ [y_0 = y] \tag{3.594}$$

is a corresponding truncated Euler operator.

Consequently, from (3.591), the Euler–Lagrange equation (3.592a) for extremals of the action is equivalent to

$$\mathrm{E}_{2n}(\mathcal{L}(x,y,y_1,\dots,y_n)) = 0.$$

Hence, the surface given by

$$F(x,y,y_1,\dots,y_n) = 0 \tag{3.595}$$

defines the stationary points of the action (3.586).

Definition 3.7.4-1. An nth-order ODE (3.585) has a variational principle given by an action functional (3.586) if there exists some function $\mathcal{L}(x,y,y',\dots,y^{(n)})$, i.e., a Lagrangian, such that the Euler–Lagrange equation (3.592b) holds for all functions $y = y(x)$. Equivalently, the surface (3.595) arises as the stationary points

$$F(x,y,y_1,\dots,y_n) = \mathrm{E}_{2n}(\mathcal{L}(x,y,y_1,\dots,y_n)) = 0 \tag{3.596}$$

for a Lagrangian $\mathcal{L}(x,y,y_1,\dots,y_n)$.

We next establish the fundamental connection between the existence of an action functional (3.586) and the self-adjointness of the linearization operator for an nth-order ODE (3.585). If relation (3.596) holds for some Lagrangian $\mathcal{L}(x,y,y_1,\dots,y_n)$, then by direct calculation one can show that $(\mathrm{L}_F - \mathrm{L}^*_F)V = 0$ is satisfied identically for arbitrary functions $V = V(x,y,y_1,\dots,y_\ell)$, $\ell \le n$. Hence, $\mathrm{L}_F = \mathrm{L}^*_F$. Conversely, if $\mathrm{L}_F = \mathrm{L}^*_F$, then using conditions (3.545a,b), one can verify that

$$\mathcal{L}(x,y,y_1,\dots,y_n) = y\int_0^1 F(x,\lambda y,\lambda y_1,\dots,\lambda y_n)\, d\lambda \tag{3.597}$$

is a Lagrangian yielding (3.596).

Theorem 3.7.4-1. *An nth-order ODE* (3.585) *admits an action functional* (3.586) *if and only if* (3.585) *is self-adjoint. An explicit Lagrangian is then given by* (3.597).

The same results hold for a self-adjoint ODE in a solved form (3.535) for y_n. In particular, the Lagrangian (3.597) then takes the form

$$\mathcal{L}(x,y,y_1,\ldots,y_n)=\tfrac{1}{2}yy_n-y\int_0^1 f(x,\lambda y,\lambda y_1,\ldots,\lambda y_{n-1})\,d\lambda.$$

When ODE (3.535) is not self-adjoint, we remark that one can also consider an action functional whose Euler–Lagrange equation yields (3.535) to within a nonconstant multiplier, i.e.,

$$F(x,y,y_1,\ldots,y_n)=(y_n-f(x,y,y_1,\ldots,y_{n-1}))h=\mathrm{E}_{2n}(\mathcal{L}(x,y,y_1,\ldots,y_n)) \qquad (3.598)$$

in terms of some function $h=h(x,y,y_1,\ldots,y_n)$. Necessary and sufficient conditions for the existence of such a multiplier $h(x,y,y_1,\ldots,y_n)$ are given by (3.545a,b). The cases $n=2,3$ are considered in Exercises 3.7-8. Theorem 3.7.4-1 corresponds to the situation when $h(x,y,y_1,\ldots,y_n)\equiv 1$.

As discussed in Section 3.5.1, each symmetry of ODE (3.535) is characterized by the infinitesimal of a one-parameter group of local transformations leaving invariant the surface (3.536). In particular, if $\hat{\eta}(x,y,y_1,\ldots,y_\ell)$ is a symmetry of order $0\le\ell\le n-1$ admitted by ODE (3.535), then the corresponding local transformation group acting on functions $y(x)$ is given by the extended infinitesimal generator

$$\hat{\mathrm{X}}^{(\infty)}=\sum_{i=0}^{\infty}\left(\frac{d^i}{dx^i}\hat{\eta}(x,y,y',\ldots,y^{(\ell)})\right)\frac{\partial}{\partial y^{(i)}}. \qquad (3.599)$$

For a self-adjoint ODE (3.535), any such local transformation group obviously leaves invariant the extremals of the action functional (3.586).

Definition 3.7.4-2. A symmetry $\hat{\eta}(x,y,y_1,\ldots,y_\ell)$ of order ℓ admitted by a self-adjoint ODE (3.535) is a *variational symmetry* if the action functional (3.586) is invariant under (3.599) to within a boundary term, i.e.,

$$\hat{\mathrm{X}}^{(\infty)}S[y(x)]=\int_a^b \hat{\mathrm{X}}^{(\infty)}\mathcal{L}(x,y,y',\ldots,y^{(n)})\,dx=B(\hat{\eta};x,y,y',\ldots,y^{(n+\ell-1)})\Big|_{x=a}^{x=b}$$

for some function $B(\hat{\eta};x,y,y',\ldots,y^{(n+\ell-1)})$ which, without loss of generality, depends on $x,y,$ and $y^{(i)}$ up to at most order $i=n+\ell-1$.

Invariance of the action functional to within a boundary term is equivalent to the invariance of the Lagrangian to within a total derivative

$$\hat{\mathrm{X}}^{(n)}\mathcal{L}(x,y,y_1,\ldots,y_n)=\mathrm{D}_kB(V;x,y,y_1,\ldots,y_{k-1}) \quad \text{with } \hat{\mathrm{X}}=V\frac{\partial}{\partial y}, \qquad (3.600)$$

where $V=\hat{\eta}(x,y,y_1,\ldots,y_\ell)$, and $k=n+\ell$. In turn, (3.600) can be expressed equivalently using the Euler operator equation given by Theorem 3.6.4-2 as follows:

Theorem 3.7.4-2. *Suppose* $\hat{\eta}(x,y,y_1,\ldots,y_\ell)$ *is a symmetry of order* ℓ *admitted by a self-adjoint ODE* (3.535). *Let* $\theta(x,y,y_1,\ldots,y_k)=\hat{\mathrm{X}}^{(n)}\mathcal{L}(x,y,y_1,\ldots,y_n)$, $k=n+\ell$, *where* $\hat{\mathrm{X}}^{(n)}$ *is the nth extension of* $\hat{\mathrm{X}}=\hat{\eta}\dfrac{\partial}{\partial y}$. *Let, inductively,* $\Psi_k=\theta_{y_k}$, $\Psi_{k-j}=\theta_{y_{k-j}}-D_k\Psi_{k-j+1}$, $j=1,\ldots,k$, *where* D_k *is the truncated total derivative operator* (3.468) *and* $\Psi_0=\mathrm{E}_k(\theta)$ *is the Euler operator* (3.594). *Then* $\hat{\eta}(x,y,y_1,\ldots,y_\ell)$ *is a variational symmetry of the self-adjoint ODE* (3.535) *if and only if*

$$\frac{\partial\Psi_{k-j}}{\partial y_k}=0,\quad j=0,1,\ldots,k-1, \tag{3.601a}$$

$$\Psi_0=0. \tag{3.601b}$$

Theorem 3.7.4-2 provides a system of $n+\ell+1$ linear determining equations for the variational symmetries $\hat{\eta}(x,y,y_1,\ldots,y_\ell)$ of a self-adjoint ODE (3.535). From Lemma 3.7.2-1 and Theorem 3.7.2-1 (in the self-adjoint case), we see that this system is equivalent to the symmetry determining equation

$$\mathbf{D}^n\hat{\eta}-\sum_{i=0}^{n-1}f_{y_i}\mathbf{D}^i\hat{\eta}=0 \tag{3.602a}$$

and the $n/2$ adjoint invariance conditions

$$\hat{\eta}_{y_{n-2m}}+\sum_{i=0}^{2m-i}(-1)^{i-1}(\mathbf{D}^{i-1}(f_{y_{n-2m+i}}\hat{\eta}))_{y_{n-1}}+(\mathbf{D}^{2m-1}\hat{\eta})_{y_{n-1}}=0,\quad m=1,\ldots,n/2, \tag{3.602b}$$

where $\mathbf{D}$ is the derivative operator (3.445b) associated with the surface (3.536). Hence, we obtain a system of $1+(n/2)$ linear determining equations that are necessary and sufficient for a symmetry of a self-adjoint ODE (3.535) to be a variational symmetry.

Theorem 3.7.4-3 (Variational Symmetry Determining Equations). *The variational symmetries of a self-adjoint nth-order ODE* (3.535) *are those symmetries* $\hat{\eta}$ *of* (3.535) *that satisfy the n/2 adjoint invariance conditions* (3.602b).

We now state the fundamental theorem of Noether for variational symmetries:

Theorem 3.7.4-4 (Noether's Theorem). *For a self-adjoint ODE* (3.535), *every variational symmetry of order* $0\le\ell\le n-1$ *is an integrating factor and, conversely, every integrating factor of order* $0\le\ell\le n-1$ *is a variational symmetry. In particular, if* $\hat{\eta}(x,y,y_1,\ldots,y_\ell)$ *is a variational symmetry of ODE* (3.535), *then*

$$\psi(x, y, y_1, \ldots, y_{n-1}) = B(\hat{\eta}; x, y, y_1, \ldots, y_{n-1}, 0, \ldots) - A(\hat{\eta}; x, y, y_1, \ldots, y_{n-1}, 0, \ldots) \tag{3.603}$$

yields a first integral of (3.535), *where* $A(V; x, y, y_1, \ldots, y_{2n-1})$ *and* $B(V; x, y, y_1, \ldots, y_{n+\ell-1})$ *are given by* (3.593) *and* (3.600).

Proof. Suppose $\hat{\eta}(x, y, y_1, \ldots, y_\ell)$ is a variational symmetry of a self-adjoint ODE (3.535). By combining the Lagrangian variations (3.600) and (3.593) for $V = \hat{\eta}$, we obtain

$$\hat{\eta}F = \mathrm{D}_{2n}(B - A) \quad \text{with } F = \mathrm{E}_{2n}(\mathcal{L}).$$

This is the characteristic equation [cf. Section 3.6.4] stating that $\hat{\eta}(x, y, y_1, \ldots, y_\ell)$ is an integrating factor of ODE (3.535) with corresponding first integral (3.603).

Conversely, suppose $\Lambda(x, y, y_1, \ldots, y_\ell)$ is an integrating factor of a self-adjoint ODE (3.535). Then we have the characteristic equation

$$\Lambda F = \mathrm{D}_n \psi \quad \text{with } F = \mathrm{E}_{2n}(\mathcal{L}).$$

By using identity (3.593) with $V = \Lambda$, we obtain

$$\hat{\mathrm{X}}^{(n)} \mathcal{L} = \mathrm{D}_{2n}(\psi + A), \tag{3.604}$$

and hence, $\hat{\mathrm{X}} = \Lambda(x, y, y_1, \ldots, y_\ell)\partial / \partial y$ is the infinitesimal generator of a one-parameter local transformation group leaving invariant the action functional (3.586) to within a boundary term. Since the extremals of (3.586) remain invariant, $\Lambda(x, y, y_1, \ldots, y_\ell)$ is a symmetry of ODE (3.535) and, hence, from (3.604) we conclude that $\Lambda(x, y, y_1, \ldots, y_\ell)$ is a variational symmetry of (3.535). □

It is common to see Noether's Theorem applied to a self-adjoint ODE (3.535) in the following way: One first finds symmetries of ODE (3.535). Next, one checks which of these symmetries are variational symmetries, i.e., if the Lagrangian is invariant to within a total derivative. Finally, one calculates the first integral (3.603) for each variational symmetry. This procedure is quite awkward computationally since it is cumbersome to verify directly the invariance (3.600). A much more effective approach is given by Theorem 3.7.4-3. In particular, one can solve the linear determining system (3.602a,b) to find only those symmetries of ODE (3.535) which are variational symmetries. Most important, one is able to mingle the *n*/2 adjoint invariance conditions (3.602b) with the symmetry determining equation (3.602a) to optimally solve the system. In practice, this provides a significant computational advantage compared to the approach through Noether's Theorem. Moreover, the calculation of first integrals can be carried out directly in terms of the variational symmetries of ODE (3.535) by the line integral formula given in Theorem 3.6.4-3.

Theorem 3.7.4-5. *For a self-adjoint ODE* (3.535), *the first integral corresponding to a variational symmetry (i.e., integrating factor)* $\hat{\eta}(x,y,y_1,\ldots,y_\ell)$ *of* (3.535) *is given by the line integral*

$$\psi =$$

$$\int_C \left[\sum_{i=1}^{n-1} \left(\sum_{j=0}^{n-i-1} (-1)^{j+1} (\mathrm{D}_{n-1})^j (f\hat{\eta})_{y_{i+j}} + (-1)^{n-i} (\mathrm{D}_{n-1})^{n-i} \hat{\eta} \right) (dy_{i-1} - y_i\, dx) + \hat{\eta}(dy_{n-1} - f\, dx) \right],$$

where C *is any path curve from a point* $(\tilde{x},\tilde{y},\tilde{y}_1,\ldots,\tilde{y}_{n-1})$ *to* $(x,y,y_1,\ldots,y_{n-1})$.

3.7.5 COMPARISON OF CALCULATIONS OF SYMMETRIES, ADJOINT-SYMMETRIES, AND INTEGRATING FACTORS

For a given nth-order ODE (3.535), the nature of the calculation of its symmetries $\hat{\eta}(x,y,y_1,\ldots,y_\ell)$, adjoint-symmetries $\omega(x,y,y_1,\ldots,y_\ell)$, and integrating factors $\Lambda(x,y,y_1,\ldots,y_\ell)$, $0 \le \ell \le n-1$, is the same. In each situation one has to solve a system of linear determining equations, given by (3.542), (3.543), and (3.552a,b), respectively. For $0 \le \ell < n-1$, these determining systems reduce to overdetermined systems of linear homogeneous PDEs in $\ell+2$ independent variables $x, y, y_1, \ldots, y_\ell$ and, consequently, there are at most a finite number of linearly independent solutions. In practice, one is typically able to find all these solutions explicitly. However, for $\ell = n-1$, the determining systems are no longer overdetermined and now possess an infinite number of solutions. In this case one can use special ansatzes (e.g., elimination of variables, separation of variables, point symmetry invariance) to seek solutions.

In the classical case of a first-order ODE, there is an explicit one-to-one relation between symmetries and integrating factors [cf. Section 3.2.2], namely, $\hat{\eta}(x,y)$ is a symmetry if and only if $\Lambda(x,y) = 1/\hat{\eta}(x,y)$ is an integrating factor. Moreover, here integrating factors are the same as adjoint-symmetries. However, for second- and higher-order ODEs, these relationships break down (i.e., there are now adjoint invariance conditions).

When $n \ge 2$, the size of the solution space for adjoint-symmetries of order $\ell = n-1$ of ODE (3.535) is always of a larger cardinality (in terms of free functions) than that for integrating factors of the same order $\ell = n-1$ since, from Theorem 3.7.2-1, it follows that not every adjoint-symmetry satisfies the $[n/2]$ adjoint invariance conditions for determining integrating factors.

The size of the solution space for symmetries of order $\ell = n-1$ of ODE (3.535) is of the same cardinality as that for adjoint-symmetries of order $\ell = n-1$, since both the symmetry determining equation and adjoint-symmetry determining equation are of the same nature, i.e., they are linear homogeneous PDEs in $n+1$ independent variables $x, y, y_1, \ldots, y_{n-1}$.

The situation for the solution spaces of symmetries, adjoint-symmetries, and integrating factors, respectively, of order $0 \le \ell < n-1$ is much more involved since, in general, a given nth-order ODE (3.535) may admit no such symmetries, adjoint-symmetries, or integrating factors. An interesting question then is how do sizes of respective classes of nth-order ODEs (3.535) compare if one considers a specific ansatz for admitted symmetries, adjoint-symmetries, and integrating factors. This is relevant for assessing a priori the utility of, say, point symmetry analysis versus point-form integrating factor analysis.

To make an explicit comparison, we classify second-order ODEs

$$y_2 = f(x, y, y_1) \tag{3.605}$$

admitting

(i) point symmetries:

$$\hat{\eta} = y_1; \tag{3.606a}$$

(ii) point-form adjoint-symmetries:

$$\omega = y_1; \tag{3.606b}$$

(iii) point-form integrating factors:

$$\Lambda = y_1. \tag{3.606c}$$

The substitution of (3.606a) into the symmetry determining equation (3.542) yields

$$\mathbf{D}^2 y_1 - f_y y_1 - f_{y_1} \mathbf{D} y_1 = f_x = 0.$$

Hence, the class of ODEs (3.605) admitting the point symmetry ansatz (3.606a) is given by

$$f = a(y, y_1) \tag{3.607}$$

(i.e., x is missing) depending on an arbitrary function $a(y, y_1)$. In contrast, the substitution of (3.606b) into the adjoint-symmetry determining equation (3.543) yields

$$\mathbf{D}^2 y_1 - f_y y_1 + \mathbf{D}(y_1 f_{y_1}) = f_x + 2 f f_{y_1} + y_1 f_{xy_1} + (y_1)^2 f_{yy_1} + y_1 f f_{y_1 y_1} = 0. \tag{3.608}$$

This is a second-order nonlinear PDE for $f(x, y, y_1)$ and, hence, in effect its general solution depends on two arbitrary functions of two independent variables. But the substitution of (3.606c) into the adjoint invariance condition (3.552b) yields

$$(f_{y_1} y_1)_{y_1} + (\mathbf{D} y_1)_{y_1} = 2 f_{y_1} + y_1 f_{y_1 y_1} = 0. \tag{3.609}$$

Solving (3.609), we obtain $f = a_1(x, y)(y_1)^{-1} + a_0(x, y)$, and then (3.608) reduces to $(a_0)_x = (a_1)_y$. Hence, the class of ODEs (3.605) admitting the point-form integrating factor ansatz (3.606c) is given by

$$f = a(x, y)(y_1)^{-1} + \int a_y(x, y)\, dx + b(y) \tag{3.610}$$

depending on arbitrary functions $a(x, y)$ and $b(y)$. We observe that both classes (3.610) and (3.607) of ODEs (3.605) involve one arbitrary function of two independent variables, while the class given by (3.608) depends on two such arbitrary functions. Moreover, the class (3.610) also involves an additional arbitrary function of one independent variable.

Therefore, a priori the utility of a point-form integrating factor ansatz is no less than that of a point symmetry ansatz for second-order ODEs. More generally, a similar conclusion holds for any particular ansatz for symmetries and integrating factors of nth-order ODEs.

EXERCISES 3.7

1. Consider the harmonic oscillator equation

$$y'' + \nu^2 y = 0, \quad \nu = \text{const.} \tag{3.611}$$

   (a) Show that the adjoint-symmetries of ODE (3.611) are the same as the symmetries of (3.611).
   (b) Show that the translation symmetry $x \to x + \varepsilon,\ y \to y$ of ODE (3.611) is a variational symmetry (i.e., an integrating factor of (3.611)) but the scaling symmetry $x \to x,\ y \to \lambda y$ of (3.611) is not. Find the first integral of ODE (3.611) arising from the translation symmetry.
   (c) Consider the ansatz (3.553) for ODE (3.611) by using the scaling symmetry and first integral corresponding to the translation symmetry. Show that this ansatz yields a variational symmetry of ODE (3.611) giving a first integral functionally independent of the previous one.
   (d) Show that these two first integrals correspond to the energy and phase of the harmonic oscillator, and obtain the quadrature of ODE (3.611).

2. Consider the KdV traveling wave ODE (3.499).
   (a) Find the first-order adjoint-symmetries of ODE (3.499) and verify that the two admitted point-form adjoint-symmetries satisfy the adjoint invariance conditions for integrating factors of (3.499).
   (b) Use the ansatz (3.553) to obtain an additional integrating factor of ODE (3.499). Show that this ansatz only yields two functionally independent first integrals corresponding to first-order integrating factors.
   (c) Look for second-order adjoint-symmetries of ODE (3.499) given by the ansatzes (3.554a,b), using the invariants of the scaling symmetry $x \to \lambda x,\ y \to \lambda^{-2} y$ admitted by (3.499). Verify which of these adjoint-symmetries satisfy the adjoint invariance conditions for integrating factors of ODE (3.499).
   (d) Obtain the quadrature of ODE (3.499) using the first integrals corresponding to its admitted first- and second-order integrating factors.

3. Consider the fourth-order ODE

$$y^{(4)} = \tfrac{4}{3}(y''')^2 / y''. \tag{3.612}$$

   (a) Show that ODE (3.612) is not self-adjoint.

(b) Find the first- and second-order adjoint-symmetries of ODE (3.612). Verify which ones satisfy the adjoint invariance conditions for integrating factors of (3.612).

(c) Find third-order adjoint-symmetries of ODE (3.612) that are given by the ansatzes (3.554a,b) using the joint invariants of the scaling symmetries $x \to \alpha x,\ y \to \beta y$ and the translation symmetries $x \to x+\varepsilon,\ y \to y+\delta$ admitted by (3.612). Verify which of these third-order adjoint-symmetries satisfy the adjoint invariance conditions for integrating factors of ODE (3.612).

(d) Obtain the quadrature of ODE (3.612) by using the first integrals arising from the admitted integrating factors.

4. Show that the fourth-order wave speed ODE (3.571) (not in solved form) is not self-adjoint.

5. Prove Lemma 3.7.1-2.

6. For an nth-order ODE (3.535), show that if $\psi(\psi_1,\ldots,\psi_n)$ is a function of first integrals ψ_i, $i=1,\ldots,k$, and $\omega(x,y,y_1,\ldots,y_\ell)$ is an adjoint-symmetry, then $\Lambda=\omega\psi$ satisfies (3.552a) while (3.552b) reduces to a first-order linear homogeneous system of PDEs for $\psi(\psi_1,\ldots,\psi_n)$.

7. Let $X=\xi(x,y)\frac{\partial}{\partial x}+\eta(x,y)\frac{\partial}{\partial y}$ be the infinitesimal generator of a point symmetry admitted by an nth-order ODE (3.535). Through the use of canonical coordinates [cf. Section 3.3.1], show that the symmetry determining equation (3.542) and the adjoint-symmetry determining equation (3.543) admit the nth-extended infinitesimal generator $X^{(n)}$ of X.

8. (a) Consider a second-order ODE $y''=f(x,y,y')$. Obtain a linear homogeneous PDE that is a necessary and sufficient condition for $F(x,y,y',y'')=h(x,y,y')(y''-f(x,y,y'))=0$ to be a self-adjoint ODE in terms of a multiplier $h(x,y,y_1)$. Show that such a multiplier exists for any $f(x,y,y_1)$.

(b) Consider a third-order ODE $y'''=f(x,y,y',y'')$. Show that the ODE $F(x,y,y',y'',y''')=h(x,y,y',y'')(y'''-f(x,y,y',y''))=0$ is not self-adjoint for any multiplier $h(x,y,y',y'')$.

9. Classify all self-adjoint fourth-order ODEs $y^{(4)}=f(x,y',y'',y''')$.

10. A second-order ODE $y''=f(x,y,y')$ admits the point-form adjoint-symmetry $\omega=y_1$ if and only if $f(x,y,y_1)$ satisfies the second-order nonlinear PDE (3.608).

(a) Show that $f=a(x,y)y_1+b(x,y)$ is a solution of (3.608) if $a_y=0$ and $b_x+2ab=0$.

(b) Show that $\omega=y_1$ is not an integrating factor for the corresponding class of ODEs $y''=a(x)y'+[a(x)]^{-2}b(y)$.

(c) Compare the class of ODEs (b) to the class given by (3.610) admitting the integrating factor $\Lambda = y_1$.

11. Classify all second-order ODEs (3.605) admitting:
    (a) the point symmetry $\hat{\eta} = y$;
    (b) the point-form adjoint-symmetry $\omega = y$;
    (c) the point-form integrating factor $\Lambda = y$;
    (d) the first-order symmetry $\hat{\eta} = 1/y_1$;
    (e) the first-order adjoint-symmetry $\omega = 1/y_1$; and
    (f) the first-order integrating factor $\Lambda = 1/y_1$.

12. Classify all third-order ODEs $y''' = f(x, y, y', y'')$ admitting:
    (a) the point symmetry $\hat{\eta} = y_1$;
    (b) the point-form adjoint-symmetry $\omega = y_1$; and
    (c) the point-form integrating factor $\Lambda = y_1$.

13. An nth-order ODE (3.535) is *skew-adjoint* if $\mathrm{L}_F = -\mathrm{L}^*_F$.
    (a) Show that for any skew-adjoint ODE, admitted symmetries are the same as admitted adjoint-symmetries.
    (b) Classify all second-, third-, and fourth-order skew-adjoint ODEs.

3.8 DIRECT CONSTRUCTION OF FIRST INTEGRALS THROUGH SYMMETRIES AND ADJOINT-SYMMETRIES

We now present two additional methods for construction of first integrals of nth-order ODEs.

The first method yields an algebraic first integral formula (i.e., without integration) from any pair consisting of a symmetry and an adjoint-symmetry admitted by a given ODE. For an nth-order ODE that admits a scaling symmetry, we show that the first integral arising from a pair consisting of the scaling symmetry and an adjoint-symmetry given by an admitted integrating factor is the same as the first integral arising from the line integral formula for the admitted integrating factor. Thus, for such an ODE, one can use the algebraic first integral formula in place of the line integral formula for constructing a first integral in terms of an integrating factor. Most important, the algebraic first integral formula yields a first integral from any adjoint-symmetry whether or not it is an integrating factor.

The second method uses a Wronskian determinant formula yielding first integrals for ODEs that admit sufficiently many symmetries or adjoint-symmetries. If an nth-order ODE in solved form does not have an explicit dependence on the $(n-1)$th derivative of the dependent variable, then either n symmetries or n adjoint-symmetries are sufficient to obtain a first integral by the Wronskian formula. More generally, for an nth-order ODE in solved form with an essential dependence on the $(n-1)$th derivative of the dependent variable, the formula requires at least either $n+1$ symmetries or $n+1$ adjoint-symmetries or, alternatively, n symmetries together with n adjoint-symmetries.

3.8.1 FIRST INTEGRALS FROM SYMMETRY AND ADJOINT-SYMMETRY PAIRS

Consider a second- or higher-order ODE

$$y^{(n)} = f(x, y, y', ..., y^{(n-1)}), \quad n \geq 2, \tag{3.613}$$

represented as a surface

$$F(x, y, y_1, ..., y_{n-1}) = y_n - f(x, y, y_1, ..., y_{n-1}) = 0. \tag{3.614}$$

Recall that the symmetries $\hat{\eta}$ and adjoint-symmetries ω of ODE (3.613) are given by the solutions of $\mathbf{L}_F \hat{\eta} = 0$ and $\mathbf{L}^*_F\, \omega = 0$, respectively, where $\mathbf{L}_F$ is the linearization operator of (3.614) and $\mathbf{L}^*_F$ is the adjoint operator [cf. (3.541a,b)]. Now, from Lemma 3.7.1-1, it follows that $\mathbf{L}_F$ and $\mathbf{L}^*_F$ satisfy the identity

$$W\mathbf{L}_F V - V\mathbf{L}^*_F\, W = \mathbf{D}S[W, V; F] \tag{3.615a}$$

with

$$S[W, V; F] = \sum_{i=0}^{n-1} \sum_{j=0}^{i} (-1)^j (\mathbf{D}^{i-j} V) \mathbf{D}^j (W F_{y_{i+1}}), \tag{3.615b}$$

for arbitrary functions $V(x, y, y_1, ..., y_{n-1}), W(x, y, y_1, ..., y_{n-1})$. Hence, if we let $V = \hat{\eta}(x, y, y_1, \ldots, y_\ell)$ and $W = \omega(x, y, y_1, \ldots, y_\ell)$, then (3.615a) yields $\mathbf{D}S[\omega, \hat{\eta}; F] = 0$. Therefore, $S[\omega, \hat{\eta}; F]$ is either a constant or a first integral of ODE (3.613), depending on whether its corresponding integrating factor is identically zero.

Theorem 3.8.1-1. *If $(\hat{\eta}, \omega)$ is a pair consisting of a symmetry $\hat{\eta}(x, y, y_1, \ldots, y_\ell)$ and an adjoint-symmetry $\omega(x, y, y_1, \ldots, y_\ell)$ of ODE* (3.613), *then*

$$\hat{\psi}(\hat{\eta}, \omega) = \sum_{j=0}^{n-1} (-1)^j (\mathbf{D}^j \omega) \mathbf{D}^{n-j-1} \hat{\eta} + \sum_{i=0}^{n-2} \sum_{j=0}^{i} (-1)^{j+1} (\mathbf{D}^j (\omega f_{y_{i+1}})) \mathbf{D}^{i-j} \hat{\eta} \tag{3.616}$$

is a first integral of (3.613) *provided that* $\partial \hat{\psi}(\hat{\eta}, \omega) / \partial y_{n-1} \not\equiv 0$.

Now suppose we have a point symmetry $\hat{\eta}(x, y, y_1) = \eta(x, y) - \xi(x, y) y_1$ of ODE (3.613) and an adjoint-symmetry $\omega(x, y, y_1, \ldots, y_\ell)$ given by an integrating factor of ODE (3.613). Then the following relation holds between the first integral (3.616) determined by the pair $(\hat{\eta}, \omega)$ and the first integral corresponding to the integrating factor $\Lambda = \omega(x, y, y_1, \ldots, y_\ell)$.

Theorem 3.8.1-2. *Let* $\Lambda = \omega(x, y, y_1, \dots, y_\ell)$ *be an integrating factor of ODE* (3.613) *with the corresponding first integral* $\psi(x, y, y_1, \dots, y_{n-1})$. *If* $\hat{\eta} = \eta(x, y) - \xi(x, y) y_1$ *is a point symmetry of* (3.613), *then the first integral* $\hat{\psi}(\hat{\eta}, \omega)$, *given by* (3.616), *satisfies*

$$\hat{\psi}(\hat{\eta}, \omega) = \mathrm{X}^{(n-1)} \psi + c, \quad c = \text{const}, \tag{3.617}$$

where $\mathrm{X}^{(n-1)}$ *is the* $(n-1)$*th extension* [cf. (2.100a,b)] *of the point symmetry generator* $X = \xi \frac{\partial}{\partial x} + \eta \frac{\partial}{\partial y}$. *In particular, the integrating factor corresponding to the first integral* (3.617) *is given by*

$$\hat{\Lambda}(\hat{\eta}, \omega) = \mathrm{X}^{(n-1)} \omega + R_{n-1} \omega, \tag{3.618}$$

where

$$R_{n-1} \equiv \frac{\partial \eta^{(n-1)}}{\partial y_{n-1}} = \eta_y - n \xi_y y_1 - (n-1) \xi_x.$$

Proof. We calculate the integrating factor $\hat{\Lambda}(\hat{\eta}, \omega) = \partial \hat{\psi}(\hat{\eta}, \omega) / \partial y_{n-1}$ from (3.616). First, since $\partial \hat{\eta} / \partial y_i = 0$ for $i \geq 2$, the terms in $\partial \hat{\psi}(\hat{\eta}, \omega) / \partial y_{n-1}$ involving differentiation $\partial / \partial y_{n-1}$ of $\mathbf{D}^k \hat{\eta}$ reduce to

$$\omega (\mathbf{D} \mathrm{D}^{n-2} \hat{\eta})_{y_{n-1}} - (\omega f_{y_{n-1}} + \mathbf{D} \omega)(\mathrm{D}^{n-2} \hat{\eta})_{y_{n-1}}, \tag{3.619}$$

where

$$\begin{aligned} (\mathrm{D}^{n-2} \hat{\eta})_{y_{n-1}} &= -\xi, \\ (\mathbf{D}\mathrm{D}^{n-2} \hat{\eta})_{y_{n-1}} &= -(n-1) \mathrm{D} \xi - f_{y_{n-1}} \xi + \eta_y - \xi_y y_1. \end{aligned}$$

Next, the terms in $\partial \hat{\psi}(\hat{\eta}, \omega) / \partial y_{n-1}$ with no differentiation of $\mathbf{D}^k \hat{\eta}$ yield

$$\sum_{k=0}^{n-2} \left((-1)^{n-k-1} (\mathbf{D}^{n-k-1} \omega)_{y_{n-1}} + \sum_{i=1}^{n-k-1} (-1)^i (\mathbf{D}^{i-1} (\omega f_{y_{k+i}}))_{y_{n-1}} \right) \mathbf{D}^k \hat{\eta} + \omega_{y_{n-1}} \mathbf{D}^{n-1} \hat{\eta}. \tag{3.620}$$

By using the adjoint invariance conditions (3.551), which hold since $\omega(x, y, y_1, \dots, y_\ell)$ is an integrating factor of ODE (3.613), we find that (3.620) simplifies to

$$\sum_{k=0}^{n-1} \omega_{y_k} \mathbf{D}^k \hat{\eta}. \tag{3.621}$$

We now combine (3.621) and (3.619), which yield

$$\frac{\partial \hat{\psi}(\hat{\eta}, \omega)}{\partial y_{n-1}} = R_{n-1} \omega + \sum_{k=0}^{n-1} \omega_{y_k} \mathrm{D}^k (\eta - \xi y_1) + \xi \mathrm{D} \omega.$$

The identities $\mathrm{D}^k (\eta - \xi y_1) = \eta^{(k)} - \xi y_{k+1}$, $0 \leq k \leq n-1$, [cf. (2.219)] then lead to

$$\frac{\partial \hat{\psi}(\hat{\eta}, \omega)}{\partial y_{n-1}} = \mathrm{X}^{(n-1)} \omega + R_{n-1} \omega.$$

Hence, we have established (3.618). Finally, (3.617) follows from Theorem 3.6.4-7. □

Now, if ODE (3.613) admits a scaling symmetry

$$\hat{\eta}_S = \eta - \xi y_1, \quad \text{with } \eta = py, \ \xi = qx, \tag{3.622a}$$

i.e., $x \to \alpha^q x, y \to \alpha^p y$ for some $p = \text{const},\ q = \text{const},$ then

$$\mathrm{X}_S^{(n)}(y_n - f(x, y, y_1, \ldots, y_{n-1})) = (p - nq)(y_n - f(x, y, y_1, \ldots, y_{n-1})), \tag{3.622b}$$

where $p - nq$ is the *scaling weight* of the surface (3.614), i.e., $F \to \alpha^{p-nq} F$ under $\mathrm{X}_S^{(n)}$. If an integrating factor $\Lambda(x, y, y_1, \ldots, y_\ell)$ of ODE (3.613) is homogeneous with respect to this scaling symmetry, i.e., $\Lambda \to \alpha^s \Lambda$ under $x \to \alpha^q x,\ y \to \alpha^p y,\ y_i \to \alpha^{p-qi} y_i$, then

$$\mathrm{X}_S^{(\ell)} \Lambda(x, y, y_1, \ldots, y_\ell) = s \Lambda(x, y, y_1, \ldots, y_\ell) \tag{3.623}$$

for some s which defines the scaling weight of $\Lambda(x, y, y_1, \ldots, y_\ell)$.

Definition 3.8.1-1. A homogeneous integrating factor (3.623) of ODE (3.613) has *critical* scaling weight with respect to the scaling symmetry (3.622a) if $s = (n-1)q - p$.

In the case of a scaling symmetry, Theorem 3.8.1-2 yields an algebraic formula for first integrals.

Theorem 3.8.1-3. *Suppose ODE* (3.613) *admits a scaling symmetry* (3.622a,b). *Let* $\Lambda(x, y, y_1, \ldots, y_\ell)$ *be a homogeneous integrating factor of* (3.613) *with scaling weight* (3.623), *and let* $\psi(x, y, y_1, \ldots, y_{n-1})$ *be its corresponding homogeneous first integral. Then*

$$\hat{\psi}(\hat{\eta}_S, \Lambda) = r\psi, \tag{3.624}$$

where

$$r = s + p - (n-1)q. \tag{3.625}$$

In particular, if the scaling weight s of $\Lambda(x, y, y_1, \ldots, y_\ell)$ *is not critical, i.e.* $r \neq 0$, *then*

$$\begin{aligned}\psi(x, y, y_1, \ldots, y_{n-1}) &= r^{-1}\hat{\psi}(\hat{\eta}_S, \Lambda) \\ &= r^{-1}((p-(n-1)q)y_{n-1} - qxf)\Lambda \\ &\quad + r^{-1}\sum_{i=0}^{n-2}((p-iq)y_i - qxy_{i+1})\left((-1)^{n-i-1}\mathbf{D}^{n-i-1}\omega + \sum_{j=1}^{n-i-1}(-1)^j \mathbf{D}^{j-1}(\omega f_{y_{i+j}})\right).\end{aligned} \tag{3.626}$$

We now illustrate the use of Theorems 3.8.1-1 to 3.8.1-3 through several examples.

For a first example, consider the nonlinear Duffing equation

$$y'' + ay' + by + y^3 = 0, \tag{3.627}$$

which is not self-adjoint if $a \neq 0$ (i.e., the damping constant is nonzero). The point symmetries of ODE (3.627) are given by

$$\hat{\eta}_1 = y_1, \tag{3.628a}$$

$$\hat{\eta}_2 = e^{(a/3)x}(\tfrac{1}{3}ay + y_1) \quad \text{if} \quad b = \tfrac{2}{9}a^2. \tag{3.628b}$$

The adjoint-symmetries of (3.627) are given by [cf. Section 3.7.3]

$$\omega_1 = e^{ax}y_1, \tag{3.629a}$$

$$\omega_2 = e^{(4a/3)x}(\tfrac{1}{3}ay + y_1) \quad \text{if} \quad b = \tfrac{2}{9}a^2. \tag{3.629b}$$

Here, we apply Theorem 3.8.1-1, using the four pairs of symmetries and adjoint-symmetries given by (3.628a,b) and (3.629a,b) to obtain first integrals of ODE (3.627) given by

$$\hat{\psi}(\hat{\eta}, \omega) = \omega \mathbf{D}\hat{\eta} - \hat{\eta}\mathbf{D}\omega + a\hat{\eta}\omega, \tag{3.630}$$

where

$$\mathbf{D} = \frac{\partial}{\partial x} + y_1 \frac{\partial}{\partial y} - (ay_1 + by + y^3)\frac{\partial}{\partial y_1}.$$

Substitution of (3.628a,b) and (3.629a,b) into (3.630) gives

$$\hat{\psi}(\hat{\eta}_2, \omega_2) = \hat{\psi}(\hat{\eta}_1, \omega_1) = 0, \tag{3.631a}$$

$$\hat{\psi}(\hat{\eta}_1, \omega_2) = -\hat{\psi}(\hat{\eta}_2, \omega_1) = -\tfrac{2}{3}ae^{(4a/3)x}((y_1 + \tfrac{1}{3}ay)^2 + \tfrac{1}{2}y^4). \tag{3.631b}$$

This yields a single first integral that is a multiple $-4a/3$ of the first integral

$$\psi_2 = e^{(4a/3)x}(\tfrac{1}{2}(y_1 + \tfrac{1}{3}ay)^2 + \tfrac{1}{4}y^4) \text{ if } b = \tfrac{2}{9}a^2, \tag{3.631c}$$

where ψ_2 is the first integral given by the integrating factor $\Lambda = \omega_2$ of ODE (3.627) [cf. Section 3.6.3]. From Theorem 3.8.1-2, note that (3.631a,b) corresponds to the action of the point symmetry generators

$$\mathrm{X}_1^{(1)} = -\frac{\partial}{\partial x}$$

and

$$\mathrm{X}_2^{(1)} = -e^{(a/3)x}\frac{\partial}{\partial x} + \tfrac{1}{3}ae^{(a/3)x}y\frac{\partial}{\partial y} + e^{(a/3)x}(\tfrac{2}{3}ay_1 + \tfrac{1}{9}a^2 y)\frac{\partial}{\partial y_1}$$

on ψ_2, i.e., $\mathrm{X}_1^{(1)}\psi_2 = \hat{\psi}(\hat{\eta}_1, \omega_2) = -\tfrac{4}{3}a\psi_2$ and $\mathrm{X}_2^{(1)}\psi_2 = \hat{\psi}(\hat{\eta}_2, \omega_2) = 0$.

As a second example, consider the third-order nonlinear ODE

$$y''' + yy' = 0 \tag{3.632}$$

that describes traveling wave solutions of the KdV equation [see Exercise 4.1-2]. Since (3.632) is a third-order ODE, it is not self-adjoint. Its admitted point symmetries are given by a translation in x and a scaling in x, y:

$$\hat{\eta}_1 = y_1, \tag{3.633a}$$

$$\hat{\eta}_2 = -2y - xy_1. \tag{3.633b}$$

Its admitted first-order adjoint-symmetries [cf. Exercise 3.7-2] are given by

$$\omega_1 = 1, \tag{3.634a}$$

$$\omega_2 = y, \tag{3.634b}$$

$$\omega_3 = \tfrac{1}{3}y^3 + (y_1)^2. \tag{3.634c}$$

The two point-form adjoint-symmetries (3.634a,b) are integrating factors of ODE (3.632) [cf. Section 3.6.5] while (3.634c) is a first-order adjoint-symmetry that is not an integrating factor of ODE (3.632). Under the action of the scaling symmetry (3.633b), the adjoint-symmetries (3.634a–c) are homogeneous with scaling weights given, respectively, by

$$s_1 = 0, \quad s_2 = -2, \quad s_3 = -6. \tag{3.635}$$

We now apply Theorem 3.8.1-1, using the symmetries (3.633a,b) and adjoint-symmetries (3.634a–c) to obtain first integrals of ODE (3.632) given by

$$\hat{\psi}(\hat{\eta}, \omega) = \omega \mathbf{D}^2 \hat{\eta} - (\mathbf{D}\omega)\mathbf{D}\hat{\eta} + \hat{\eta}\mathbf{D}^2\omega + y\omega\hat{\eta}, \tag{3.636}$$

where

$$\mathbf{D} = \frac{\partial}{\partial x} + y_1 \frac{\partial}{\partial y} + y_2 \frac{\partial}{\partial y_1} - yy_1 \frac{\partial}{\partial y_2}.$$

First we use the scaling symmetry $\hat{\eta} = \hat{\eta}_2$. Substituting (3.634a,b) into (3.636), we obtain

$$\hat{\psi}(\hat{\eta}_2, \omega_1) = -4(y_2 + \tfrac{1}{2}y^2) = -4\psi_1, \tag{3.637a}$$

$$\hat{\psi}(\hat{\eta}_2, \omega_2) = -6(yy_2 + \tfrac{1}{3}y^3 - \tfrac{1}{2}(y_1)^2) = -6\psi_2, \tag{3.637b}$$

which are scaling multiples $r_1 = -4$ and $r_2 = -6$ of the first integrals ψ_1 and ψ_2 given by the integrating factors (3.634a,b). Note that these scaling factors are in accordance with Theorem 3.8.1-3 where, from (3.625), $r_i = s_i - 4$, $i = 1,2$. Next, the substitution of (3.634c) into (3.636) yields the functionally dependent first integral

$$\hat{\psi}(\hat{\eta}_2, \omega_3) = -4\psi_1\psi_2. \tag{3.638}$$

Finally, if we use the translation symmetry $\hat{\eta} = \hat{\eta}_1$ in (3.363) with (3.634a–c), then we obtain

$$\hat{\psi}(\hat{\eta}_1, \omega_1) = \hat{\psi}(\hat{\eta}_1, \omega_2) = \hat{\psi}(\hat{\eta}_1, \omega_3) = 0.$$

As a third example, consider the fourth-order ODE

$$(yy'(y(y')^{-1})'')' = 0$$

or, equivalently,

$$y^{(4)} = f(y, y', y'', y''') = -\frac{(y')^2 y'''}{y^2} + \frac{4(y'')^2}{y} - \frac{4(y'')^3}{(y')^2} - \frac{3y'y'''}{y} + \frac{5y''y'''}{y'}, \quad (3.639)$$

arising in the study of the symmetry properties of the wave equation with wave speed $y(x)$. The ODE (3.639) is not self-adjoint [cf. Section 3.7.3]. Its admitted point symmetries consist of translations in x and independent scalings in x and y given by

$$\hat{\eta}_1 = y_1, \quad (3.640a)$$

$$\hat{\eta}_2 = y, \quad (3.640b)$$

$$\hat{\eta}_3 = -xy_1. \quad (3.640c)$$

It does not admit any contact symmetries or second-order symmetries [cf. Section 3.5.2]. The adjoint-symmetries up to second order of (3.639) [cf. Section 3.7.3] are given by

$$\omega_1 = \frac{y^2}{y_1}, \quad (3.641a)$$

$$\omega_2 = \frac{y^2}{(y_1)^3}, \quad (3.641b)$$

$$\omega_3 = y^4 \frac{(y_2)^2}{(y_1)^5}, \quad (3.641c)$$

where (3.641a,b) are integrating factors but (3.641c) is not [cf. Section 3.6.5]. The first integrals corresponding to (3.641a,b) are given by

$$\psi_1 = y^2 (y_1)^{-1} y_3 + yy_2 - 2y^2 (y_1)^{-2} (y_2)^2, \quad (3.642a)$$

$$\psi_2 = y^2 (y_1)^{-3} y_3 + y(y_1)^{-2} y^2 - y^2 (y_1)^{-4} (y_2)^2. \quad (3.642b)$$

Here, we apply Theorems 3.8.1-1 and 3.8.1-3, using pairs of symmetries (3.640a–c) and adjoint-symmetries (3.641a–c) to obtain the first integrals (3.642a,b). First, using the translation symmetry (3.640a), we find that the first integral formula (3.616) trivially yields $\hat{\psi}(\hat{\eta}_1, \omega_1) = \hat{\psi}(\hat{\eta}_1, \omega_2) = \hat{\psi}(\hat{\eta}_1, \omega_3) = 0$. Next consider the scaling symmetries (3.640b,c). The scaling weights of (3.641a–c) are $s_1 = s_3 = -s_2 = 1$ with respect to (3.640b), and $s_1 = s_3 = 1, s_2 = 3$ with respect to (3.640c). From the first integral formula

(3.616), we obtain (3.616)

$$\hat{\psi}(\hat{\eta}_2,\omega_1) = -\hat{\psi}(\hat{\eta}_3,\omega_1) = 2\psi_1, \tag{3.643a}$$

$$\hat{\psi}(\hat{\eta}_2,\omega_2) = \hat{\psi}(\hat{\eta}_3,\omega_2) = 0, \tag{3.643b}$$

$$\hat{\psi}(\hat{\eta}_2,\omega_3) = -\hat{\psi}(\hat{\eta}_3,\omega_3) = 2\psi_1\psi_2, \tag{3.643c}$$

where (3.643a,b) are scaling multiples $r_1 = 2$ and $r_2 = 0$ of (3.642a,b), respectively. Note that $r_2 = 0$ reflects the x and y scaling invariance of ψ_2, i.e., correspondingly, ω_2 has critical scaling weight with respect to both (3.640b,c).

3.8.2 FIRST INTEGRALS FROM A WRONSKIAN FORMULA USING SYMMETRIES OR ADJOINT-SYMMETRIES

Consider again a second- or higher-order ODE

$$y^{(n)} = f(x, y, y', \ldots, y^{(n-1)}), \quad n \geq 2, \tag{3.644}$$

represented as a surface

$$y_n = f(x, y, y_1, \ldots, y_{n-1}). \tag{3.645}$$

Recall that the determining equations for symmetries and adjoint-symmetries of ODE (3.644) are nth-order linear homogeneous equations in terms of the operator

$$\mathbf{D} = \frac{\partial}{\partial x} + y_1\frac{\partial}{\partial y} + \cdots + y_{n-1}\frac{\partial}{\partial y_{n-2}} + f(x, y, y_1, \ldots, y_{n-1})\frac{\partial}{\partial y_{n-1}}.$$

In particular, both of the determining equations have the form

$$\mathbf{D}^n\rho + \mathrm{A}_{n-1}\mathbf{D}^{n-1}\rho + \cdots + \mathrm{A}_1\mathbf{D}\rho + \mathrm{A}_0\rho = 0, \tag{3.646}$$

where, for symmetries $\hat{\eta}(x, y, y_1, \ldots, y_\ell)$ of ODE (3.644), in (3.646) we have

$$\rho = \hat{\eta}, \quad \mathrm{A}_i = f_{y_i}, \quad i = 0, 1, \ldots, n-1, \tag{3.647a}$$

with $y_0 = y$; for adjoint-symmetries $\omega(x, y, y_1, \ldots, y_\ell)$ of (3.644), in (3.646) we have

$$\rho = \omega, \quad \mathrm{A}_i = \sum_{j=0}^{n-i-1} (-1)^{n+i+j}\frac{(i+j)!}{i!\,j!}\mathbf{D}^j f_{y_{i+j}}, \quad i = 0, 1, \ldots, n-1, \tag{3.647b}$$

with $y_0 = y$.

Definition 3.8.2-1. If $\{\rho_i(x, y, y_1, \ldots, y_\ell)\}$, $i = 1, 2, \ldots, n$, is a set of n solutions of the nth-order linear homogeneous PDE (3.646), then the *Wronskian* of $\{\rho_i\}$ is given by the $n \times n$ determinant

$$W(\rho_1,\ldots,\rho_n)=\begin{vmatrix}\rho_1 & \cdots & \rho_n\\ \mathbf{D}\rho_1 & \cdots & \mathbf{D}\rho_n\\ \vdots & & \vdots\\ \mathbf{D}^{n-1}\rho_1 & \cdots & \mathbf{D}^{n-1}\rho_n\end{vmatrix}. \tag{3.648}$$

The determinant (3.648) possesses properties completely analogous to the classical Wronskian for nth-order linear ODEs [Coddington (1961)].

Definition 3.8.2-2. A set of k solutions $\{\rho_i(x,y,y_1,\ldots,y_\ell)\}$, $i=1,2,\ldots,k$, of the determining equation (3.646) is *linearly dependent with respect to the surface* (3.645) if $\sum_{i=1}^{k} c_i\rho_i=0$ holds for some functions $c_i(x,y,y_1,\ldots,y_{n-1})$ satisfying $\mathbf{D}c_i=0$. Otherwise, the set of k solutions $\{\rho_i(x,y,y_1,\ldots,y_\ell)\}$, $i=1,2,\ldots,k$, of (3.646) is *linearly independent with respect to the surface* (3.645).

Definition 3.8.2-2 is stronger than the usual strict linear dependence of functions. In particular, a set of solutions of determining equation (3.646) can be linearly independent in the strict sense (i.e., none is a constant coefficient linear combination of the others) but may still be linearly dependent with respect to the surface (3.645).

Lemma 3.8.2-1. *Let* $\{\rho_i(x,y,y_1,\ldots,y_\ell)\}$, $i=1,2,\ldots,n$, *be a set of* n *solutions of determining equation* (3.646). *Then* $W(\rho_1,\ldots,\rho_n)=0$ *if and only if the set is linearly dependent with respect to the surface* (3.645).

Proof. Let $\{\rho_i\}$ be a set of n strictly linearly independent solutions of (3.646). If $\{\rho_i\}$ is linearly dependent with respect to the surface (3.645), then $\sum_{i=1}^{n} c_i\rho_i=0$ holds with $\mathbf{D}c_i=0$. Hence, by repeated differentiation with respect to $\mathbf{D}$, one has

$$\sum_{i=1}^{n} c_i\mathbf{D}^k\rho_i=0,\quad k\geq 0.$$

Thus, the column vectors of the Wronskian (3.648) are linearly dependent, i.e.,

$$\sum_{i=1}^{n} c_i\begin{bmatrix}\rho_i\\ \mathbf{D}\rho_i\\ \vdots\\ \mathbf{D}^{n-1}\rho_i\end{bmatrix}=0, \tag{3.649}$$

and, consequently, the Wronskian $W(\rho_1,\ldots,\rho_n)$ vanishes.

Conversely, if the Wronskian satisfies $W(\rho_1,\ldots,\rho_n)=0$, then its column vectors must be linearly dependent for some coefficient functions c_i that are not all zero. Clearly, without loss of generality, we may assume that $c_n \neq 0$. We now use an induction argument on n. Suppose $n = 2$. Then (3.649) yields

$$\rho_2 = \widetilde{c}\rho_1, \tag{3.650a}$$

$$\mathbf{D}\rho_2 = \widetilde{c}\mathbf{D}\rho_1, \tag{3.650b}$$

with $\widetilde{c} = -c_1 / c_2$. Substitution of (3.650a) into (3.650b) yields $\mathbf{D}\widetilde{c} = 0$, and thus, the set $\{\rho_i\}$ is linearly dependent with respect to the surface (3.645). Now suppose $n > 2$. To proceed inductively, we assume that if the Wronskian of a proper subset of $\{\rho_i\}$ vanishes, then the subset is linearly dependent with respect to the surface (3.645). Consider (3.649) with $n > 2$ and divide this equation by c_n. The second row minus the derivative $\mathbf{D}$ of the first row yields the equation

$$\sum_{i=1}^{n-1} \hat{c}_i \rho_i = 0, \quad \hat{c}_i = \mathbf{D}\left(\frac{c_i}{c_n}\right). \tag{3.651}$$

Similarly, the other rows of (3.649) lead to

$$\sum_{i=1}^{n-1} \hat{c}_i \mathbf{D}^k \rho_i = 0, \quad k = 0,1,\ldots,n-2. \tag{3.652}$$

There are now two cases to consider. If $\hat{c}_i = 0$ for all $i \leq n-1$, then the first row of (3.649) yields $\sum_{i=1}^{n} \widetilde{c}_i \rho_i = 0$ with $\widetilde{c}_i = c_i / c_n$, where $\mathbf{D}\widetilde{c}_i = 0$. Otherwise, if $\hat{c}_i \neq 0$ for some $i \leq n-1$, then from (3.651) and (3.652) the determinant $W(\rho_1,\ldots,\rho_{n-1})$ has linearly dependent column vectors, and thus, $W(\rho_1,\ldots,\rho_{n-1}) = 0$. Hence, $\sum_{i=1}^{n} \widetilde{c}_i \rho_i = 0$ holds with $\widetilde{c}_n = 0$ and $\widetilde{c}_i = \hat{c}_i$, $i = 1,\ldots,n-1$, where $\mathbf{D}\widetilde{c}_i = 0$ by the induction assumption. In either case, we conclude that the set $\{\rho_i\}$ is linearly dependent with respect to the surface (3.645). □

The following result holds analogously to the situation for the classical Wronskian case. The proof is left to Exercise 3.8-6:

Lemma 3.8.2-2. *Let* $\{\rho_i(x, y, y_1,\ldots, y_\ell)\}$, $i = 1,2,\ldots,n$, *be a set of* n *solutions of determining equation* (3.646). *Then the Wronskian* (3.648) *satisfies the first-order linear PDE*

$$\mathbf{D}W(\rho_1,\ldots,\rho_n) = \mathrm{A}_{n-1}W(\rho_1,\ldots,\rho_n). \tag{3.653}$$

In (3.653), the coefficient A_{n-1} is given by $f_{y_{n-1}}$ or $-f_{y_{n-1}}$ when (3.646) is, respectively, the symmetry determining equation or adjoint-symmetry determining equation. Hence, if ODE (3.645) does not involve y_{n-1}, then $A_{n-1} = 0$, and thus, (3.653) yields $\mathbf{D}W(\rho_1,\ldots,\rho_n) = 0$. This establishes the following theorem:

Theorem 3.8.2-1. *Suppose an nth-order ODE* (3.645) *does not involve* y_{n-1}, *i.e.,* $f_{y_{n-1}} = 0$. *Let* $\rho_i(x, y, y_1,\ldots, y_\ell)$, $i = 1,2,\ldots,n$, *be either n symmetries or n adjoint-symmetries of* (3.645). *If this set* $\{\rho_i\}$ *is linearly independent with respect to the surface* (3.645), *then*

$$\hat{\psi}(\rho_1,\ldots,\rho_n) = \begin{vmatrix} \rho_1 & \cdots & \rho_n \\ \mathbf{D}\rho_1 & \cdots & \mathbf{D}\rho_n \\ \vdots & & \vdots \\ \mathbf{D}^{n-1}\rho_1 & \cdots & \mathbf{D}^{n-1}\rho_n \end{vmatrix} \tag{3.654}$$

yields a first integral of ODE (3.645) *provided that* $\partial\hat{\psi}(\rho_1,\ldots,\rho_n)/\partial y_{n-1} \not\equiv 0$.

If y_{n-1} appears in ODE (3.645), then the coefficient A_{n-1} in (3.653) is nonzero. In this case, one can still obtain first integrals from (3.653) provided that one has at least $n+1$ linearly independent solutions of the determining equation (3.646) for either admitted symmetries or admitted adjoint-symmetries or, alternatively, n admitted symmetries together with n admitted adjoint-symmetries.

Theorem 3.8.2-2. *Suppose an nth-order ODE* (3.645) *has an essential dependence on* y_{n-1}, i.e., $f_{y_{n-1}} \neq 0$, *and thus adjoint-symmetries of* (3.645) *are not symmetries of* (3.645).

(i) *Let both* $\rho_i(x, y, y_1,\ldots, y_\ell)$ and $\tilde{\rho}_i(x, y, y_1,\ldots, y_\ell)$, $i = 1,2,\ldots,n$, *be either n symmetries or n adjoint-symmetries of* (3.645). *If each set* $\{\rho_i\}$ *and* $\{\tilde{\rho}_i\}$ *is linearly independent with respect to the surface* (3.645), *then*

$$\hat{\psi}(\rho_1,\ldots\rho_n,\tilde{\rho}_1,\ldots,\tilde{\rho}_n) = W(\rho_1,\ldots,\rho_n)/W(\tilde{\rho}_1,\ldots,\tilde{\rho}_n) \tag{3.655a}$$

yields a first integral of ODE (3.645) *provided that* $\partial\hat{\psi}(\rho_1,\ldots,\rho_n,\tilde{\rho}_1,\ldots,\tilde{\rho}_n)/\partial y_{n-1} \not\equiv 0$.

(ii) *Let* $\rho_i(x, y, y_1,\ldots, y_\ell)$, $i = 1,2,\ldots,n$, *be n symmetries of* (3.645) *and let* $\tilde{\rho}_i(x, y, y_1,\ldots, y_\ell)$, $i = 1,2,\ldots,n$, *be n adjoint-symmetries of* (3.645). *If each set* $\{\rho_i\}$ *and* $\{\tilde{\rho}_i\}$ *is linearly independent with respect to the surface* (3.645), *then*

$$\hat{\psi}(\rho_1,\ldots,\rho_n,\tilde{\rho}_1,\ldots,\tilde{\rho}_n) = W(\rho_1,\ldots,\rho_n)W(\tilde{\rho}_1,\ldots,\tilde{\rho}_n) \tag{3.655b}$$

yields a first integral of ODE (3.645) *provided that* $\partial\hat{\psi}(\rho_1,\ldots,\rho_n,\tilde{\rho}_1,\ldots,\tilde{\rho}_n)/\partial y_{n-1} \not\equiv 0$.

Proof. Left to Exercise 3.8-7. □

In particular, if one has a set of at least $n+1$ symmetries (or $n+1$ adjoint-symmetries) which is linearly independent with respect to the surface (3.645), then by considering the $n+1$ distinct subsets of n symmetries (or n adjoint-symmetries), one obtains $n+1$ first integrals (3.655a) of ODE (3.644). This yields a reduction of (3.644) to an $(n-r)$th-order ODE when r of the $n+1$ first integrals are functionally independent. If $r=n$, then one obtains the quadrature of ODE (3.644), i.e., its general solution depending on n essential constants.

We note that for an nth-order ODE (3.644), if one knows a set of $1 \le k \le n$ symmetries (or adjoint-symmetries) which is linearly dependent with respect to the surface (3.645), i.e., $\sum_{i=1}^{k} c_i \rho_i = 0$ where $\mathbf{D}c_i = 0$, then it follows that each coefficient $c_i(x, y, y_1, \ldots, y_{n-1})$, $i = 1, 2, \ldots, k$, yields a first integral of (3.644) provided that $\partial c_i / \partial y_{n-1} \not\equiv 0$.

We now illustrate the use of Theorems 3.8.2-2 through several examples. Examples illustrating Theorem 3.8.2-1 will be considered in the next section and in Exercise 3.8-2.

For a first example, consider the second-order ODE [Stephani (1989)]

$$y'' = 2(y')^2 \cot y + \sin y \cos y, \tag{3.656}$$

which describes the geodesics, i.e., great circles, on a unit sphere (x is the polar angle or longitude and y is the azimuthal angle or latitude). The point symmetries admitted by (3.656) are given by [cf. Exercise 3.5-3]

$$\hat{\eta}_1 = y_1, \tag{3.657a}$$

$$\hat{\eta}_2 = \sin x - y_1 \cot y \cos x, \tag{3.657b}$$

$$\hat{\eta}_3 = \cos x + y_1 \cot y \sin x. \tag{3.657c}$$

The point symmetries (3.657a–c) form the Lie algebra SO(3) with generators

$$\mathrm{X}_1 = \frac{\partial}{\partial x}, \quad \mathrm{X}_2 = \cot y \cos x \frac{\partial}{\partial x} + \sin x \frac{\partial}{\partial y}, \quad \mathrm{X}_3 = \cot y \sin x \frac{\partial}{\partial x} - \cos x \frac{\partial}{\partial y}.$$

We now apply Theorem 3.8.2-2(i), using (3.657a–c) to obtain first integrals of ODE (3.656) that yield its complete quadrature. The Wronskians (3.648) arising for the three pairs of symmetries obtained from (3.657a–c) are given by

$$W(\hat{\eta}_1, \hat{\eta}_2) = \begin{vmatrix} \hat{\eta}_1 & \hat{\eta}_2 \\ \mathbf{D}\hat{\eta}_1 & \mathbf{D}\hat{\eta}_2 \end{vmatrix} = \frac{((y_1)^2 + \sin^2 y)(y_1 \cos x - \sin x \sin y \cos y)}{\sin^2 y}, \tag{3.658a}$$

$$W(\hat{\eta}_1, \hat{\eta}_3) = \begin{vmatrix} \hat{\eta}_1 & \hat{\eta}_3 \\ \mathbf{D}\hat{\eta}_1 & \mathbf{D}\hat{\eta}_3 \end{vmatrix} = -\frac{((y_1)^2 + \sin^2 y)(y_1 \sin x + \cos x \sin y \cos y)}{\sin^2 y}, \tag{3.658b}$$

$$W(\hat{\eta}_2, \hat{\eta}_3) = \begin{vmatrix} \hat{\eta}_2 & \hat{\eta}_3 \\ \mathbf{D}\hat{\eta}_2 & \mathbf{D}\hat{\eta}_3 \end{vmatrix} = -((y_1)^2 + \sin^2 y), \tag{3.658c}$$

where

$$\mathbf{D}=\frac{\partial}{\partial x}+y_1\frac{\partial}{\partial y}+(2(y_1)^2\cot y+\sin y\cos y)\frac{\partial}{\partial y_1}.$$

Hence, the ratios (3.655a) arising from (3.658a–c) yield the three first integrals

$$\psi_1=\frac{W(\hat{\eta}_1,\hat{\eta}_2)}{W(\hat{\eta}_2,\hat{\eta}_3)}=\frac{-y_1\cos x+\sin x\sin y\cos y}{\sin^2 y}, \tag{3.659a}$$

$$\psi_2=\frac{W(\hat{\eta}_1,\hat{\eta}_3)}{W(\hat{\eta}_2,\hat{\eta}_3)}=\frac{y_1\sin x+\cos x\sin y\cos y}{\sin^2 y}, \tag{3.659b}$$

$$\psi_3=\frac{W(\hat{\eta}_1,\hat{\eta}_2)}{W(\hat{\eta}_1,\hat{\eta}_3)}=\frac{-y_1\cos x+\sin x\sin y\cos y}{y_1\sin x+\cos x\sin y\cos y}. \tag{3.659c}$$

Clearly, any two of (3.659a–c) are functionally independent and thus yield the quadrature of the second-order ODE (3.656). In particular, $\psi_1=\text{const}=c_1$, $\psi_2=\text{const}=c_2$ yield its general solution

$$c_1\sin x+c_2\cos x=\cot y. \tag{3.660}$$

As a second example, we return to the nonlinear Duffing equation (3.627), which admits two point symmetries (3.628a,b) and two point-form adjoint-symmetries (3.629a,b). Here, we apply Theorem 3.8.2-2(ii), using together the pairs (3.628a,b) and (3.629a,b) to obtain the first integral (3.631c). The Wronskians corresponding to (3.628a,b) and (3.629a,b) are given by

$$W(\hat{\eta}_1,\hat{\eta}_2)=\begin{vmatrix}\hat{\eta}_1 & \hat{\eta}_2\\ \mathbf{D}\hat{\eta}_1 & \mathbf{D}\hat{\eta}_2\end{vmatrix}=\tfrac{4}{3}ae^{(a/3)x}(\tfrac{1}{2}(y_1+\tfrac{1}{3}ay)^2+\tfrac{1}{4}y^4), \tag{3.661a}$$

$$W(\omega_1,\omega_2)=\begin{vmatrix}\omega_1 & \omega_2\\ \mathbf{D}\omega_1 & \mathbf{D}\omega_2\end{vmatrix}=\tfrac{4}{3}ae^{(7a/3)x}(\tfrac{1}{2}(y_1+\tfrac{1}{3}ay)^2+\tfrac{1}{4}y^4). \tag{3.661b}$$

Hence, the corresponding product (3.655b) yields the first integral

$$\psi=(\tfrac{4}{3}ae^{(4a/3)x}(\tfrac{1}{2}(y_1+\tfrac{1}{3}ay)^2+\tfrac{1}{4}y^4)^2, \tag{3.662}$$

which is a multiple of the square of (3.631c).

For a third example, we return to the third-order nonlinear ODE (3.632) represented by the surface

$$y_3=-yy_1=f(y,y_1). \tag{3.663}$$

ODE (3.663) admits the first-order adjoint-symmetries (3.634a–c). Since $f_{y_2}=0$, from Theorem 3.8.2-1 it follows that three adjoint-symmetries could yield a first integral (3.654) of ODE (3.663). However, we find that the Wronskian arising from (3.634a–c) vanishes:

$$W(\omega_1,\omega_2,\omega_3)=\begin{vmatrix}\omega_1 & \omega_2 & \omega_3\\ \mathbf{D}\omega_1 & \mathbf{D}\omega_2 & \mathbf{D}\omega_3\\ \mathbf{D}^2\omega_1 & \mathbf{D}^2\omega_2 & \mathbf{D}\omega_3\end{vmatrix}=\begin{vmatrix}1 & y & \frac{1}{3}y^3+(y_1)^2\\ 0 & y_1 & (y^2+2y_2)y_1\\ 0 & y_2 & y^2y_2+2(y_2)^2\end{vmatrix}=0. \tag{3.664}$$

Consequently, from Lemma 3.8.2-1, it follows that the adjoint-symmetries (3.634a–c) are linearly dependent with respect to surface (3.663) and, hence, we cannot apply Theorem 3.8.2-1 in this situation. However, the linear dependence among (3.634a–c) directly leads to two first integrals as follows. From (3.664), we have

$$c_1\begin{bmatrix}1\\0\\0\end{bmatrix}+c_2\begin{bmatrix}y\\y_1\\y_2\end{bmatrix}+c_3\begin{bmatrix}\frac{1}{3}y^3+(y_1)^2\\ y^2y_1+2y_1y_2\\ y^2y_2+2(y_2)^2\end{bmatrix}=0 \tag{3.665}$$

for some functions c_i satisfying

$$\mathbf{D}c_i=0 \quad \text{where } \mathbf{D}=\frac{\partial}{\partial x}+y_1\frac{\partial}{\partial y}+y_2\frac{\partial}{\partial y_1}-yy_1\frac{\partial}{\partial y_2}.$$

The linear system (3.665) yields

$$\frac{c_2}{c_3}=-2(y_2+\tfrac{1}{2}y^2)=\psi_1, \tag{3.666a}$$

$$\frac{c_1}{c_3}=2(yy_2+\tfrac{1}{3}y^3-\tfrac{1}{2}(y_1)^2)=\psi_2. \tag{3.666b}$$

Thus, since c_2/c_3 and c_1/c_3 are not constants, the expressions (3.666a,b) give two first integrals ψ_1 and ψ_2 of ODE (3.663) which clearly are functionally independent. Note that ψ_1 and ψ_2 are multiples of the first integrals arising from integrating factors (3.634a,b) [cf. Section 3.6.5].

For a fourth example, consider the third-order ODE

$$y'''=6x\frac{(y'')^3}{(y')^2}+6\frac{(y'')^2}{y'}$$

represented by the surface

$$y_3=6x(y_2)^3(y_1)^{-2}+6(y_2)^2(y_1)^{-1}=f(x,y_1,y_2) \tag{3.667}$$

in (x,y,y_1,y_2,y_3)-space. As shown in Section 3.5.2, ODE (3.667) admits three point symmetries and seven contact symmetries. It also admits three second-order adjoint-symmetries that are integrating factors, as shown in Section 3.6.5. Here, we obtain the three corresponding first integrals of (3.667) through Theorem 3.8.2-2(i) by using the three admitted point symmetries

$$\hat{\eta}_1=y,\quad \hat{\eta}_2=xy_1,\quad \hat{\eta}_3=1, \tag{3.668}$$

and the two admitted contact symmetries

$$\hat{\eta}_4 = (y_1)^{-1}, \tag{3.669a}$$

$$\hat{\eta}_5 = (y_1)^{-2}. \tag{3.669b}$$

Note that, since $f_{y_2} \neq 0$, we need at least four symmetries in order to obtain two independent Wronskians (3.648). Now from (3.668) and (3.669a,b), we obtain $5!/(3!2!) = 10$ Wronskians using sets of three symmetries each. The corresponding ratios (3.655a) of these Wronskians yield 45 first integrals, which can be shown to include three functionally independent ones. In particular, the following are functionally independent:

$$\psi_1 = \frac{W(\hat{\eta}_1, \hat{\eta}_2, \hat{\eta}_3)}{W(\hat{\eta}_1, \hat{\eta}_3, \hat{\eta}_4)} = x(y_1)^2 + \tfrac{1}{3}(y_1)^2(y_2)^{-1}, \tag{3.670a}$$

$$\psi_2 = \frac{W(\hat{\eta}_1, \hat{\eta}_2, \hat{\eta}_3)}{W(\hat{\eta}_1, \hat{\eta}_3, \hat{\eta}_5)} W = \tfrac{1}{2}x(y_1)^3 + \tfrac{1}{4}(y_1)^4(y_2)^{-1}, \tag{3.670b}$$

$$\psi_3 = \frac{W(\hat{\eta}_1, \hat{\eta}_2, \hat{\eta}_4)}{W(\hat{\eta}_1, \hat{\eta}_3, \hat{\eta}_4)} = \tfrac{3}{4}\frac{W(\hat{\eta}_1, \hat{\eta}_2, \hat{\eta}_5)}{W(\hat{\eta}_1, \hat{\eta}_3, \hat{\eta}_5)} = -\tfrac{2}{3}y + 2xy_1 + \tfrac{1}{3}(y_1)^2(y_2)^{-1}. \tag{3.670c}$$

From the 45 first integrals, one can show that among those connected with just (3.668) and either (3.669a) or (3.669b), there are only two functionally independent ones. Hence, both of the contact symmetries (3.669a,b) are needed to obtain three functionally independent first integrals of (3.667).

As a final example, we consider the fourth-order ODE

$$y^{(4)} = \frac{4}{3}\frac{(y''')^2}{y''}$$

represented by the surface

$$y_4 = \frac{4}{3}\frac{(y_3)^2}{y_2} = f(y_2, y_3) \tag{3.671}$$

in $(x, y, y_1, y_2, y_3, y_4)$ – space. In Section 3.5.2, it was shown that ODE (3.671) admits 12 second-order symmetries. Similarly, one can show that ODE (3.671) admits 17 second-order adjoint-symmetries [cf. Exercise 3.7-3]. Since $f_{y_3} \neq 0$, at least five symmetries or adjoint-symmetries are needed to obtain two independent Wronskians (3.648) yielding a first integral of ODE (3.671) from Theorem 3.8.2-2(i). Here we obtain four functionally independent first integrals of the fourth-order ODE (3.671) by using five of its admitted symmetries

$$\hat{\eta}_1 = (y_2)^{1/3}, \quad \hat{\eta}_2 = x(y_2)^{1/3}, \quad \hat{\eta}_3 = y(y_2)^{1/3}, \quad \hat{\eta}_4 = xy(y_2)^{1/3}, \quad \hat{\eta}_5 = x^2(y_2)^{1/3}, \tag{3.672}$$

and five of its admitted adjoint-symmetries

$$\omega_1 = (y_2)^{-4/3}, \quad \omega_2 = x(y_2)^{-4/3}, \quad \omega_3 = y(y_2)^{-4/3}, \quad \omega_4 = y_1(y_2)^{-4/3}, \quad \omega_5 = xy_1(y_2)^{-4/3}. \tag{3.673}$$

From (3.672), we obtain five Wronskians using sets of four symmetries each, yielding 10 first integrals given by the ratios (3.655a). Likewise, we obtain 10 first integrals from (3.673). Each set of first integrals can be shown to lead to four functionally independent first integrals of (3.671). In particular, we obtain

$$\psi_1 = \frac{W(\hat{\eta}_1,\hat{\eta}_2,\hat{\eta}_3,\hat{\eta}_4)}{W(\hat{\eta}_1,\hat{\eta}_2,\hat{\eta}_3,\hat{\eta}_5)} = \frac{1}{2}\frac{W(\omega_1,\omega_3,\omega_4,\omega_5)}{W(\omega_1,\omega_2,\omega_3,\omega_4)} = y_1 - \tfrac{3}{2}(y_2)^2(y_3)^{-1}, \tag{3.674a}$$

$$\psi_2 = \frac{W(\hat{\eta}_1,\hat{\eta}_2,\hat{\eta}_4,\hat{\eta}_5)}{W(\hat{\eta}_1,\hat{\eta}_2,\hat{\eta}_3,\hat{\eta}_5)} = \frac{W(\omega_1,\omega_2,\omega_3,\omega_5)}{W(\omega_1,\omega_2,\omega_3,\omega_4)} = x + 3y_2(y_3)^{-1}, \tag{3.674b}$$

$$\psi_3 = \frac{W(\hat{\eta}_1,\hat{\eta}_3,\hat{\eta}_4,\hat{\eta}_5)}{W(\hat{\eta}_1,\hat{\eta}_2,\hat{\eta}_3,\hat{\eta}_5)} = \frac{W(\omega_2,\omega_3,\omega_4,\omega_5)}{W(\omega_1,\omega_2,\omega_3,\omega_4)} = 2xy_1 - y + 3(y_1y_2 - x(y_2)^2)(y_3)^{-1}, \tag{3.674c}$$

$$\psi_4 = \frac{W(\hat{\eta}_2,\hat{\eta}_3,\hat{\eta}_4,\hat{\eta}_5)}{W(\hat{\eta}_1,\hat{\eta}_2,\hat{\eta}_3,\hat{\eta}_5)} = \frac{W(\omega_2,\omega_3,\omega_4,\omega_5)W(\omega_1,\omega_2,\omega_3,\omega_5)}{[W(\omega_1,\omega_2,\omega_4,\omega_5)]^2}$$
$$= -xy + x^2y_1 + (3xy_1y_2 - \tfrac{3}{2}x^2(y_2)^2 - 3yy_2)(y_3)^{-1}. \tag{3.674d}$$

The same first integrals arise from Theorem 3.8.2-2(ii) through products of Wronskians involving (3.672) and (3.673). Then, from (3.674a–d), we have the quadrature of ODE (3.671) given by $\psi_i = \text{const} = c_i$, $i = 1,2,3,4$, which yields

$$y = \frac{c_4 - c_3x + c_1x^2}{x - c_2}. \tag{3.675}$$

3.8.3 FIRST INTEGRALS FOR SELF-ADJOINT ODEs

Here we briefly specialize the results of Sections 3.8.1 and 3.8.2 to self-adjoint ODEs. Recall that for an nth-order ODE given by the surface (3.645), the conditions for self-adjointness are that n is even and that $f(x,y,y_1,\ldots,y_{n-1})$ satisfies (3.540b). In particular, it is necessary that $f_{y_{n-1}} = 0$, and thus,

$$y_n = f(x,y,y_1,\ldots,y_{n-2}), \; n \geq 2. \tag{3.676}$$

For a self-adjoint ODE (3.676), adjoint-symmetries are symmetries. If one knows at least two linearly independent symmetries, then Theorem 3.8.1-1 yields a first integral of (3.676) for each such pair of admitted symmetries $\hat{\eta}_1(x,y,y_1,\ldots,y_\ell)$ and $\hat{\eta}_2(x,y,y_1,\ldots,y_\ell)$, given by the algebraic formula

$$\psi(\hat{\eta}_1,\hat{\eta}_2)=\sum_{j=0}^{n-1}(-1)^j(\mathbf{D}^j\hat{\eta}_1)\mathbf{D}^{n-j-1}\hat{\eta}_2+\sum_{i=0}^{n-3}\sum_{j=0}^{i}(-1)^{j+1}(\mathbf{D}^j(\hat{\eta}_1 f_{y_{i+1}}))\mathbf{D}^{i-j}\hat{\eta}_2 \tag{3.677}$$

provided that $\partial\hat{\psi}(\hat{\eta}_1,\hat{\eta}_2)/\partial y_{n-1}\not\equiv 0$. We emphasize that the symmetries used in (3.677) need *not* be variational symmetries.

Now suppose $\hat{\eta}_1(x,y,y_1,\ldots,y_\ell)$ and $\hat{\eta}_2(x,y,y_1,\ldots,y_\ell)$ are variational symmetries of order ℓ of a self-adjoint ODE (3.676). Then one can show [Exercise 3.8-6] that the first integral (3.677) is yielded by the integrating factor

$$\hat{\Lambda}(\hat{\eta}_1,\hat{\eta}_2)=\sum_{i=0}^{\ell}((\hat{\eta}_1)_{y_i}\mathbf{D}^i\hat{\eta}_2-(\hat{\eta}_2)_{y_i}\mathbf{D}^i\hat{\eta}_1)=\hat{\mathrm{X}}_1^{(\ell)}\hat{\eta}_2-\hat{\mathrm{X}}_2^{(\ell)}\hat{\eta}_1, \tag{3.678}$$

where $\hat{\mathrm{X}}_i=\hat{\eta}_i\dfrac{\partial}{\partial y}$ is the infinitesimal generator corresponding to $\hat{\eta}_i, i=1,2$. Since an integrating factor is a variational symmetry for a self-adjoint ODE [cf. Section 3.7.4], it follows that the integrating factor (3.678) of a self-adjoint ODE (3.676) is a symmetry of (3.676). In particular,

$$\hat{\mathrm{X}}=\hat{\Lambda}(\hat{\eta}_1,\hat{\eta}_2)\frac{\partial}{\partial y}=[\hat{\mathrm{X}}_1,\hat{\mathrm{X}}_2] \tag{3.679}$$

is equal to the commutator symmetry of the symmetries $\hat{\eta}_1$ and $\hat{\eta}_2$. Correspondingly, if ψ_1 and ψ_2 are first integrals arising from integrating factors given by the variational symmetries $\hat{\eta}_1$ and $\hat{\eta}_2$, then

$$\hat{\psi}(\hat{\eta}_1,\hat{\eta}_2)=\tfrac{1}{2}(\hat{\mathrm{X}}_1^{(n-1)}\psi_2-\hat{\mathrm{X}}_2^{(n-1)}\psi_1). \tag{3.680}$$

Now suppose a self-adjoint ODE (3.676) admits a scaling symmetry

$$\hat{\eta}_S=py-qxy_1,\quad p=\text{const},\quad q=\text{const}, \tag{3.681}$$

i.e., $x\to\alpha^q x,\ y\to\alpha^p y$. If one knows a variational symmetry $\hat{\eta}(x,y,y_1,\ldots,y_\ell)$ of order ℓ of (3.676) with noncritical scaling weight $s\neq(n-1)q-p$ with respect to (3.681), then (3.677) yields an algebraic formula for the corresponding first integral of ODE (3.676) given by

$$\hat{\psi}(\hat{\eta}_S,\hat{\eta})=r^{-1}(p-(n-1)q)y_{n-1}-qxf)\hat{\eta}$$
$$+r^{-1}\sum_{i=0}^{n-2}((p-iq)y_i-qxy_{i+1})\left((-1)^{n-i-1}\mathbf{D}^{n-i-1}\hat{\eta}+\sum_{j=1}^{n-i-2}(-1)^j\mathbf{D}^j(\hat{\eta}f_{y_{i+j}})\right), \tag{3.682a}$$

where

$$r=s+p-(n-1)q. \tag{3.682b}$$

Finally, since $f_{y_{n-1}} = 0$ holds for a self-adjoint ODE (3.676), Theorem 3.8.2-1 yields a first integral of (3.676) if one knows a set of n symmetries $\hat{\eta}_i(x, y, y_1, \ldots, y_\ell)$, $i = 1,2,\ldots, n$, that is linearly independent with respect to the surface (3.676). In particular, the first integral is given by

$$\hat{\psi}(\hat{\eta}_1,\ldots,\hat{\eta}_n) = \begin{vmatrix} \hat{\eta}_1 & \cdots & \hat{\eta}_n \\ \mathbf{D}\hat{\eta}_1 & \cdots & \mathbf{D}\hat{\eta}_n \\ \vdots & & \vdots \\ \mathbf{D}^{n-1}\hat{\eta}_1 & \cdots & \mathbf{D}^{n-1}\hat{\eta}_n \end{vmatrix} \tag{3.683}$$

provided that $\partial\hat{\psi}(\hat{\eta}_1,\ldots,\hat{\eta}_n)/\partial y_{n-1} \not\equiv 0$. The symmetries used in (3.683) need not be variational symmetries.

For $n = 2$, we note that (3.683) and (3.677) both yield the same first integral formula $\hat{\psi}(\hat{\eta}_1, \hat{\eta}_2) = \hat{\eta}_1\mathbf{D}\hat{\eta}_2 - \hat{\eta}_2\mathbf{D}\hat{\eta}_1$. However, for $n > 2$, the corresponding formulas are different.

We now consider two examples.

As a first example, we consider the nonlinear Duffing equation (3.627) with no damping, i.e., the self-adjoint ODE $y'' + by + y^3 = 0$. This second-order ODE admits the point symmetry (3.628a). In the case $b = 0$, the fully nonlinear Duffing equation

$$y'' + y^3 = 0 \tag{3.684}$$

admits a second point symmetry given by the scaling symmetry $x \to \alpha x$, $y \to \alpha^{-1}y$. We now apply the Wronskian (3.683) to obtain a first integral of the self-adjoint ODE (3.684) from the two admitted point symmetries

$$\hat{\eta}_1 = y_1, \quad \hat{\eta}_2 = y + xy_1. \tag{3.685}$$

This yields

$$\hat{\psi}(\hat{\eta}_1, \hat{\eta}_2) = 2(y_1)^2 + y^4. \tag{3.686}$$

The corresponding integrating factor is given by the variational point symmetry

$$\hat{\Lambda}(\hat{\eta}_1, \hat{\eta}_2) = 4y_1 = 4\hat{\eta}_1. \tag{3.687}$$

Consequently, the first integral (3.686) can also be obtained directly in terms of $\hat{\eta}_1$ from the scaling formula (3.682a,b) since the scaling weight $s = -2$ of (3.687) is noncritical (i.e., $q = 1$, $p = -1$, and hence, $r = -4 \neq 0$). Note that the first integral (3.686) is a multiple of the energy of the nonlinear oscillator described by (3.684).

As our final example, we consider the fourth-order nonlinear ODE [Sheftel (1997)]

$$y^{(4)} = y^{-5/3}. \tag{3.688}$$

Its admitted point symmetries are given by [cf. Exercise 3.5-7]

$$\hat{\eta}_1 = y_1, \tag{3.689a}$$

$$\hat{\eta}_2 = 3y - 2xy_1, \tag{3.689b}$$

$$\hat{\eta}_3 = 3xy - x^2 y_1. \tag{3.689c}$$

Hence, there are an insufficient number of symmetries to apply the first integral Wronskian formula (3.683). However, we can apply the first integral formula (3.677) by using pairs of symmetries (3.689a–c). This yields three functionally independent first integrals

$$\hat{\psi}_1(\hat{\eta}_1, \hat{\eta}_2) = 2y_1 y_3 - (y_2)^2 + 3y^{-2/3}, \tag{3.690a}$$

$$\hat{\psi}_2(\hat{\eta}_1, \hat{\eta}_3) = (2xy_1 - 3y)y_3 + y_1 y_2 - x(y_2)^2 + 3xy^{-2/3}, \tag{3.690b}$$

$$\hat{\psi}_3(\hat{\eta}_2, \hat{\eta}_3) = (-2x^2 y_1 + 6xy)y_3 - 6yy_2 - 2xy_1 y_2 + x^2(y_2)^2 + 4(y_1)^2 - 3x^2 y^{-2/3}, \tag{3.690c}$$

with corresponding integrating factors given by

$$\hat{\Lambda}_1(\hat{\eta}_1, \hat{\eta}_2) = 2y_1 = 2\hat{\eta}_1, \tag{3.691a}$$

$$\hat{\Lambda}_2(\hat{\eta}_1, \hat{\eta}_3) = 2xy_1 - 3y = -\hat{\eta}_2, \tag{3.691b}$$

$$\hat{\Lambda}_3(\hat{\eta}_2, \hat{\eta}_3) = -2x^2 y_1 + 6xy = 2\hat{\eta}_3. \tag{3.691c}$$

Hence, the three point symmetries, including the scaling symmetry, are variational symmetries of ODE (3.688). Note that the two first integrals (3.690a,c) can also be obtained using the scaling formula (3.682a) in terms of (3.689a,c). We also note that the commutators (3.679) arising from the three point symmetries (3.689a–c) are given by

$$[\hat{\mathrm{X}}_1, \hat{\mathrm{X}}_2] = 2\hat{\mathrm{X}}_1, \quad [\hat{\mathrm{X}}_1, \hat{\mathrm{X}}_3] = -\hat{\mathrm{X}}_2, \quad [\hat{\mathrm{X}}_2, \hat{\mathrm{X}}_3] = 2\hat{\mathrm{X}}_3,$$

in accordance with the results (3.678) and (3.691a–c).

Finally, from (3.690a–c), we obtain three quadratures $\psi_i = \text{const} = c_i$, $i = 1,2,3$, yielding a reduction of ODE (3.688) to a first-order ODE given by

$$\tfrac{1}{9}(x^2c_1 - 2xc_2 - c_3)(y_1)^2 - \tfrac{2}{3}(xc_1 - c_2)yy_1 + c_1 y^2 - 3y^{4/3} + \tfrac{1}{36}(x^2c_1 - 2xc_2 - c_3)^2 = 0. \tag{3.692}$$

EXERCISES 3.8

1. Consider the harmonic oscillator equation $y'' + \nu^2 y = 0$, $\nu = \text{const}$. Find first integrals from its admitted point symmetries by using the scaling formula (3.682a,b) and the Wronskian formula (3.683).

2. Consider the third-order ODE $y''' = y^{-2}(y')^3$, which admits the translation symmetry $x \to x + \varepsilon,\ y \to y$ and the scaling symmetries $x \to \alpha x,\ y \to \beta y$. Use the Wronskian formula (3.654) with these three admitted point symmetries to find a first integral and corresponding integrating factor.

3. Consider the third-order ODE (3.667), which admits the scaling symmetries $x \to \alpha x,\ y \to \beta y$.
   (a) Find first integrals from the admitted second-order adjoint-symmetries (3.513) and (3.515) by using the scaling formula (3.626). Compare the results obtained from invariance under the x and y scalings, respectively.
   (b) Classify which of the second-order adjoint-symmetries (3.513) and (3.515) of ODE (3.667) have noncritical scaling dimension with respect to the x and y scalings. Determine if there is a combination of the x and y scalings such that *all* the second-order adjoint-symmetries (3.513) and (3.515) have noncritical scaling dimensions.

4. Consider the fourth-order ODE (3.671), which admits the scaling symmetries $x \to \alpha x,\ y \to \beta y$.
   (a) Find first integrals from the admitted second-order adjoint-symmetries [cf. Exercise 3.7-3] by using the scaling formula (3.626) for the x and y scalings.
   (b) Find first integrals from the admitted second-order symmetries and adjoint-symmetries by using the Wronskian formula (3.655a,b).
   (c) Compare the results obtained from the scaling formula and the Wronskian formula.

5. Consider the fourth-order wave speed ODE (3.639). Find first integrals by using the Wronskian formulas (3.655a,b) for:
   (a) admitted first- and second-order symmetries [cf. Section 3.5.2]; and
   (b) admitted first- and second-order adjoint-symmetries [cf. Section 3.7.3].

6. Show that the first integral formula (3.677) for a pair of variational symmetries reduces to the expression (3.680). Show that the corresponding integrating factor is given by the commutator expression (3.678), (3.679).

7. Prove Lemma 3.8.2-2.

8. Prove Theorem 3.8.2-2.

9. Here we consider another algebraic formula for first integrals that applies to the special class of nth-order ODEs

$$y_n = f(x, y_1, \dots, y_{n-1}), \quad f_y = 0, \tag{3.693}$$

   i.e., f has no dependence on y. For an ODE (3.693), the adjoint-symmetry determining equation (3.543) takes the form

$$\mathbf{D}\psi = 0, \quad \psi = (-1)^n \mathbf{D}^{n-1}\omega - \sum_{i=1}^{n-1} (-1)^i \mathbf{D}^{i-1}(f_{y_i}\omega). \tag{3.694}$$

Hence, (3.694) yields a first integral of ODE (3.693) provided that $\psi_{y_{n-1}} \not\equiv 0$.

(a) Calculate first integrals given by (3.694) for the third-order ODE (3.506) by using its admitted second-order adjoint-symmetries [cf. Section 3.7.3].

(b) Calculate first integrals given by (3.694) for the fourth-order ODE (3.612) by using its admitted adjoint-symmetries determined in Exercise 3.7-3.

3.9 APPLICATIONS TO BOUNDARY VALUE PROBLEMS

We show how reduction of order is applied to boundary value problems for ODEs. Since reduction of order holds for essentially all solutions of a given ODE, it follows that a posed boundary value problem for the given ODE will map into a boundary value problem for the reduced order ODE. We illustrate this through an example.

Consider again the Prandtl–Blasius problem for a flat plate discussed in Section 1.3.1. The boundary value problem (1.62a–e) reduces to solving the Blasius equation

$$y''' + \tfrac{1}{2} y y'' = 0, \quad 0 < x < \infty, \tag{3.695a}$$

with the boundary conditions

$$y(0) = y'(0) = 0, \tag{3.695b}$$

$$y'(\infty) = 1. \tag{3.695c}$$

We wish to determine the value of $\sigma = y''(0)$.

In Section 3.4.2, we saw that the invariance of (3.695a) under the two-parameter Lie group of point transformations (3.139a,b) reduced this ODE to the first-order ODE

$$\frac{dV}{dU} = \frac{V}{U}\left[\frac{\frac{1}{2} + V + U}{2U - V}\right] \tag{3.696}$$

plus the quadratures (3.171) and (3.172), where

$$V = \frac{y''}{yy'}, \quad U = \frac{y'}{y^2}.$$

Let $V = \phi(U; C_1)$ be the general solution of (3.696). Consider the phase plane diagram in the UV-plane associated with (3.696). At some point on the solution curve of the boundary value problem (3.695a–c), one must have $y' > 0$. Then, from the phase plane diagram, it follows that $U > 0$ along the whole solution curve of the boundary value problem (3.695a–c). Consequently, $y > 0, y' > 0$ for $0 < x < \infty$. Then, at some point along the solution curve of (3.695a–c), one must have $y'' > 0$. Hence, the solution curve of the boundary value problem (3.695a–c) must lie entirely in the first quadrant [cf. Figure 3.3]. It then follows that along the solution curve one has $y > 0, y' > 0, y'' > 0$ for $0 < x < \infty$. Thus, $y''(0) = \sigma > 0$. Then,

$$\frac{dU}{dx} = \frac{y'}{y}[V - 2U] \gtrless 0 \quad \text{if } V \gtrless 2U,$$

leads to the direction of increasing x indicated by the arrows in Figure 3.3. Thus, as $x \to \infty$, $(U,V) \to (0,0)$. As $x \to 0$, three cases can arise: $(U,V) \to (0,\infty)$; $U \to \infty$ with $V \ll U$; and $(U,V) \to (\infty,\infty)$ with $V = O(U)$. We examine each of these three cases in terms of the boundary conditions (3.695b,c):

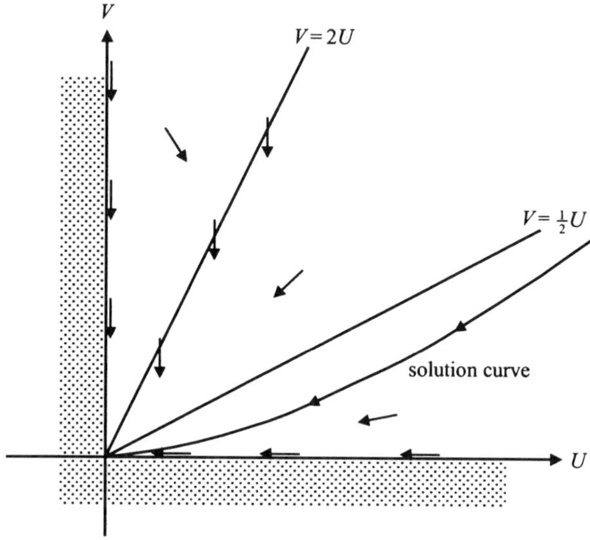

Figure 3.3

Case I. $(U,V) \to (0,\infty)$ as $x \to 0$.
Here $dV/dU \sim -V/U$ as $x \to 0$. Then $y''/y^3 = VU \sim \text{const} = C_1$ as $x \to 0$, which is impossible if $y''(0) = \sigma = \text{const} \neq 0$.

Case II. $U \to \infty$ with $V \ll U$ as $x \to 0$.
Here $dV/dU \sim V/2U$ as $x \to 0$. Then $y''/(y')^{3/2} \sim \text{const} = C_2$ as $x \to 0$, which again is impossible.

Thus, the following case must hold:

Case III. $(U,V) \to (\infty,\infty)$ with $V = O(U)$ as $x \to 0$.
From ODE (3.696), it follows that the solution must lie along a *separatrix* (*exceptional path*) [see Section 3.10]

$$V \sim \tfrac{1}{2}U \quad \text{as } x \to 0,$$

i.e., $y'' \sim \frac{1}{2}(y')^2/y$ as $x \to 0$. Then ODE (3.170) becomes

$$\frac{ds}{dr} \sim -\frac{1}{3r} \quad \text{as} \quad x \to 0,$$

which leads to

$$\frac{(y')^2}{y} \sim \text{const} = C_\infty \quad \text{as} \quad x \to 0^+.$$

Then $y''(0) = \frac{1}{2}C_\infty$.

As $x \to \infty$, the boundary condition (3.695c) leads to $(U,V) \to (0,0)$ with $V << U$. Consequently, as $x \to \infty$, we have

$$\frac{dV}{dU} \sim \frac{V}{4U^2},$$

so that

$$V \sim d_0 e^{-1/4U}$$

for some constant d_0. Thus, from ODE (3.170), we get $y' = s = \text{const} = C_0$ as $x \to \infty$. As shown in Section 1.3.1, it then follows that $\sigma = C_\infty / 2(C_0)^{3/2}$.

The solution of the boundary value problem (3.695a–c) is now obtained by starting from the exceptional path as $U \to \infty$, then integrating out to $V \sim d_0 e^{-1/4U}$ as $U \to 0$, and finally determining constants C_0 and C_∞. Then we obtain $\sigma = C_\infty / 2(C_0)^{3/2}$. See Dresner (1983) for further details.

Further examples of applications to boundary value problems appear in Bluman and Cole (1974) and Dresner (1983, 1999).

EXERCISES 3.9

1. Consider the nonlinear diffusion equation

$$u_t = (uu_x)_x, \quad 0 < x < \infty, \quad 0 < t < \infty,$$

with the boundary conditions

$$u(x,0) = 0, \quad x > 0,$$
$$u(0,t) = 1, \quad t > 0.$$

From its invariance under scalings, its solution is of the form $u = y(\eta)$ with $\eta = x / \sqrt{t}$.

(a) Derive the second-order ODE satisfied by $y(\eta)$ and the corresponding boundary conditions.

(b) Show that this ODE admits a one-parameter Lie group of point transformations. Reduce this ODE to a first-order ODE plus a quadrature.
(c) Study the phase plane of this first-order ODE and discuss which path yields the solution of the posed boundary value problem.
(d) Sketch $u(x,t) = y(\eta)$.

2. Consider the Thomas–Fermi equation

$$y'' = x^{-1/2} y^{3/2}. \tag{3.697}$$

(a) Find a scaling symmetry admitted by (3.697).
(b) Use this symmetry to reduce ODE (3.697) to a first-order ODE.
(c) Find the curve in the phase plane that corresponds to the physically interesting boundary conditions

$$y(0) = 1, \quad y(\infty) = 0.$$

For a full discussion of this problem, see Bluman and Cole (1974) and Dresner (1999).

3. In a geometrically nonlinear theory of axisymmetric deformation of a membrane under pressure loading, one obtains the ODE

$$(x^3 y')' = x^3 \nu(x) - \frac{x^3}{y^2} q(x), \quad 0 < x < 1, \tag{3.698}$$

where y is the deflection from the original shape, x is a radial spherical coordinate, $\nu(x)$ is a shape function with $\nu(x) = \text{const}$ for a sphere, and $q(x)$ is a load function with $q(x) \sim x^4$ for uniform pressure. Assume that the membrane is spherical and loaded with constant pressure near $x = 0$. The boundary conditions are

$$y(0) \text{ is finite (regularity at the axis)}, \tag{3.699a}$$

$$y(1) = 0 \text{ (membrane fixed at the edge)}. \tag{3.699b}$$

(a) Show that ODE (3.698), with $\nu(x)$ and $q(x)$ having the desired properties and with the solution satisfying the regularity condition, admits a one-parameter Lie group of point transformations if and only if $\nu(x)$ and $q(x)$ are of the form

$$\nu(x) = \nu_0 (1 + \alpha x^2)^{-3}, \quad q(x) = q_0 x^4 (1 + \alpha x^2)^{-5}, \tag{3.700}$$

for arbitrary constants α, ν_0, q_0.
(b) Show that the infinitesimal generator of the point symmetry admitted by ODE (3.698), when its coefficients satisfy (3.700), is given by

$$\mathrm{X} = (x + \alpha x^3) \frac{\partial}{\partial x} + 2y \frac{\partial}{\partial y}. \tag{3.701}$$

(c) Accordingly, reduce ODE (3.698) to a first-order ODE and isolate the particular path in the phase plane that solves the boundary value problem (3.698), (3.699a,b).

For a full discussion of this problem, see Bluman and Cole (1974).

3.10 INVARIANT SOLUTIONS

Consider an nth-order ODE

$$y^{(n)} = f(x, y, y', \ldots, y^{(n-1)}) \tag{3.702}$$

or, equivalently, the surface $y_n = f(x, y, y_1, \ldots, y_n)$, that is assumed to admit a one-parameter Lie group of point transformations (point symmetry) with the infinitesimal generator

$$\mathrm{X} = \xi(x, y)\frac{\partial}{\partial x} + \eta(x, y)\frac{\partial}{\partial y}. \tag{3.703}$$

Definition 3.10-1. $y = \phi(x)$ is an *invariant solution* of ODE (3.702) resulting from its invariance under the point symmetry (3.703) if and only if:

(i) $y = \phi(x)$ is an invariant curve of (3.703), i.e., $\mathrm{X}(y - \phi(x)) = 0$ when $y = \phi(x)$;

(ii) $y = \phi(x)$ solves (3.702).

It follows that $y = \phi(x)$ is an invariant solution of ODE (3.702) resulting from its invariance under the point symmetry (3.703) if and only if $y = \phi(x)$ satisfies

(i)
$$\xi(x, \phi(x))\phi'(x) = \eta(x, \phi(x)); \tag{3.704a}$$

(ii)
$$\phi^{(n)}(x) = f(x, \phi(x), \phi'(x), \ldots, \phi^{(n-1)}(x)). \tag{3.704b}$$

More generally, $\Phi(x, y) = 0$ defines an invariant solution of ODE (3.702) resulting from its invariance under the point symmetry (3.703) if and only if:

(i) $\Phi(x, y) = 0$ is an invariant curve of (3.703), i.e., $\mathrm{X}\Phi = 0$ when $\Phi = 0$;

(ii) $\Phi(x, y) = 0$ solves (3.702).

In particular, here (i) and (ii) are equivalent to:

(i) $\Phi(x, y) = 0$ is a solution of the first-order ODE $y' = \eta(x, y)/\xi(x, y)$;

(ii) $\Phi(x, y) = 0$ is a solution of the ODE $y^{(n)} = f(x, y, y', \ldots, y^{(n-1)})$.

An obvious (naive) procedure to find invariant solutions of ODE (3.702), resulting from its invariance under the point symmetry (3.703), is to first try to solve the ODE $y' = \eta(x, y)/\xi(x, y)$ to obtain its general solution $g(x, y; C) = 0$. If one is able to do this, the values of C that yield invariant solutions, if any exist, are then determined by substituting this general solution into the given ODE (3.702). Any such value of $C = C^*$

determines an invariant solution $\Phi(x,y)=g(x,y;C^*)=0$ of ODE (3.702) resulting from its invariance under (3.703).

A better alternative procedure that avoids determining the general solution of $y'=\eta(x,y)/\xi(x,y)$ will now be given. In particular, we will see that usually *it is not necessary to solve the ODE* $y'=\eta(x,y)/\xi(x,y)$ *or any other ODE in order to find the invariant solutions of ODE* (3.702) *resulting from its invariance under the point symmetry* (3.408).

Theorem 3.10-1 [Bluman (1990c)]. *Suppose an nth-order ODE* (3.702) *admits the point symmetry* (3.703) *with* $\xi \not\equiv 0$. *Let* $\mathrm{Y}=\frac{\partial}{\partial x}+\Psi(x,y)\frac{\partial}{\partial y}$ *where* $\Psi(x,y)\equiv\eta(x,y)/\xi(x,y)$. *Consider the function* $Q(x,y)$ *defined by*

$$Q(x,y)=\left(y_n-f(x,y,y_1,\ldots,y_{n-1})\right)\Big|_{y_k=\mathrm{Y}^{k-1}\Psi}=\mathrm{Y}^{n-1}\Psi-f(x,y,\Psi,\mathrm{Y}\Psi,\ldots,\mathrm{Y}^{n-2}\Psi). \tag{3.705}$$

Three cases arise for the algebraic equation $Q(x,y)=0$:

(i) $Q(x,y)=0$ *defines no curves in the xy-plane*;

(ii) $Q(x,y)=0$ *is identically satisfied for all values of* x *and* y;

(iii) $Q(x,y)=0$ *defines curves in the xy-plane.*

In Case (i), *the ODE* (3.702) *has* **no** *solutions resulting from its invariance under the point symmetry* (3.703).

In Case (ii), **any** *solution of the ODE* $y'=\eta(x,y)/\xi(x,y)$ *is an invariant solution of the ODE* (3.702) *resulting from its invariance under the point symmetry* (3.703).

In Case (iii), *an invariant solution of ODE* (3.702), *resulting from its invariance under the point symmetry* (3.703), *is a curve satisfying* $Q(x,y)=0$ *and, conversely,* **any curve** *satisfying* $Q(x,y)=0$ *is an invariant solution of ODE* (3.702) *resulting from its invariance under* (3.703).

Proof. If $y_1=y'=\eta(x,y)/\xi(x,y)=\Psi(x,y)$, then we successively obtain $y_k=y^{(k)}=d^{k-1}y_1/dx^{k-1}=\mathrm{Y}^{k-1}\Psi$, for $k=1,2,\ldots,n$. Hence, any invariant solution of ODE (3.702), resulting from its invariance under the point symmetry (3.703), must satisfy the algebraic equation

$$Q(x,y)=0.$$

From this it immediately follows that:

(i) if $Q(x,y)=0$ defines no curves in the xy-plane, then ODE (3.702) has no solutions resulting from its invariance under the point symmetry (3.703); and

(ii) if $Q(x,y)\equiv 0$ for all x,y, then *any* solution of the ODE $y'=\eta(x,y)/\xi(x,y)$ is an invariant solution of ODE (3.702) resulting from its invariance under (3.703).

In Case (iii), consider any curve satisfying $Q(x,y)=0$. This curve is an invariant solution of ODE (3.702), resulting from its invariance under the point symmetry (3.703), if and only if its differential consequence

$$Q_x + Q_y y' = 0$$

satisfies ODE $y' = \eta(x,y)/\xi(x,y)$. This is equivalent to

$$\mathrm{Y}Q = 0 \quad \text{when } Q(x,y) = 0. \tag{3.706}$$

We now show that (3.706) holds for any curve defined by $Q(x,y) = 0$.

Since ODE (3.702) admits the point symmetry (3.703), it follows that the point symmetry determining equation

$$\eta^{(n)} = \xi \frac{\partial f}{\partial x} + \eta \frac{\partial f}{\partial y} + \eta^{(1)} \frac{\partial f}{\partial y_1} + \cdots + \eta^{(n-1)} \frac{\partial f}{\partial y_{n-1}} \tag{3.707a}$$

must hold for all values of $x, y, y_1, \ldots, y_n$ satisfying

$$y_n = f(x, y, y_1, \ldots, y_{n-1}), \tag{3.707b}$$

with $\eta^{(k)}$ given by (2.100a,b) for $k = 1, 2, \ldots, n$. Given any values of x and y, from (3.705) we see that the values $x, y, y_k = \mathrm{Y}^{k-1}\Psi$, $k = 1,2,\ldots,n$, satisfy (3.707b) if $Q(x,y) = 0$. Then it follows that $\mathrm{Y} = \mathrm{D}$, where D is the total derivative operator (2.96). Hence, in (3.707a), we have

$$\eta^{(1)} = \mathrm{D}\eta - y_1 \mathrm{D}\xi = \mathrm{Y}\eta - \Psi\mathrm{Y}\xi = \mathrm{Y}(\xi\Psi) - \Psi\mathrm{Y}\xi = \xi\mathrm{Y}\Psi$$

and, by induction,

$$\eta^{(k+1)} = \mathrm{D}\eta^{(k)} - y_{k+1}\mathrm{D}\xi = \xi\mathrm{Y}^{k+1}\Psi, \quad k = 1, 2, \ldots, n-1.$$

Thus, the point symmetry determining equation (3.707a) yields

$$\mathrm{Y}^n\Psi = \frac{\partial f}{\partial x} + \Psi \frac{\partial f}{\partial y} + (\mathrm{Y}\Psi)\frac{\partial f}{\partial y_1} + \cdots + (\mathrm{Y}^{(n-1)}\Psi)\frac{\partial f}{\partial y_{n-1}} \tag{3.708}$$

evaluated at $y_k = \mathrm{Y}^{k-1}\Psi$, $k = 1,2,\ldots,n-1$, for any curve satisfying $Q(x,y) = 0$. On the other hand, by applying Y to (3.705), we have

$$\mathrm{Y}Q \equiv \mathrm{Y}^n\Psi - \left[\frac{\partial f}{\partial x} + \Psi \frac{\partial f}{\partial y} + (\mathrm{Y}\Psi)\frac{\partial f}{\partial y_1} + \cdots + (\mathrm{Y}^{n-1}\Psi)\frac{\partial f}{\partial y_{n-1}}\right]$$

evaluated at $y_k = \mathrm{Y}^{k-1}\Psi$, $k = 1,2,\ldots,n-1$. Hence, from (3.708), we obtain

$$\mathrm{Y}Q = 0 \quad \text{when } Q(x,y) = 0. \qquad \square$$

A similar result holds for invariant solutions of ODE (3.702) resulting from an admitted point symmetry (3.703) with $\xi \equiv 0$.

Theorem 3.10-2. *Suppose an nth-order ODE* (3.702) *admits a point symmetry* (3.703) *with* $\xi \equiv 0$. *Consider the algebraic equation* $\eta(x, y) = 0$. *Then two cases arise:*

(i) $\eta(x, y) = 0$ *defines* **no** *curves in the xy-plane; and*

(ii) $\eta(x, y) = 0$ *defines curves in the xy-plane.*

In Case (i), *ODE* (3.702) *has* **no** *invariant solutions resulting from its invariance under* (3.703).

In Case (ii), *a curve* $y = \phi(x)$ *is an invariant solution of ODE* (3.702) *resulting from its invariance under* (3.703) *if and only if this curve satisfies* $\eta(x, y) = 0$.

Proof. Left to Exercise 3.10-13. □

As a first example, consider the *n*th-order linear homogeneous ODE with constant coefficients,

$$y^{(n)} + a_1 y^{(n-1)} + \cdots + a_{n-1} y' + a_n y = 0. \tag{3.709}$$

We find all invariant solutions of ODE (3.709) from its invariance under both translations in x and scalings in y generated by the infinitesimal generator

$$\mathrm{X} = \alpha \frac{\partial}{\partial x} + \beta y \frac{\partial}{\partial y}$$

for arbitrary constants α, β. Let $\lambda = \beta / \alpha$ with $\alpha \neq 0$. The corresponding invariant solutions $y = \phi(x)$ satisfy

$$y' = \frac{\eta}{\xi} = \lambda y. \tag{3.710}$$

[Hence, we have $\Psi = \lambda y, \mathrm{Y} = \frac{\partial}{\partial x} + \lambda y \frac{\partial}{\partial y}, y^{(k)} = \mathrm{Y}^{k-1}\Psi = \lambda^k y$.] Substituting (3.710) into ODE (3.709), we obtain

$$Q(x, y) = [\lambda^n + a_1 \lambda^{n-1} + \cdots + a_{n-1}\lambda + a_n] y = 0.$$

This yields the well-known characteristic polynomial equation

$$p(\lambda) = \lambda^n + a_1 \lambda^{n-1} + \cdots + a_{n-1}\lambda + a_n = 0, \tag{3.711}$$

that arises for determining solutions $y = Ce^{\lambda x}, C = \text{const}$, of ODE (3.709). Here these solutions are the invariant solutions of the given ODE (3.709) resulting from invariance under $\mathrm{X} = \alpha \frac{\partial}{\partial x} + \beta y \frac{\partial}{\partial y}$, i.e., $y = Ce^{\lambda x}$ satisfies $[\mathrm{X}(y - Ce^{\lambda x})]\big|_{y = Ce^{\lambda x}} = 0$ when $p(\lambda) = 0$. In terms of Theorem 3.10-1, Case (ii) arises when λ is a root of $p(\lambda) = 0$, Case (iii) arises when λ is not a root of $p(\lambda) = 0$, and here $y = 0$ is the corresponding (trivial) invariant solution. In Case (ii), an invariant solution is any solution of ODE (3.710) of the form $y = Ce^{\lambda x}$, for arbitrary constant C.

As a second example, consider again the Blasius equation

$$y''' + \tfrac{1}{2}yy'' = 0, \tag{3.712}$$

which admits

$$\mathrm{X} = (\alpha x + \beta)\frac{\partial}{\partial x} - \alpha y\frac{\partial}{\partial y},$$

for arbitrary constants α, β. When $\alpha = 0$, the corresponding invariant solution is $y = \text{const} = C$, for any constant C. When $\alpha \neq 0$, let $\lambda = \beta / \alpha$.

We first consider obtaining invariant solutions through the obvious procedure. Then an invariant solution $y = \phi(x)$ satisfies the ODE

$$y' = \frac{\eta}{\xi} = -\frac{y}{x+\lambda}, \tag{3.713a}$$

which has the general solution,

$$y = \frac{C}{x+\lambda}, \tag{3.713b}$$

for an arbitrary constant C. After substituting (3.713b) into ODE (3.712), we find that $C = 0$ or $C = 6$, which leads to the corresponding invariant solutions

$$y = \phi_1(x) = 0, \quad y = \phi_2(x) = \frac{6}{x+\lambda}, \tag{3.714}$$

of the Blasius equation (3.712) for an arbitrary constant λ.

Now we obtain the invariant solutions (3.714) through the procedure arising from Theorem 3.10-1. Here

$$\mathrm{Y} = \frac{\partial}{\partial x} - \frac{y}{x+\lambda}\frac{\partial}{\partial y}, \quad \Psi = -\frac{y}{x+\lambda}, \quad \mathrm{Y}\Psi = \frac{2y}{(x+\lambda)^2}, \quad \mathrm{Y}^2\Psi = -\frac{6y}{(x+\lambda)^3}.$$

Then

$$Q(x,y) = \frac{y^2}{(x+\lambda)^2} - \frac{6y}{(x+\lambda)^3}.$$

Consequently, $Q(x, y) = 0$ yields the invariant solutions (3.714).

For both examples, there is little difference in the amount of computation for the two approaches. However, in general, the procedure outlined in Theorem 3.10-1 is clearly superior since it avoids an unnecessary integration of the ODE $y' = \eta / \xi$.

Invariant solutions are especially interesting for first-order ODEs.

3.10.1 INVARIANT SOLUTIONS FOR FIRST-ORDER ODEs: SEPARATRICES AND ENVELOPES

In the case of a first-order ODE

$$y' = f(x,y), \tag{3.715}$$

one only considers invariant solutions resulting from invariance under a nontrivial infinitesimal generator (3.703), admitted by (3.715), for which $\eta(x,y)/\xi(x,y) \not\equiv f(x,y)$, i.e., $Q(x,y) = \eta(x,y)/\xi(x,y) - f(x,y) \not\equiv 0$ for all values of x, y. This clearly follows from Theorem 3.10-1, since if $\eta(x,y)/\xi(x,y) \equiv f(x,y)$, then $Q(x,y) \equiv 0$ and hence, trivially, all solutions of ODE (3.715) are invariant solutions.

Consider the set of all solution curves of ODE (3.715) in the *xy*-plane (phase plane). This set may include *separatrices* (*exceptional paths*), e.g., limit cycles, which are solution curves that behave topologically "abnormally" in relation to neighboring solution curves, i.e., "separate" topologically distinct solution curves [Lefschetz (1963)]. By the following argument we show that a separatrix is an invariant solution of ODE (3.715) for *any* admitted Lie group of transformations.

A one-parameter (ε) Lie group of transformations admitted by ODE (3.715) induces a continuous deformation of the solutions of (3.715) to other solutions of (3.715) through the parameter (ε). But two solutions of ODE (3.715) that are topologically distinct cannot be continuously deformed to each other and, hence, cannot be mapped into each other under the group. Thus, it follows that a separatrix must be an invariant solution of ODE (3.715) for all admitted Lie groups of transformations.

By a similar argument as that for separatrices, it follows that any singular solution, in particular any *envelope solution* (if one exists), for a first-order ODE

$$F(x,y,y') = 0, \tag{3.716}$$

must be an invariant solution for any admitted Lie group of transformations. If ODE (3.716) admits a nontrivial one-parameter Lie group of transformations with the infinitesimal generator $\mathrm{X} = \xi(x,y)\dfrac{\partial}{\partial x} + \eta(x,y)\dfrac{\partial}{\partial y}$, then

$$\mathrm{X}^{(1)}F = \xi F_x + \eta F_y + [\eta_x + (\eta_y - \xi_x)y' - \xi_y (y')^2]F_{y'} = 0 \quad \text{when } F(x,y,y') = 0,$$

and

$$F\left(x, y, \frac{\eta(x,y)}{\xi(x,y)}\right) \not\equiv 0 \quad \text{for all values of } x, y.$$

By a simple extension of Theorem 3.10-1, it follows that the invariant solutions of ODE (3.716), resulting from its invariance under $\mathrm{X} = \xi(x,y)\dfrac{\partial}{\partial x} + \eta(x,y)\dfrac{\partial}{\partial y}$, are the curves defined by the algebraic equation

$$F\left(x, y, \frac{\eta(x,y)}{\xi(x,y)}\right) = 0. \tag{3.717}$$

Hence, an envelope solution of ODE (3.421) satisfies the algebraic equation (3.717) for *any* point symmetry $\mathrm{X} = \xi(x,y)\frac{\partial}{\partial x} + \eta(x,y)\frac{\partial}{\partial y}$ admitted by (3.716).

As a first example, consider the first-order ODE

$$y' = y^2, \tag{3.718}$$

which obviously admits

$$\mathrm{X}_1 = \frac{\partial}{\partial x}, \quad \mathrm{X}_2 = x\frac{\partial}{\partial x} - y\frac{\partial}{\partial y}.$$

From its invariance under X_1, it follows that a separatrix solution $y = \phi(x)$ of ODE (3.718) must satisfy

$$Q(x,y) = \frac{\eta(x,y)}{\xi(x,y)} - f(x,y) = -y^2 = 0.$$

Hence, the only possible separatrix solution could be $y = 0$.

From its invariance under X_2, it follows that a separatrix solution $y = \phi(x)$ of ODE (3.718) must also satisfy

$$Q(x,y) = \frac{\eta(x,y)}{\xi(x,y)} - f(x,y) = -\left(\frac{y}{x} + y^2\right) = -y\left(y + \frac{1}{x}\right) = 0,$$

which leads to possible separatrices $y = 0$ and $y = -1/x$. Since $y = -1/x$ is not an invariant solution of (3.718), resulting from its invariance under X_1, it cannot be a separatrix of ODE (3.718). This solution is a particular solution of ODE (3.718) arising from its general solution $y = -1/(x + C)$, $C =$ const, when $C = 0$. The solution curves are illustrated in Figure 3.4. Clearly, $y = 0$ is a separatrix.

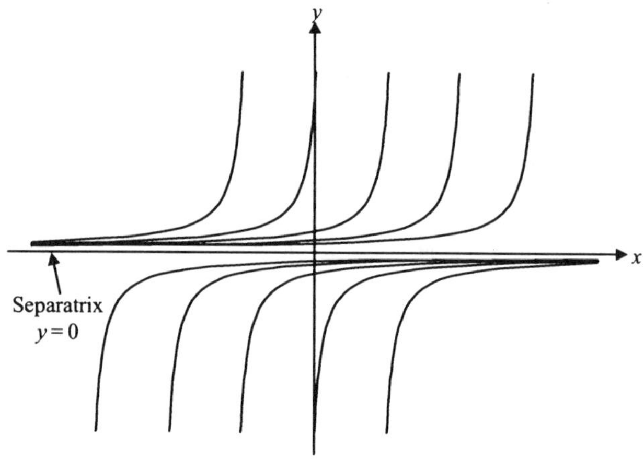

Figure 3.4. Solution curves of $y' = y^2$.

As a second example, consider the first-order ODE

$$y' = \frac{x\sqrt{x^2 + y^2} + y(x^2 + y^2 - 1)}{x(x^2 + y^2 - 1) - y\sqrt{x^2 + y^2}}, \tag{3.719}$$

which admits the rotation group

$$\mathrm{X} = y\frac{\partial}{\partial x} - x\frac{\partial}{\partial y}. \tag{3.720}$$

Invariant curves of (3.720) are the circles

$$x^2 + y^2 = c^2. \tag{3.721}$$

Thus, a separatrix of ODE (3.719) must satisfy (3.721). After substituting (3.721) into ODE (3.719), we get

$$-\frac{x}{y} = \frac{xc + y(c^2 - 1)}{x(c^2 - 1) - yc},$$

so that $c = 1$. Hence, the only possible separatrix is the circle

$$x^2 + y^2 = 1. \tag{3.722}$$

One can show that the circle (3.722) is indeed a limit cycle of ODE (3.719). The situation is illustrated in Figure 3.5.

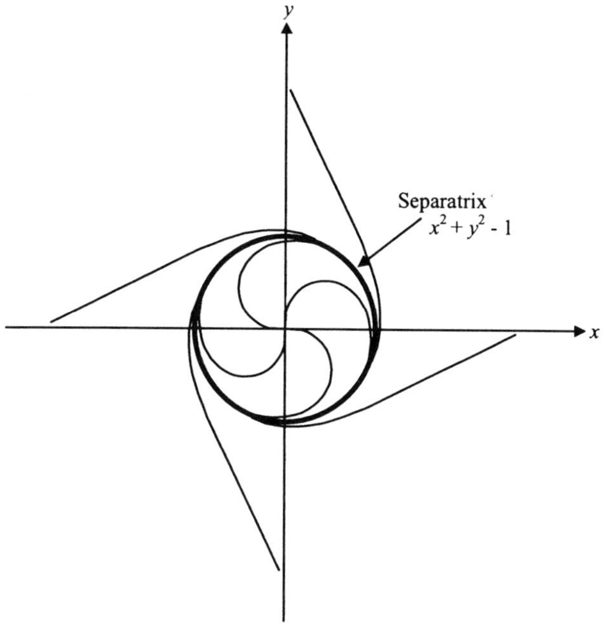

Figure 3.5. Solution curves of ODE (3.719).

As a third example, consider Clairaut's equation

$$y = xy' + \frac{m}{y'}, \tag{3.723}$$

where m is a constant.

Clearly, ODE (3.723) admits the scaling group

$$\mathrm{X} = 2x\frac{\partial}{\partial x} + y\frac{\partial}{\partial y}. \tag{3.724}$$

An envelope solution of ODE (3.723) must satisfy the algebraic equation (3.717) with $\eta / \xi = y / 2x$, i.e.,

$$y = \tfrac{1}{2}y + \frac{2mx}{y}$$

and so

$$y^2 = 4mx. \tag{3.725}$$

Hence, $y^2 = 4mx$ is the only possible envelope solution of ODE (3.723).

The invariance of ODE (3.723) under (3.724) yields its general solution

$$y = cx + \frac{m}{c} \tag{3.726}$$

for an arbitrary constant c. Clearly, the parabola defined by (3.725) is the envelope of the family of straight lines (3.726). The situation is illustrated in Figure 3.6 for $m = 1$.

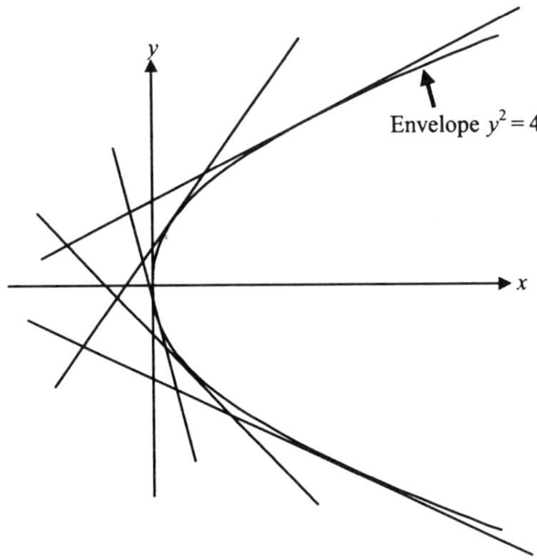

Figure 3.6. Typical solution curves of ODE (3.723) for $m = 1$.

EXERCISES 3.10

1. (a) Suppose that $\lambda = r$ is a double root of the characteristic polynomial equation (3.711).
   (i) Show that $\mathrm{X} = \alpha \frac{\partial}{\partial x} + (\beta y + \gamma e^{rx}) \frac{\partial}{\partial y}$ is admitted by ODE (3.709) for arbitrary constants α, β, γ.
   (ii) Find the corresponding invariant solutions of ODE (3.709).
   (b) What is the situation if $\lambda = r$ is a root of multiplicity k of the characteristic polynomial equation (3.711)?

2. Consider the Euler equation

$$x^2 y'' + Axy' + By = 0, \quad A = \text{const}, \quad B = \text{const}.$$

Find its general solution in terms of invariant solutions. Note that the scalings $\mathrm{X} = \alpha x \frac{\partial}{\partial x} + \beta y \frac{\partial}{\partial y}$ are admitted by the Euler equation for arbitrary constants α, β.

3. Find all invariant solutions of ODE (3.194) resulting from its invariance under the three-parameter Lie group of point transformations (3.195a,b).

4. Find the general solution of ODE (3.719) and sketch several of the solution curves.

5. Use the invariance of ODE (3.723) under (3.724) to derive its general solution (3.726).

6. Find necessary conditions for a first-order ODE $y' = f(x,y)$ so that it admits the rotation group $\mathrm{X} = y\frac{\partial}{\partial x} - x\frac{\partial}{\partial y}$ and has the circle $x^2 + y^2 = 1$ as a limit cycle (separatrix).

7. Consider the ODE

$$y' = A\frac{y}{x}, \quad A = \text{const} \neq 0. \tag{3.727}$$

(a) Find invariant solutions resulting from the invariance of ODE (3.727) under

(i) $\mathrm{X}_1 = x\frac{\partial}{\partial x}$;

(ii) $\mathrm{X}_2 = y\frac{\partial}{\partial y}$; and

(iii) $\mathrm{X}_3 = x\frac{\partial}{\partial x} + y\frac{\partial}{\partial y}$.

(b) Determine the separatrices of ODE (3.727).

(c) Sketch typical solution curves of ODE (3.727).

8. Use the invariance of the ODE $y' = x/y$ under $\mathrm{X} = x\frac{\partial}{\partial x} + y\frac{\partial}{\partial y}$ to find its separatrices.

9. Find the separatrices and sketch typical solution curves for the ODE

$$y' = \frac{y(y-2x)}{x(x-2y)}.$$

10. Show that the ODE

$$y' = -\frac{x^3}{y}$$

has no separatrices, without explicitly solving it.

11. Consider the ODE

$$y + \tfrac{2}{3}(y')^3 - (x + (y')^2) = 0. \tag{3.728}$$

(a) Find a one-parameter Lie group of transformations admitted by ODE (3.728).

(b) Sketch typical solution curves of ODE (3.728).

(c) Find the envelope of the solution curves of ODE (3.728).

12. Do the same as in the Exercise 3.10-11 for the Clairaut ODE

$$y = xy' + (1 + (y')^2)^{1/2}.$$

13. Prove Theorem 3.10-2.

3.11 DISCUSSION

In this chapter, we have shown how to:

(i) use invariance under Lie groups of point transformations to construct solutions to a given ODE;
(ii) use invariance under one-parameter groups of local transformations (point, contact, higher-order) to reduce the order of a given ODE;
(iii) find infinitesimal generators of local transformations admitted by a given ODE through solving determining equations arising from Lie's algorithm;
(iv) find the most general nth-order ODE that admits a given Lie group of transformations;
(v) use first integrals to reduce the order of a given ODE;
(vi) find first integrals through corresponding integrating factors admitted by a given ODE; and
(vii) construct first integrals algebraically through use of symmetries and adjoint-symmetries admitted by a given ODE.

If a given nth-order ODE admits a one-parameter Lie group of point transformations, then its order can be reduced, constructively, to an $(n - 1)$th-order ODE through the use of canonical coordinates or differential invariants associated with the group. Moreover, the solution of the given ODE can be found by quadrature after the reduced ODE is solved.

The invariance of a first-order ODE under a nontrivial one-parameter Lie group of transformations is equivalent to the existence of an integrating factor for the ODE. In general, this equivalence does not hold for higher-order ODEs. However, if a higher-order ODE possesses a variational formulation (Lagrangian), then its invariance under a one-parameter Lie group of local transformations admitted by the variational principle for the ODE (variational symmetry) is equivalent to the existence of an integrating factor of the ODE (the classical Noether's Theorem). An nth-order ODE $(n \geq 1)$ possesses a variational principle if the linearized equation of the ODE is self-adjoint (which is only possible if the ODE is of even order). For a self-adjoint ODE, a one-parameter Lie group of point transformations admitted by its variational principle leads, constructively, to a reduction of order by two [Olver (1986)]. A geometrical reduction procedure for such an ODE in a Hamiltonian setting is discussed in Marsden and Ratiu (1999).

Any first-order ODE admits a nontrivial infinite-parameter Lie group of point transformations. A second-order ODE admits at most an eight-parameter Lie group of point transformations. Moreover, if a second-order ODE does admit an eight-parameter

Lie group of point transformations, then there exists a point transformation mapping the ODE into a linear ODE and, in particular, to the ODE $y'' = 0$. An nth-order ODE $(n \geq 3)$ admits at most an $(n + 4)$-parameter Lie group of point transformations.

Olver (1986) showed that if an nth-order ODE admits an r-parameter solvable Lie group of point transformations $(r \leq n)$, then it can be reduced to an $(n - r)$th-order ODE plus r quadratures. The reduction algorithm presented in Section 3.4 appeared in Bluman (1990b). The use of solvable Lie groups (called *integrable groups* in earlier literature) to reduce the order of a system of first-order ODEs appears to have been first considered by Bianchi (1918, Section 167) [Eisenhart (1933, Section 36)].

One can extend Lie's work on the invariance of ODEs under Lie groups of point transformations (point symmetries) to invariance under more general local transformations characterized by infinitesimal generators depending on derivatives of dependent variables. Such extensions include invariance under Lie groups of contact transformations (contact symmetries) relevant for second- or higher-order ODEs or, more generally, higher-order transformations (higher-order symmetries) relevant for third- and higher-order ODEs. In making such generalizations, it is more convenient to consider the invariance of an ODE from the point of view of directly mapping its solutions into solutions by local transformations that keep the independent variable of the ODE fixed. Here, a symmetry (point, contact, or higher-order) of an ODE is generated by any solution of the linear determining equation arising from linearization about *all* solutions of the given ODE. Contact symmetries for ODEs are considered in Abraham-Shrauner et al. (1995), Stephani (1989), and Hydon (2000).

Geometrically, symmetries of an nth order ODE describe motions on its space of solutions. Such motions are most naturally formulated in the setting of jet spaces [c.f. Section 2.8] whose coordinates consist of the independent variable, and the dependent variable and its derivatives up to at least nth-order. Here, a given nth-order ODE corresponds to a surface of co-dimension one, and its symmetries (one-parameter local transformation groups) in characteristic form represent integral curves of vector fields that are tangent to the surface and involve no motion with respect to the coordinate given by the independent variable, while preserving the derivative relations (contact ideal) among the remaining coordinates. Point symmetries and contact symmetries are distinguished as local transformations that arise from vector fields well-defined on the entire jet space (i.e., irrespective of the ODE and its corresponding surface), whereas higher-order symmetries cannot be so defined except on the infinite-order extension (prolongation) of the jet space to include coordinates given by derivatives of the dependent variable to all orders

Any second-order ODE admits an infinite number of contact symmetries, i.e., an infinite-parameter Lie group of contact transformations. Any nth-order ODE $(n \geq 3)$ admits an infinite number of $(n - 1)$th-order symmetries, i.e., an infinite-parameter group of local transformations of order $n-1$. However, an nth-order ODE $(n \geq 3)$ admits at most a finite number of symmetries of order strictly less than $n-1$, i.e., any admitted group of local transformations of order at most $n-2$ has finite dimension. The complete symmetry group of an ODE naturally has the structure of an abstract infinite-dimensional Lie group [cf. Section 2.8]. Subgroups of admitted point symmetries and contact symmetries correspond to finite-dimensional abstract Lie groups in the cases of ODEs of orders $n \geq 2$ and $n \geq 3$, respectively.

Similarly, any second- or nth-order ODE ($n \geq 2$) admits an infinite number of linearly independent integrating factors of order $n-1$ (i.e., with an essential dependence on up to $(n-1)$th-order derivatives of the dependent variable) but admits at most a finite number with order strictly less than $n-1$. In general, an integrating factor of an ODE has no obvious relation to any underlying invariance or geometrical motion other than in the case of the classical Noether's Theorem where integrating factors are symmetries that leave invariant a variational principle for the ODE.

An integrating factor multiplying an ODE transforms it into an total derivative (exact) form. Consequently, any integrating factor admitted by an nth-order ODE satisfies a system of linear determining equations arising from annihilation by an Euler operator [Olver (1986)] applied to the product of the ODE and the integrating factor. The basic framework for exact nth-order ODEs and integrating factors is given in Kamke (1943) and Kaplan (1958). For a first-order ODE, the classical formulation of integrating factors involves transforming the ODE to an exact differential form $d\psi(x,y) = \Lambda(x,y)(dy - f(x,y)dx) = 0$ in the independent and dependent variables. This generalizes naturally in the context of jet space to higher-order ODEs, since one can write an exact nth-order ODE $d\psi(x,y,y_1,\ldots,y_{n-1})/dx = 0$ as a system of n exact differential forms $d\psi(x,y,y_1,\ldots,y_{n-1}) = \Lambda(x,y,y_1,\ldots,y_{n-1})(dy_{n-1} - f(x,y,y_1,\ldots,y_{n-1})dx) = 0$, and $dy_i = y_{i+1}dx$, $i = 0,1,\ldots,n-2$, $[y_0 \equiv y]$ using the jet space coordinates. Here, the determining equations for an integrating factor $\Lambda(x,y,y_1,\ldots,y_{n-1})$ can be derived from the fact that any exact differential form is closed. This leads to a system of $1+(n(n-1)/2)$ determining equations that is essentially the same as the system obtained from the standard Euler operator framework. Moreover, the line integral formula for first integrals $\psi(x,y,y_1,\ldots,y_{n-1})$ is the same as the Poincaré homotopy formula [Olver (1986)] used for showing that a closed differential form is locally exact.

In contrast, our framework for exact nth-order ODEs, presented in Section 3.6.4, uses a truncation of the standard Euler operator (defined on the infinite-order jet space) to the finite-order jet space naturally associated with a given ODE. Most signifcantly, this leads to a simplification of the integrating factor determining equations into an equivalent smaller system of $1+[n/2]$ determining equations, which has not, to our knowledge, appeared elsewhere in the literature.

Any integrating factor also satisfies the adjoint equation of the linear determining equation for symmetries of a given ODE. In particular, the $1+[n/2]$ determining equations for integrating factors have a natural splitting into this adjoint equation plus $[n/2]$ extra determining equations. Use of the adjoint equation appears in related work by Gordon (1986) and Sarlet et al. (1987, 1990), where solutions of the adjoint equation are named "adjoint symmetries." Keeping in mind that, in general, the solutions of the adjoint equation are not themselves generators of symmetries unless the given ODE is self-adjoint, we call them adjoint-symmetries. The splitting of the integrating factor determining equations leads to the useful alternative characterization of integrating factors as adjoint-symmetries that satisfy extra adjoint invariance conditions, as given in Section 3.7.2. Moreover, from this point of view, the classical equivalence between integrating factors and variational symmetries in the case of an ODE with a variational principle is obviously generalized to an equivalence between integrating factors and those adjoint-symmetries satisfying the adjoint invariance conditions in the case of an ODE without any variational principle. Finally, this provides a more geometrically appealing formulation of the symplified system of $1+[n/2]$ determining equations for integrating factors from Section 3.6.4, where the total derivative operator on jet space is now replaced by a tangential derivative with respect to the surface defined by the ODE.

The local existence theory for nth-order ODEs guarantees that a given ODE admits n functionally independent first integrals (i.e., none is a function combination of the others). Any admitted first integral is equal to a constant for each solution of the ODE and hence its total derivative with respect to the independent variable of the ODE yields an integrating factor multiplying the ODE. The correspondence between a first integral and an integrating factor can be thought of as being similar to the correspondence between a group of local transformations and the infinitesimal of its infinitesimal generator in characteristic form.

Any integrating factor admitted by an ODE leads, constructively, to a single reduction of order of the ODE through the corresponding first integral given by the line integral formula. This reduction of order method is complementary to Lie's symmetry method for reducing the order of a given ODE. Moreover, the integrating factor procedure is computationally no more complex than Lie's algorithm. Most important, the integrating factor approach yields first integrals which are reductions of order in terms of the given variables, unlike the situation for reduction of order under Lie groups of transformation where the reduction of order does not usually hold in terms of the original variables. A full discussion of the integrating factor approach appears in Anco and Bluman (1997a, 1998). Related work also appears in Sheftel (1997, Section 3.5), Cheb-Terrab and Roche (1999), and Hydon (2000).

The nature of the calculation of symmetries, adjoint-symmetries, and integrating factors of an ODE is similar, as in each situation there is a linear determining system to solve. Moreover, for an nth-order ODE, the determining system reduces to an overdetermined system of linear homogeneous PDEs when one calculates symmetries, adjoint-symmetries, or integrating factors of order strictly less than $n-1$. Typically, in these cases, all solutions of the respective determining systems can be obtained explicitly by obvious extensions of the standard calculational algorithm for solving the overdetermined linear system for point symmetries. One can also use effective ansatzes to seek particular solutions by directly splitting the determining equations into an overdetermined system of linear homogeneous PDEs. Such ansatzes include a point-form ansatz, an elimination of variables ansatz, and an invariant-solution ansatz [see Chapter 4] arising naturally from any admitted Lie group of point transformations of a given ODE. Available powerful computer algebra systems, e.g., REDUCE, MATHEMATICA, MAPLE (which was used for the calculations for the ODE examples in Sections 3.5–3.8), are readily adapted for solving the resulting overdetermined systems.

Most important, the cardinality of the class of ODEs admitting symmetries of a given ansatz is, in general, the same as that of the class of ODEs admitting integrating factors of the same ansatz. Consequently, the use of the integrating factor method for reducing the order of ODEs should be viewed as having an a priori utility no less than that of symmetry methods.

Another procedure for reducing the order of an ODE is through direct construction of first integrals by special formulas using either symmetries or adjoint-symmetries. The scaling symmetry formula in Section 3.8.1 is a counterpart for ODEs of a conservation law formula in the case of linear PDEs appearing in works of Olver (1986) and Anco and Bluman (1996, 1997a). The Wronskian formula in Section 3.8.2 is a generalization of a related procedure given in Hydon (2000) for obtaining a first integral of any nth-order ODE that admits at least $n+1$ point symmetries.

In the case of a boundary value problem for an nth-order ODE, invariance of the ODE under an r-parameter solvable Lie group of transformations ($r \leq n$) reduces the given boundary value problem to a boundary value problem for an ($n - r$)th-order ODE. This can be especially useful for obtaining qualitative results about the solution of the boundary value problem. Moreover, finding a Lie group of transformations admitted by the ODE of a boundary value problem may lead to an effective numerical method for solving the ODE [Dresner (1983, 1999)]. In addition, if the boundary conditions are of the right type, then the shooting method may be reduced to a single shooting when used in combination with Lie group invariance (which yields a parameter-dependent family of solutions from a given solution).

If an ODE admits a Lie group of transformations, then one can construct special solutions that are invariant under the admitted transformations. Such solutions are also invariant curves of the group. The construction of invariant solutions for an nth-order ODE can be extended to invariance with respect to an admitted Lie group of local transformations of any order less than n. For a second- or higher-order ODE, invariant solutions, resulting from invariance under point symmetries, can be found without explicitly solving the given ODE. For a first-order ODE, invariant solutions are found by solving related algebraic equations. Moreover, separatrices and singular envelope solutions, if they exist, are invariant solutions for any admitted Lie group of transformations. Consequently, such solutions can be found without determining the general solution of the given first-order ODE. The results presented in Section 3.10 appeared in Bluman (1990c). Wulfman (1979) considered group aspects of separatrices that are limit cycles. The construction of separatrices in the case of scaling invariance is discussed in Dresner (1983, 1999). Discussions of envelope solutions from invariance considerations appear in Page (1896, 1897), Cohen (1911), and Dresner (1999).

Some known results for systems of ODEs are now summarized. Using the ideas developed in Section 3.2, one can show that a system of first-order ODEs always admits an infinite-parameter Lie group of nontrivial point transformations and an infinite-parameter Lie group of trivial point transformations. But there is no constructive procedure for finding the groups [Ovsiannikov (1982, Section 8)]. Gonzalez-Gascon and Gonzalez-Lopez (1983) showed that a system of m nth-order ODEs can admit at most a $[2(m+1)^2]$-parameter Lie group of point transformations if $n = 2$, and at most a $(2m^2+mn+2)$-parameter Lie group of point transformations if $n > 2$. If a given system of m first-order ODEs admits an r-parameter solvable Lie group of point transformations ($r \leq m$), then it can be reduced to a system of $m - r$ first-order ODEs plus r quadratures. The latter two results appear in Olver (1986). The framework for constructing integrating factors for systems of ODEs appears in Anco and Bluman (1998). Interesting examples appear in Senthilvelan and Lakshmanan (1995).

In Chapter 4, we consider the invariance of PDEs under Lie groups of point transformations. In general, unlike the case for an ODE where invariance under a one-parameter Lie group of transformations leads to a reduced ODE whose solution includes *all* solutions of the given ODE, the invariance of a PDE does not lead to a reduced PDE whose solution includes all solutions of the given PDE. However, in the same way as for ODEs, we can define and construct *invariant solutions* for PDEs resulting from invariance under Lie groups of transformations. For ODEs, such special solutions are obtained by solving reduced algebraic equations; for PDEs, such special solutions arise from solving reduced PDEs with fewer independent variables.

4

Partial Differential Equations (PDEs)

4.1 INTRODUCTION

In this chapter, we show how to construct solutions of partial differential equations (PDEs) from invariance under Lie groups of point transformations (point symmetries). We will consider both scalar PDEs and systems of PDEs.

As is the situation for an ordinary differential equation (ODE), we will see that the infinitesimal criterion for the invariance of a PDE leads directly to an algorithm to determine the infinitesimal generators of the Lie group of point transformations admitted by a given PDE. The invariant surfaces of the Lie group of point transformations lead to *invariant solutions* (*similarity solutions*). These solutions are obtained by solving PDEs with fewer independent variables than appear in the given PDE.

We will discuss how one can use invariance under Lie groups of point transformations to solve boundary value problems for PDEs. If a one-parameter Lie group of point transformations admitted by a PDE also leaves invariant both the domain and boundary conditions of a posed boundary value problem, then the solution of the boundary value problem is an invariant solution. Hence, the boundary value problem is reduced constructively to a posed boundary value problem with one less independent variable. The situation is less restrictive in the case of a linear PDE. Here, one need not leave invariant the boundary conditions of a posed boundary value problem. A superposition of invariant solutions, corresponding to an eigenfunction expansion, could yield the solution of the boundary value problem, where the eigenvalue arises from the invariance of a linear homogeneous PDE under scalings of its dependent variables. We will also consider the invariance of boundary value problems under multiparameter Lie groups of point transformations.

4.1.1 INVARIANCE OF A PDE

First we consider a scalar PDE. We represent a kth-order scalar PDE by

$$F(x,u,\partial u,\partial^2 u,\ldots,\partial^k u)=0, \tag{4.1}$$

where $x=(x_1,x_2,\ldots,x_n)$ denotes the coordinates corresponding to its n independent variables, u denotes the coordinate corresponding to its dependent variable, and $\partial^j u$ denotes the coordinates with components $\partial^j u/\partial x_{i_1}\partial x_{i_2}\cdots\partial x_{i_j} = u_{i_1 i_2\cdots i_j}$, $i_j=1,2,\ldots,n,$ for $j=1,2,\ldots,k,$ corresponding to all jth-order partial derivatives of u with respect to x.

In terms of the coordinates $x,u,\partial u,\partial^2 u,\ldots,\partial^k u,$ the PDE (4.1) becomes an algebraic equation that defines a hypersurface in $(x,u,\partial u,\partial^2 u,\ldots,\partial^k u)$ – space. [Here,

$(x,u,\partial u)-$space is of dimension $2n+1$, $(x,u,\partial u,\partial^2 u)-$space is of dimension $\frac{1}{2}(n^2+5n+2)$, etc.] For any solution $u=\Theta(x)$ of PDE (4.1), the equation

$$(x,u,\partial u,\partial^2 u,\ldots,\partial^k u)=(x,\Theta(x),\partial\Theta(x),\partial^2\Theta(x),\ldots,\partial^k\Theta(x))$$

defines a solution surface that lies on the surface $F(x,u,\partial u,\partial^2 u,\ldots,\partial^k u)=0$.

We assume that PDE (4.1) can be written in solved form in terms of some specific component of the ℓth-order partial derivatives of u:

$$F(x,u,\partial u,\partial^2 u,\ldots,\partial^k u)=u_{i_1 i_2\cdots i_\ell}-f(x,u,\partial u,\partial^2 u,\ldots,\partial^k u)=0, \tag{4.2}$$

where $f(x,u,\partial u,\partial^2 u,\ldots,\partial^k u)$ does not depend explicitly on $u_{i_1 i_2\cdots i_\ell}$.

Definition 4.1.1-1. The one-parameter Lie group of point transformations

$$x^* = X(x,u;\varepsilon), \tag{4.3a}$$

$$u^* = U(x,u;\varepsilon), \tag{4.3b}$$

leaves invariant the PDE (4.1), i.e., *is a point symmetry admitted by PDE* (4.1), if and only if its kth extension, defined by (2.115a–d) and (2.116a,b), leaves invariant the surface (4.1).

A solution $u=\Theta(x)$ of PDE (4.1) lies on the surface (4.1) with $u_{i_1 i_2\cdots i_j}=\partial^j\Theta(x)/\partial x_{i_1}\partial x_{i_2}\cdots\partial x_{i_j}$, $i_j=1,2,\ldots,n$, for $j=1,2,\ldots,k$. The invariance of surface (4.1) under the kth-extension of (4.3a,b) means that any solution $u=\Theta(x)$ of PDE (4.1) maps into another solution $u=\Phi(x;\varepsilon)$ of (4.1) under the action of the one-parameter group (4.3a,b) for any ε. Moreover, if a transformation (4.3a,b) maps any solution $u=\Theta(x)$ of PDE (4.1) into another solution $u=\Phi(x;\varepsilon)$ of (4.1), then the surface (4.1) is invariant under (4.3a,b) with $u_{i_1 i_2\cdots i_j}=\partial^j\Phi(x;\varepsilon)/\partial x_{i_1}\partial x_{i_2}\cdots\partial x_{i_j}$, $i_j=1,2,\ldots,n$, for $j=1,2,\ldots,k$. Consequently, *the set of all solutions of PDE* (4.1) *is invariant under the one-parameter Lie group of point transformations* (4.3a,b) *if and only if* (4.1) *admits* (4.3a,b).

The following theorem arises from Definition 4.1.1-1, Theorem 2.6.7-1 on the criterion for an invariant surface in terms of an infinitesimal generator, and Theorem 2.4.4-1 on extended infinitesimals. [For the rest of this section, we assume the summation convention for repeated indices.]

Theorem 4.1.1-1 (Infinitesimal Criterion for the Invariance of a PDE). *Let*

$$\mathrm{X}=\xi_i(x,u)\frac{\partial}{\partial x_i}+\eta(x,u)\frac{\partial}{\partial u} \tag{4.4}$$

be the infinitesimal generator of the Lie group of point transformations (4.3a,b). *Let*

$$\mathrm{X}^{(k)} = \xi_i(x,u)\frac{\partial}{\partial x_i} + \eta(x,u)\frac{\partial}{\partial u} + \eta_i^{(1)}(x,u,\partial u)\frac{\partial}{\partial u_i} + \cdots$$
$$+ \eta_{i_1 i_2 \cdots i_k}^{(k)}(x,u,\partial u,\partial^2 u,\ldots,\partial^k u)\frac{\partial}{\partial u_{i_1 i_2 \cdots i_k}} \tag{4.5}$$

be the kth-extended infinitesimal generator of (4.4), *where* $\eta_i^{(1)}$ *is given by* (2.119a) *and* $\eta_{i_1 i_2 \cdots i_j}^{(j)}$ *by* (2.119b), $i_j = 1,2,\ldots,n$, *for* $j = 1,2,\ldots,k$, *in terms of* $\xi(x,u) = (\xi_1(x,u),\xi_2(x,u),\ldots,\xi_n(x,u))$, $\eta(x,u)$. *Then the one-parameter Lie group of point transformations* (4.3a,b) *is admitted by PDE* (4.1), *i.e., is a point symmetry of PDE* (4.1), *if and only if*

$$\mathrm{X}^{(k)}F(x,u,\partial u,\partial^2 u,\ldots,\partial^k u) = 0 \quad \textit{when } F(x,u,\partial u,\partial^2 u,\ldots,\partial^k u) = 0. \tag{4.6}$$

Proof. Left to Exercise 4.1-3. □

4.1.2 ELEMENTARY EXAMPLES

(1) *Group of Translations*
The second-order PDE

$$u_{xx} = f(u_{xt},u_{tt},u_x,u_t,u,x) \tag{4.7}$$

admits the one-parameter (ε) Lie group of translations

$$x^* = x, \tag{4.8a}$$

$$t^* = t + \varepsilon, \tag{4.8b}$$

$$u^* = u, \tag{4.8c}$$

since under (4.8a–c) we have

$$u^*_{x^*x^*} = u_{xx}, \quad u^*_{x^*t^*} = u_{xt}, \quad u^*_{t^*t^*} = u_{tt}, \quad u^*_{x^*} = u_x, \quad u^*_{t^*} = u_t,$$

so that the surface defined by (4.7) is invariant in $(x,u,\partial u,\partial^2 u)$ – space with $x_1 = x, x_2 = t.$ Then

$$u = \phi(x) \tag{4.9}$$

is invariant under (4.8a–c) and defines a solution (*invariant solution*) of PDE (4.7) provided that $\phi(x)$ solves the second-order ODE

$$\phi''(x) = f(0,0,\phi'(x),0,\phi(x),x).$$

Note that

$$u = \Theta(x,t)$$

defines an *invariant surface* of (4.8a–c) [cf. Theorem 2.6.7-1] if and only if

$$X(u-\Theta(x,t)) = -\frac{\partial\Theta(x,t)}{\partial t} = 0 \quad \text{when} \quad u = \Theta(x,t),$$

where $X = \frac{\partial}{\partial t}$ is the infinitesimal generator of the Lie group of translations (4.8a–c). This leads to the *invariant form* (*similarity form*) (4.9) for an invariant solution resulting from the invariance of PDE (4.7) under (4.8a–c).

Under the action of the Lie group of translations (4.8a–c), a solution $u = \Theta(x,t)$ of PDE (4.7) is mapped into a *one-parameter family of solutions* $u = \Phi(x,t;\varepsilon) = \Theta(x,t+\varepsilon)$ provided that $u = \Theta(x,t)$ is not an invariant solution of PDE (4.7) resulting from its invariance under (4.8a–c), i.e., $\Theta(x,t)$ depends essentially on t.

(2) *Group of Scalings*

The wave equation

$$u_{xx} = u_{tt} \tag{4.10}$$

admits the one-parameter (α) Lie group of scalings

$$x^* = \alpha x, \tag{4.11a}$$

$$t^* = \alpha t, \tag{4.11b}$$

$$u^* = u, \tag{4.11c}$$

since $u^*_{x^*x^*} = \alpha^{-2}u_{xx}$, $u^*_{t^*t^*} = \alpha^{-2}u_{tt}$ and, consequently, $u^*_{x^*x^*} = u^*_{t^*t^*}$ when $u_{xx} = u_{tt}$.

If one chooses *canonical coordinates* $r = x/t$, $s = \log t$, u, so that the Lie group of scalings (4.11a–c) becomes $r^* = r$, $s^* = s + \log\alpha$, $u^* = u$, then PDE (4.10) transforms into the PDE

$$(1-r^2)u_{rr} - u_{ss} + 2ru_{rs} + u^s - 2ru_r = 0. \tag{4.12}$$

Correspondingly,

$$u = \phi(r) = \phi\left(\frac{x}{t}\right) \tag{4.13}$$

defines an invariant solution of PDE (4.12), and hence of the wave equation (4.10), provided that $\phi(r)$ satisfies the ODE

$$(1-r^2)\phi''(r) - 2r\phi'(r) = 0. \tag{4.14}$$

The infinitesimal generator of the group of scalings (4.11a–c) is given by $X = x\frac{\partial}{\partial x} + t\frac{\partial}{\partial t}$. Hence, $u = \Theta(x,t)$ defines an invariant surface of (4.11a–c) if and only if

$$X(u-\Theta(x,t)) = 0 \quad \text{when} \quad u = \Theta(x,t),$$

i.e., if and only if

$$x\Theta_x + t\Theta_t = 0. \tag{4.15}$$

The solution of the corresponding characteristic equations

$$\frac{dx}{x} = \frac{dt}{t} = \frac{d\Theta}{0},$$

leads to the invariant form (4.13). The substitution of (4.13) into the wave equation (4.10) yields the ODE (4.14). We see that it is unnecessary to find canonical coordinates of the Lie group of scalings (4.11a–c) in order to find the resulting invariant solutions.

Moreover, under the action of the Lie group of scalings (4.11a–c), a solution $u = \Theta(x,t)$ of the wave equation (4.10) maps into the one-parameter family of solutions $u = \Phi(x,t;\alpha) = \Theta(\alpha x, \alpha t)$ of (4.10) provided that $u = \Theta(x,t)$ is not an invariant solution of (4.10) resulting from its invariance under (4.11a–c), i.e., $u = \Theta(x,t)$ is not of the form (4.13).

(3) *Superposition of Invariant Solutions for Linear PDEs*

The wave equation (4.10) admits the one-parameter (ε) Lie group of point transformations

$$x^* = x, \tag{4.16a}$$

$$t^* = t + \varepsilon, \tag{4.16b}$$

$$u^* = e^{\varepsilon\lambda}u, \tag{4.16c}$$

for *any* constant $\lambda \in \mathrm{C}$. The infinitesimal generator of (4.16a–c) is given by

$$\mathrm{X} = \frac{\partial}{\partial t} + \lambda u\frac{\partial}{\partial u}.$$

Actually, the Lie group of point transformations (4.16a–c) defines a two-parameter (ε, λ) Lie group of transformations corresponding to the invariance of the wave equation under translations in t and scalings in u. The resulting invariant solutions $u = \Theta(x,t)$ must satisfy the invariant surface condition

$$\mathrm{X}(u - \Theta(x,t)) = \lambda u - \Theta_t = 0 \quad \text{when } u = \Theta(x,t),$$

i.e.,

$$u_t = \lambda u. \tag{4.17}$$

As is the situation for ODEs [cf. Section 3.10], we can find invariant solutions through two procedures:

Method (I) (Invariant Form Method). The solution of the invariant surface condition (4.17) is given by the *invariant form*

$$u = \phi(x)e^{\lambda t}, \tag{4.18}$$

for an arbitrary function $\phi(x)$. Then the substitution of (4.18) into the wave equation (4.10) yields the ODE

$$\phi''(x) = \lambda^2\phi(x).$$

Hence, after solving this simple ODE, we obtain invariant solutions

$$u = \Theta(x,t) = Ce^{\lambda(t\pm x)} \tag{4.19}$$

of PDE (4.10), where C is an arbitrary constant.

Method (II) (Direct Substitution Method). Here we directly substitute the invariant surface condition (4.17) into PDE (4.10) and avoid solving explicitly (4.17). Then $u_{tt} = \lambda u_t = \lambda^2 u$. Hence, $u_{xx} = \lambda^2 u$, so that

$$u = \psi(t)e^{\pm\lambda x} \tag{4.20}$$

for an arbitrary function $\psi(t)$. Then the substitution of (4.20) into the invariant surface condition (4.17) leads to $\psi(t)$ satisfying the ODE $\psi'(t) = \lambda\psi(t)$, and hence to the invariant solutions (4.19).

Since the wave equation (4.10) is a linear homogeneous PDE, it follows that superpositions of invariant solutions

$$\sum_{\lambda} C(\lambda)e^{\lambda(t\pm x)}, \quad \sum_{\lambda}[C_1(\lambda)e^{\lambda(t-x)} + C_2(\lambda)e^{\lambda(t+x)}], \quad \int_{\gamma} C(\lambda)e^{\lambda(t\pm x)}d\lambda, \text{ etc.,}$$

define solutions of (4.10) where $\lambda \in \mathbf{C}$ is an "eigenvalue," and γ defines a path in the complex λ–plane. Special superpositions correspond to Fourier series, the Laplace transform, and the Fourier transform representations of solutions.

As a general remark, we note that the Invariant Form Method is useful if one can find the general solution of the invariant surface condition whereas the Direct Substitution Method must be used if one is unable to solve explicitly the invariant surface condition.

EXERCISES 4.1

1. Find invariant solutions for the wave equation (4.10) resulting from its invariance under the one-parameter (ε) Lie group of translations

$$x^* = x + \varepsilon,$$
$$t^* = t + \alpha\varepsilon,$$

for any fixed constant $\alpha \in \mathbf{R}$. How do these solutions relate to the general solution of PDE (4.10)?

2. (a) Show that the most general second-order PDE that admits the two-parameter $(\varepsilon_1, \varepsilon_2)$ Lie group of translations $x^* = x + \varepsilon_1$, $t^* = t + \varepsilon_2$, is of the form

$$F(u_{xx}, u_{xt}, u_{tt}, u_x, u_t, u) = 0.$$

(b) Find the invariant solutions of this PDE resulting from its invariance under the one-parameter (ε) Lie group of translations $x^* = x + c\varepsilon$, $t^* = t + \varepsilon$, where $c \in \mathbf{R}$ is a fixed constant. Interpret these solutions.

(c) As an example, find invariant solutions of the Korteweg–de Vries (KdV) equation $u_t + uu_x + u_{xxx} = 0$. In particular, show that $u = \Theta(x,t) = U(z) + c$ satisfies the KdV equation where $z = ct - x$, $c = \text{const} > 0$, and $U(z)$ satisfies the ODE $U''' + UU' = 0$. Verify that $U = c(3\operatorname{sech}^2(\frac{1}{2}\sqrt{c}z) - 1)$ is a particular solution of this ODE. Show that the corresponding traveling wave solution $u = \Theta(x,t)$ of the KdV equation, the well-known KdV soliton solution, satisfies $u \to 0$ as $x \to \pm\infty$.

3. Prove Theorem 4.1.1-1.

4.2 INVARIANCE FOR SCALAR PDEs

4.2.1 INVARIANT SOLUTIONS

Consider a kth-order PDE (4.1) $(k \geq 2)$ that admits a one-parameter Lie group of point transformations with the infinitesimal generator (4.4). We assume that $\xi(x,u) \not\equiv 0$.

Definition 4.2.1-1. $u = \Theta(x)$ is an *invariant solution* of PDE (4.1) resulting from its admitted point symmetry with the infinitesimal generator (4.4) if and only if:

(i) $u = \Theta(x)$ is an invariant surface of (4.4); and

(ii) $u = \Theta(x)$ solves (4.1).

It follows that $u = \Theta(x)$ is an invariant solution of PDE (4.1) resulting from its invariance under the point symmetry (4.4) if and only if $u = \Theta(x)$ satisfies both:

(i) $\mathrm{X}(u - \Theta(x)) = 0$ when $u = \Theta(x)$, i.e.,

$$\xi_i(x, \Theta(x)) \frac{\partial \Theta(x)}{\partial x_i} = \eta(x, \Theta(x)); \tag{4.21a}$$

and

(ii) $F(x, u, \partial u, \partial^2 u, \ldots, \partial^k u) = 0$ when $u = \Theta(x)$, i.e.,

$$F(x, \Theta(x), \partial\Theta(x), \partial^2\Theta(x), \ldots, \partial^k\Theta(x)) = 0. \tag{4.21b}$$

Equation (4.21a) is called the *invariant surface condition* for the invariant solutions resulting from invariance under the point symmetry (4.4). Invariant solutions for PDEs were first considered by Lie (1881). They can be determined by two procedures.

(I) Invariant Form Method. Here, we first solve the invariant surface condition, i.e., the first-order PDE (4.21a), by solving the corresponding characteristic equations for $u = \Theta(x)$:

$$\frac{dx_1}{\xi_1(x,u)} = \frac{dx_2}{\xi_2(x,u)} = \cdots = \frac{dx_n}{\xi_n(x,u)} = \frac{du}{\eta(x,u)}. \tag{4.22}$$

If $y_1(x,u), y_2(x,u), \ldots, y_{n-1}(x,u), v(x,u)$ are n functionally independent constants arising from solving the system of n first order ODEs (4.22) with $\partial v / \partial u \not\equiv 0$, then the general solution $u = \Theta(x)$ of PDE (4.21a) is given, implicitly, by the invariant form

$$v(x,u) = \Phi(y_1(x,u), y_2(x,u), \ldots, y_{n-1}(x,u)), \tag{4.23}$$

where Φ is an arbitrary differentiable function of $y_1(x,u), y_2(x,u), \ldots, y_{n-1}(x,u)$. Note that $y_1(x,u), y_2(x,u), \ldots, y_{n-1}(x,u), v(x,u)$ are n functionally independent group invariants of the point symmetry (4.4) and thus are n canonical coordinates for the Lie group of point transformations (4.3a,b). Let $y_n(x,u)$ be the $(n+1)$th canonical coordinate satisfying

$$\mathrm{X}y_n = 1.$$

If PDE (4.1) is transformed to another kth-order PDE in terms of independent variables $y_1, y_2, \ldots, y_n$ and dependent variable v, then the transformed PDE would admit the one-parameter Lie group of translations

$$y^*_i = y_i, \quad i = 1,2,\ldots,n-1, \tag{4.24a}$$

$$y^*_n = y_n + \varepsilon, \tag{4.24b}$$

$$v^* = v. \tag{4.24c}$$

Thus, the variable y_n would not appear explicitly in the transformed PDE, and hence, the transformed PDE would have solutions of the form (4.23). Consequently, the PDE (4.1) has invariant solutions given implicitly by the invariant form (4.23). Such solutions are found by solving a reduced differential equation with $n-1$ independent variables $y_1, y_2, \ldots, y_{n-1}$ and a dependent variable v. The variables $y_1, y_2, \ldots, y_{n-1}$ are commonly called *similarity variables*. The reduced differential equation is found by substituting the invariant form (4.23) into PDE (4.1). We assume that this substitution does not lead to a singular differential equation for v. Note that if $\partial \xi / \partial u \equiv 0$, as is usually the case, then $y_i = y_i(x), i = 1, 2, \ldots, n-1$; if $n = 2$, then the reduced differential equation is an ODE and we denote the similarity variable by $\zeta = y_1$.

(II) Direct Substitution Method. This procedure is especially useful, in fact, it is necessary to use it, if one cannot solve explicitly the invariant surface condition (4.21a), i.e., the characteristic equations (4.22). We can assume that $\xi_n(x,u) \not\equiv 0$. [If $\xi_i(x,u) \equiv 0, i = 1,2,\ldots,n,$ then the solutions $u = \Theta(x)$ of the algebraic equation $\eta(x,u) = 0$ define the invariant surfaces satisfying (4.21a). Any such $u = \Theta(x)$ is an invariant solution of (4.1) if and only if it satisfies the given PDE (4.1).] Hence, the first-order PDE (4.21a) can be written as

$$\frac{\partial u}{\partial x_n} = \frac{\eta(x,u)}{\xi_n(x,u)} - \sum_{i=1}^{n-1} \frac{\xi_i(x,u)}{\xi_n(x,u)} \frac{\partial u}{\partial x_i}. \tag{4.25}$$

From (4.25) and its differential consequences, it follows that any term involving derivatives of u with respect to x_n can be expressed in terms of x, u, and derivatives of u with respect to the variables $x_1, x_2, \ldots, x_{n-1}$. Hence, after directly substituting (4.25) and its differential consequences into the given PDE (4.1), for all terms in (4.1) that involve derivatives of u with respect to x_n, we obtain a reduced differential equation of order at most k involving the dependent variable u, the $n - 1$ independent variables $x_1, x_2, \ldots, x_{n-1}$, and the parameter x_n. Any solution of this reduced differential equation defines an invariant solution of PDE (4.1), resulting from its invariance under the Lie group of point transformations with infinitesimal generator (4.4), provided that the invariant surface condition (4.25) or, equivalently, the given PDE (4.1) itself, is also satisfied. If $n = 2$, the reduced differential equation is an ODE. The constants appearing in the general solution of this ODE are arbitrary functions of the parameter x_n. These arbitrary functions are then determined by substituting this general solution into either (4.25) or the given PDE (4.1). Note that the Direct Substitution Method is computationally better than the Invariant Form Method since it avoids integration of the characteristic ODEs.

In Section 4.4.1, we extend the Invariant Form Method and the Direct Substitution Method to obtain invariant solutions from admitted multiparameter groups of point symmetries.

4.2.2 DETERMINING EQUATIONS FOR SYMMETRIES OF A *k*th-ORDER PDE

Consider a *k*th-order PDE $(k \geq 2, \ell \leq k)$

$$u_{i_1 i_2 \cdots i_\ell} = f(x,u,\partial u,\partial^2 u,\ldots,\partial^k u), \tag{4.26}$$

where $f(x,u,\partial u,\partial^2 u,\ldots,\partial^k u)$ does not depend explicitly on $u_{i_1 i_2 \cdots i_\ell}$. From Theorem 4.1.1-1, we see that PDE (4.26) admits the one-parameter Lie group of point transformations with the infinitesimal generator

$$\mathrm{X} = \xi_i(x,u)\frac{\partial}{\partial x_i} + \eta(x,u)\frac{\partial}{\partial u}, \tag{4.27}$$

and with its kth extension given by (4.5), if and only if $\xi(x,u)$ and $\eta(x,u)$ satisfy the symmetry determining equation

$$\eta^{(\ell)}_{i_1i_2\cdots i_\ell} = \xi_j \frac{\partial f}{\partial x_j} + \eta \frac{\partial f}{\partial u} + \eta^{(1)}_j \frac{\partial f}{\partial u_j} + \cdots + \eta^{(k)}_{j_1j_2\cdots j_k} \frac{\partial f}{\partial u_{j_1j_2\cdots j_k}} \quad \text{when } u \text{ satisfies (4.26).} \tag{4.28}$$

It is easy to show that:

(i) $\eta^{(p)}_{j_1j_2\cdots j_p}$ is linear in the components of the coordinates $\partial^p u$ if $p \geq 2$; and

(ii) $\eta^{(p)}_{j_1j_2\cdots j_p}$ is a polynomial in the components of the coordinates $\partial u, \partial^2 u, \ldots, \partial^p u$, with coefficients that are linear homogeneous in $\xi(x,u), \eta(x,u)$ and their derivatives with respect to x and u to order p.

From (i) and (ii), it follows that if $f(x,u,\partial u,\partial^2 u,\ldots,\partial^k u)$ is a polynomial in the components of $\partial u,\partial^2 u,\ldots,\partial^k u$, then the symmetry determining equation (4.28) is a polynomial equation in the components of $\partial u,\partial^2 u,\ldots,\partial^k u$ with coefficients that are linear homogeneous in $\xi(x,u),\eta(x,u)$, and their derivatives to order k. Observe that at any point x, we can assign arbitrary values to each component of $u,\partial u,\partial^2 u,\ldots,\partial^k u$ provided that PDE (4.26) is satisfied. In particular, after $u_{i_1i_2\cdots i_\ell}$ is replaced by $f(x,u,\partial u,\partial^2 u,\ldots,\partial^k u)$ in the symmetry determining equation (4.28), the resulting expression must hold for arbitrary values of the remaining components of coordinates $x,u,\partial u,\partial^2 u,\ldots,\partial^k u$. Moreover, the resulting expression is a polynomial equation in the remaining components of $\partial u,\partial^2 u,\ldots,\partial^k u$. Consequently, the coefficients of this polynomial equation must vanish separately. This yields a system of *linear homogeneous* PDEs for the $n + 1$ functions $\xi(x,u),\eta(x,u)$. This system of linear PDEs is called the set of *determining equations* for the infinitesimal generators (4.27) admitted by PDE (4.26). The set of determining equations is an *overdetermined* system of PDEs for $\xi(x,u)$ and $\eta(x,u)$ since, in general, there are more than $n+1$ determining equations.

When PDE (4.26) is not a polynomial equation in the components of $\partial u,\partial^2 u,\ldots,\partial^k u$, one can still split the symmetry determining equation (4.28) into a system of linear homogeneous PDEs for $\xi(x,u)$ and $\eta(x,u)$ by using the independence of the components of $\partial u,\partial^2 u,\ldots,\partial^k u$ after substitution for the component $u_{i_1i_2\cdots i_\ell}$. Typically, the resulting set of determining equations will be an overdetermined system for $\xi(x,u)$, $\eta(x,u)$.

When the set of determining equations is overdetermined, it often happens that its only solution is the trivial solution $\xi(x,u) = \eta(x,u) = 0$. In this case, the given PDE (4.26) does not admit point symmetries (although (4.26) could still admit contact symmetries, higher-order symmetries, or nonlocal symmetries, that result from considering a more general infinitesimal generator than (4.27)).

When the general solution of the set of determining equations is nontrivial, two cases arise: If the general solution contains at most a finite number r of essential arbitrary constants, then it yields an r-parameter Lie group of point transformations admitted by

PDE (4.26); if the general solution cannot be expressed in terms of a finite number of essential constants (e.g., when it contains an infinite number of essential constants or involves arbitrary functions of components of x, u), then it yields an infinite-parameter Lie group of point transformations admitted by PDE (4.26).

One can easily verify that any linear nonhomogeneous PDE, defined by a linear operator L,

$$\mathrm{L}u = g(x), \tag{4.29}$$

admits a trivial infinite-parameter Lie group of point transformations

$$x^* = x, \tag{4.30a}$$

$$u^* = u + \varepsilon\omega(x), \tag{4.30b}$$

where $\omega(x)$ is any solution of the associated linear homogeneous PDE

$$\mathrm{L}\omega = 0. \tag{4.31}$$

[The group (4.30a,b) is important when considering the problem of mapping nonlinear PDEs to linear PDEs [Kumei and Bluman (1982); Bluman and Kumei (1990a)].] To within this trivial infinite-parameter Lie group of point transformations, the Lie group of point transformations admitted by a linear PDE typically has at most a finite number of parameters.

We now state some useful results on the forms of admitted point symmetries for wide classes of scalar PDEs (4.1) that arise in applications. For a given PDE (4.1), these results significantly simplify the many calculations involved in setting up and solving the set of determining equations for the admitted infinitesimals $\xi(x,u)$, $\eta(x,u)$. Suppose PDE (4.1) is such that $F(x,u,\partial u,\partial^2 u,\ldots,\partial^k u)$ is linear in the components of $\partial^k u$ and, in addition, suppose the coefficients of the components of $\partial^k u$ depend at most on x and u. Then PDE (4.1) is of the form

$$A_{i_1 i_2 \cdots i_k}(x,u)u_{i_1 i_2 \cdots i_k} = g(x,u,\partial u,\ldots,\partial^{k-1}u), \tag{4.32}$$

with coefficients $A_{i_1 i_2 \cdots i_k}(x,u)$ that are symmetric with respect to their indices. The following theorems hold:

Theorem 4.2.2-1. *Suppose a kth-order PDE* (4.26) *is of the form*

$$B_{i_1 i_2 \cdots i_k}(x)u_{i_1 i_2 \cdots i_k} = g(x,u,\partial u,\ldots,\partial^{k-1}u), \tag{4.33}$$

$k \geq 2$, *and admits a Lie group of point transformations with the infinitesimal generator* (4.27). *If there* does not *exist a change of independent variables* x *under which PDE* (4.33) *is equivalent to a PDE*

$$\frac{\partial^k u}{\partial x_1^k} = G(x,u,\partial u,\ldots,\partial^{k-1}u) \tag{4.34}$$

for some function $G(x,u,\partial u,\ldots,\partial^{k-1}u)$, *then*

$$\frac{\partial \xi_i}{\partial u} = 0, \quad i = 1,2,\ldots,n.$$

Theorem 4.2.2-2. *Suppose a PDE of the form* (4.34) *is of order* $k \geq 2$ *and admits a Lie group of point transformations with the infinitesimal generator* (4.27). *Then*

$$\frac{\partial \xi_i}{\partial u} = 0, \quad i = 2,\ldots,n.$$

Theorem 4.2.2-3. *Suppose a PDE* (4.26), *of order* $k \geq 3$, *is of the form*

$$A_{i_1 i_2 \cdots i_k}(x,u) u_{i_1 i_2 \cdots i_k} = B_{j_1 j_2 \cdots j_{k-1}}(x,u,\partial u) u_{j_1 j_2 \cdots j_{k-1}} + h(x,u,\partial u,\ldots,\partial^{k-2} u), \tag{4.35}$$

and admits a Lie group of point transformations with the infinitesimal generator (4.27). *Then*

$$\frac{\partial \xi_i}{\partial u} = 0, \quad i = 1,2,\ldots,n.$$

Theorem 4.2.2-4. *Suppose a PDE* (4.26), *of order* $k \geq 3$, *is of the form*

$$A_{i_1 i_2 \cdots i_k}(x,u) u_{i_1 i_2 \cdots i_k} = C_{j_1 j_2 \cdots j_{k-1}}(x,u) u_{j_1 j_2 \cdots j_{k-1}} + h(x,u,\partial u,\ldots,\partial^{k-2} u), \tag{4.36}$$

and admits a Lie group of point transformations with the infinitesimal generator (4.27). *Then*

$$\frac{\partial \xi_i}{\partial u} = 0, \quad i = 1,2,\ldots,n,$$

and

$$\frac{\partial^2 \eta}{\partial u^2} = 0.$$

Theorem 4.2.2-5. *Suppose a second-order PDE* (4.26) *is of the form*

$$A_{ij}(x,u) u_{ij} = C_k(x,u) u_k + h(x,u),$$

and admits a Lie group of point transformations with the infinitesimal generator (4.27) *such that*

$$\frac{\partial \xi_i}{\partial u} = 0, \quad i = 1,2,\ldots,n.$$

Then

$$\frac{\partial^2 \eta}{\partial u^2} = 0.$$

Theorem 4.2.2-6. *Suppose a PDE* (4.26), *of order* $k \geq 2$, *is a* linear *PDE that admits a Lie group of point transformations with the infinitesimal generator* (4.27). *Then*

$$\frac{\partial \xi_i}{\partial u} = 0, \quad i = 1,2,\ldots,n,$$

$$\frac{\partial^2 \eta}{\partial u^2} = 0.$$

Theorems 4.2.2-1 to 4.2.2-5 are proved in Bluman (1990a). Theorem 4.2.2-6 is proved in Ovsiannikov (1962, Chapter 6; 1982, Section 27) for $k = 2$, and in Bluman (1990a) for $k > 2$. Further classification results for special cases of PDEs of the form (4.32) appear in Heredero and Olver (1996).

For $n = 2$, introduce the notations

$$x_1 = x, \quad x_2 = t, \quad \xi_1(x_1, x_2) = \xi(x,t), \quad \xi_2(x_1,x_2) = \tau(x,t),$$

and

$$u_1 = u_x = \frac{\partial u}{\partial x}, \quad u_2 = u_t = \frac{\partial u}{\partial t}, \quad \eta_1^{(1)} = \eta_x^{(1)}, \quad \eta_2^{(1)} = \eta_t^{(1)}, \quad \text{etc.}$$

If $\partial \xi / \partial u = 0$, $\partial \tau / \partial u = 0$, $\partial^2 \eta / \partial u^2 = 0$, then an admitted infinitesimal generator for a point symmetry is of the form

$$\mathrm{X} = \xi(x,t)\frac{\partial}{\partial x} + \tau(x,t)\frac{\partial}{\partial t} + [f(x,t)u + g(x,t)]\frac{\partial}{\partial u}. \tag{4.37}$$

It follows that for an infinitesimal generator of the form (4.37), we have [cf. (2.123)–(2.137)]

$$\eta = fu + g, \tag{4.38}$$

$$\eta_x^{(1)} = \frac{\partial g}{\partial x} + \frac{\partial f}{\partial x}u + \left[f - \frac{\partial \xi}{\partial x}\right]u_x - \frac{\partial \tau}{\partial x}u_t, \tag{4.39}$$

$$\eta_t^{(1)} = \frac{\partial g}{\partial t} + \frac{\partial f}{\partial t}u + \left[f - \frac{\partial \xi}{\partial t}\right]u_x - \frac{\partial \tau}{\partial t}u_t, \tag{4.40}$$

$$\eta_{xx}^{(2)} = \frac{\partial^2 g}{\partial x^2} + \frac{\partial^2 f}{\partial x^2}u + \left[2\frac{\partial f}{\partial x} - \frac{\partial^2 \xi}{\partial x^2}\right]u_x - \frac{\partial^2 \tau}{\partial x^2}u_t + \left[f - 2\frac{\partial \xi}{\partial x}\right]u_{xx} - 2\frac{\partial \tau}{\partial x}u_{xt}, \tag{4.41}$$

$$\eta_{xt}^{(2)} = \frac{\partial^2 g}{\partial x \partial t} + \frac{\partial^2 f}{\partial x \partial t}u + \left[\frac{\partial f}{\partial x} - \frac{\partial^2 \xi}{\partial x \partial t}\right]u_x + \left[\frac{\partial f}{\partial t} - \frac{\partial^2 \tau}{\partial x \partial t}\right]u_t - \frac{\partial \xi}{\partial t}u_{xx}$$
$$+\left[f - \frac{\partial \xi}{\partial x} - \frac{\partial \tau}{\partial t}\right]u_{xt} - \frac{\partial \tau}{\partial x}u_{tt}, \tag{4.42}$$

$$\eta_{tt}^{(2)}=\frac{\partial^2 g}{\partial t^2}+\frac{\partial^2 f}{\partial t^2}u-\frac{\partial^2 \xi}{\partial t^2}u_x+\left[2\frac{\partial f}{\partial t}-\frac{\partial^2 \tau}{\partial t^2}\right]u_t-2\frac{\partial \xi}{\partial t}u_{xt}+\left[f-2\frac{\partial \tau}{\partial t}\right]u_{tt}. \quad (4.43)$$

4.2.3 EXAMPLES

(1) *Heat Equation*
Consider the heat equation

$$u_{xx}=u_t. \quad (4.44)$$

From Theorem 4.2.2-6, it immediately follows that the infinitesimal generator of a point symmetry

$$\mathrm{X}=\xi(x,t,u)\frac{\partial}{\partial x}+\tau(x,t,u)\frac{\partial}{\partial t}+\eta(x,t,u)\frac{\partial}{\partial u} \quad (4.45)$$

admitted by PDE (4.44) must be of the form (4.37). We now find all infinitesimal generators of point symmetries (4.37) admitted by the heat equation (4.44). For PDE (4.44), the symmetry determining equation (4.28) becomes

$$\eta_{xx}^{(2)}=\eta_t^{(1)} \quad \text{when } u_{xx}=u_t. \quad (4.46)$$

Substituting (4.40) and (4.41) into (4.46), and then eliminating u_{xx} through (4.44), we obtain

$$\left[\frac{\partial^2 g}{\partial x^2}-\frac{\partial g}{\partial t}\right]+\left[\frac{\partial^2 f}{\partial x^2}-\frac{\partial f}{\partial t}\right]u+\left[2\frac{\partial f}{\partial x}-\frac{\partial^2 \xi}{\partial x^2}+\frac{\partial \xi}{\partial t}\right]u_x$$
$$+\left[\frac{\partial \tau}{\partial t}-\frac{\partial^2 \tau}{\partial x^2}-2\frac{\partial \xi}{\partial x}\right]u_t-2\frac{\partial \tau}{\partial x}u_{xt}=0. \quad (4.47)$$

The symmetry determining equation (4.47) must hold for all values of x,t,u,u_x,u_t,u_{xt}. Hence, from setting to zero the coefficients of u_{xt},u_t,u_x,u and the first bracketed term of (4.47) we obtain the following set of five *determining equations* for $\xi(x,t),\tau(x,t),f(x,t),g(x,t)$:

$$\frac{\partial \tau}{\partial x}=0, \quad (4.48a)$$

$$\frac{\partial \tau}{\partial t}-\frac{\partial^2 \tau}{\partial x^2}-2\frac{\partial \xi}{\partial x}=0, \quad (4.48b)$$

$$2\frac{\partial f}{\partial x}-\frac{\partial^2 \xi}{\partial x^2}+\frac{\partial \xi}{\partial t}=0, \quad (4.48c)$$

$$\frac{\partial^2 f}{\partial x^2}-\frac{\partial f}{\partial t}=0, \quad (4.48d)$$

$$\frac{\partial^2 g}{\partial x^2} - \frac{\partial g}{\partial t} = 0. \tag{4.48e}$$

The solution of PDE (4.48e) corresponds to the trivial infinite-parameter Lie group of point symmetries (4.30a,b) with $\omega(x) = g(x,t)$. Nontrivial point symmetries arise from solving the system of linear PDEs (4.48a–d). One can show that the solution of the determining equations (4.48a–d) is given by

$$\xi(x,t) = \kappa + \beta x + \gamma xt + \delta t, \tag{4.49a}$$

$$\tau(x,t) = \tau(t) = \alpha + 2\beta t + \gamma t^2, \tag{4.49b}$$

$$f(x,t) = -\gamma(\tfrac{1}{4}x^2 + \tfrac{1}{2}t) - \tfrac{1}{2}\delta x + \lambda, \tag{4.49c}$$

where $\alpha, \beta, \gamma, \delta, \kappa, \lambda$ are six arbitrary parameters [Bluman and Cole (1969)]. Hence, the point symmetry generators admitted by the heat equation (4.44) are given by

$$\mathrm{X}_1 = \frac{\partial}{\partial x}, \quad \mathrm{X}_2 = \frac{\partial}{\partial t}, \quad \mathrm{X}_3 = x\frac{\partial}{\partial x} + 2t\frac{\partial}{\partial t}, \quad \mathrm{X}_4 = xt\frac{\partial}{\partial x} + t^2\frac{\partial}{\partial t} - (\tfrac{1}{4}x^2 + \tfrac{1}{2}t)u\frac{\partial}{\partial u},$$
$$\mathrm{X}_5 = t\frac{\partial}{\partial x} - \tfrac{1}{2}xu\frac{\partial}{\partial u}, \quad \mathrm{X}_6 = u\frac{\partial}{\partial u}. \tag{4.50}$$

The infinitesimal generators (4.50) correspond to a six-parameter Lie group of nontrivial point transformations acting on (x,t,u) – space.

The commutator table for the Lie algebra arising from the infinitesimal generators (4.50) is given by

	X_1	X_2	X_3	X_4	X_5	X_6
X_1	0	0	X_1	X_5	$-\tfrac{1}{2}\mathrm{X}_6$	0
X_2	0	0	$2\mathrm{X}_2$	$\mathrm{X}_3 - \tfrac{1}{2}\mathrm{X}_6$	X_1	0
X_3	$-\mathrm{X}_1$	$-2\mathrm{X}_2$	0	$2\mathrm{X}_4$	X_5	0
X_4	$-\mathrm{X}_5$	$-\mathrm{X}_3 + \tfrac{1}{2}\mathrm{X}_6$	$-2\mathrm{X}_4$	0	0	0
X_5	$\tfrac{1}{2}\mathrm{X}_6$	$-\mathrm{X}_1$	$-\mathrm{X}_5$	0	0	0
X_6	0	0	0	0	0	0

From the form of the infinitesimals (4.49a,b), we see that the infinitesimal generators (4.50) induce a five-parameter $(\alpha, \beta, \gamma, \delta, \kappa)$ Lie group of point transformations acting on (x,t)–space

$$\mathrm{Y}_1 = \frac{\partial}{\partial x}, \quad \mathrm{Y}_2 = \frac{\partial}{\partial t}, \quad \mathrm{Y}_3 = x\frac{\partial}{\partial x} + 2t\frac{\partial}{\partial t}, \quad \mathrm{Y}_4 = xt\frac{\partial}{\partial x} + t^2\frac{\partial}{\partial t}, \quad \mathrm{Y}_5 = t\frac{\partial}{\partial x}. \tag{4.51}$$

This five-parameter Lie group is a subgroup of the eight-parameter Lie group of projective transformations in $\mathbf{R}^2$ defined by (2.168a,b) with infinitesimal generators given by (2.169).

Consider the infinitesimal generator X_4 (parameter γ) in (4.50). The corresponding one-parameter Lie group of point transformations is obtained by solving the initial value problem for the first order system of ODEs,

$$\frac{dx^*}{d\varepsilon} = x^* t^*,$$

$$\frac{dt^*}{d\varepsilon} = (t^*)^2,$$

$$\frac{du^*}{d\varepsilon} = -[\tfrac{1}{4}(x^*)^2 + \tfrac{1}{2}t^*]u^*,$$

with $u^* = u,\ x^* = x,\ t^* = t$ at $\varepsilon = 0$. This yields

$$x^* = X(x,t,u;\varepsilon) = \frac{x}{1-\varepsilon t},$$

$$t^* = T(x,t,u;\varepsilon) = \frac{t}{1-\varepsilon t},$$

$$u^* = U(x,t,u;\varepsilon) = \sqrt{1-\varepsilon t}\exp\left[-\frac{\varepsilon x^2}{4(1-\varepsilon t)}\right]u.$$

Now we find the invariant solutions $u = \Theta(x,t)$ of the heat equation (4.44) resulting from its invariance under X_4 by both methods outlined in Section 4.2.1.

(I) Invariant Form Method. Here, the invariant surface condition (4.21a) becomes

$$xtu_x + t^2u_t = -(\tfrac{1}{4}x^2 + \tfrac{1}{2}t)u. \tag{4.52}$$

The corresponding characteristic equations are given by

$$\frac{dx}{xt} = \frac{dt}{t^2} = \frac{du}{-(\frac{1}{4}x^2 + \frac{1}{2}t)u}.$$

This solution of the characteristic equations yields two invariants of X_4:

$$\zeta = \frac{x}{t}, \quad v = \sqrt{t}\, e^{x^2/4t}u.$$

Thus, the solution of the invariant surface condition (4.52) is given by the invariant form

$$\sqrt{t}e^{x^2/4t}u = \phi\left(\frac{x}{t}\right)$$

or, after we solve for u,

$$u = \Theta(x,t) = \frac{1}{\sqrt{t}}e^{-x^2/4t}\phi(\zeta) \tag{4.53}$$

in terms of the similarity variable $\zeta = x/t$. Substitution of (4.53) into the heat equation (4.44) leads to $\phi(\zeta)$ satisfying the reduced ODE

$$\phi''(\zeta) = 0.$$

Hence, the invariant solutions of PDE (4.44), resulting from its invariance under X_4, are given by

$$u = \Theta(x,t) = \frac{1}{\sqrt{t}}\left[C_1 + C_2\frac{x}{t}\right]e^{-x^2/4t}, \tag{4.54}$$

where C_1 and C_2 are arbitrary constants.

(II) Direct Substitution Method. Here, we first express the invariant surface condition in a solved form for u_t:

$$u_t = -\frac{x}{t}u_x - \left[\frac{x^2}{4t^2} + \frac{1}{2t}\right]u. \tag{4.55}$$

After substituting (4.55) into the heat equation (4.44), we obtain the following ODE with t playing the role of a parameter:

$$u_{xx} + \frac{x}{t}u_x + \left[\frac{x^2}{4t^2} + \frac{1}{2t}\right]u = 0. \tag{4.56}$$

The general solution of the parametric ODE (4.56) is given by

$$u = [A(t) + B(t)x]e^{-x^2/4t}, \tag{4.57}$$

where $A(t)$ and $B(t)$ are arbitrary functions. Substitution of (4.57) into the invariant surface condition (4.55) yields

$$\left[\left(A'(t) + \frac{1}{2t}A(t)\right) + \left(B'(t) + \frac{3}{2t}B(t)\right)x\right]e^{-x^2/4t} = 0.$$

Hence,

$$A'(t) + \frac{1}{2t}A(t) = 0,$$

$$B'(t) + \frac{3}{2t}B(t) = 0,$$

which in turn yields the invariant solutions (4.54).

We now find the one-parameter (ε) family of solutions $u = \Phi(x,t;\varepsilon)$, resulting from the invariance of the heat equation (4.44) under the point symmetry X_4, obtained from any solution $u = \Theta(x,t)$ that is not of the form (4.54). Let

$$\hat{x} = X(x,t,u;\varepsilon) = \frac{x}{1-\varepsilon t},$$

$$\hat{t} = T(x,t,u;\varepsilon) = \frac{t}{1-\varepsilon t},$$

$$\hat{u} = \Theta(\hat{x},\hat{t}).$$

Then

$$u = \Phi(x,t;\varepsilon) = U(\hat{x},\hat{t},\hat{u};-\varepsilon) = \frac{1}{\sqrt{1-\varepsilon t}}\exp\left[\frac{\varepsilon x^2}{4(1-\varepsilon t)}\right]\Theta\left(\frac{x}{1-\varepsilon t},\frac{t}{1-\varepsilon t}\right).$$

Lie (1881) found the admitted group of point transformations of the heat equation (4.44). Bluman (1967, Chapter II) [see also Bluman and Cole (1969, 1974 (Section 2.7))] constructed all corresponding invariant solutions of the heat equation (4.44).

(2) *Nonlinear Heat Conduction Equation*
For a second example, we consider a group classification problem. In particular, we completely classify the admitted point symmetries for the nonlinear heat conduction equation

$$u_t = (K(u)u_x)_x, \tag{4.58}$$

in terms of the conductivity $K(u)$. Since PDE (4.58) is of the form (4.34), from Theorem 4.2.2-2 it immediately follows that for any $K(u)$, an admitted infinitesimal generator for a point symmetry (4.45) must be of the form

$$\mathrm{X} = \xi(x,t,u)\frac{\partial}{\partial x} + \tau(x,t)\frac{\partial}{\partial t} + \eta(x,t,u)\frac{\partial}{\partial u}.$$

For PDE (4.58), the symmetry determining equation (4.28) becomes

$$\eta_t^{(1)} = [K'(u)u_{xx} + K''(u)(u_x)^2]\eta + K(u)\eta_{xx}^{(2)} + 2K'(u)u_x\eta_x^{(1)} \tag{4.59a}$$

with

$$u_t = (K(u)u_x)_x = K(u)u_{xx} + K'(u)(u_x)^2, \tag{4.59b}$$

where $\eta_x^{(1)}, \eta_t^{(1)}, \eta_{xx}^{(2)}$ are given by (2.123)–(2.125). After using the given PDE (4.59b) to eliminate u_t from (4.59a), we obtain a polynomial equation in powers of u_{xx}, u_{xt}, and u_x that must hold for arbitrary values of x,t,u,u_x,u_{xx},u_{xt}. From the coefficients of u_{xt} and $u_{xx}u_x$, we find that

$$\frac{\partial \xi}{\partial u} = 0, \quad \frac{\partial \tau}{\partial x} = 0, \tag{4.60a}$$

so that $\xi = \xi(x,t)$, $\tau = \tau(t)$. Using (4.60a) we get, from the coefficients of $u_x, u_{xx}, (u_x)^2$, respectively,

$$\frac{\partial \xi}{\partial t} + 2K'(u)\frac{\partial \eta}{\partial x} + K(u)\left[2\frac{\partial^2 \eta}{\partial x \partial u} - \frac{\partial^2 \xi}{\partial x^2}\right] = 0, \tag{4.60b}$$

$$K(u)\left[\tau'(t) - 2\frac{\partial \xi}{\partial x}\right] + K'(u)\eta = 0, \tag{4.60c}$$

$$K(u)\frac{\partial^2 \eta}{\partial u^2} + K'(u)\left[\tau'(t) - 2\frac{\partial \xi}{\partial x} + \frac{\partial \eta}{\partial u}\right] + K''(u)\eta = 0. \tag{4.60d}$$

The terms not involving u_{xx}, u_{xt}, u_x, lead to the final determining equation

$$K(u)\frac{\partial^2 \eta}{\partial x^2} - \frac{\partial \eta}{\partial t} = 0. \tag{4.60e}$$

Solving (4.60c) for η, and then substituting this result into (4.60e), we find that

$$\xi(x,t) = \rho x^2 + [\tfrac{1}{2}\tau'(t) + \beta]x + \gamma(t), \quad \eta(x,t,u) = \frac{K(u)}{K'(u)}[4\rho x + 2\beta], \tag{4.61}$$

where ρ, β are arbitrary constants and $\gamma(t)$ is an arbitrary function of t. After substituting (4.61) into the determining equation (4.60d), we find that if one of ρ, β is nonzero then it is necessary that the conductivity $K(u)$ satisfy the ODE

$$\left(\frac{K(u)}{K'(u)}\right)'' = 0$$

whose solution is given by $K(u) = \lambda(u+\kappa)^{\nu}$ (with the limiting case $K(u) = \lambda e^{\nu u}$), where λ, κ, ν are arbitrary constants. Finally, after substituting (4.61) into the determining equation (4.60b), we obtain

$$2\gamma'(t) + \tau''(t)x + 4\rho\left[7 - 4\frac{K(u)K''(u)}{[K'(u)]^2}\right]K(u) = 0.$$

Hence, for any $K(u) \neq \text{const}$, it immediately follows that $\gamma'(t) = \tau''(t) = 0$, so that $\gamma = \text{const}$, $\tau(t) = \delta t + \sigma$. Thus, there are five possible parameters $\beta, \rho, \gamma, \delta, \sigma$. The parameters γ, δ, σ exist for any $K(u)$ but the existence of the parameters β, ρ depends on the form of $K(u)$. Three cases arise:

Case I. $K(u)$ arbitrary.
Here, $\rho = \beta = 0$, and the nonlinear heat conduction equation (4.58) admits a three-parameter Lie group of point transformations with its infinitesimal generators given by

$$\mathrm{X}_1 = \frac{\partial}{\partial x}, \quad \mathrm{X}_2 = \frac{\partial}{\partial t}, \quad \mathrm{X}_3 = x\frac{\partial}{\partial x} + 2t\frac{\partial}{\partial t}. \tag{4.62}$$

Case II. $K(u) = \lambda(u+\kappa)^{\nu}$.
Here, $\rho = 0$, and PDE (4.58) admits a four-parameter Lie group of point transformations with its infinitesimal generators given by (4.62) and

$$\mathrm{X}_4 = x\frac{\partial}{\partial x} + \frac{2}{\nu}(u+\kappa)\frac{\partial}{\partial u}. \tag{4.63}$$

In the limiting case, where $K(u) = \lambda e^{\nu u}$, the infinitesimal generator (4.63) becomes

$$\mathrm{X}_4 = x\frac{\partial}{\partial x} + \frac{2}{\nu}\frac{\partial}{\partial u}.$$

Case III. $K(u) = \lambda(u+\kappa)^{-4/3}$.
Here, PDE (4.58) admits a five-parameter Lie group of point transformations with its infinitesimal generators given by (4.62), (4.63) $[\nu = -4/3]$, and

$$\mathrm{X}_5 = x^2\frac{\partial}{\partial x} - 3x(u+\kappa)\frac{\partial}{\partial u}.$$

Ovsiannikov (1959, 1962) derived the above results by considering PDE (4.58) as a system of PDEs $v = K(u)u_x$, $v_x = u_t$. The classification presented here appeared in Bluman (1967) [see also Bluman and Cole (1974); Ovsiannikov (1982)]. The group classification problem for the n-dimensional radially symmetric nonlinear heat conduction equation, $u_t = x^{1-n}(x^{n-1}K(u)u_x)_x$, is presented in Sophocleous (1992).

(3) *Wave Equation for an Inhomogeneous Medium*
As a third example, we consider the complete group classification, with respect to admitted point symmetries, for the wave equation with a variable wave speed $c(x)$:

$$u_{tt} = c^2(x)u_{xx}. \tag{4.64}$$

Since PDE (4.64) is a linear PDE, from Theorem 4.2.2-6 it follows that (4.64) only admits point symmetries with infinitesimal generators of the form (4.37). The symmetry determining equation (4.28) is given by

$$\eta_{tt}^{(2)} = c^2(x)\eta_{xx}^{(2)} + 2c(x)c'(x)u_{xx}\xi$$

with

$$u_{tt} = c^2(x)u_{xx},$$

where $\eta_{xx}^{(2)}$ and $\eta_{tt}^{(2)}$ are given by (4.41) and (4.43), respectively (without loss of generality, we can set $g = 0$). For an arbitrary wave speed $c(x)$, after eliminating u_{tt} through use of the given PDE (4.64), and then using the independence of $u_{xt}, u_{xx}, u_t, u_x, u$, we obtain the following set of five determining equations for $\xi(x,t), \tau(x,t), f(x,t)$:

$$\frac{\partial \xi}{\partial t} - c^2(x)\frac{\partial \tau}{\partial x} = 0, \tag{4.65a}$$

$$c(x)\left[\frac{\partial \tau}{\partial t} - \frac{\partial \xi}{\partial x}\right] + c'(x)\xi = 0, \tag{4.65b}$$

$$\frac{\partial^2 \tau}{\partial t^2} - c^2(x)\frac{\partial^2 \tau}{\partial x^2} - 2\frac{\partial f}{\partial t} = 0, \tag{4.65c}$$

$$\frac{\partial^2 \xi}{\partial t^2} + c^2(x)\left[2\frac{\partial f}{\partial x} - \frac{\partial^2 \xi}{\partial x^2}\right] = 0, \tag{4.65d}$$

$$\frac{\partial^2 f}{\partial t^2} - c^2(x)\frac{\partial^2 f}{\partial x^2} = 0. \tag{4.65e}$$

Solving (4.65a) for $\partial\tau / \partial x$, (4.65b) for $\partial\tau / \partial t$, and then using $\partial^2\tau / \partial x\partial t = \partial^2\tau / \partial t\partial x$, we obtain the equation

$$\frac{\partial^2 \xi}{\partial x^2} - c^{-2}(x)\frac{\partial^2 \xi}{\partial t^2} - \frac{\partial}{\partial x}[H(x)\xi] = 0, \tag{4.66}$$

where

$$H(x) = \frac{c'(x)}{c(x)}.$$

The solution of PDE (4.66) and the determining equation (4.65d) lead to

$$f(x,t) = \tfrac{1}{2}H(x)\xi(x,t) + S(t), \tag{4.67}$$

where $S(t)$ is an arbitrary function of t. After substituting (4.67) into the determining equation (4.65c), solving (4.65a) for $\partial\xi / \partial t$ and (4.65b) for $\partial\xi / \partial x$, and then using $\partial^2\xi / \partial x\partial t = \partial^2\xi / \partial t\partial x$, we find that $S(t) = \text{const} = s$. Hence, we obtain

$$f(x,t) = \tfrac{1}{2}H(x)\xi(x,t) + s. \tag{4.68}$$

After substituting (4.68) into the determining equation (4.65e) and then using the determining equation (4.65d), we obtain

$$H''(x)\xi + 2H'(x)\frac{\partial \xi}{\partial x} + H(x)\frac{\partial}{\partial x}(H(x)\xi) = 0$$

or, equivalently,

$$\frac{\partial}{\partial x}[(2H'(x)+H^2(x))\xi^2]=0. \tag{4.69}$$

Three cases then arise:

Case I. $2H'(x)+H^2(x)=0.$
Here, it is easy to show that

$$c(x)=(Ax+B)^2, \tag{4.70}$$

where A, B are arbitrary constants, with $H(x)=2A/(Ax+B)$. For *any* solution $\xi(x,t)$ of the corresponding PDE (4.66), we find that the functions $\tau(x,t), f(x,t)$, solving the set of determining equations (4.65a–e), are given by

$$\tau(x,t)=\int\left[\frac{\partial\xi}{\partial x}-H(x)\xi\right]dt, \tag{4.71a}$$

$$f(x,t)=\frac{A}{Ax+B}\xi(x,t). \tag{4.71b}$$

The set of functions $\{\xi,\tau,f\}$, determined by any solution $\xi(x,t)$ of PDE (4.66) and by (4.71a,b), corresponds to the invariance of the wave equation (4.64) under a nontrivial infinite-parameter Lie group of point transformations when its wave speed is given by (4.70). One can show that for $A\neq 0$, the wave equation

$$u_{tt}=(Ax+B)^4u_{xx} \tag{4.72}$$

can be transformed to the wave equation

$$w_{XT}=0$$

by the point transformation [Bluman (1983b)]

$$X=[Ax+B]^{-1}+At,$$
$$T=[Ax+B]^{-1}-At,$$
$$w=[Ax+B]^{-1}u.$$

Hence, the general solution of (4.72) is given by

$$u=(Ax+B)[F(X)+G(T)],$$

where $F(X), G(T)$ are arbitrary twice differentiable functions of their respective arguments.

Case II. $2H'(x)+H^2(x)\neq 0, \xi(x,t)\neq 0.$
Here, from (4.69) it follows that $\xi(x,t)$ can be expressed in the separable form

$$\xi(x,t)=\alpha(x)\beta(t), \tag{4.73}$$

where

$$\alpha^2(x) = \rho[2H'(x) + H^2(x)]^{-1} \tag{4.74}$$

for some constant ρ; $\beta(t)$ is to be determined. After substituting (4.68) and (4.73) into the determining equation (4.65d), we find that

$$\frac{\beta''(t)}{\beta(t)} = \frac{c^2(x)[\alpha'(x) - H(x)\alpha(x)]'}{\alpha(x)} = \text{const} = \sigma^2, \tag{4.75}$$

where σ is a real or imaginary constant. We distinguish the subcases $\sigma = 0, \sigma \neq 0$.

Case IIa. $\sigma = 0$.
Here, the wave speed $c(x)$ must satisfy the fourth-order ODE

$$[\alpha'(x) - H(x)\alpha(x)]' = 0. \tag{4.76}$$

Correspondingly, from (4.75),

$$\beta(t) = p + qt,$$

where p, q are arbitrary constants. Substitution of (4.68) and (4.73) into the determining equation (4.65e) leads to the equation

$$[\alpha(x)H(x)]'' = 0. \tag{4.77}$$

The wave speed $c(x)$ must satisfy (4.76), (4.77), and (4.74). This leads to

$$\begin{aligned} \alpha(x) &= Bx^2 + Cx + D, \\ \alpha(x)H(x) &= A + 2Bx, \\ \rho &= 4BD + A^2 - 2AC, \end{aligned}$$

where A, B, C, D are arbitrary constants. Consequently,

$$\frac{c'(x)}{c(x)} = H(x) = \frac{A + 2Bx}{Bx^2 + Cx + D},$$

i.e.,

(i) $\quad c(x) = [Bx^2 + Cx + D]\exp[(A - C)\int [Bx^2 + Cx + D]^{-1}dx.$

The corresponding $\xi(x,t)$ and $f(x,t)$ are obtained, respectively, from (4.73) and (4.68); $\tau(x,t)$ is obtained from the determining equations (4.65a,b). This yields a four-parameter Lie group of point transformations admitted by the corresponding wave equation (4.64). The infinitesimal generators are given by

$$\mathrm{X}_1 = \frac{\partial}{\partial t}, \quad \mathrm{X}_2 = [Bx^2 + Cx + D]\frac{\partial}{\partial x} + [C - A]t\frac{\partial}{\partial t} + [\tfrac{1}{2}A + Bx]u\frac{\partial}{\partial u},$$

$$X_3 = [Bx^2 + Cx + D]t\frac{\partial}{\partial x} + \left[\tfrac{1}{2}(C-A)t^2 + \int \frac{Bx^2 + Cx + D}{c^2(x)}dx\right]\frac{\partial}{\partial t} + [\tfrac{1}{2}A + Bx]tu\frac{\partial}{\partial u},$$

$$X_4 = u\frac{\partial}{\partial u}.$$

The nonzero commutators of the corresponding Lie algebra are given by $[X_1, X_2] = (C-A)X_1, [X_1, X_3] = X_2, [X_2, X_3] = (C-A)X_3$. One can show that the Lie algebra with basis generators X_1, X_2, X_3 is isomorphic to the Lie algebra $SO(2,1)$ when $A \neq C$. When $A = C$, one has $c(x) = Bx^2 + Cx + D$.

It is easy to see that to within arbitrary scalings and translations in x, a wave speed $c(x)$, given by (i), is equivalent to one of the following five canonical forms:

(a) $c(x) = x^A$ $[B = D = 0, C = 1]$;
(b) $c(x) = e^x$ $[B = C = 0, A = D = 1]$;
(c) $c(x) = (1+x^2)e^{A \arctan x}$ $[C = 0, B = D = 1]$;
(d) $c(x) = (1+x)^{1+(A/2)}(1-x)^{1-(A/2)}$ $[C = 0, B = -1, D = 1]$; and
(e) $c(x) = x^2 e^{1/x}$ $[C = D = 0, A = -1, B = 1]$.

We now list special cases of the wave speed (i), together with the infinitesimal generators (constants A, B, C, D are renamed) admitted by the corresponding wave equations (4.64).

(ii) $c(x) = (Ax+B)^C, \quad C \neq 0, 1, 2.$

$$X_1 = \frac{\partial}{\partial t}, \quad X_2 = (Ax+B)\frac{\partial}{\partial x} + A(1-C)t\frac{\partial}{\partial t} + \tfrac{1}{2}ACu\frac{\partial}{\partial u},$$

$$X_3 = (Ax+B)t\frac{\partial}{\partial x} + \tfrac{1}{2}\left[A(1-C)t^2 + \frac{(Ax+B)^{2-2C}}{A(1-C)}\right]\frac{\partial}{\partial t} + \tfrac{1}{2}ACtu\frac{\partial}{\partial u}, \quad X_4 = u\frac{\partial}{\partial u}.$$

The commutator table is the same as for (i) with $(C-A)$ replaced by $A(1-C)$.

(iii) $c(x) = Ax + B.$

$$X_1 = \frac{\partial}{\partial t}, \quad X_2 = (Ax+B)\frac{\partial}{\partial x} + \tfrac{1}{2}Au\frac{\partial}{\partial u},$$

$$X_3 = (Ax+B)t\frac{\partial}{\partial x} + \tfrac{1}{2}\log(Ax+B)\frac{\partial}{\partial t} + \tfrac{1}{2}Atu\frac{\partial}{\partial u}, \quad X_4 = u\frac{\partial}{\partial u}.$$

The nonzero commutators of the corresponding Lie algebra are given by $[X_1, X_3] = X_2$, $[X_2, X_3] = X_1$.

(iv) $c(x) = Ae^{Bx}.$

$$X_1 = \frac{\partial}{\partial t}, \quad X_2 = \frac{\partial}{\partial x} - Bt\frac{\partial}{\partial t} + \tfrac{1}{2}Bu\frac{\partial}{\partial u},$$

$$\mathrm{X}_3 = At\frac{\partial}{\partial x} - \tfrac{1}{2}[ABt^2 + (AB)^{-1}e^{-2Bx}]\frac{\partial}{\partial t} + \tfrac{1}{2}ABtu\frac{\partial}{\partial u}, \quad \mathrm{X}_4 = u\frac{\partial}{\partial u}.$$

The commutator table is the same as for (i) with $C - A$ replaced by $-AB$.

Case IIb. $\sigma \neq 0$.
Here, (4.75) leads to the wave speed $c(x)$ solving the fourth-order ODE

$$c^2(x)[\alpha'(x) - H(x)\alpha(x)]' = \sigma^2\alpha(x), \tag{4.78}$$

where $H(x) = c'(x)/c(x)$, and $\alpha(x)$ is given by (4.74). [Without loss of generality, $\rho = 1$ in (4.74).] One can show that if $c(x)$ satisfies ODE (4.78), then the corresponding wave equation (4.64) admits a four-parameter Lie group of point transformations with its infinitesimal generators given by

$$\mathrm{X}_1 = \frac{\partial}{\partial t}, \quad \mathrm{X}_2 = e^{\sigma t}\left[\alpha(x)\frac{\partial}{\partial x} + \sigma^{-1}(\alpha'(x) - H(x)\alpha(x))\frac{\partial}{\partial t} + \tfrac{1}{2}\alpha(x)H(x)u\frac{\partial}{\partial u}\right],$$
$$\mathrm{X}_3 = e^{-\sigma t}\left[\alpha(x)\frac{\partial}{\partial x} - \sigma^{-1}(\alpha'(x) - H(x)\alpha(x))\frac{\partial}{\partial t} + \tfrac{1}{2}\alpha(x)H(x)u\frac{\partial}{\partial u}\right], \quad \mathrm{X}_4 = u\frac{\partial}{\partial u}.$$

The nonzero commutators of the corresponding Lie algebra are given by

$$[\mathrm{X}_1, \mathrm{X}_2] = \sigma\mathrm{X}_2, \quad [\mathrm{X}_1, \mathrm{X}_3] = -\sigma\mathrm{X}_3,$$
$$[\mathrm{X}_2, \mathrm{X}_3] = 2\sigma^{-1}[(\alpha'(x) - H(x)\alpha(x))^2 - (\sigma\alpha(x)/c(x))^2]\mathrm{X}_1.$$

It immediately follows that

$$(\alpha'(x) - H(x)\alpha(x))^2 \text{-} (\sigma\alpha(x)/c(x))^2 = \text{const} = K. \tag{4.79}$$

Hence, (4.79) is a quadrature of ODE (4.78), i.e., the commutator $[\mathrm{X}_2, \mathrm{X}_3]$ yields a first integral of ODE (4.78)! The third-order ODE (4.79) for the wave speed $c(x)$ admits two point symmetries. Using the methods of Chapter 3, it can be reduced to a first-order ODE. If the reduced ODE can be solved, then the general solution of ODE (4.79) is obtained through three quadratures. [When σ is imaginary, then appropriate linear combinations of X_2 and X_3 yield the corresponding infinitesimal generators.] When $K \neq 0$, one can show that the Lie algebra with basis generators $\mathrm{X}_1, \mathrm{X}_2, \mathrm{X}_3$ is isomorphic to the Lie algebra of SO(2,1).

Case III. $\xi = 0$.
From the determining equations (4.65a–e), it immediately follows that $\tau = \text{const} = r$, $f = \text{const} = s$ and, hence, the wave equation (4.64) here only admits translations in t and scalings in u. In particular, if the wave speed $c(x)$ does not satisfy (4.74) and (4.79) for any values of the constants ρ, σ, K, then the wave equation (4.64) only admits the two-parameter Lie group of point transformations with infinitesimal generators $\mathrm{X}_1 = \dfrac{\partial}{\partial t}$, $\mathrm{X}_2 = u\dfrac{\partial}{\partial u}$.

In summary, we have the following theorem:

Theorem 4.2.3-1. *The wave equation* (4.64), *whose wave speed* $c(x)$ *is a solution of the system* (4.74) *and* (4.79) *for some constants* ρ, σ, K, *admits a four-parameter Lie group of point transformations. This group becomes an infinite-parameter group if and only if* $c(x) = (Ax + B)^C, C = 0, 2$. *For all other wave speeds* $c(x)$, *the wave equation* (4.64) *only admits the two-parameter Lie group of translations in* t *and scalings in* u.

The group classification of the wave equation (4.64) appeared in Bluman and Kumei (1987). This paper includes the corresponding invariant solutions.

(4) *Biharmonic Equation*
As a final example, we find the Lie group of point transformations admitted by the fourth-order biharmonic equation

$$\Delta u = \nabla^2 \nabla^2 u = 0,$$

where $\nabla^2 = \partial^2/\partial x^2 + \partial^2/\partial y^2$ or, equivalently, the PDE

$$u_{yyyy} = -2u_{xxyy} - u_{xxxx}. \tag{4.80}$$

From Theorem 4.2.2-6 $[x_1 = x, x_2 = y]$, we see that for the PDE (4.80) an admitted nontrivial point symmetry has an infinitesimal generator of the form

$$\mathrm{X} = X(x, y)\frac{\partial}{\partial x} + Y(x, y)\frac{\partial}{\partial y} + f(x, y)u\frac{\partial}{\partial u},$$

where the symmetry determining equation (4.28) is given by

$$\eta^{(4)}_{yyyy} = -2\eta^{(4)}_{xxyy} - \eta^{(4)}_{xxxx} \tag{4.81a}$$

with

$$u_{yyyy} = -2u_{xxyy} - u_{xxxx}. \tag{4.81b}$$

Then it is straightforward to derive the following set of determining equations for $X(x, y)$, $Y(x, y)$, $f(x, y)$:

$$\frac{\partial Y}{\partial x} + \frac{\partial X}{\partial y} = 0, \tag{4.82a}$$

$$\frac{\partial X}{\partial x} - \frac{\partial Y}{\partial y} = 0, \tag{4.82b}$$

$$2\frac{\partial f}{\partial x} - 3\frac{\partial^2 X}{\partial x^2} - \frac{\partial^2 X}{\partial y^2} = 0, \tag{4.82c}$$

$$2\frac{\partial f}{\partial y} - 4\frac{\partial^2 X}{\partial x\,\partial y} - 3\frac{\partial^2 Y}{\partial x^2} - \frac{\partial^2 Y}{\partial y^2} = 0, \tag{4.82d}$$

$$2\frac{\partial f}{\partial x} - 4\frac{\partial^2 Y}{\partial x\,\partial y} - 3\frac{\partial^2 X}{\partial y^2} - \frac{\partial^2 X}{\partial x^2} = 0, \tag{4.82e}$$

$$2\frac{\partial f}{\partial y} - 3\frac{\partial^2 Y}{\partial y^2} - \frac{\partial^2 Y}{\partial x^2} = 0, \tag{4.82f}$$

$$3\frac{\partial^2 f}{\partial x^2} + \frac{\partial^2 f}{\partial y^2} - 2\left[\frac{\partial^3 X}{\partial x^3} + \frac{\partial^3 X}{\partial x\,\partial y^2}\right] = 0, \tag{4.82g}$$

$$2\frac{\partial^2 f}{\partial x\,\partial y} - \frac{\partial^3 X}{\partial y^3} - \frac{\partial^3 X}{\partial x^2\,\partial y} - \frac{\partial^3 Y}{\partial x^3} - \frac{\partial^3 Y}{\partial x\,\partial y^2} = 0, \tag{4.82h}$$

$$3\frac{\partial^2 f}{\partial y^2} + \frac{\partial^2 f}{\partial x^2} - 2\left[\frac{\partial^3 Y}{\partial y^3} + \frac{\partial^3 Y}{\partial x^2\,\partial y}\right] = 0, \tag{4.82i}$$

$$4\left[\frac{\partial^3 f}{\partial x^3} + \frac{\partial^3 f}{\partial x\,\partial y^2}\right] - \frac{\partial^4 X}{\partial x^4} - 2\frac{\partial^4 X}{\partial x^2\,\partial y^2} - \frac{\partial^4 X}{\partial y^4} = 0, \tag{4.82j}$$

$$4\left[\frac{\partial^3 f}{\partial y^3} + \frac{\partial^3 f}{\partial x^2\,\partial y}\right] - \frac{\partial^4 Y}{\partial x^4} - 2\frac{\partial^4 Y}{\partial x^2\,\partial y^2} - \frac{\partial^4 Y}{\partial y^4} = 0, \tag{4.82k}$$

$$\Delta f = 0. \tag{4.82ℓ}$$

From the determining equations (4.82a,b), it follows that

$$\nabla^2 X = 0, \quad \nabla^2 Y = 0. \tag{4.83}$$

After substituting (4.83) into the determining equations (4.82c,f), we find that

$$f(x,y) = \frac{\partial X}{\partial x} + s, \quad s = \text{const.} \tag{4.84}$$

Then the determining equations (4.82d,e) are also satisfied. After substituting (4.84), (4.82a,b) into the determining equations (4.82g–i), we find that third-order derivatives of X and Y vanish. Consequently, the remaining determining equations (4.82j–ℓ) are automatically satisfied. Hence, we obtain

$$X(x,y) = \alpha_1 x^2 + \beta_1 xy + \gamma_1 y^2 + \delta_1 x + \kappa_1 y + \rho_1, \tag{4.85a}$$

$$Y(x,y) = \alpha_2 x^2 + \beta_2 xy + \gamma_2 y^2 + \delta_2 x + \kappa_2 y + \rho_2, \tag{4.85b}$$

and, after renaming s in (4.84),

$$f(x,y) = 2\alpha_1 x + \beta_1 y + s, \tag{4.85c}$$

where the indicated constants are to be determined. From the determining equations (4.82a,b), it follows that

$$2\alpha_2 x + \beta_2 y + \delta_2 = -\beta_1 x - 2\gamma_1 y - \kappa_1, \quad 2\alpha_1 x + \beta_1 y + \delta_1 = \beta_2 x + 2\gamma_2 y + \kappa_2.$$

Hence,

$$\beta_1 = -2\alpha_2, \quad \gamma_2 = -\alpha_2, \quad \beta_2 = 2\alpha_1, \quad \gamma_1 = -\alpha_1, \quad \kappa_2 = \delta_1, \quad \kappa_1 = -\delta_2.$$

Consequently, after renaming the constants $\delta_1, \delta_2, \rho_1, \rho_2, s$, we see that the point symmetry generators

$$\mathrm{X} = X(x,y)\frac{\partial}{\partial x} + Y(x,y)\frac{\partial}{\partial y} + f(x,y)u\frac{\partial}{\partial u}$$

admitted by the biharmonic equation (4.80) are given by the infinitesimals

$$X(x,y) = \alpha_1(x^2 - y^2) - 2\alpha_2 xy + \alpha_3 x - \alpha_4 y + \alpha_5, \tag{4.86a}$$

$$Y(x,y) = 2\alpha_1 xy + \alpha_2(x^2 - y^2) + \alpha_3 y + \alpha_4 x + \alpha_6, \tag{4.86b}$$

$$f(x,y) = 2\alpha_1 x - 2\alpha_2 y + \alpha_7. \tag{4.86c}$$

It is left to Exercise 4.2-5 to show that, in terms of the complex variable $z = x + iy$, these infinitesimal generators determine a seven-parameter $(\alpha_1, \alpha_2, \ldots, \alpha_7)$ Lie group of point transformations given by

$$z^* = \frac{az+b}{cz+d}, \tag{4.87a}$$

$$u^* = \lambda\left|\frac{dz^*}{dz}\right|u, \tag{4.87b}$$

where a,b,c,d are arbitrary complex constants such that $ad - bc \neq 0$, and λ is an arbitrary real constant. Equation (4.87a) is a general Möbius (bilinear) transformation. This example was considered in Bluman and Gregory (1985).

Reid (1990, 1991) showed that the distinguished cases in group classification problems for admitted point symmetries of a given PDE can be determined *without* solving the determining equations. Lisle (1992) modified Reid's algorithm by introducing a method based on moving frames to show how to solve complex classification problems that involve two or more classifying functions. He applied his method to give the complete group classification, with respect to admitted point symmetries, for the scalar diffusion convection equation

$$u_t = (D(u)u_x - K(u))_x,$$

in terms of two classifying functions: the diffusion $D(u)$ and convection $K(u)$.

The point symmetries admitted by the porous medium equation, $u_t = (u^p)_{xx} + g(x)u^q + f(x)(u^r)_x$, for various functions $g(x), f(x)$ and constants p,q,r, are considered in Gandarias (1996).

EXERCISES 4.2

1. Consider the heat equation (4.44).
   (a) Find the invariant solutions of PDE (4.44) resulting from its invariance under X_5, using both methods of Section 4.2.1.
   (b) For any solution $u = \Theta(x,t)$ that is not an invariant solution related to invariance under X_5, find the generated one-parameter family of solutions $\Phi(x,t;\varepsilon)$ of PDE (4.44).

2. (a) For which solutions $u = \Theta(x,t)$ of the heat equation (4.44) do the infinitesimal generators $X_1, X_2, \ldots, X_6$, given by (4.50), yield a six-parameter family of solutions $u = \Phi(x,t;\varepsilon_1,\varepsilon_2,\ldots,\varepsilon_6)$ of (4.44)?
   (b) Determine $\Phi(x,t;\varepsilon_1,\varepsilon_2,\ldots,\varepsilon_6)$.

3. Consider the wave equation $u_{tt} = e^{2x}u_{xx}$.
   (a) Find invariant solutions resulting from its admitted infinitesimal generators:
      (i) $X_2 + sX_1$; and
      (ii) $X_2 + sX_1$,
      where s is an arbitrary constant.
   (b) Given any solution $u = \Theta(x,t)$ of this wave equation, find the four-parameter family of solutions generated by X_1, X_2, X_3, X_4. What condition must $\Theta(x,t)$ satisfy?

4. Show that (4.49a–c) yield the general solution of the set of determining equations (4.48a–d).

5. Show that (4.87a,b) define a seven-parameter Lie group of point transformations with its infinitesimals given by (4.86a–c).

6. Consider the heat equation in n spatial dimensions.
   (a) Find the nine-parameter Lie group of point transformations admitted by $u_t = u_{xx} + u_{yy}$ $[n = 2]$.
   (b) Find the 13-parameter Lie group of point transformations admitted by $u_t = u_{xx} + u_{yy} + u_{zz}$ $[n = 3]$.
   (c) Generalize to the case of the n-dimensional heat equation $u_t = \sum_{j=1}^{n} u_{x_j x_j}$.

7. Consider the axisymmetric wave equation

$$u_{tt} = u_{rr} + \frac{1}{r}u_r. \tag{4.88}$$

   (a) Show that the Lie group of point transformations admitted by PDE (4.88) has its infinitesimal generators given by

$$X_1 = r\frac{\partial}{\partial r} + t\frac{\partial}{\partial t}, \quad X_2 = 2rt\frac{\partial}{\partial r} + (r^2 + t^2)\frac{\partial}{\partial t} - tu\frac{\partial}{\partial u}, \quad X_3 = u\frac{\partial}{\partial u}, \quad X_4 = \frac{\partial}{\partial t}.$$

(b) Find invariant solutions of PDE (4.88) resulting from the admitted infinitesimal generators:
   (i) $X_1 + sX_3$; and
   (ii) $X_2 + sX_3$,
   where s is an arbitrary constant.

8. Consider the nonlinear wave equation

$$u_{tt} = c^2(u)u_{xx}, \tag{4.89}$$

$c(u) \neq$ const. In terms of infinitesimal generators, show that the group classification for the invariance of PDE (4.89), with respect to point symmetries, is given by:
(a) $c(u)$ arbitrary:

$$X_1 = x\frac{\partial}{\partial x} + t\frac{\partial}{\partial t}, \quad X_2 = \frac{\partial}{\partial x}, \quad X_3 = \frac{\partial}{\partial t}.$$

(b) $c(u) = A(u+B)^C$, where A, B, C are arbitrary constants:

$$X_1,\ X_2,\ X_3,\ X_4 = Cx\frac{\partial}{\partial x} + (u+B)\frac{\partial}{\partial u}.$$

(c) $c(u) = A(u+B)^2$, where A and B are arbitrary constants:

$$X_1,\ X_2,\ X_3,\ X_4(C=2), \quad X_5 = x^2\frac{\partial}{\partial x} + x(u+B)\frac{\partial}{\partial u}.$$

9. Consider Laplace's equation in $n \geq 3$ dimensions,

$$\sum_{j=1}^{n} u_{x_j x_j} = 0. \tag{4.90}$$

(a) Show that the $[1+\frac{1}{2}(n+1)(n+2)]$ – parameter Lie group of point transformations admitted by PDE (4.90) has the infinitesimal generator

$$X = \sum_{j=1}^{n} \xi_j(x)\frac{\partial}{\partial x_j} + f(x)u\frac{\partial}{\partial u}$$

with the infinitesimals given by

$$\xi_j(x) = \alpha_j + \sum_{k=1}^{n} \beta_{jk}x_k - \gamma_j\sum_{k=1}^{n}(x_k)^2 + 2x_j\sum_{k=1}^{n}\gamma_k x_k + \lambda x_j, \quad j = 1,2,\ldots,n,$$

$$f(x) = (2-n)\sum_{k=1}^{n}\gamma_k x_k + \delta,$$

where $\alpha_j, \delta, \lambda, \gamma_j$, and $\beta_{jk} = -\beta_{kj}$, $j,k = 1,2,\ldots,n$, are $1+\frac{1}{2}(n+1)(n+2)$ arbitrary constants. The subgroup corresponding to $\delta = 0$ is called the conformal group.

One can show that the conformal group is isomorphic to $\mathrm{SO}(n+1,1)$ [Bluman (1967)].

(b) Find the corresponding global seven-parameter Lie group of point transformations admitted by PDE (4.90) when $n=2$.

10. Consider the nonlinear diffusion equation

$$u_{xx} = (u_x)^2 u_t. \tag{4.91}$$

(a) Find the infinite-parameter Lie group of point transformations admitted by PDE (4.91).

(b) Compare the Lie algebra for the infinitesimal generators admitted by PDE (4.91) with the Lie algebra for the infinitesimal generators admitted by the linear heat equation (4.44).

11. Find the five-parameter Lie group of point transformations admitted by Burgers' equation $u_t = u_{xx} - uu_x$. The admitted point symmetries and corresponding invariant solutions for the two- and three-dimensional Burgers' equations, $u_t = \nabla^2 u - uu_x$, appears in Edwards and Broadbridge (1995).

12. Find the four-parameter Lie group of point transformations admitted by the Kortweg–de Vries (KdV) equation

$$u_{xxx} + uu_x + u_t = 0. \tag{4.92}$$

13. Find the four-parameter Lie group of point transformations admitted by the cylindrical KdV equation

$$u_{xxx} + uu_x + \frac{1}{2t}u + u_t = 0. \tag{4.93}$$

See Bluman and Kumei (1989b, Chapter 6) on relating PDEs (4.92) and (4.93) through their group properties.

14. The motion of an incompressible two-dimensional constant-property fluid is described by the stream-function equation

$$\nabla^2 u_t + u_y \nabla^2 u_x - u_x \nabla^2 u_y = \nu \nabla^4 u, \tag{4.94}$$

where $u(x,y,t)$ is the stream function for the flow, $\nu = \text{const}$ is the kinematic viscosity, and $\nabla^2 = \partial^2/\partial x^2 + \partial^2/\partial y^2$.

(a) If $\nu = 0$, show that the infinite-parameter Lie group of point transformations admitted by PDE (4.94) can be represented by the infinitesimal generator

$$\mathrm{X} = X(x,y,t)\frac{\partial}{\partial x} + Y(x,y,t)\frac{\partial}{\partial y} + T(t)\frac{\partial}{\partial t} + \eta(x,y,t,u)\frac{\partial}{\partial u}$$

with the infinitesimals given by

$$X(x,y,t) = ax + by + cyt + f_1(t),$$

$$Y(x,y,t) = ay - bx - cxt + f_2(t),$$
$$T(t) = ht + k,$$
$$\eta(x,y,t,u) = (2a-h)u + \tfrac{1}{2}c(x^2+y^2) + f_1'(t)y - f_2'(t)x + f_3(t),$$

where a,b,c,h,k are arbitrary constants and $f_1(t), f_2(t), f_3(t)$ are arbitrary differentiable functions of the time coordinate t.

(b) If $\nu \neq 0$, show that the admitted group is the same as for (a) except that $h = 2a$ [Cantwell (1978)].

15. Consider the nonlinear reaction–diffusion equation

$$u_t = u_{xx} + F(u). \tag{4.95}$$

For an arbitrary reaction function $F(u)$, one can show that PDE (4.95) is only invariant under translations in x and t. Show that PDE (4.95) admits a three-parameter Lie group of point transformations only if $F(u)$ is given by one of the three forms: $Au^B, u(A + B\log u), Ae^{Bu}$, to within translations in u, where A and B are arbitrary constants. [Liu and Fang (1986). For generalizations, see Galaktionov et al. (1988).]

16. Consider the nonlinear wave equation

$$u_{tt} = (c^2(u)u_x)_x, \tag{4.96}$$

$c(u) \neq$ const. In terms of the infinitesimal generators for admitted Lie groups of point transformations, show that the group classification of PDE (4.96) is given by:

(a) $c(u)$ arbitrary:

$$\mathrm{X}_1 = x\frac{\partial}{\partial x} + t\frac{\partial}{\partial t}, \quad \mathrm{X}_2 = \frac{\partial}{\partial x}, \quad \mathrm{X}_3 = \frac{\partial}{\partial t}.$$

(b) $c(u) = A(u+B)^C$, where A,B,C are arbitrary constants:

$$\mathrm{X}_1,\ \mathrm{X}_2,\ \mathrm{X}_3,\ \ \mathrm{X}_4 = Cx\frac{\partial}{\partial x} + (u+B)\frac{\partial}{\partial u}.$$

(c) $c(u) = A(u+B)^{-2}$, where A and B are arbitrary constants:

$$\mathrm{X}_1,\ \mathrm{X}_2,\ \mathrm{X}_3,\ \mathrm{X}_4(C=-2),\ \ \mathrm{X}_5 = t^2\frac{\partial}{\partial t} + t(u+B)\frac{\partial}{\partial u}.$$

(d) $c(u) = A(u+B)^{-2/3}$, where A and B are arbitrary constants:

$$\mathrm{X}_1,\ \mathrm{X}_2,\ \mathrm{X}_3,\ \mathrm{X}_4(C=-\tfrac{2}{3}),\ \ \mathrm{X}_5 = x^2\frac{\partial}{\partial x} - 3x(u+B)\frac{\partial}{\partial u}.$$

[Ames, Lohner, and Adams (1981). A group classification for PDE (4.96) with $c(u)$ replaced by $c(x,u)$ has been investigated by Torrisi and Valenti (1985). Generalizations to higher-dimensional nonlinear wave equations of the form

$u_{tt} = (f(u)u_x)_x + (g(u)u_y)_y + (h(u)u_z)_z$ are given in Baikov, Gazizov, and Ibragimov (1990, 1991).]

17. Show that the two-dimensional nonlinear Schrödinger equation

$$u_{xx} + u_{yy} + r\,|\,u^2\,|\,u = iu_t, \quad r = \text{const},$$

admits an eight-parameter Lie group of point transformations with its infinitesimal generators given by

$$\mathrm{X}_1 = \frac{\partial}{\partial x}, \quad \mathrm{X}_2 = \frac{\partial}{\partial y}, \quad \mathrm{X}_3 = \frac{\partial}{\partial t}, \quad \mathrm{X}_4 = x\frac{\partial}{\partial x} + y\frac{\partial}{\partial y} + 2t\frac{\partial}{\partial t}, \quad \mathrm{X}_5 = y\frac{\partial}{\partial x} - x\frac{\partial}{\partial y},$$

$$\mathrm{X}_6 = t\frac{\partial}{\partial x} - \tfrac{1}{2}ixu\frac{\partial}{\partial u}, \quad \mathrm{X}_7 = t\frac{\partial}{\partial y} - \tfrac{1}{2}iyu\frac{\partial}{\partial u},$$

$$\mathrm{X}_8 = xt\frac{\partial}{\partial x} + yt\frac{\partial}{\partial y} + t^2\frac{\partial}{\partial t} - [t + \tfrac{1}{4}i(x^2 + y^2)]u\frac{\partial}{\partial u}$$

[Tajiri (1983)].

18. Show that the PDE $u_{xx} + u_{yy} + (e^u)_{zz} = 0$, which arises in Riemannian geometry, admits

$$\mathrm{X}_1 = \frac{\partial}{\partial x}, \quad \mathrm{X}_2 = \frac{\partial}{\partial y}, \quad \mathrm{X}_3 = \frac{\partial}{\partial z}, \quad \mathrm{X}_4 = x\frac{\partial}{\partial x} + y\frac{\partial}{\partial y} + z\frac{\partial}{\partial z},$$

and the infinite-parameter Lie group of point transformations represented by the infinitesimal generator

$$\mathrm{X}_\infty = \alpha(x,y)x\frac{\partial}{\partial x} + \alpha(x,y)y\frac{\partial}{\partial y} - 2\alpha(x,y)\frac{\partial}{\partial u},$$

where $\nabla^2\alpha(x,y) = 0$. [See Drew, Kloster, and Gegenberg (1989) where the cylindrically symmetric case is also considered.]

19. Show that the most general second-order scalar PDE $u_{xx} = f(x,t,u,u_x,u_t,u_{xt},u_{tt})$ that admits the group of the heat equation with the six infinitesimal generators (4.53) is given by

$$u_{xx} = u_t + \frac{(u_x)^2}{u}K(\varsigma),$$

where

$$\varsigma = \frac{u^2 u_{xt} - 3uu_t u_x}{(u_x)^3}$$

and $K(\varsigma)$ is any solution of the first-order ODE

$$(K - \tfrac{3}{2}\varsigma - 3)K' + K = 0.$$

Find a point symmetry admitted by this first-order ODE and show that its general solution is given by

$$\frac{K^3}{(2K-2-\varsigma)} = \text{const.}$$

20. Consider the nonlinear second-order PDE

$$u_x + 2uu_z + zu_{xz} + z^2u_{yz} + zuu_{zz} = 0,$$

which arises in the classification of ODEs that admit point-form adjoint-symmetries [cf. Section 3.7.5]. Find the two-parameter Lie group of scalings admitted by this PDE and the corresponding invariant solutions.

4.3 INVARIANCE FOR A SYSTEM OF PDEs

Consider a system of N PDEs ($N > 1$) with n independent variables $x = (x_1, x_2, \ldots, x_n)$ and m dependent variables $u = (u^1, u^2, \ldots, u^m)$, given by

$$F^\mu(x, u, \partial u, \partial^2 u, \ldots, \partial^k u) = 0, \quad \mu = 1, 2, \ldots, N. \tag{4.97}$$

Definition 4.3-1. The one-parameter Lie group of point transformations

$$x^* = X(x, u; \varepsilon), \tag{4.98a}$$

$$u^* = U(x, u; \varepsilon), \tag{4.98b}$$

leaves invariant the system of PDEs (4.97), i.e., *is a point symmetry admitted by* (4.97), if and only if its kth extension, defined by (2.134a–d), (2.130)–(2.132), leaves invariant the N surfaces in $(x, u, \partial u, \partial^2 u, \ldots, \partial^k u)$-space, defined by (4.97).

In analogy to the situation for a scalar PDE, it is easy to prove the following theorem. [For the rest of this section, we assume the summation convention for repeated indices.]

Theorem 4.3-1 (Infinitesimal Criterion for the Invariance of a System of PDEs). *Let*

$$\mathrm{X} = \xi_i(x, u)\frac{\partial}{\partial x_i} + \eta^\nu(x, u)\frac{\partial}{\partial u^\nu} \tag{4.99}$$

be the infinitesimal generator of the Lie group of point transformations (4.98a,b). *Let*

$$\mathrm{X}^{(k)} = \xi_i(x, u)\frac{\partial}{\partial x_i} + \eta^\nu(x, u)\frac{\partial}{\partial u^\nu} + \eta_i^{(1)\nu}(x, u, \partial u)\frac{\partial}{\partial u_i^\nu} + \cdots$$

$$+\eta_{i_1 i_2 \cdots i_k}^{(k)\nu}(x,u,\partial u,\partial^2 u,\ldots,\partial^k u)\frac{\partial}{\partial u_{i_1 i_2 \cdots i_k}^{\nu}} \tag{4.100}$$

be the kth-extended infinitesimal generator of (4.99) *where* $\eta_i^{(1)\nu}$ *is given by* (2.135) *and* $\eta_{i_1 i_2 \cdots i_j}^{(j)\nu}$ *by* (2.136), $\nu = 1,2,\ldots,m$, *and* $i_j = 1,2,\ldots,n$, *for* $j = 1,2,\ldots,k$, *in terms of* $\xi(x,u) = (\xi_1(x,u), \xi_2(x,u), \ldots, \xi_n(x,u))$, $\eta(x,u) = (\eta^1(x,u), \eta^2(x,u), \ldots, \eta^m(x,u))$. *Then the one-parameter Lie group of point transformations* (4.98a,b) *is admitted by the system of PDEs* (4.97) *if and only if*

$$\mathrm{X}^{(k)} F^{\sigma}(x,u,\partial u,\partial^2 u,\ldots,\partial^k u) = 0 \quad \textit{when } u \textit{ satisfies } (4.97) \tag{4.101}$$

for each $\sigma = 1,2,\ldots,N$.

Proof. Left to Exercise 4.3-1. □

Note that the invariance criterion (system of symmetry determining equations (4.101)) involves the substitution of the N PDEs (4.97) and their differential consequences into each of the N determining equations given by (4.101).

4.3.1 INVARIANT SOLUTIONS

Consider a system of PDEs (4.97) that admits a one-parameter Lie group of point transformations with the infinitesimal generator (4.99). We assume that $\xi(x,u) \not\equiv 0$.

Definition 4.3.1-1. $u = \Theta(x)$, with components $u^\nu = \Theta^\nu(x)$, $\nu = 1,2,\ldots,m$, is an *invariant solution* of the system of PDEs (4.97) resulting from an admitted point symmetry with infinitesimal generator (4.99) if and only if:

(i) $u^\nu = \Theta^\nu(x)$ is an invariant surface of (4.99) for each $\nu = 1,2,\ldots,m$;

(ii) $u = \Theta(x)$ solves (4.97).

It follows that $u = \Theta(x)$ is an invariant solution of the system of PDEs (4.97), resulting from its invariance under the Lie group of point transformations (4.98a,b), if and only if $u = \Theta(x)$ satisfies:

(i) $\mathrm{X}(u^\nu - \Theta^\nu(x)) = 0$ when $u = \Theta(x)$, $\nu = 1,2,\ldots,m$, i.e.,

$$\xi_i(x,\Theta(x))\frac{\partial \Theta^\nu(x)}{\partial x_i} = \eta^\nu(x,\Theta(x)), \quad \nu = 1,2,\ldots,m; \tag{4.102}$$

(ii) $F^\mu(x,u,\partial u,\partial^2 u,\ldots,\partial^k u) = 0$ when $u = \Theta(x)$, $\mu = 1,2,\ldots,N$, i.e.,

$$F^\mu(x,\Theta(x),\partial\Theta(x),\partial^2\Theta(x),\ldots,\partial^k\Theta(x)) = 0, \quad \mu = 1,2,\ldots,N. \tag{4.103}$$

where $\partial^j\Theta(x)$ denotes $\partial^j\Theta^\nu(x)/\partial x_{i_1}\partial x_{i_2}\ldots\partial x_{i_j}$, $\nu=1,2,\ldots,m$, and $i_j=1,2,\ldots,n$, for $j=1,2,\ldots,k$.

Equations (4.102) are the *invariant surface conditions* for the invariant solutions of the system of PDEs (4.97) resulting from its invariance under the point symmetry (4.99). As is the situation for a scalar PDE, invariant solutions can be determined by two procedures.

(I) Invariant Form Method. Here, we solve the invariant surface conditions (4.102) by explicitly solving the corresponding characteristic equations for $u=\Theta(x)$ given by

$$\frac{dx_1}{\xi_1(x,u)}=\frac{dx_2}{\xi_2(x,u)}=\cdots=\frac{dx_n}{\xi_n(x,u)}=\frac{du^1}{\eta^1(x,u)}=\frac{du^2}{\eta^2(x,u)}=\cdots=\frac{du^m}{\eta^m(x,u)}. \tag{4.104}$$

If $y_1(x,u),y_2(x,u),\ldots,y_{n-1}(x,u),v^1(x,u),v^2(x,u),\ldots,v^m(x,u)$, are $n+m-1$ functionally independent constants that arise from solving the system of $n+m-1$ first-order ODEs (4.104) with the Jacobian $\partial(v^1,v^2,\ldots,v^m)/\partial(u^1,u^2,\ldots,u^m)\neq 0$, then the general solution $u=\Theta(x)$ of the system of PDEs (4.102) is given implicitly by the invariant form

$$v^\nu(x,u)=\Phi^\nu(y_1(x,u),y_2(x,u),\ldots,y_{n-1}(x,u)), \tag{4.105}$$

where Φ^ν is an arbitrary differentiable function of $y_1(x,u),y_2(x,u),\ldots,y_{n-1}(x,u)$, for $\nu=1,2,\ldots,m$. Note that $y_1(x,u),y_2(x,u),\ldots,y_{n-1}(x,u),v^1(x,u),v^2(x,u),\ldots,v^m(x,u)$ are $n+m-1$ functionally independent group invariants of (4.99) and hence are $n+m-1$ canonical coordinates for the Lie group of point transformations (4.98a,b). Let $y_n(x,u)$ be the $(n+m)$th canonical coordinate satisfying

$$\mathrm{X}y_n=1.$$

If the system of PDEs (4.97) is transformed into a system of PDEs in terms of independent variables $y_1,y_2,\ldots,y_n$ and dependent variables $v^1,v^2,\ldots,v^m$, then the transformed system of PDEs admits the one-parameter Lie group of translations

$$\begin{aligned} y^*_i&=y_i, \quad i=1,2,\ldots,n-1,\\ y^*_n&=y_n+\varepsilon,\\ v^{*\nu}&=v^\nu, \quad \nu=1,2,\ldots,m. \end{aligned}$$

Thus, the variable y_n does not appear explicitly in the transformed system of PDEs and, hence, the transformed system of PDEs has solutions of the form (4.105). Consequently, the system of PDEs (4.97) has invariant solutions given implicitly by the invariant form (4.105). Such solutions are found by solving a reduced system of differential equations with $n-1$ independent variables $y_1,y_2,\ldots,y_{n-1}$ and m dependent variables $v^1,v^2,\ldots,v^m$. The variables $y_1,y_2,\ldots,y_{n-1}$ are commonly called similarity variables. The reduced system of differential equations is found by substituting the invariant form (4.105) into the given system of PDEs (4.97). We assume that this substitution does not lead to a singular differential equation. Note that if $\partial\xi/\partial u \equiv 0$, as is typically the case, then $y_i = y_i(x)$, $i = 1,2,\ldots,n-1$. If $n = 2$, then the reduced system of differential equations is a system of ODEs and we denote the similarity variable by $\zeta = y_1$.

(II) Direct Substitution Method. This procedure is essential if one is unable to solve explicitly the invariant surface conditions (4.102), i.e., the characteristic equations (4.104). We assume that $\xi_n(x,u) \not\equiv 0$. Then the first-order system of PDEs (4.102) can be written as

$$\frac{\partial u^\nu}{\partial x_n} = \frac{\eta^\nu(x,u)}{\xi_n(x,u)} - \sum_{i=1}^{n-1} \frac{\xi_i(x,u)}{\xi_n(x,u)} \frac{\partial u^\nu}{\partial x_i}, \quad \nu = 1,2,\ldots,m. \tag{4.106}$$

From (4.106) and its differential consequences, it follows that any term involving derivatives of u with respect to x_n can be expressed in terms of x, u, and derivatives of u with respect to the variables $x_1, x_2, \ldots, x_{n-1}$. Hence, after directly substituting (4.106) and its differential consequences into the given system of PDEs (4.97) for all terms in (4.97) that involve derivatives of u with respect to x_n, we obtain a reduced system of differential equations involving m dependent variables $u^1, u^2, \ldots, u^m$, $n-1$ independent variables $x_1, x_2, \ldots, x_{n-1}$, and parameter x_n. Any solution of this reduced system of differential equations defines an invariant solution of the system of PDEs (4.97) resulting from its invariance under the Lie group of point transformations with the infinitesimal generator (4.99), provided that the invariant surface conditions (4.102) or, equivalently, the given system of PDEs (4.97) itself, are also satisfied. If $n = 2$, the reduced system of differential equations is a system of ODEs. The constants appearing in the general solution of this system of ODEs are arbitrary functions of the parameter x_n. These arbitrary functions are then determined by substituting this general solution into either the invariant surface conditions (4.102) or the given system of PDEs (4.97).

4.3.2 DETERMINING EQUATIONS FOR SYMMETRIES OF A SYSTEM OF PDEs

Consider a system of PDEs (4.97) with each of its PDEs given in a solved form

$$u^{\nu_\mu}_{i_1 i_2 \cdots i_{\ell_\mu}} = f^\mu(x,u,\partial u,\partial^2 u,\ldots,\partial^k u), \tag{4.107}$$

in terms of some specific ℓ_μth-order partial derivative of u^{ν_μ} for some $\nu_\mu = 1,2,\ldots,m$, where $f^\mu(x,u,\partial u,\partial^2 u,\ldots,\partial^k u)$ does not depend explicitly on any of the components $u^{\nu_\sigma}_{i_1 i_2 \cdots i_{\ell_\mu}}$, $\sigma = 1,2,\ldots,N$, for each $\mu = 1,2,\ldots,N$. From Theorem 4.3-1, we see that the system of PDEs (4.107) admits the point symmetry

$$\mathrm{X} = \xi_i(x,u)\frac{\partial}{\partial x_i} + \eta^\nu(x,u)\frac{\partial}{\partial u^\nu}, \tag{4.108}$$

with the kth extension of (4.108) given by (4.100), if and only if

$$\eta^{(\ell_\mu)\nu_\mu}_{i_1 i_2 \cdots i_{\ell_\mu}} = \xi_j \frac{\partial f^\mu}{\partial x_j} + \eta^\nu \frac{\partial f^\mu}{\partial u^\nu} + \eta^{(1)\nu}_j \frac{\partial f^\mu}{\partial u^\nu_j} + \cdots + \eta^{(k)\nu}_{j_1 j_2 \cdots j_k} \frac{\partial f^\mu}{\partial u^\nu_{j_1 j_2 \cdots j_k}}, \quad \mu = 1,2,\ldots,N, \tag{4.109a}$$

with

$$u^{\nu_\sigma}_{i_1 i_2 \cdots i_{\ell_\sigma}} = f^\sigma(x,u,\partial u,\partial^2 u,\ldots,\partial^k u), \quad \sigma = 1,2,\ldots,N. \tag{4.109b}$$

It is easy to see that $\eta^{(p)\nu}_{j_1 j_2 \cdots j_p}$ is a polynomial in the components of $\partial u, \partial^2 u, \ldots, \partial^p u$, with coefficients that are linear homogeneous in the components of $\xi(x,u)$, $\eta(x,u)$ and their derivatives to order p. Thus, ξ and η appear linearly in (4.109a). As is the situation for a given scalar PDE, the system of symmetry determining equations (4.109a,b) leads to a *system of linear homogeneous PDEs* for ξ and η. First, we eliminate the components $u^{\nu_\sigma}_{i_1 i_2 \cdots i_{\ell_\sigma}}$ and their differential consequences from (4.109a) by substitution from (4.109b) and the differential consequences of (4.109b), $\sigma = 1,2,\ldots,N$. Consequently, the components of x, u and the remaining components of $\partial u, \partial^2 u, \ldots, \partial^k u$ that appear in the resulting system of symmetry determining equations (4.109a) are themselves independent variables, i.e., take on arbitrary values. Since the resulting expression for (4.109a) holds for any values of these independent variables, one obtains a system of linear homogeneous PDEs for ξ and η that constitutes a set of *determining equations* for the infinitesimal generators X admitted by the given system of PDEs (4.97). In particular, if each $f^\mu(x,u,\partial u,\partial^2 u,\ldots,\partial^k u)$, $\mu = 1,2,\ldots,N$, is a polynomial in the components of $\partial u, \partial^2 u, \ldots, \partial^k u$, then the system of symmetry determining equations (4.109a) yields polynomial equations in the independent components of $\partial u, \partial^2 u, \ldots, \partial^k u$. Consequently, the coefficients of these polynomial equations must vanish separately. This yields the set of linear determining equations for ξ and η. Typically, the number of determining equations is far greater than $n+m$, so that the set of determining equations is very overdetermined.

A linear system of nonhomogeneous PDEs,

$$\mathrm{L}u = g(x), \tag{4.110}$$

admits a trivial infinite-parameter Lie group of point transformations

$$x^* = x, \tag{4.111a}$$

$$u^* = u + \varepsilon\omega(x), \tag{4.111b}$$

where $\omega(x)$ is any solution of the associated linear homogeneous system of PDEs

$$\mathrm{L}u = 0.$$

To within this trivial infinite-parameter Lie group of point transformations, the Lie group of point transformations admitted by a linear system of PDEs usually has at most a finite number of parameters.

Unlike the case for scalar PDEs, there still appears to be very little known about the forms of admitted point symmetries for systems of PDEs. It is conjectured that for a linear system of PDEs (4.110), an admitted infinitesimal generator for a point symmetry (modulo the admitted trivial infinite-parameter Lie group of point transformations (4.111a,b)) is such that ξ has no dependence on u, and η is linear in u, i.e.,

$$\xi_i = \xi_i(x), \quad i = 1, 2, \ldots, n, \tag{4.112a}$$

$$\eta^\nu = k^\nu_\sigma(x) u^\sigma, \tag{4.112b}$$

for some functions $k^\nu_\sigma(x)$ for $\sigma, \nu = 1, 2, \ldots, m$. We will assume that the conditions (4.112a,b) hold for a Lie group of point transformations admitted by a linear system of PDEs. [It is easy to check that the conditions (4.112a,b) hold for all examples treated in this book.]

For a linear system with n = 2 and m = 2, write $x_1 = x,\ x_2 = t,\ \xi_1 = \xi(x,t)$, $\xi_2 = \tau(x,t),\ u^1 = u,\ u^2 = v,\ \eta^1 = \eta^u = f(x,t)u + g(x,t)v,\ \eta^2 = \eta^v = k(x,t)v + \ell(x,t)u$. Here an admitted infinitesimal generator for a point symmetry is of the form

$$\mathrm{X} = \xi(x,t)\frac{\partial}{\partial x} + \tau(x,t)\frac{\partial}{\partial t} + [f(x,t)u + g(x,t)v]\frac{\partial}{\partial u} + [k(x,t)v + \ell(x,t)u]\frac{\partial}{\partial v}, \tag{4.113}$$

and the once-extended infinitesimals are given by

$$\eta^{(1)u}_x = \frac{\partial f}{\partial x}u + \frac{\partial g}{\partial x}v + \left[f - \frac{\partial \xi}{\partial x}\right]u_x - \frac{\partial \tau}{\partial x}u_t + g v_x, \tag{4.114}$$

$$\eta^{(1)v}_x = \frac{\partial \ell}{\partial x}u + \frac{\partial k}{\partial x}v + \ell u_x + \left[k - \frac{\partial \xi}{\partial x}\right]v_x - \frac{\partial \tau}{\partial x}v_t, \tag{4.115}$$

$$\eta^{(1)u}_t = \frac{\partial f}{\partial t}u + \frac{\partial g}{\partial t}v - \frac{\partial \xi}{\partial t}u_x + \left[f - \frac{\partial \tau}{\partial t}\right]u_t + g v_t, \tag{4.116}$$

$$\eta^{(1)v}_t = \frac{\partial \ell}{\partial t}u + \frac{\partial k}{\partial t}v + \ell u_t - \frac{\partial \xi}{\partial t}v_x + \left[k - \frac{\partial \tau}{\partial t}\right]v_t. \tag{4.117}$$

4.3.3 EXAMPLES

(1) *System of Wave Equations*

Consider the linear system of first-order wave equations

$$v_t = u_x, \tag{4.118a}$$

$$u_t = x^4 v_x. \tag{4.118b}$$

Note that if the pair $(u(x,t), v(x,t))$ solves (4.118a,b), then $u(x,t)$ solves the wave equation

$$u_{tt} = x^4 u_{xx},$$

and $v(x,t)$ solves the wave equation

$$v_{tt} = (x^4 v_x)_x.$$

The system of symmetry determining equations (4.102a,b) for the system of PDEs (4.118a,b) is given by

$$\eta_t^{(1)v} = \eta_x^{(1)u}, \tag{4.119a}$$

$$\eta_t^{(1)u} = 4x^3 v_x \xi + x^4 \eta_x^{(1)v}, \tag{4.119b}$$

with $v_t = u_x, u_t = x^4 v_x$. After substituting (4.114)–(4.117) into (4.119a,b), and then eliminating v_t and u_t through substitutions from the given system of PDEs (4.118a,b), we obtain the system of symmetry determining equations given by

$$\left[\frac{\partial \ell}{\partial t} - \frac{\partial f}{\partial x}\right]u + \left[\frac{\partial k}{\partial t} - \frac{\partial g}{\partial x}\right]v + \left[x^4\left(\ell + \frac{\partial \tau}{\partial x}\right) - g - \frac{\partial \xi}{\partial t}\right]v_x + \left[k - f + \frac{\partial \xi}{\partial x} - \frac{\partial \tau}{\partial t}\right]u_x = 0, \tag{4.120a}$$

$$\left[\frac{\partial f}{\partial t} - x^4\frac{\partial \ell}{\partial x}\right]u + \left[\frac{\partial g}{\partial t} - x^4\frac{\partial k}{\partial x}\right]v + \left[g - \frac{\partial \xi}{\partial t} + x^4\left(\frac{\partial \tau}{\partial x} - \ell\right)\right]u_x + \left[x^4\left(f - k + \frac{\partial \xi}{\partial x} - \frac{\partial \tau}{\partial t}\right) - 4x^3\xi\right]v_x = 0. \tag{4.120b}$$

Each of the equations (4.120a,b) must hold for arbitrary values of x, t, u, v, u_x, v_x. Consequently, we obtain a set of eight *symmetry determining equations* for ξ, τ, f, g, k, ℓ, that simplify to the equations

$$\frac{\partial k}{\partial t} - \frac{\partial g}{\partial x} = 0, \tag{4.121a}$$

$$\frac{\partial \ell}{\partial t} - \frac{\partial f}{\partial x} = 0, \tag{4.121b}$$

$$x^4 \ell - g = 0, \tag{4.121c}$$

$$x^4\frac{\partial \tau}{\partial x} - \frac{\partial \xi}{\partial t} = 0, \tag{4.121d}$$

$$x^4\frac{\partial k}{\partial x} - \frac{\partial g}{\partial t} = 0, \tag{4.121e}$$

$$x^4\frac{\partial \ell}{\partial x}-\frac{\partial f}{\partial t}=0, \tag{4.121f}$$

$$x\left[\frac{\partial \tau}{\partial t}-\frac{\partial \xi}{\partial x}\right]+2\xi=0, \tag{4.121g}$$

$$k-f+\frac{\partial \xi}{\partial x}-\frac{\partial \tau}{\partial t}=0. \tag{4.121h}$$

It is left to Exercise 4.3-3 to show that the solution of the symmetry determining equations (4.121a–h) is given by

$$\xi(x,t)=\alpha x+2\beta xt, \tag{4.122a}$$

$$\tau(x,t)=-\alpha t-\beta(x^{-2}+t^2)+\gamma, \tag{4.122b}$$

$$f(x,t)=3\beta t+\delta, \tag{4.122c}$$

$$g(x,t)=-\beta x, \tag{4.122d}$$

$$k(x,t)=-2\alpha-\beta t+\delta, \tag{4.122e}$$

$$\ell(x,t)=-\beta x^{-3}, \tag{4.122f}$$

where $\alpha,\beta,\gamma,\delta$ are four arbitrary constants. Hence, the point symmetry generators admitted by the system of wave equations (4.118a,b) are given by

$$\mathrm{X}_1=\frac{\partial}{\partial t},\quad \mathrm{X}_2=x\frac{\partial}{\partial x}-t\frac{\partial}{\partial t}-2v\frac{\partial}{\partial v},$$

$$\mathrm{X}_3=2xt\frac{\partial}{\partial x}-(x^{-2}+t^2)\frac{\partial}{\partial t}+(3tu-xv)\frac{\partial}{\partial u}-(tv+x^{-3}u)\frac{\partial}{\partial v},\quad \mathrm{X}_4=u\frac{\partial}{\partial u}+v\frac{\partial}{\partial v}.$$

These infinitesimal generators determine a nontrivial four-parameter Lie group of point transformations acting on (x,t,u,v) – space. The nonzero commutators of the corresponding Lie algebra are given by

$$[\mathrm{X}_1,\mathrm{X}_2]=-\mathrm{X}_1,\quad [\mathrm{X}_1,\mathrm{X}_3]=2\mathrm{X}_2+3\mathrm{X}_4,\quad [\mathrm{X}_2,\mathrm{X}_3]=-\mathrm{X}_3.$$

One can show that the Lie algebra with basis generators $\mathrm{Y}_1=\mathrm{X}_1$, $\mathrm{Y}_2=\mathrm{X}_2+\frac{3}{2}\mathrm{X}_4$, $\mathrm{Y}_3=\mathrm{X}_3$, is isomorphic to the Lie algebra of SO(2,1).

Consider the infinitesimal generator X_3 (parameter β). We find the resulting invariant solutions $(u,v)=(\Theta_1(x,t),\Theta_2(x,t))$ by both procedures outlined in Section 4.3.1.

(I) Invariant Form Method. Here, the characteristic equations (4.104) become

$$\frac{dx}{2xt}=-\frac{dt}{x^{-2}+t^2}=\frac{du}{3tu-xv}=-\frac{dv}{tv+x^{-3}u}. \tag{4.123}$$

The integration of the first ODE in (4.123), i.e., $dx/dt = -2xt/(x^{-2}+t^2)$, yields the similarity variable (invariant)

$$y_1 = \zeta = \text{const} = x^{-1} - xt^2. \tag{4.124}$$

To determine the other invariants of (4.123), we consider the corresponding system of first-order characteristic ODEs

$$\frac{dx}{d\varepsilon} = 2xt, \tag{4.125a}$$

$$\frac{dt}{d\varepsilon} = -(x^{-2}+t^2), \tag{4.125b}$$

$$\frac{du}{d\varepsilon} = 3tu - xv, \tag{4.125c}$$

$$\frac{dv}{d\varepsilon} = -(tv + x^{-3}u). \tag{4.125d}$$

After substituting the constant of integration (4.124) into the system of ODEs (4.125a,b), from (4.125b) we obtain

$$\zeta^{-1}xt + \varepsilon = \text{const} = E. \tag{4.126}$$

The constant E is related to the invariance of (4.125a–d) under translations in ε. Without loss of generality, we can set $E = 0$. From (4.125a–d), we get

$$\frac{d^2v}{d\varepsilon^2} + 4t\frac{dv}{d\varepsilon} + 2(t^2 - x^{-2})v = 0.$$

Then using (4.124), one can show that this ODE simplifies to

$$\frac{d^2}{d\varepsilon^2}(xv) = 0.$$

Hence,

$$xv = v^1\varepsilon + v^2, \tag{4.127a}$$

where v^1 and v^2 are constants of integration. Equation (4.125d) now yields

$$u = x^2[t(v^1\varepsilon + v^2) - v^1]. \tag{4.127b}$$

Using $\varepsilon = -\zeta^{-1}xt$ [cf. (4.126)], we can eliminate ε from (4.127a,b), and thus obtain

$$u = x^2[-xt^2\zeta^{-1}v^1 + tv^2 - v^1],$$
$$v = x^{-1}[-xt\zeta^{-1}v^1 + v^2].$$

The constants ζ, v^1, v^2 are independent invariants of (4.123); ζ is the similarity variable for the invariant solutions resulting from X_3. These invariant solutions are now found by replacing v^1, v^2 by functions of ζ, i.e., $v^1 = F(\zeta), v^2 = G(\zeta)$ [$\Phi^1 = F, \Phi^2 = G$ in the invariant form (4.105)]. Then

$$u = x^2[-xt^2\zeta^{-1}F(\zeta) + tG(\zeta) - F(\zeta)], \tag{4.128a}$$

$$v = x^{-1}[-xt\zeta^{-1}F(\zeta) + G(\zeta)]. \tag{4.128b}$$

We now substitute (4.128a,b) into the wave equations (4.118a,b) to determine $F(\zeta)$ and $G(\zeta)$. Equations (4.118a,b), respectively, lead to the system

$$xt[2G(\zeta) + \zeta G'(\zeta)] + [F'(\zeta) - \zeta^{-1}F(\zeta)] = 0,$$

$$[2G(\zeta) + \zeta G'(\zeta)] + \frac{x^2 t}{1 - x^2 t^2}[\zeta F'(\zeta) - F(\zeta)] = 0.$$

Consequently,

$$2G(\zeta) + \zeta G'(\zeta) = 0,$$

$$\zeta F'(\zeta) - F(\zeta) = 0,$$

and thus,

$$G(\zeta) = a\zeta^{-2}, \quad F(\zeta) = b\zeta,$$

where a, b are arbitrary constants. This yields the two linearly independent solutions

$$(u, v) = (x, t), \tag{4.129a}$$

and

$$(u, v) = \left(\frac{x^4 t}{[1 - x^2 t^2]^2}, \frac{x}{[1 - x^2 t^2]^2}\right), \tag{4.129b}$$

of the system of wave equations (4.118a,b).

(II) Direct Substitution Method. Here, the invariant surface conditions (4.106) become

$$u_x = \tfrac{1}{2}[[x^{-3}t^{-1} + x^{-1}t]u_t + 3x^{-1}u - t^{-1}v], \tag{4.130a}$$

$$v_x = \tfrac{1}{2}[[x^{-3}t^{-1} + x^{-1}t]v_t - x^{-1}v - x^{-4}t^{-1}u]. \tag{4.130b}$$

Using (4.130a,b), we now eliminate derivatives of u and v with respect to x from (4.118a,b) so that these equations become

$$v_t = \tfrac{1}{2}[[x^{-3}t^{-1} + x^{-1}t]u_t + 3x^{-1}u - t^{-1}v], \tag{4.131a}$$

$$u_t = \tfrac{1}{2}[[xt^{-1} + x^3 t]v_t - x^3 v - t^{-1}u], \tag{4.131b}$$

which is a system of first-order ODEs with independent variable t and parameter x.

Expressing (4.131a,b) in solved form in terms of v_t, u_t, and setting $\sigma = xt$, we obtain

$$v_\sigma = \frac{\sigma(3+\sigma^2)v + (1-5\sigma^2)x^{-2}u}{(1-\sigma^2)^2}, \tag{4.132a}$$

$$u_\sigma = \frac{-\sigma(1+3\sigma^2)u + (3\sigma^2-1)x^2 v}{(1-\sigma^2)^2}. \tag{4.132b}$$

Next we take $\partial/\partial\sigma$ of (4.132a) and eliminate u_σ through (4.132b). Finally, we eliminate u through expressing (4.132a) in the form

$$u = x^2\left(\frac{\sigma(3+\sigma^2)v + (\sigma^2-1)v_\sigma}{5\sigma^2-1}\right). \tag{4.133}$$

This leads to the second-order ODE

$$(1-5\sigma^2)(\sigma^2-1)\frac{\partial^2 v}{\partial\sigma^2} - 4\sigma(5\sigma^2+1)\frac{\partial v}{\partial\sigma} + 4(5\sigma^2+1)v = 0. \tag{4.134}$$

Linearly independent solutions of (4.134) are given by

$$v = \sigma, \quad v = (1-\sigma^2)^{-2}.$$

Hence,

$$v = A(x)\sigma + B(x)(1-\sigma^2)^{-2} \tag{4.135a}$$

is the general solution of ODE (4.134) where $A(x), B(x)$ are arbitrary functions. From (4.133), we then get

$$u = x^2 A(x) + x^2 B(x)\sigma(1-\sigma^2)^{-2}. \tag{4.135b}$$

After substituting (4.135a,b) into the given PDE (4.118a), we find that

$$[A(x) + xA'(x)] = \frac{\sigma}{(1-\sigma^2)^2}[B(x) - xB'(x)]. \tag{4.136}$$

Since (4.136) must hold for all values of x and σ, we get

$$A(x) = ax^{-1}, \quad B(x) = bx,$$

where a and b are arbitrary constants. In turn, this leads to the solutions (4.129a,b).

Note that the Direct Substitution Method avoids integration of the characteristic equations (4.104) and, hence, it is more adaptable to automatic computation through symbolic manipulation programs.

(2) *Nonlinear Heat Conduction Equation*
Consider again the nonlinear heat conduction equation

$$u_t = (K(u)u_x)_x. \tag{4.137}$$

We form an associated system of PDEs

$$v_t = K(u)u_x, \tag{4.138a}$$

$$v_x = u. \tag{4.138b}$$

Note that if the pair $(u(x,t), v(x,t))$ solves the system of PDEs (4.138a,b), then $u(x,t)$ solves the nonlinear heat conduction equation (4.137), and $v(x,t)$ solves the nonlinear diffusion equation

$$v_t = K(v_x)v_{xx}.$$

The Lie group of point transformations admitted by the system of PDEs (4.138a,b) can yield a symmetry admitted by the scalar PDE (4.137) that is neither a point transformation nor even a local transformation. For a full discussion of how to find and use such *nonlocal symmetries* for a given system of PDEs, see Bluman, Kumei, and Reid (1988), Bluman and Kumei (1989b, Chapter 7)), Bluman and Doran-Wu (1995), and Anco and Bluman (1996, 1997b).

We now completely classify the invariance properties of the system of PDEs (4.138a,b) in terms of its admitted point symmetries. We leave many details to the reader. Suppose the system of PDEs (4.138a,b) admits an infinitesimal generator of the form

$$\mathrm{X} = \xi(x,t,u,v)\frac{\partial}{\partial x} + \tau(x,t,u,v)\frac{\partial}{\partial t} + \eta^u(x,t,u,v)\frac{\partial}{\partial u} + \eta^v(x,t,u,v)\frac{\partial}{\partial v}. \tag{4.139}$$

Here, the system of symmetry determining equations (4.102) becomes

$$\eta_t^{(1)v} = K'(u)u_x\eta^u + K(u)\eta_x^{(1)u}, \tag{4.140a}$$

$$\eta_x^{(1)v} = \eta^u, \tag{4.140b}$$

with

$$v_t = K(u)u_x, \quad v_x = u, \tag{4.140c}$$

where $\eta_t^{(1)v}, \eta_x^{(1)u}, \eta_x^{(1)v}$ are given by (2.135). After eliminating v_x and v_t through substitution from the given system of PDEs (4.138a,b), we obtain

$$\left[\frac{\partial\eta^v}{\partial t} - u\frac{\partial\tau}{\partial t} - K(u)\left(\frac{\partial\eta^u}{\partial x} + u\frac{\partial\eta^u}{\partial v}\right)\right] + \left[K(u)\left(\frac{\partial\eta^v}{\partial v} - \frac{\partial\tau}{\partial t} - \frac{\partial\eta^u}{\partial u} + \frac{\partial\xi}{\partial x}\right) - K'(u)\eta^u\right]u_x$$
$$+\left[\frac{\partial\eta^v}{\partial u} - u\frac{\partial\xi}{\partial t} + K(u)\left(\frac{\partial\tau}{\partial x} + u\frac{\partial\tau}{\partial v}\right)\right]u_t + K(u)\left[\frac{\partial\xi}{\partial u} - K(u)\frac{\partial\tau}{\partial v}\right](u_x)^2 = 0, \tag{4.141a}$$

$$\left[\frac{\partial\eta^v}{\partial x} + u\frac{\partial\eta^v}{\partial v} - u\frac{\partial\xi}{\partial x} - u^2\frac{\partial\xi}{\partial v} - \eta^u\right] + \left[\frac{\partial\eta^v}{\partial u} - u\frac{\partial\xi}{\partial u} - K(u)\left(\frac{\partial\tau}{\partial x} + u\frac{\partial\tau}{\partial v}\right)\right]u_x$$
$$- K(u)\frac{\partial\tau}{\partial u}(u_x)^2 = 0. \tag{4.141b}$$

Each of the symmetry determining equations (4.141a,b) must hold for arbitrary values of x,t,u,v,u_x,u_t. Consequently, we obtain a set of seven symmetry determining equations for ξ,τ,η^u,η^v, that simplify to the equations

$$\frac{\partial \tau}{\partial u} = 0, \tag{4.142a}$$

$$\frac{\partial \tau}{\partial x} + u\frac{\partial \tau}{\partial v} = 0, \tag{4.142b}$$

$$\frac{\partial \eta^v}{\partial u} - u\frac{\partial \xi}{\partial u} = 0, \tag{4.142c}$$

$$\frac{\partial \xi}{\partial u} - K(u)\frac{\partial \tau}{\partial v} = 0, \tag{4.142d}$$

$$\frac{\partial \eta^v}{\partial t} - u\frac{\partial \xi}{\partial t} - K(u)\left(\frac{\partial \eta^u}{\partial x} + u\frac{\partial \eta^u}{\partial v}\right) = 0, \tag{4.142e}$$

$$\frac{\partial \eta^v}{\partial v} - \frac{\partial \tau}{\partial t} - \frac{\partial \eta^u}{\partial u} + \frac{\partial \xi}{\partial x} - \frac{K'(u)}{K(u)}\eta^u = 0, \tag{4.142f}$$

$$\frac{\partial \eta^v}{\partial x} + u\frac{\partial \eta^v}{\partial v} - u\frac{\partial \xi}{\partial x} - u^2\frac{\partial \xi}{\partial v} - \eta^u = 0. \tag{4.142g}$$

The solution of the set of symmetry determining equations (4.142a–g) is left to Exercise 4.3-5. The results can be summarized as follows [Bluman, Kumei, and Reid (1988)].

Case I. $K(u)$ arbitrary.
Here, the given system of PDEs (4.138a,b) admits a four-parameter Lie group of point transformations with its infinitesimal generators given by

$$\mathrm{X}_1 = \frac{\partial}{\partial x}, \quad \mathrm{X}_2 = \frac{\partial}{\partial t}, \quad \mathrm{X}_3 = x\frac{\partial}{\partial x} + 2t\frac{\partial}{\partial t} + v\frac{\partial}{\partial v}, \quad \mathrm{X}_4 = \frac{\partial}{\partial v}. \tag{4.143}$$

Case II. $K(u) = \lambda(u+\kappa)^\nu$.
Here, the system of PDEs (4.138a,b) admits a five-parameter Lie group of point transformations with its infinitesimal generators given by (4.143) and

$$\mathrm{X}_5 = x\frac{\partial}{\partial x} + 2\nu^{-1}(u+\kappa)\frac{\partial}{\partial u} + [(1+2\nu^{-1})v + 2\nu^{-1}\kappa x]\frac{\partial}{\partial v}. \tag{4.144}$$

Case III. $K(u) = \lambda(u+\kappa)^{-2}$.
Here, the system of PDEs (4.138a,b) admits an infinite-parameter Lie group of point transformations with its infinitesimal generators given by (4.143), (4.144) [with $\nu = -2$],

$$X_6 = -x(v+\kappa x)\frac{\partial}{\partial x} + (u+\kappa)[v + x(u+2\kappa)]\frac{\partial}{\partial u} + [2\lambda t + \kappa x(v+\kappa x)]\frac{\partial}{\partial v},$$

$$X_7 = -x[(v+\kappa x)^2 + 2\lambda t]\frac{\partial}{\partial x} + 4\lambda t^2\frac{\partial}{\partial t} + (u+\kappa)[6\lambda t + (v+\kappa x)^2 + 2x(u+\kappa)(v+\kappa x)]\frac{\partial}{\partial u}$$
$$+ [\kappa x(v+\kappa x)^2 + 2\lambda t(2v+3\kappa x)\frac{\partial}{\partial v},$$

$$X_\infty = \phi(z,t)\frac{\partial}{\partial x} - (u+\kappa)^2\frac{\partial \phi(z,t)}{\partial z}\frac{\partial}{\partial u} - \kappa\phi(z,t)\frac{\partial}{\partial v}, \tag{4.145}$$

where $z = v + \kappa x$, and $w = \phi(z,t)$ is any solution of the linear heat equation

$$w_t = \lambda w_{zz}.$$

The use of the infinitesimal generator X_∞ to map the nonlinear heat conduction equation

$$u_t = \lambda((u+\kappa)^{-2}u_x)_x$$

to a linear PDE is discussed in Kumei and Bluman (1982) and Bluman and Kumei (1990a) [see also Bluman and Kumei (1989b, Chapter 6)].

Case IV. $K(u) = \dfrac{1}{u^2+pu+q}\exp\left[r\displaystyle\int\frac{du}{u^2+pu+q}\right]$, where p,q,r are arbitrary constants such that $p^2 - 4q - r^2 \neq 0$.

Here, the system of PDEs (4.138a,b) admits a five-parameter Lie group of point transformations with its infinitesimal generators given by (4.143) and

$$X_5 = v\frac{\partial}{\partial x} + (r-p)t\frac{\partial}{\partial t} - (u^2+pu+q)\frac{\partial}{\partial u} - (qx+pv)\frac{\partial}{\partial v}. \tag{4.146}$$

(3) *Wave Equation for an Inhomogeneous Medium*

Consider again the wave equation in an inhomogeneous medium with a variable wave speed $c(x)$:

$$u_{tt} = c^2(x)u_{xx}. \tag{4.147}$$

We form an associated system of first-order PDEs

$$v_t = u_x, \tag{4.148a}$$

$$u_t = c^2(x)v_x. \tag{4.148b}$$

If the pair $(u(x,t), v(x,t))$ solves (4.148a,b), then $u(x,t)$ solves the wave equation (4.147) and $v(x,t)$ solves the hyperbolic equation

$$v_{tt} = (c^2(x)v_x)_x.$$

We now give a complete group classification of the system of PDEs (4.148a,b) with respect to its invariance under a Lie group of point transformations. Suppose (4.148a,b) admits an infinitesimal generator of the form (4.113). It is left to Exercise 4.3-6 to show that the set of symmetry determining equations for $\xi(x,t)$, $\tau(x,t)$, $f(x,t)$, $g(x,t)$, $k(x,t)$, $\ell(x,t)$ is given by

$$\frac{\partial k}{\partial t}-\frac{\partial g}{\partial x}=0, \tag{4.149a}$$

$$\frac{\partial \ell}{\partial t}-\frac{\partial f}{\partial x}=0, \tag{4.149b}$$

$$c^2(x)\ell-g=0, \tag{4.149c}$$

$$c^2(x)\frac{\partial \tau}{\partial x}-\frac{\partial \xi}{\partial t}=0, \tag{4.149d}$$

$$c^2(x)\frac{\partial k}{\partial x}-\frac{\partial g}{\partial t}=0, \tag{4.149e}$$

$$c^2(x)\frac{\partial \ell}{\partial x}-\frac{\partial f}{\partial t}=0, \tag{4.149f}$$

$$c(x)\left[\frac{\partial \tau}{\partial t}-\frac{\partial \xi}{\partial x}\right]+c'(x)\xi=0, \tag{4.149g}$$

$$k-f+\frac{\partial \xi}{\partial x}-\frac{\partial \tau}{\partial t}=0. \tag{4.149h}$$

The integrability conditions arising from the determining equations (4.149a–c,f) lead to $g(x,t)$ satisfying

$$\frac{\partial g}{\partial x}H(x)+gH'(x)=0, \tag{4.150}$$

where

$$H(x)=\frac{c'(x)}{c(x)}. \tag{4.151}$$

Consequently,

$$g(x,t)=-\frac{a(t)}{2H(x)}, \tag{4.152}$$

in terms of an arbitrary function $a(t)$. Two cases arise that depend on whether or not the wave speed $c(x)$ satisfies the ODE

$$cc'\left(\frac{c}{c'}\right)''=\text{const}=\mu, \tag{4.153}$$

for some constant μ. If $c(x)$ satisfies ODE (4.153), then $a(t)$ satisfies the ODE

$$a''(t) = \mu a(t).$$

If $c(x)$ does not satisfy ODE (4.153) for any constant μ, then $a(t) \equiv 0$, and the corresponding system of PDEs (4.148a,b) only admits the two obvious infinitesimal generators

$$\mathrm{X}_1 = \frac{\partial}{\partial t}, \quad \mathrm{X}_2 = u\frac{\partial}{\partial u} + v\frac{\partial}{\partial v}. \tag{4.154}$$

If the wave speed $c(x)$ satisfies ODE (4.153), then one can show that the corresponding system of PDEs (4.148a,b) admits a four-parameter Lie group of point transformations. In terms of the solutions of ODE (4.153), the group classification is summarized as follows, modulo scalings and translations in x [Bluman and Kumei (1987, 1988)]:

(I) $\mu = 0$.
In this case,

$$c(x) = e^x \text{ or } x^C, \tag{4.155}$$

where C is an arbitrary constant.

(II) $\mu \neq 0$.
Here, the ODE (4.153) cannot be solved explicitly but reduces to one of the following first-order ODEs:

$$c' = \upsilon^{-1}\sin(\upsilon \log c); \tag{4.156a}$$

$$c' = \upsilon^{-1}\sinh(\upsilon \log c); \tag{4.156b}$$

$$c' = \log c; \tag{4.156c}$$

$$c' = \upsilon^{-1}\cosh(\upsilon \log c); \tag{4.156d}$$

$\upsilon \neq 0$ is an arbitrary constant. If $c(x) = \phi(x, \upsilon)$ is a solution of any one of the ODEs (4.156a–d), then the corresponding general solution of ODE (4.153) is given by

$$c(x) = K\phi(Lx + M, \upsilon),$$

where $K^2L^2 = |\mu|$ for arbitrary constants L, M, υ. The admitted infinitesimal generators for the various subcases include:

Case I. $\mu = 0$.

Case Ia. $c(x) = x^C, C \neq 0,1$.

$$\mathrm{X}_1 = \frac{\partial}{\partial t}, \quad \mathrm{X}_2 = x\frac{\partial}{\partial x} + (1-C)t\frac{\partial}{\partial t} - Cv\frac{\partial}{\partial v},$$

$$X_3 = 2xt\frac{\partial}{\partial x} + \left[(1-C)t^2 + \frac{x^{2-2C}}{1-C}\right]\frac{\partial}{\partial t} + [(2C-1)tu - xv]\frac{\partial}{\partial u} - [tv + x^{1-2C}u]\frac{\partial}{\partial v},$$

$$X_4 = u\frac{\partial}{\partial u} + v\frac{\partial}{\partial v}. \tag{4.157}$$

Case Ib. $c(x) = x$.

$$X_1 = \frac{\partial}{\partial t}, \quad X_2 = x\frac{\partial}{\partial x} - v\frac{\partial}{\partial v}, \quad X_3 = 2xt\frac{\partial}{\partial x} + 2\log x\frac{\partial}{\partial t} + [tu - xv]\frac{\partial}{\partial u} - [tv + x^{-1}u]\frac{\partial}{\partial v},$$

$$X_4 = u\frac{\partial}{\partial u} + v\frac{\partial}{\partial v}. \tag{4.158}$$

Case Ic. $c(x) = e^x$.

$$X_1 = \frac{\partial}{\partial t}, \quad X_2 = \frac{\partial}{\partial x} - t\frac{\partial}{\partial t} - v\frac{\partial}{\partial v},$$

$$X_3 = -4t\frac{\partial}{\partial x} + 2[t^2 + e^{-2x}]\frac{\partial}{\partial t} + 2[-2tu + v]\frac{\partial}{\partial u} + 2e^{-2x}u\frac{\partial}{\partial v}, \quad X_4 = u\frac{\partial}{\partial u} + v\frac{\partial}{\partial v}. \tag{4.159}$$

Case II. $\mu \neq 0$.
If the wave speed $c(x)$ satisfies either of the ODEs (4.156a or b), then the given system of PDEs (4.148a,b) admits

$$X_1 = \frac{\partial}{\partial t},$$

$$X_2 = e^t\left\{2c(c')^{-1}\frac{\partial}{\partial x} + 2[(c(c')^{-1})' - 1]\frac{\partial}{\partial t} + [(2 - (c(c')^{-1})')u - c(c')^{-1}v]\frac{\partial}{\partial u} - [(c(c')^{-1})'v + (cc')u\frac{\partial}{\partial v}\right\},$$

$$X_3 = e^{-t}\left\{2c(c')^{-1}\frac{\partial}{\partial x} + 2[1 - (c(c')^{-1})']\frac{\partial}{\partial t} + [(2 - (c(c')^{-1})')u + c(c')^{-1}v]\frac{\partial}{\partial u} - [(c(c')^{-1})'v - (cc')u\frac{\partial}{\partial v}\right\},$$

$$X_4 = u\frac{\partial}{\partial u} + v\frac{\partial}{\partial v}. \tag{4.160}$$

The resulting invariant solutions appear in Bluman and Kumei (1987, 1988). Special classes of invariant solutions will be considered in Section 4.4.3.

Lisle (1992) gave a partial group classification of the diffusion convection system of PDEs

$$v_x = u,$$

$$v_t = D(u)u_x - K(u),$$

by delineating those classes of functions $D(u), K(u)$ for which the system admits extra point symmetries (including those that induce nonlocal symmetries of the scalar diffusion convection equation).

Akhatov, Gazizov, and Ibragimov (1988) [see also Ibragimov (1995)] present the group classification of the one-dimensional system of adiabatic gas equations given by

$$\rho_t + v\rho_x + \rho v_x = 0,$$

$$\rho(v_t + vv_x) + p_x = 0,$$

$$\rho(p_t + vp_x) + B(p,\rho)v_x = 0,$$

where $\rho(x,t)$ is the density of the gas, $p(x,t)$ is the pressure, $v(x,t)$ is the velocity, and $B(p,\rho)$ is the constitutive law.

Many further examples are exhibited in Ibragimov (1995).

EXERCISES 4.3

1. Prove Theorem 4.3-1.
2. Show that the infinitesimal generators for point symmetries admitted by the system of PDEs (4.118a,b) are of the form (4.113).
3. Show that (4.122a–f) is the general solution of the set of symmetry determining equations (4.121a–h).
4. The linear system of wave equations (4.118a,b) admits the infinitesimal generator

$$\mathrm{X} = \mathrm{X}_3 + s\mathrm{X}_4 = 2xt\frac{\partial}{\partial x} - [x^{-2} + t^2]\frac{\partial}{\partial t} + [(3t+s)u - xv]\frac{\partial}{\partial u} + [(s-t)v - x^{-3}u]\frac{\partial}{\partial v}.$$

   (a) For invariant solutions resulting from invariance under X, show that the invariant form is given by

$$u = x^2 e^{-sxt\zeta^{-1}}[-xt^2\zeta^{-1}F(\zeta;s) + tG(\zeta;s) - F(\zeta;s)],$$

$$v = x^{-1}e^{-sxt\zeta^{-1}}[-xt^2\zeta^{-1}F(\zeta;s) + G(\zeta;s)],$$

   where $F(\zeta;s)$ and $G(\zeta;s)$ are arbitrary functions of ζ and s. The similarity variable ζ is given by (4.124).

   (b) Determine the coupled system of ODEs that are satisfied by $F(\zeta;s), G(\zeta;s)$. Simplify and express the solution in terms of special functions.

   (c) Derive these invariant solutions by the Direct Substitution Method.

5. Complete the group classification of the system of PDEs (4.138a,b) and derive (4.143)–(4.146).

6. Derive the set of symmetry determining equations (4.149a–h).

7. Consider the two-dimensional nonstationary boundary layer equations

$$u_t + uu_x + vu_y + p_x = u_{yy}, \quad p_y = 0, \quad u_x + v_y = 0, \tag{4.161}$$

[$u(x,y,t), v(x,y,t)$ are components of the velocity vector; $p(x,t)$ is the pressure; without loss of generality, the viscosity and density constants are set to equal one]. Show that the admitted point symmetries of (4.161) are given by

$$\mathrm{X}_1 = \frac{\partial}{\partial t}, \quad \mathrm{X}_2 = 2x\frac{\partial}{\partial x} + y\frac{\partial}{\partial y} + 2t\frac{\partial}{\partial t} - v\frac{\partial}{\partial v}, \quad \mathrm{X}_3 = x\frac{\partial}{\partial x} + u\frac{\partial}{\partial u} + 2p\frac{\partial}{\partial p},$$

$$\mathrm{X}_{\infty_1} = \alpha(t)\frac{\partial}{\partial x} + \alpha'(t)\frac{\partial}{\partial u} - x\alpha''(t)\frac{\partial}{\partial p}, \quad \mathrm{X}_{\infty_2} = \beta(t)\frac{\partial}{\partial y} + \beta'(t)\frac{\partial}{\partial v}, \quad \mathrm{X}_{\infty_3} = \gamma(t)\frac{\partial}{\partial p},$$

where $\alpha(t)$, $\beta(t)$, and $\gamma(t)$ are arbitrary sufficiently smooth functions of t [Ovsiannikov (1982)].

8. Show that the two-dimensional steady-state boundary layer equations [$u_t = 0$ in the system of PDEs (4.161)] admit

$$\mathrm{X}_1 = \frac{\partial}{\partial t}, \quad \mathrm{X}_2 = 2x\frac{\partial}{\partial x} + y\frac{\partial}{\partial y} - v\frac{\partial}{\partial v}, \quad \mathrm{X}_3 = x\frac{\partial}{\partial x} + u\frac{\partial}{\partial u} + 2p\frac{\partial}{\partial p}, \quad \mathrm{X}_4 = \frac{\partial}{\partial p},$$

$$\mathrm{X}_\infty = \phi(x)\frac{\partial}{\partial y} + u\phi'(x)\frac{\partial}{\partial v},$$

where $\phi(x)$ is an arbitrary differentiable function [Ovsiannikov (1982)].

9. Show that the three-dimensional incompressible Navier–Stokes equations [x_1, x_2, x_3 are spatial variables; t is time; u^1, u^2, u^3 are components of the velocity vector; p is pressure; $\nabla^2 = \sum_{i=1}^3 \partial^2/\partial x_i^2$; without loss of generality, the viscosity is set to equal 1],

$$\sum_{i=1}^{3} u^i_{x_i} = 0, \quad u^j_t + \sum_{i=1}^{3} u^i u^j_{x_i} + p_{x_j} = \nabla^2 u^j, \quad j = 1,2,3,$$

admit

$$\mathrm{X}_1 = \frac{\partial}{\partial t}, \quad \mathrm{X}_2 = \sum_{i=1}^{3}\left[x_i\frac{\partial}{\partial x_i} - u^i\frac{\partial}{\partial u^i}\right] + 2\left[t\frac{\partial}{\partial t} - p\frac{\partial}{\partial p}\right],$$

$$\mathrm{X}_3 = x_2\frac{\partial}{\partial x_1} - x_1\frac{\partial}{\partial x_2} + u^2\frac{\partial}{\partial u^1} - u^1\frac{\partial}{\partial u^2}, \quad \mathrm{X}_4 = x_3\frac{\partial}{\partial x_1} - x_1\frac{\partial}{\partial x_3} + u^3\frac{\partial}{\partial u^1} - u^1\frac{\partial}{\partial u^3},$$

$$\mathrm{X}_5 = x_3\frac{\partial}{\partial x_2} - x_2\frac{\partial}{\partial x_3} + u^3\frac{\partial}{\partial u^2} - u^2\frac{\partial}{\partial u^3},$$

$$X_{\infty_j} = \alpha_j(t)\frac{\partial}{\partial x_j} + \alpha_j'(t)\frac{\partial}{\partial u^j} - x_j\alpha_j''(t)\frac{\partial}{\partial p}, \quad j = 1,2,3,$$

$$X_{\infty_4} = \beta(t)\frac{\partial}{\partial p},$$

where $\beta(t), \alpha_j(t), j = 1,2,3,$ are arbitrary functions. [See Boisvert, Ames, and Srivastava (1983). In this paper various invariant solutions are given.]

10. If the complex-valued wave function $\psi(x,t)$ satisfies the cubic nonlinear Schrödinger equation

$$i\psi_t = -\psi_{xx} + V(x)\psi + |\psi|^2\psi \tag{4.162}$$

for an external potential $V(x)$, then the canonical transformation

$$\psi(x,t) = \sqrt{v}e^{-iu/2},$$

where $u(x,t)$ and $v(x,t)$ are real-valued functions, transforms PDE (4.162) into the nonlinear system of PDEs, representing a Madelung fluid, given by

$$u_t + \tfrac{1}{2}(u_x)^2 + 2v + 2V(x) = 2v^{-1/2}(v^{1/2})_{xx}, \quad v_t + (vu_x)_x = 0. \tag{4.163}$$

Show that if $V(x) = -x$, then the system of PDEs (4.163) admits the infinitesimal generators

$$X_1 = (x+3t^2)\frac{\partial}{\partial x} + 2t\frac{\partial}{\partial t} + (6xt + 2t^2)\frac{\partial}{\partial u} - 2v\frac{\partial}{\partial v}, \quad X_2 = \frac{\partial}{\partial t},$$

$$X_3 = t\frac{\partial}{\partial x} + (x+t^2)\frac{\partial}{\partial u}, \quad X_4 = \frac{\partial}{\partial u}, \quad X_5 = \frac{\partial}{\partial x} + 2t\frac{\partial}{\partial u}.$$

[See Baumann and Nonnenmacher (1987), where these infinitesimal generators and resulting invariant solutions are given.]

11. Show that the coupled two-dimensional nonlinear system of Schrödinger equations

$$iu_t - u_{xx} + u_{yy} + |u|^2 u - 2uv = 0, \quad v_{xx} + v_{yy} - (|u|^2)_{xx} = 0,$$

where $u(x,t)$ and $v(x,t)$ are complex-valued functions, admits the infinitesimal generators

$$X_1 = \frac{\partial}{\partial x}, \quad X_2 = \frac{\partial}{\partial y}, \quad X_3 = \frac{\partial}{\partial t}, \quad X_4 = x\frac{\partial}{\partial x} + y\frac{\partial}{\partial y} + 2t\frac{\partial}{\partial t} - u\frac{\partial}{\partial u} - 2v\frac{\partial}{\partial v},$$

$$X_5 = -t\frac{\partial}{\partial x} + \tfrac{1}{2}ixu\frac{\partial}{\partial u}, \quad X_6 = t\frac{\partial}{\partial y} + \tfrac{1}{2}iyu\frac{\partial}{\partial u},$$

$$X_7 = xt\frac{\partial}{\partial x} + yt\frac{\partial}{\partial y} + t^2\frac{\partial}{\partial t} - [t + \tfrac{1}{4}i(x^2 - y^2)]u\frac{\partial}{\partial u} - 2tv\frac{\partial}{\partial v}, \quad X_8 = iu\frac{\partial}{\partial u}.$$

[Tajiri and Hagiwara (1983).]

12. Show that the shallow water wave equations for two-dimensional flow over a flat bottom, given by

$$u_t + uu_x + vu_y + gH_x = 0, \quad v_t + vu_x + vu_y + gH_y = 0,$$

$$H_t + uH_x + vH_y + H(u_x + v_y) = 0,$$

where $u(x,y,t), v(x,y,t)$ are the components of the velocity vector, $H(x,y,t)$ is the depth of the water, and g = const is the acceleration due to gravity, admit the point symmetries

$$\mathrm{X}_1 = \frac{\partial}{\partial t}, \quad \mathrm{X}_2 = \frac{\partial}{\partial x}, \quad \mathrm{X}_3 = \frac{\partial}{\partial y}, \quad \mathrm{X}_4 = t\frac{\partial}{\partial x} + \frac{\partial}{\partial u}, \quad \mathrm{X}_5 = t\frac{\partial}{\partial y} + \frac{\partial}{\partial v},$$

$$\mathrm{X}_6 = y\frac{\partial}{\partial x} - x\frac{\partial}{\partial y} + v\frac{\partial}{\partial u} - u\frac{\partial}{\partial v}, \quad \mathrm{X}_7 = t\frac{\partial}{\partial t} + x\frac{\partial}{\partial x} + y\frac{\partial}{\partial y},$$

$$\mathrm{X}_8 = x\frac{\partial}{\partial x} + y\frac{\partial}{\partial y} + u\frac{\partial}{\partial u} + v\frac{\partial}{\partial v} + 2H\frac{\partial}{\partial H},$$

$$\mathrm{X}_9 = t^2\frac{\partial}{\partial t} + tx\frac{\partial}{\partial x} + ty\frac{\partial}{\partial y} + (x - tu)\frac{\partial}{\partial u} + (x - tv)\frac{\partial}{\partial v} - 2tH\frac{\partial}{\partial H}.$$

[Ibragimov (1983).]

13. Consider the mapping from $(x, u, \partial u, \partial^2 u, \ldots, \partial^\ell u)$ – space to $(y, v, \partial v, \partial^2 v, \ldots, \partial^\ell v)$ – space given by $x = X(y,v)$, $u = U(y,v)$, where $x = (x_1, x_2, \ldots, x_n)$, $u = (u^1, u^2, \ldots, u^m)$, $y = (y_1, y_2, \ldots, y_n)$, and $v = (v^1, v^2, \ldots, v^m)$; $\partial^j v$ denotes jth-order partial derivatives of the components of v with respect to the components of y. A function $u = \Theta(x)$ transforms to a function $v = \Phi(y)$ through solving the equations $\Theta^\alpha(X(y,v)) = U^\alpha(y,v)$, $\alpha = 1,2,\ldots,m$. Let

$$\frac{DX_k}{Dy_j} = \frac{\partial X_k}{\partial y_j} + \frac{\partial X_k}{\partial v^\mu}\frac{\partial v^\mu}{\partial y_j}, \quad \frac{DU^\alpha}{Dy_j} = \frac{\partial U^\alpha}{\partial y_j} + \frac{\partial U^\alpha}{\partial v^\mu}\frac{\partial v^\mu}{\partial y_j},$$

$$\frac{D^2U^\alpha}{Dy_jDy_k} = \frac{\partial^2 U^\alpha}{\partial y_j \partial y_k} + \frac{\partial^2 U^\alpha}{\partial y_j \partial v^\mu}\frac{\partial v^\mu}{\partial y_k} + \frac{\partial^2 U^\alpha}{\partial v^\mu \partial v^\lambda}\frac{\partial v^\mu}{\partial y_k}\frac{\partial v^\lambda}{\partial y_j} + \frac{\partial U^\alpha}{\partial v^\mu}\frac{\partial^2 v^\mu}{\partial y_j \partial y_k}.$$

Assume that under the above mapping, the Jacobian determinant satisfies

$$\frac{DX}{Dy} = \frac{D(X_1, X_2, \ldots, X_n)}{D(y_1, y_2, \ldots, y_n)} = \det\left\|\frac{DX_i}{Dy_j}\right\| = \begin{vmatrix} \dfrac{DX_1}{Dy_1} & \cdots & \dfrac{DX_1}{Dy_n} \\ \vdots & & \vdots \\ \dfrac{DX_n}{Dy_1} & \cdots & \dfrac{DX_n}{Dy_n} \end{vmatrix} \neq 0.$$

(a) Show that

$$\frac{DU^\alpha}{Dy_j} = \frac{DX_k}{Dy_j}\frac{\partial u^\alpha}{\partial x_k},$$

and hence,

$$\frac{\partial u^\alpha}{\partial x_i} = \frac{\dfrac{D(X_1, X_2, \ldots, X_{i-1}, U^\alpha, X_{i=1}, \ldots, X_n)}{Dy}}{\dfrac{DX}{Dy}}.$$

(b) Show that

$$\frac{D^2U^\alpha}{Dy_j Dy_k} = \frac{D^2 X_p}{Dy_j Dy_k}\frac{\partial u^\alpha}{\partial x_p} + \frac{DX_p}{Dy_j}\frac{DX_q}{Dy_k}\frac{\partial^2 u^\alpha}{\partial y_p \partial y_q}.$$

(c) One can show that

$$\det\left\|\frac{DX_p}{Dy_j}\frac{DX_q}{Dy_k}\right\| = \det\left\|\frac{DX_i}{Dy_j}\otimes\frac{DX_i}{Dy_j}\right\| = \left(\frac{DX}{DY}\right)^{2n}$$

[Greub (1967, p. 26)]. Hence, as an example, show that

$$\frac{\partial^2 u^\alpha}{\partial x_1^2} = \frac{\begin{vmatrix} \dfrac{D^2U^\alpha}{Dy_1^2} - \dfrac{D^2X_k}{Dy_1^2}\dfrac{\partial u^\alpha}{\partial x_k} & \dfrac{DX_1}{Dy_1}\dfrac{DX_2}{Dy_1} & \cdots & \dfrac{DX_n}{Dy_1}\dfrac{DX_n}{Dy_1} \\ \dfrac{D^2U^\alpha}{Dy_1 Dy_2} - \dfrac{D^2X_k}{Dy_1 Dy_2}\dfrac{\partial u^\alpha}{\partial x_k} & \dfrac{DX_1}{Dy_1}\dfrac{DX_2}{Dy_2} & \cdots & \dfrac{DX_n}{Dy_1}\dfrac{DX_n}{Dy_2} \\ \vdots & \vdots & & \vdots \\ \dfrac{D^2U^\alpha}{Dy_n^2} - \dfrac{D^2X_k}{Dy_n^2}\dfrac{\partial u^\alpha}{\partial x_k} & \dfrac{DX_1}{Dy_n}\dfrac{DX_2}{Dy_n} & \cdots & \dfrac{DX_n}{Dy_n}\dfrac{DX_n}{Dy_n} \end{vmatrix}}{\left(\dfrac{DX}{DY}\right)^{2n}}.$$

4.4 APPLICATIONS TO BOUNDARY VALUE PROBLEMS

In Sections 4.1 to 4.3, we have shown how to find point symmetries admitted by given PDEs and how to use them to find resulting invariant solutions. Now we consider the problem of using invariance to solve boundary value problems posed for PDEs. The application of Lie symmetries to boundary value problems for PDEs is much more restrictive than is the situation for ODEs.

In the case of an ODE, an admitted integrating factor or point symmetry (or, more generally, a higher-order symmetry) leads to a reduction in the order of the ODE. In terms of the original variables (integrating factor reduction) or in terms of the corresponding differential invariants (point symmetry reduction), any posed boundary value problem for the ODE is *automatically* reduced to a boundary value problem for a lower order ODE.

In the case of a PDE, an invariant solution arising from an admitted point symmetry solves a given boundary value problem provided that the symmetry leaves invariant all boundary conditions. This means that the domain of the boundary value problem or, equivalently, its boundary as well as the conditions (boundary conditions) imposed on the boundary must be invariant.

The situation is not so restrictive in the case of boundary value problems posed for linear PDEs. Here a boundary value problem need not be completely invariant (*incomplete invariance*) since one can use an appropriate superposition of invariant solutions in the following situations:

(i) For a linear *nonhomogeneous* PDE with linear homogeneous boundary conditions, an infinitesimal generator $\mathrm{X} \neq u\frac{\partial}{\partial u}$, admitted by the associated linear homogeneous PDE, is useful if X is also admitted by the homogeneous boundary conditions. Then the boundary value problem can be solved by a superposition (i.e., eigenfunction expansion, integral transform representation) of invariant form functions arising from the infinitesimal generator $\mathrm{X} + \lambda u\frac{\partial}{\partial u}$, where λ is an arbitrary constant, since $\mathrm{X} + \lambda u\frac{\partial}{\partial u}$ is admitted by both the associated linear homogeneous PDE and the homogeneous boundary conditions [λ plays the role of an eigenvalue].

(ii) For a linear *homogeneous* PDE with $p \geq 1$ linear homogeneous boundary conditions and one linear nonhomogeneous boundary condition, an infinitesimal generator $\mathrm{X} \neq u\frac{\partial}{\partial u}$, admitted by the PDE, is useful if X is also admitted by the p homogeneous boundary conditions. Consequently, for any complex constant λ, the infinitesimal generator $\mathrm{X} + \lambda u\frac{\partial}{\partial u}$ is admitted by the PDE and its p homogeneous boundary conditions. Here, one solves the boundary value problem by first constructing the invariant solutions, resulting from $\mathrm{X} + \lambda u\frac{\partial}{\partial u}$, that satisfy the PDE and its homogeneous boundary conditions. Then one finds a superposition of these invariant solutions that solves the nonhomogeneous boundary condition. Note that in this case (unlike in (i)), X does not necessarily leave invariant the domain of the boundary value problem.

The results to be presented in Sections 4.4.1 and 4.4.2 first appeared in a more rudimentary form in Bluman (1967, 1974) and Bluman and Cole (1969, 1974). [For the rest of this section, we assume the summation convention for repeated indices.]

4.4.1 FORMULATION OF INVARIANCE OF A BOUNDARY VALUE PROBLEM FOR A SCALAR PDE

Consider a boundary value problem for a kth-order $(k \geq 2)$ scalar PDE that can be written in a solved form

$$F(x,u,\partial u,\partial^2 u,\ldots,\partial^k u) = u_{i_1 i_2 \cdots i_\ell} - f(x,u,\partial u,\partial^2 u,\ldots,\partial^k u) = 0 \tag{4.164a}$$

$[f(x,u,\partial u,\partial^2 u,\ldots,\partial^k u)$ does not depend explicitly on $u_{i_1 i_2 \cdots i_\ell}]$, defined on a domain Ω_x in $x-$space $[x = (x_1,x_2,\ldots,x_n)]$ with boundary conditions

$$B_\alpha(x,u,\partial u,\ldots,\partial^{k-1} u) = 0 \tag{4.164b}$$

prescribed on boundary surfaces

$$\omega_\alpha(x) = 0, \quad \alpha = 1,2,\ldots,s. \tag{4.164c}$$

We assume that the boundary value problem (4.164a–c) has a unique solution. Consider an infinitesimal generator of the form

$$\mathrm{X} = \xi_i(x)\frac{\partial}{\partial x_i} + \eta(x,u)\frac{\partial}{\partial u}, \tag{4.165}$$

which defines a point symmetry acting on both $(x,u)-$space as well as on its projection to $x-$space.

Definition 4.4.1-1. The point symmetry X of the form (4.165) is *admitted by the* boundary value problem (4.164a–c) if and only if:

(i) $\mathrm{X}^{(k)} F(x,u,\partial u,\partial^2 u,\ldots,\partial^k u) = 0$ when $F(x,u,\partial u,\partial^2 u,\ldots,\partial^k u) = 0;$ (4.166a)

(ii) $\mathrm{X}\omega_\alpha(x) = 0$ when $\omega_\alpha(x) = 0, \alpha = 1,2,\ldots,s;$ (4.166b)

(iii) $\mathrm{X}^{(k-1)} B_\alpha(x,u,\partial u,\ldots,\partial^{k-1} u) = 0$
when $B_\alpha(x,u,\partial u,\ldots,\partial^{k-1} u) = 0$ on $\omega_\alpha(x) = 0, \alpha = 1,2,\ldots,s.$ (4.166c)

Theorem 4.4.1-1. *Suppose the boundary value problem* (4.164a–c) *admits the Lie group of point transformations with infinitesimal generator* (4.165). *Let* $y = (y_1(x), y_2(x),\ldots,y_{n-1}(x))$ *be* $n-1$ *functionally independent group invariants of* (4.165) *that depend only on x. Let* $v(x,u)$ *be a group invariant of* (4.165) *such that* $\partial v/\partial u \neq 0$. *Then the boundary value problem* (4.164a–c) *reduces to*

$$G(y,v,\partial v,\partial^2 v,\ldots,\partial^k v) = 0 \tag{4.167a}$$

defined on some domain Ω_y *in* $y-$*space with boundary conditions*

$$C_\alpha(y,v,\partial v,\partial^2 v,\ldots,\partial^{k-1} v) = 0 \tag{4.167b}$$

prescribed on boundary surfaces

$$v_\alpha(y) = 0, \tag{4.167c}$$

for some $G(y, v, \partial v, \partial^2 v, \ldots, \partial^k v)$, $C_\alpha(y, v, \partial v, \partial^2 v, \ldots, \partial^{k-1} v)$, $v_\alpha(y)$, $\alpha = 1, 2, \ldots, s$. *Moreover, in the boundary value problem* (4.167a,b), $\partial^j v$ *represents the components of jth-order partial derivatives of* v *with respect to* $y = (y_1(x), y_2(x), \ldots, y_{n-1}(x))$, $j = 1, 2, \ldots, k$; *and* (4.167a) *can be written in a solved form in terms of some specific* ℓ*th-order partial derivative of* v *with respect to* y.

Proof. Left to Exercise 4.4-1. □

Note that the surfaces $y_j(x) = 0$, $j = 1, 2, \ldots, n-1$, are invariant surfaces of the point symmetry (4.165). The invariance condition (4.166b) means that each boundary surface $\omega_\alpha(x) = 0$ is an invariant surface $v_\alpha(y) = 0$ of the projected point symmetry

$$\xi_i(x)\frac{\partial}{\partial x_i} \tag{4.168}$$

given by the restriction of point symmetry (4.165) to x-space. From the invariance of the boundary value problem under the point symmetry (4.165), the number of independent variables in (4.164a–c) is reduced by one. In particular, the solution of the boundary value problem (4.164a–c) is an invariant solution

$$v = \Phi(y_1, y_2, \ldots, y_{n-1}) \tag{4.169}$$

of the PDE (4.167a) resulting from its invariance under point symmetry (4.165). In terms of the dependent variable u and independent variables x appearing in PDE (4.164a), the corresponding invariant solution $u = \Theta(x)$ of PDE (4.164a) must satisfy

$$\mathrm{X}(u - \Theta(x)) = 0 \quad \text{when } u = \Theta(x), \tag{4.170}$$

i.e.,

$$\xi_i(x)\frac{\partial \Theta(x)}{\partial x_i} = \eta(x, \Theta(x)). \tag{4.171}$$

Theorem 4.4.1-2. *If the infinitesimal generator* X, *given by* (4.165), *is of the form*

$$\mathrm{X} = \xi_i(x)\frac{\partial}{\partial x_i} + f(x)u\frac{\partial}{\partial u}, \tag{4.172}$$

then the group invariant $v(x,u)$ *is of the form* $v(x,u) = u / g(x)$ *for some specific function* $g(x)$ *and hence the invariant form related to invariance under* X *can be expressed in the separable form*

$$u = \Theta(x) = g(x)\Phi(y), \tag{4.173}$$

in terms of an arbitrary function $\Phi(y)$ of $y = (y_1(x), y_2(x), \ldots, y_{n-1}(x))$.

Proof. Left to Exercise 4.4-2. □

If the boundary value problem (4.164a–c) admits an r-parameter Lie group of point transformations with infinitesimal generators of the form

$$\mathrm{X}_i = \xi_{ij}(x)\frac{\partial}{\partial x_j} + \eta_i(x,u)\frac{\partial}{\partial u}, \quad i = 1,2,\ldots,r, \tag{4.174}$$

then the unique solution $u = \Theta(x)$ of the boundary value problem (4.164a–c) is an invariant solution satisfying

$$\mathrm{X}_i(u - \Theta(x)) = 0 \quad \text{when } u = \Theta(x), \quad i = 1,2,\ldots,r.$$

The proof of the following theorem is left to Exercise 4.4-3:

Theorem 4.4.1-3 (Invariance of a Boundary Value Problem Under a Multiparameter Lie Group of Point Transformations). *Suppose the boundary value problem* (4.164a–c) *admits an r-parameter Lie group of point transformations with infinitesimal generators of the form*

$$\mathrm{X}_i = \xi_{ij}(x)\frac{\partial}{\partial x_j} + \eta_i(x,u)\frac{\partial}{\partial u}, \quad i = 1,2,\ldots,r. \tag{4.175}$$

Let R be the rank of the $r \times n$ *matrix*

$$\Xi(x) = \begin{bmatrix} \xi_{11}(x) & \xi_{12}(x) & \cdots & \xi_{1n}(x) \\ \xi_{21}(x) & \xi_{22}(x) & \cdots & \xi_{2n}(x) \\ \vdots & \vdots & & \vdots \\ \xi_{r1}(x) & \xi_{r2}(x) & \cdots & \xi_{rn}(x) \end{bmatrix}. \tag{4.176}$$

Let $q = n - R$, *and let* $z_1(x), z_2(x), \ldots, z_q(x)$ *be a complete set of functionally independent invariants of* (4.175), *satisfying*

$$\xi_{ij}(x)\frac{\partial z_\ell(x)}{\partial x_j} = 0, \quad i = 1,2,\ldots,r, \ \ \ell = 1,2,\ldots,q. \tag{4.177}$$

Let

$$v = \frac{u}{g(x)} \tag{4.178}$$

be an invariant of (4.175) *satisfying*

$$\mathrm{X}_i v = 0, \quad i = 1,2,\ldots,r.$$

Then the boundary value problem (4.164a–c) *reduces to a boundary value problem with* $q = n - R$ *independent variables* $z = (z_1(x), z_2(x), \ldots, z_q(x))$ *and dependent variable* v *given by* (4.178). *The solution of the boundary value problem* (4.164a–c) *is an invariant solution that can be expressed in terms of a separable form*

$$u = g(x)\Phi(z), \tag{4.179}$$

where the function $\Phi(z)$ *is to be determined.*

The following examples are illustrative:

(1) *Fundamental Solutions of the Heat Equation*
Consider again the heat equation (4.47) defined on the domain $t > 0, a < x < b$. Recall that PDE (4.47) admits the six-parameter $(\alpha, \beta, \gamma, \delta, \kappa, \lambda)$ Lie group of point transformations given by infinitesimal generators of the form

$$\mathrm{X} = \xi(x,t)\frac{\partial}{\partial x} + \tau(t)\frac{\partial}{\partial t} + f(x,t)u\frac{\partial}{\partial u},$$

with its infinitesimals given by the equations

$$\xi(x,t) = \kappa + \beta x + \gamma xt + \delta t, \tag{4.180a}$$

$$\tau(t) = \alpha + 2\beta t + \gamma t^2, \tag{4.180b}$$

$$f(x,t) = -\gamma(\tfrac{1}{4}x^2 + \tfrac{1}{2}t) - \tfrac{1}{2}\delta x + \lambda. \tag{4.180c}$$

The boundary curves of the domain are $t = 0$, $x = a$, $x = b$. The invariance of $t = 0$ leads to

$$\tau(0) = 0,$$

and, hence, $\alpha = 0$. If $a = -\infty$ and $b = \infty$, then there is no further parameter reduction resulting from invariance of the boundary curves. If $a \neq -\infty$, then the invariance of $x = a$ leads to

$$\xi(a,t) = 0$$

for any $t > 0$, and hence,

$$\kappa = -\beta a, \quad \delta = -\gamma a.$$

Similarly, if $b \neq \infty$, then the invariance of $x = b$ yields

$$\kappa = -\beta b, \quad \delta = -\gamma b.$$

Consequently, if $a \neq -\infty$ and $b \neq \infty$, then $\beta = \gamma = \delta = \kappa = 0$, and so there is no nontrivial Lie group of point transformations admitted by both the heat equation and the boundary of a boundary value problem for the heat equation (4.47) defined on the domain $t > 0, a < x < b$. However, since PDE (4.47) is linear, it is not necessary to leave all boundary curves of the domain invariant, as mentioned in the introductory remarks of Section 4.4, and as will be shown in Section 4.4.2.

If $a = -\infty$ and $b = \infty$, then a four-parameter Lie group of point transformations is admitted by the boundary of a boundary value problem posed for the heat equation (4.47) on the domain $t > 0$, $a < x < b$, and hence, a boundary value problem could admit at most a five-parameter $(\beta, \gamma, \delta, \kappa, \lambda)$ Lie group of point transformations.

If $a \neq -\infty$ (without loss of generality $a = 0$) and $b = \infty$, then a two-parameter Lie group of point transformations is admitted by the boundary of a posed boundary value problem for the heat equation (4.47), and hence, a boundary value problem could admit at most a three-parameter (β, γ, λ) Lie group of point transformations with infinitesimals given by

$$\xi(x,t) = \beta x + \gamma xt, \tag{4.181a}$$

$$\tau(t) = 2\beta t + \gamma t^2, \tag{4.181b}$$

$$f(x,t) = -\gamma(\tfrac{1}{4}x^2 + \tfrac{1}{2}t) + \lambda. \tag{4.181c}$$

We now derive fundamental solutions for the heat equation (4.47) when

$$u(x,0) = \delta(x - x_0),$$

where $\delta(x - x_0)$ is the Dirac delta function centered at x_0, $a < x_0 < b$, for an infinite domain $(a = -\infty, b = \infty)$ or a semi-infinite domain $(a = 0, b = \infty)$.

(i) *Infinite Domain* $(a,b) = (-\infty, \infty)$. Consider the boundary value problem

$$u_t = u_{xx}, \tag{4.182a}$$

on the domain $t > 0$, $-\infty < x < \infty$, with boundary conditions

$$u(\pm\infty, t) = 0, \quad t > 0,$$

and

$$u(x,0) = \delta(x). \tag{4.182b}$$

Without loss of generality, one can set $x_0 = 0$.

The Lie group of point transformations with infinitesimals (4.180a–c) is admitted by the boundary value problem (4.182a,b) provided that

$$f(x,0)u(x,0) = \xi(x,0)\delta'(x) \quad \text{when} \quad u(x,0) = \delta(x),$$

i.e.,

$$f(x,0)\delta(x) = \xi(x,0)\delta'(x). \tag{4.183}$$

From the properties of the Dirac delta function (a generalized function [Lighthill (1958)]), we see that (4.183) is satisfied if

$$\xi(0,0) = 0 \tag{4.184a}$$

and

$$f(0,0) = -\xi_x(0,0). \tag{4.184b}$$

Thus, in the infinitesimal equations (4.180a–c), we have

$$\kappa = 0, \quad \lambda = -\beta.$$

Consequently, a three-parameter (β,γ,δ) Lie group of point transformations is admitted by the boundary value problem (4.182a,b). The infinitesimal generators of this group are given by

$$\mathrm{X}_1 = x\frac{\partial}{\partial x} + 2t\frac{\partial}{\partial t} - u\frac{\partial}{\partial u}, \quad \mathrm{X}_2 = t\frac{\partial}{\partial x} - \tfrac{1}{2}xu\frac{\partial}{\partial u}, \quad \mathrm{X}_3 = xt\frac{\partial}{\partial x} + t^2\frac{\partial}{\partial t} - [\tfrac{1}{4}x^2 + \tfrac{1}{2}t]u\frac{\partial}{\partial u}. \tag{4.185}$$

The corresponding matrix

$$\Xi(x,t) = \begin{bmatrix} x & 2t \\ t & 0 \\ xt & t^2 \end{bmatrix}$$

has rank $R = 2$, so that group invariance reduces the boundary value problem completely in the sense that the number of independent variables is reduced from two to zero. Note that

$$\mathrm{X}_3 = \tfrac{1}{2}[t\mathrm{X}_1 + x\mathrm{X}_2]. \tag{4.186}$$

Hence, an invariant solution resulting from joint invariance under X_1 and X_2 must also be an invariant solution resulting from X_3. Let $u = \Theta(x,t)$ be an invariant solution resulting from joint invariance under X_1 and X_2. Then

$$\mathrm{X}_1(u - \Theta(x,t)) = 0 \quad \text{when} \quad u = \Theta(x,t)$$

yields the invariant form

$$u = \Theta(x,t) = \frac{1}{\sqrt{t}}\Phi_1(\zeta_1), \tag{4.187}$$

with the similarity variable given by

$$\zeta_1 = \frac{x}{\sqrt{t}}.$$

The equation

$$\mathrm{X}_2(u - \Theta(x,t)) = 0 \quad \text{when} \quad u = \Theta(x,t)$$

leads to the invariant form

$$u = \Theta(x,t) = e^{-x^2/4t}\Phi_2(\zeta_2), \tag{4.188}$$

with the similarity variable given by

$$\zeta_2 = t.$$

From the uniqueness of the solution to the boundary value problem (4.182a,b), it follows that

$$\frac{1}{\sqrt{t}}\Phi_1(\zeta_1) = e^{-x^2/4t}\Phi_2(\zeta_2).$$

After expressing the variables x and t in terms of the similarity variables ζ_1 and ζ_2, we see that

$$e^{(\zeta_1)^2/4}\Phi_1(\zeta_1) = \sqrt{\zeta_2}\Phi_2(\zeta_2) = \text{const} = c.$$

Hence, the solution of the boundary value problem (4.182a,b) is given by the well-known expression

$$u = \Theta(x,t) = \frac{c}{\sqrt{t}}e^{-x^2/4t} \tag{4.189}$$

for some constant c. The initial condition (4.182b) yields

$$c = \frac{1}{\sqrt{4\pi}}.$$

Note that from relation (4.186) it must automatically follow that

$$\mathrm{X}_3\left(u - \frac{c}{\sqrt{t}}e^{-x^2/4t}\right) = 0$$

for any value of the constant c.

(ii) *Semi-Infinite Domain* $(a,b) = (0,\infty)$. Consider the boundary value problem

$$u_t = u_{xx}, \tag{4.190a}$$

on the domain $t > 0$, $x > 0$, with boundary conditions

$$u(0,t) = 0, \quad t > 0, \tag{4.190b}$$

and

$$u(x,0) = \delta(x - x_0), \quad 0 < x_0 < \infty. \tag{4.190c}$$

The three-parameter Lie group of point transformations with infinitesimals given by (4.181a–c) is admitted by PDE (4.190a), the boundary curves $t = 0$ and $x = 0$, and the boundary condition (4.190b). The invariance of the initial condition (4.190c) leads to the restriction

$$f(x,0)u(x,0) = \xi(x,0)\delta'(x - x_0) \quad \text{when} \quad u(x,0) = \delta(x - x_0),$$

i.e.,

$$f(x,0)\delta(x - x_0) = \xi(x,0)\delta'(x - x_0).$$

Hence,

$$\xi(x_0,0) = 0,$$

$$f(x_0,0) = -\xi_x(x_0,0).$$

Consequently, in the infinitesimal equations (4.181a–c), we must have

$$\beta = 0, \quad \lambda = \tfrac{1}{4}(x_0)^2\gamma.$$

Thus, the boundary value problem (4.190a–c) admits the point symmetry

$$\mathrm{X} = xt\frac{\partial}{\partial x} + t^2\frac{\partial}{\partial t} + [\tfrac{1}{4}(x_0)^2 - (\tfrac{1}{4}x^2 + \tfrac{1}{2}t)]\,u\frac{\partial}{\partial u}.$$

The corresponding invariant solution has the invariant form

$$u = \Theta(x,t) = \frac{e^{-[x^2+(x_0)^2]/4t}}{\sqrt{t}}\Phi(\zeta), \tag{4.191}$$

where $\Phi(\zeta)$ is an arbitrary function of the similarity variable

$$\zeta = \frac{x}{t}.$$

After substituting the invariant form (4.191) into the heat equation (4.190a), we find that $\Phi(\zeta)$ satisfies the ODE

$$\Phi'' = \tfrac{1}{4}(x_0)^2\Phi.$$

Hence, the solution of the boundary value problem (4.190a–c) has the form

$$u = \Theta(x,t) = \frac{1}{\sqrt{t}}[Ce^{-(x-x_0)^2/4t} + De^{-(x+x_0)^2/4t}]$$

for some constants C and D. The boundary condition (4.190b) leads to $D = -C$, and from the initial condition (4.190c) we find that $C = 1/\sqrt{4\pi}$. This yields the well-known solution of the boundary value problem (4.190a–c), usually obtained by the method of images, given by

$$u = \Theta(x,t) = G(x - x_0,t) - G(x + x_0,t),$$

where

$$G(x,t) = \frac{e^{-x^2/4t}}{\sqrt{4\pi t}}.$$

(2) *Fundamental Solution of the Axisymmetric Wave Equation*
The fundamental solution of the axisymmetric wave equation (4.88) is the solution of the boundary value problem

$$\mathrm{L}u = u_{tt} - u_{rr} - \frac{1}{r}u_r = \frac{1}{2\pi r}\delta(r)\delta(t),$$

i.e.,

$$r\mathrm{L}u = \frac{1}{2\pi}\delta(r)\delta(t), \tag{4.192a}$$

on the domain $r > 0, t > 0$, with the causality condition that

$$u \equiv 0 \quad \text{if} \quad r > t. \tag{4.192b}$$

It is left to Exercise 4.2-7 to show that the linear homogeneous equation

$$\mathrm{L}u = 0$$

admits a four-parameter $(\alpha, \beta, \gamma, \lambda)$ Lie group of point transformations represented by the infinitesimal generator

$$\mathrm{X} = \rho(r,t)\frac{\partial}{\partial r} + \tau(r,t)\frac{\partial}{\partial t} + f(t)u\frac{\partial}{\partial u}, \tag{4.193}$$

with its infinitesimals given by

$$\rho(r,t) = \alpha r + 2\beta rt, \tag{4.194a}$$

$$\tau(r,t) = \alpha t + \beta(r^2 + t^2) + \gamma, \tag{4.194b}$$

$$f(t) = -\beta t + \lambda. \tag{4.194c}$$

The invariance of the wavefront $r = t$ leads to

$$\rho(t,t) = \tau(t,t),$$

and hence, $\gamma = 0$.

Under the action of (4.193), we have

$$r^* \mathrm{L}^* u^* = \left[1 + \varepsilon\left[f(t) - 2\tau_t(r,t) + \frac{\rho(r,t)}{r}\right] + O(\varepsilon^2)\right] r\mathrm{L}u,$$

$$\delta(r^*)\delta(t^*) = \delta(r)\delta(t) + \varepsilon[\rho(r,t)\delta'(r)\delta(t) + \tau(r,t)\delta(r)\delta'(t)] + O(\varepsilon^2).$$

Hence, (4.193), (4.194a–c) is admitted by (4.192a) if

$$\left[f(t) - 2\tau_t(r,t) + \frac{\rho(r,t)}{r}\right]\delta(r)\delta(t) = \rho(r,t)\delta'(r)\delta(t) + \tau(r,t)\delta(r)\delta'(t). \tag{4.195}$$

Since $z\delta'(z) \equiv -\delta(z)$, after using this in (4.195) with $z = r,\ t$, we see that (4.192a,b) admits (4.193) and (4.194a–c) if

$$\left[f(t) - 2\tau_t(r,t) + \frac{2}{r}\rho(r,t) + \frac{1}{t}\tau(r,t)\right]\delta(r)\delta(t) = 0. \tag{4.196}$$

Equation (4.196) reduces to

$$\left[\lambda + \alpha + \beta\frac{r^2}{t}\right]\delta(r)\delta(t) = 0. \tag{4.197}$$

Since (4.197) needs to be satisfied only on the wavefront $r = t$ when $t = 0$, it follows that β remains arbitrary and $\lambda = -\alpha$. Thus, the boundary value problem (4.192a,b) admits a two-parameter Lie group of point transformations given by the infinitesimal generators

$$\mathrm{X}_1 = r\frac{\partial}{\partial r} + t\frac{\partial}{\partial t} - u\frac{\partial}{\partial u}, \quad \mathrm{X}_2 = 2rt\frac{\partial}{\partial r} + (r^2 + t^2)\frac{\partial}{\partial t} - tu\frac{\partial}{\partial u}. \tag{4.198}$$

Let $u = \Theta(r,t)$ be an invariant solution resulting from joint invariance under X_1 and X_2. Then

$$\mathrm{X}_1(u - \Theta(r,t)) = 0 \quad \text{when} \quad u = \Theta(r,t)$$

yields the invariant form

$$u = \Theta(r,t) = \frac{1}{t}\Phi_1(\zeta_1), \tag{4.199}$$

with the similarity variable

$$\zeta_1 = \frac{r}{t};$$

and

$$\mathrm{X}_2(u - \Theta(r,t)) = 0 \quad \text{when} \quad u = \Theta(r,t)$$

leads to the invariant form

$$u = \Theta(r,t) = \frac{1}{\sqrt{r}}\Phi_2(\zeta_2), \tag{4.200}$$

with the similarity variable

$$\zeta_2 = \frac{1}{r}(t^2 - r^2).$$

The uniqueness of the solution of the boundary value problem (4.192a,b) leads to

$$\frac{1}{\sqrt{r}}\Phi_2(\zeta_2) = \frac{1}{t}\Phi_1(\zeta_1). \tag{4.201}$$

Since

$$t = \frac{\zeta_2\zeta_1}{1-(\zeta_1)^2}, \quad r = \frac{\zeta_2(\zeta_1)^2}{1-(\zeta_1)^2},$$

it follows that after making these substitutions in (4.201), we have

$$\sqrt{\zeta_2}\Phi_2(\zeta_2) = \sqrt{1-(\zeta_1)^2}\Phi_1(\zeta_1) = \text{const.}$$

Hence,

$$u = \Theta(r,t) = \frac{c}{\sqrt{t^2 - r^2}} \tag{4.202}$$

for some constant c. One can show that $c = 1/2\pi$.

(3) *Fundamental Solutions of Fokker–Planck Equations*
As a third example, we consider probability distributions that arise as fundamental solutions of Fokker–Planck equations with drifts $\phi(x)$. In particular, we consider the boundary value problem

$$u_t = u_{xx} + (\phi(x)u)_x, \tag{4.203a}$$

on the domain $t > 0, a < x < b$, with the initial condition

$$u(x,0) = \delta(x - x_0), \quad a < x_0 < b, \tag{4.203b}$$

and reflecting boundaries $x = a$ and $x = b$ on which

$$\lim_{x\to a^+,b^-} [u_x + \phi(x)u] = 0. \tag{4.203c}$$

Let $u = G(x,t;x_0)$ be the solution of the boundary value problem (4.203a–c). Then it follows that for *any* x_0, $a < x_0 < b$, one must have

$$\int_a^b G(x,t;x_0)\,dx = 1. \tag{4.204}$$

We consider the group classification problem with respect to drifts $\phi(x)$ for the boundary value problem (4.203a,b) for the physically interesting situation where $\phi(x)$ is an odd function of x. Complete details appear in Bluman (1967, 1971) and Bluman and Cole (1974).

One can show that the linear PDE (4.203a) admits a point symmetry

$$\xi(x,t)\frac{\partial}{\partial x} + \tau(x,t)\frac{\partial}{\partial t} + f(x,t)u\frac{\partial}{\partial u} \tag{4.205}$$

if and only if

$$\tau(x,t) = \tau(t),$$
$$\xi(x,t) = \tfrac{1}{2}x\tau(t) + A(t),$$
$$f(x,t) = -\tfrac{1}{4}x\phi(x)\tau'(t) - \tfrac{1}{8}x^2\tau''(t) - \tfrac{1}{2}\phi(x)A(t) - \tfrac{1}{2}xA'(t) + B(t),$$

where for a given drift $\phi(x)$, the functions $A(t), B(t), \tau(t)$ must satisfy

$$P(x,t) + Q(x,t) = 0,$$

with

$$P(x,t) = \tfrac{1}{4}[\phi^2(x) + x\phi(x)\phi'(x) - 2\phi'(x) - x\phi''(x)]\tau'(t) + \tfrac{1}{4}\tau''(t) - \tfrac{1}{8}x^2\tau'''(t) + B'(t),$$

$$Q(x,t) = \tfrac{1}{2}[\phi(x)\phi'(x) - \phi''(x)]A(t) - \tfrac{1}{2}xA''(t).$$

After imposing the restriction that $\phi(x)$ is an odd function of x, we see that

$$P(x,t) \equiv 0, \tag{4.206a}$$

$$Q(x,t) \equiv 0. \tag{4.206b}$$

From the identity (2.406a), it follows that if $\tau'(t) \not\equiv 0$, then the drift $\phi(x)$ must satisfy the fifth-order ODE

$$[\phi^2(x) + x\phi(x)\phi'(x) - 2\phi'(x) - x\phi''(x)]''' = 0 \tag{4.207}$$

for the Fokker–Planck equation (4.203a) to admit nontrivial point symmetries (other than the obvious invariance under translations in t).

One can show that ODE (4.207) (with the restriction that $\phi(x)$ is odd) reduces to the Riccati equation

$$2\phi'(x) - \phi^2(x) + \beta^2 x^2 - \gamma + \frac{16\nu^2 - 1}{x^2} = 0, \tag{4.208}$$

where $\beta, \gamma,$ and ν are arbitrary constants. After substituting (4.208) into (4.206a), we see that $\tau(t)$ and $B(t)$ must satisfy the system of ODEs

$$\tau'''(t) = 4\beta^2\tau'(t), \tag{4.209a}$$

$$B'(t) = \tfrac{1}{4}[\gamma\tau'(t) - \tau''(t)]. \tag{4.209b}$$

From the identity (4.206b), it now follows that if $A(t) \not\equiv 0$, then an odd drift $\phi(x)$ must satisfy the Riccati equation

$$2\phi'(x) - \phi^2(x) + \beta^2 x^2 - \gamma = 0. \tag{4.210}$$

After substituting (4.210) into (4.206b), we see that $A(t)$ must satisfy the ODE

$$A''(t) = \beta^2 A(t). \tag{4.211}$$

Hence, a drift $\phi(x)$ simultaneously satisfies ODEs (4.208) and (4.210) if and only if $\phi(x)$ satisfies ODE (4.210). From (4.209a,b) and (4.211), we see that a Fokker–Planck PDE (4.203a) with an odd drift $\phi(x)$ admits a six-parameter Lie group of point transformations if and only if $\phi(x)$ satisfies ODE (4.210). Moreover, we see that if an odd drift $\phi(x)$ satisfies ODE (4.208) with $\nu^2 \neq \frac{1}{16}$, then the corresponding Fokker–Planck equation (4.203a) admits a four-parameter Lie group of point transformations.

The invariance of the boundary curve $t = 0$ leads to $\tau(0) = 0$ and thus reduces the number of parameters by one. The invariance of the initial condition (4.203b) requires that

$$\xi(x_0, 0) = 0,$$

$$f(x_0, 0) = -\xi_x(x_0, 0).$$

Hence, a three-parameter Lie group is admitted by (4.203a,b) if an odd drift $\phi(x)$ satisfies ODE (4.210); a one-parameter Lie group is admitted by (4.203a,b) if an odd drift $\phi(x)$ satisfies ODE (4.208) with $\nu^2 \neq \frac{1}{16}$.

The standard substitution

$$\phi(x) = -2\frac{V'(x)}{V(x)}$$

transforms any solution of the second-order linear ODE

$$4V''(x) + \left[\gamma - \beta^2 x^2 - \frac{16\nu^2 - 1}{x^2}\right] V(x) = 0 \tag{4.212}$$

to a solution of the Riccati ODE (4.208). Since ODE (4.212) is invariant under reflections in x, its general solution can be expressed in the form

$$V(x) = c_1 V_1(x) + c_2 V_2(x),$$

where $V_1(x), V_2(x)$ are, respectively, even and odd functions of x. Then $\phi(x)$ is an odd function of x that satisfies ODE (4.208) if and only if either

$$\phi(x) = -2\frac{V_1'(x)}{V_1(x)}$$

or

$$\phi(x) = -2\frac{V_2'(x)}{V_2(x)}.$$

Only $V_1(x)$ leads to a physically interesting drift. One can show that

$$V_1(x) = [\tfrac{1}{2}\beta x^2]^{\nu+(1/4)} e^{-\beta x^2/4} M(c, d, \tfrac{1}{2}\beta x^2), \tag{4.213}$$

where $M(c,d,z)$ denotes Kummer's hypergeometric function of the first kind with $c = \frac{1}{2} + \nu - \gamma/8\beta$, $d = 1 + 2\nu$, $\nu > -\frac{1}{2}$, $\beta \neq 0$. The properties of $M(c,d,z)$ are well-known [Abramowitz and Stegun (1970, Chapter 13)]:

As $z \to 0$,

$$M(c,d,z) = 1 + \frac{c}{d}z + O(z^2). \tag{4.214a}$$

As $z \to \infty$,

$$M(c,d,z) = \frac{\Gamma(d)}{\Gamma(c)} z^{c-d} e^z \left[1 + O\left(\frac{1}{z}\right)\right]. \tag{4.214b}$$

From the asymptotic properties (4.214a,b), it follows that

$$\lim_{x\to 0} x\phi(x) = -(4\nu+1), \quad \lim_{x\to\infty} \frac{\phi(x)}{x} = -\beta.$$

The following cases arise:

Case I. $\nu^2 = \frac{1}{16}$.
Here, a three-parameter Lie group of point transformations is admitted by (4.203a,b) with its infinitesimal generators given by

$$\mathrm{X}_1 = 2\beta x \sinh 2\beta t \frac{\partial}{\partial x} + 4\sinh^2 \beta t \frac{\partial}{\partial t}$$
$$+ [\gamma \sinh^2 \beta t - \beta \sinh 2\beta t(1 + x\phi(x)) + \beta^2((x_0)^2 - x^2 \cosh 2\beta t)]u\frac{\partial}{\partial u}, \tag{4.215a}$$

$$\mathrm{X}_2 = 2\sinh \beta t \frac{\partial}{\partial x} + [\beta(x_0 - x\cosh \beta t) - \phi(x)\sinh \beta t]u\frac{\partial}{\partial u}, \tag{4.215b}$$

$$\mathrm{X}_3 = 4\beta(x_0 \cosh \beta t - x\cosh 2\beta t)\frac{\partial}{\partial x} - 4\sinh 2\beta t \frac{\partial}{\partial t} + [4\beta \cosh^2 \beta t + 2\beta\phi(x)$$
$$\times (x\cosh 2\beta t - x_0 \cosh \beta t) + 2\beta^2 x(x\sinh 2\beta t - x_0 \sinh \beta t) - \gamma \sinh 2\beta t]u\frac{\partial}{\partial u}. \tag{4.215c}$$

Note that

$$\mathrm{X}_3 = -2\coth \beta t \mathrm{X}_1 + 2\beta(x \operatorname{cosech} \beta t + x_0 \coth \beta t)\mathrm{X}_2.$$

Let $u = \Theta(x,t)$ be a resulting invariant solution for the three-parameter Lie group of point transformations (4.215a–c). Then

$$\mathrm{X}_1(u - \Theta(x,t)) = 0 \quad \text{when} \quad u = \Theta(x,t)$$

leads to the invariant form

$$u = g_1(x,t)\Phi_1(\zeta_1), \tag{4.216}$$

where

$$g_1(x,t) = \frac{V_1(x)}{\sqrt{\sinh \beta t}} \exp\left[\frac{\gamma t}{4} + \frac{\beta (x_0)^2}{2(1-e^{2\beta t})} - \frac{\beta x^2 \coth \beta t}{4}\right],$$

with the similarity variable given by

$$\zeta_1 = \frac{x}{2\sinh \beta t}.$$

On the other hand,

$$\mathrm{X}_2(u - \Theta(x,t)) = 0 \quad \text{when } u = \Theta(x,t)$$

leads to the invariant form

$$u = g_2(x,t)\Phi_2(\zeta_2), \tag{4.217}$$

where

$$g_1(x,t) = V_1(x) \exp\left[\frac{\beta x_0 x}{2\sinh \beta t} - \frac{\beta x^2 \coth \beta t}{4}\right],$$

with the similarity variable

$$\zeta_2 = t.$$

Assuming uniqueness of the solution to the boundary value problem (4.203a,b), we equate the invariant forms (4.216) and (4.217) to obtain

$$\Phi_2(\zeta_2) = \Phi_2(t) = \frac{D}{\sqrt{\sinh \beta t}} \exp\left[\frac{\gamma t}{4} + \frac{\beta (x_0)^2}{2(1-e^{2\beta t})}\right],$$

where D is an arbitrary constant.

We now consider separately the subcases $\nu = \pm\frac{1}{4}$.

Case I(a). $\nu = -\frac{1}{4}$.
Here,

$$V_1(x) = e^{-\beta x^2/4} M(c, \tfrac{1}{2}, \tfrac{1}{2}\beta x^2), \quad c = \tfrac{1}{4} - \frac{\gamma}{8\beta}.$$

If there is no reflecting boundary, i.e., $a = -\infty, b = \infty$, then the solution of the boundary value problem (4.203a,b) is given by

$$u = G_1(x,t;x_0) = \frac{D}{\sqrt{\sinh \beta t}} M(c, \tfrac{1}{2}, \tfrac{1}{2}\beta x^2) \exp[\tfrac{1}{4}\gamma t - \tfrac{1}{4}\beta(1 + \coth \beta t)(x - x_0 e^{-\beta t})^2], \tag{4.218a}$$

with

$$D = \sqrt{\frac{\beta}{4\pi}} \times [M(c, \tfrac{1}{2}, \tfrac{1}{2}\beta(x_0)^2)]^{-1}, \tag{4.218b}$$

on the domain $t > 0, -\infty < x < \infty$.

If $a = 0$ is a reflecting boundary and $b = \infty$, then the solution of the boundary value problem (4.203a–c) is given by

$$u = G_1(x, t; x_0) + G_1(x, t; -x_0), \quad t > 0, 0 < x < \infty, \tag{4.219}$$

which is an even function of x with the extension of the domain of x in (4.219) to $-\infty < x < \infty$. The solution (4.219) can be interpreted as representing the response to sources located at $\pm x_0$ when $t = 0$.

Note that in the limiting case $c = 0$, the drift becomes $\phi(x) = \beta x$, and here the solutions (4.218a,b) and (4.219) become the well-known probability distributions for a free particle in a Brownian motion.

Case I(b). $\nu = \frac{1}{4}$.
Here,

$$V_1(x) = xe^{-\beta x^2/4} M(c, \tfrac{3}{2}, \tfrac{1}{2}\beta x^2), \quad c = \tfrac{3}{4} - \frac{\gamma}{8\beta},$$

$x > 0$, with $a = 0$ as a reflecting boundary and $b = \infty$. The resulting solution of the corresponding boundary value problem (4.203a–c) is given by

$$u = G_2(x, t; x_0) + G_2(x, t; -x_0), \quad t > 0, \ 0 < x < \infty,$$

where

$$G_2(x, t; x_0) = \frac{E}{\sqrt{\sinh \beta t}} xM(c, \tfrac{3}{2}, \tfrac{1}{2}\beta x^2) \exp[\tfrac{1}{4}\gamma t - \tfrac{1}{4}\beta(1 + \coth \beta t)(x - x_0 e^{-\beta t})^2],$$

with

$$E = \frac{1}{x_0} \sqrt{\frac{\beta}{4\pi}} \times [M(c, \tfrac{3}{2}, \tfrac{1}{2}\beta(x_0)^2)]^{-1}.$$

Case II. $\nu > -\frac{1}{2}$.
Here, only a one-parameter Lie group of point transformations is admitted by the boundary value problem (4.203a,b), with its infinitesimal generator given by (4.215a). The resulting invariant solution arises from the invariant form (4.216). After substituting (4.216) into PDE (4.203a) and letting $\zeta_1 = \zeta$, $\Phi_1(\zeta_1) = \Phi(\zeta)$, we find that $\Phi(\zeta)$ satisfies a second-order linear ODE for which the general solution can be expressed in terms of modified Bessel functions:

$$\Phi(\zeta) = \zeta^{1/2}[A_1 I_{2\nu}(\beta x_0 \zeta) + A_2 I_{-2\nu}(\beta x_0 \zeta)] \qquad \text{for } x > 0,$$
$$\Phi(\zeta) = |\zeta|^{1/2} [B_1 K_{2\nu}(\beta x_0 |\zeta|) + B_2 I_{2\nu}(\beta x_0 |\zeta|)] \quad \text{for } x < 0,$$

where A_1, A_2, B_1, B_2 are arbitrary constants to be determined from boundary and continuity conditions. As $t \to 0, \zeta \to \infty$. From Watson (1922, Section 7.23), we find that, as $z \to \infty$,

$$K_{2\nu}(z) = \left(\frac{\pi}{2z}\right)^{1/2} e^{-z}\left[1 + O\left(\frac{1}{z}\right)\right],$$

$$I_{2\nu}(z) = \left(\frac{1}{2\pi z}\right)^{1/2} e^{z}\left[1 + O\left(\frac{1}{z}\right)\right].$$

One can show that with $A_2 = B_1 = B_2 = 0$, we obtain a solution valid for $\nu \neq -\frac{1}{4}$, $t > 0$, $x > 0$ (i.e., $x = 0$ is a reflecting boundary) with the constant

$$A_1 = 2(x_0)^{-2\nu}(\tfrac{1}{2}\beta)^{(3/4)-\nu}[M(c,d,\tfrac{1}{2}\beta(x_0)^2)]^{-1}.$$

One can show that the reflecting boundary condition (4.203c) admits the point symmetry (4.215a) when $x = a$. An important consequence is that the conservation law equation

$$\int_0^\infty G(x,t;x_0)\,dx = 1$$

admits (4.215a), and hence, all moments of the probability distribution can be computed from this invariance and invariance of successive higher-order moments [Bluman and Cole (1974, pp. 272–274)].

An interesting special case, when the drift $\phi(x) = \beta x + \alpha / x$, is considered in Exercise 4.4-8.

The problem of finding fundamental solutions for wider classes of Fokker–Planck equations with time-dependent coefficients through mappings to the heat equation (mappings to a Wiener process) is considered in Bluman (1980) and Bluman and Shtelen (1998). Point symmetries are found for various examples of Fokker–Planck equations in Stognii and Shtelen (1991), Cicogna and Vitali (1990), and Rudra (1990).

The use of point symmetries to find fundamental solutions of linear scalar PDEs is also considered in Aksenov (1995). Rosinger and Walus (1994) give a general setting for considering the group invariance of generalized solutions.

King (1989, 1991) gives many examples of using group invariance to solve boundary value problems for nonlinear diffusion equations of the form (4.58).

4.4.2 INCOMPLETE INVARIANCE FOR A LINEAR SCALAR PDE

Consider a boundary value problem for a kth-order $(k \geq 2)$ linear scalar PDE

$$\mathrm{L}u = g(x) \tag{4.220a}$$

defined on a domain Ω, with linear boundary conditions

$$\mathrm{L}_\alpha u = h_\alpha(x) \tag{4.220b}$$

prescribed on boundary surfaces

$$\omega_\alpha(x) = 0, \tag{4.220c}$$

where L is a kth-order linear operator and L_α is a linear operator of order at most $k-1$, $\alpha = 1, 2, \ldots, s$. We assume that the boundary value problem (4.220a–c) has a unique solution. Formally, the solution of the boundary value problem (4.220a–c) can be represented as a superposition

$$u = u_0 + \sum_{\beta=1}^{s} u_\beta,$$

where u_0 satisfies

$$\mathrm{L}u_0 = g(x), \quad x \in \Omega,$$

$$\mathrm{L}_\alpha u_0 = 0 \text{ on } \omega_\alpha(x) = 0, \quad \alpha = 1, 2, \ldots, s,$$

and u_β satisfies

$$\mathrm{L}u_\beta = 0, \quad x \in \Omega,$$

$$\mathrm{L}_\alpha u_\beta = \delta_{\alpha\beta} h_\alpha(\mathrm{x}) \text{ on } \omega_\alpha(x) = 0, \quad \alpha, \beta = 1, 2, \ldots, s.$$

[$\delta_{\alpha\beta}$ is the Kronecker symbol.]

The solution of the boundary value problem (2.220a–c) reduces to the solutions of two types of boundary value problems:

(i) a linear nonhomogeneous PDE with s linear homogeneous boundary conditions:

$$\mathrm{L}u = g(x), \quad x \in \Omega, \tag{4.221a}$$

$$\mathrm{L}_\alpha u = 0 \text{ on } \omega_\alpha(x) = 0, \quad \alpha = 1,2,\ldots, s. \tag{4.221b}$$

(ii) a linear homogeneous PDE with $s - 1$ linear homogeneous boundary conditions and one linear nonhomogeneous boundary condition (without loss of generality the sth one):

$$\mathrm{L}u = 0, \quad x \in \Omega, \tag{4.222a}$$

$$\mathrm{L}_\alpha u = 0 \text{ on } \omega_\alpha(x) = 0, \quad \alpha = 1,2,\ldots, s-1, \tag{4.222b}$$

$$\mathrm{L}_s u = h(x) \text{ on } \omega_s(x) = 0. \tag{4.222c}$$

Now consider the homogeneous boundary value problem

$$\mathrm{L}u = 0, \quad x \in \Omega, \tag{4.223a}$$

$$\mathrm{L}_\alpha u = 0 \text{ on } \omega_\alpha(x) = 0, \quad \alpha = 1, 2, \ldots, s, \tag{4.223b}$$

associated with the given boundary value problem (4.220a–c). Suppose the nontrivial infinitesimal generator

$$\mathrm{X}_1 = \xi_i(x)\frac{\partial}{\partial x_i} + f(x)u\frac{\partial}{\partial u} \tag{4.224}$$

is admitted by $\mathrm{L}u = 0$ $[\xi(x) \not\equiv 0]$. Clearly, the homogeneous boundary value problem (4.223a,b) admits

$$\mathrm{X}_2 = u\frac{\partial}{\partial u}. \tag{4.225}$$

For the solution of (i), let

$$u = \Phi(x;\lambda) \tag{4.226}$$

be the *invariant form* related to the invariance of the boundary value problem (4.223a,b) under the infinitesimal generator $\mathrm{X}_\lambda = \mathrm{X}_1 + \lambda \mathrm{X}_2$ where λ is an arbitrary complex constant. Then the superposition of invariant forms

$$u = \sum_\lambda \Phi(x;\lambda) \tag{4.227}$$

solves the boundary value problem (4.221a,b) if

$$\sum_\lambda \mathrm{L}\Phi(x;\lambda) = g(x), \tag{4.228}$$

and if

$$\mathrm{L}_\alpha \Phi(x;\lambda) = 0 \text{ on } \omega_\alpha(x) = 0, \quad \alpha = 1, 2, \ldots, s, \tag{4.229}$$

for each λ in the sum (4.227). In (4.227) the superposition $\sum_\lambda$ could also represent $\int_\Gamma d\lambda$ for some curve Γ in the complex λ-plane. Typically, one solves (4.228) for $g(x) = \delta(x - x_0)$ for any $x_0 \in \Omega$. Then a superposition over the resulting Green's function is used to solve the boundary value problem (4.221a,b) for an arbitrary $g(x)$.

For the solution of (ii), let

$$u = \Theta(x;\lambda) \tag{4.230}$$

be the most general *invariant solution* of (4.222a,b) resulting from its invariance under X_λ, which usually exists for only certain complex eigenvalues λ. A superposition over such invariant solutions given by

$$u = \sum_\lambda \Theta(x;\lambda) \tag{4.231}$$

solves the boundary value problem (4.222a–c) provided that

$$\sum_\lambda \mathrm{L}_s \Theta(x;\lambda) = h(x) \text{ on } \omega_s(x) = 0. \tag{4.232}$$

Again the superposition $\sum_\lambda$ in (4.231) could also represent $\int_\Gamma d\lambda$ for some curve Γ in the complex λ-plane. The appropriate superposition is then found so as to satisfy (4.232). Typically, one first solves (4.232) for $h(x) = \delta(x-x_0)$ for any $x_0 \in \Omega$, and then uses a superposition over the resulting Green's function to solve the boundary value problem (4.222a–c) for an arbitrary $h(x)$.

The following examples are illustrative:

(1) *Fundamental Solutions of the Heat Equation for a Finite Spatial Domain*

(i) *Nonhomogeneous Heat Equation with Linear Homogeneous Boundary Conditions.* Consider the boundary value problem for the nonhomogeneous heat equation

$$\mathrm{L}u = u_t - u_{xx} = \delta(x-x_0)\delta(t), \tag{4.233a}$$

defined on the domain $t>0, 0<x<1$, where $0<x_0<1$, with linear homogeneous boundary conditions

$$u(0,t) = u(1,t) = 0. \tag{4.233b}$$

Clearly, the point symmetry (invariance under translations in t)

$$\mathrm{X}_1 = \frac{\partial}{\partial t}$$

is admitted by $\mathrm{L}u = 0$ and the homogeneous boundary conditions (4.233b). The invariant form resulting from invariance under

$$\mathrm{X}_\lambda = \frac{\partial}{\partial t} + \lambda u \frac{\partial}{\partial u}$$

is given by

$$u = \Phi(x,t;\lambda) = y(x;\lambda)e^{\lambda t}. \tag{4.234}$$

Now consider the superposition of invariant forms

$$u = \sum_\lambda y(x;\lambda)e^{\lambda t}. \tag{4.235}$$

After substituting (4.235) into PDE (4.233a), we formally find that

$$\sum_\lambda (y_{xx} - \lambda y)e^{\lambda t} = -\delta(x-x_0)\delta(t). \tag{4.236}$$

To satisfy the homogeneous boundary condition (4.233b), we require that for any λ in the superposition (4.235), we have

$$y(0;\lambda) = y(1;\lambda) = 0. \tag{4.237}$$

A natural superposition (4.235) arising from (4.234) is the inverse Laplace transform representation of the solution of the boundary value problem (4.233a,b) given by

$$u(x,t) = \frac{1}{2\pi i} \int_{\gamma - i\infty}^{\gamma + i\infty} y(x;\lambda) e^{\lambda t} \, d\lambda, \tag{4.238}$$

where $\gamma \in \mathrm{R}$ lies to the right of all singularities of $y(x;\lambda)$ in the complex λ-plane. [This superposition integral is the well-known Bromwich contour.] Formally,

$$\delta(t) = \frac{1}{2\pi i} \int_{\gamma - i\infty}^{\gamma + i\infty} e^{\lambda t} \, d\lambda.$$

Hence, $y(x;\lambda)$ satisfies the ODE

$$y_{xx} - \lambda y = -\delta(x - x_0) \tag{4.239}$$

together with the boundary conditions (4.237). Consequently,

$$y(x;\lambda) = \begin{cases} \dfrac{\sinh\sqrt{\lambda}x \sinh\sqrt{\lambda}(x_0 - 1)}{\sqrt{\lambda}\sinh\sqrt{\lambda}}, & 0 < x < x_0, \\[2ex] \dfrac{\sinh\sqrt{\lambda}x_0 \sinh\sqrt{\lambda}(x - 1)}{\sqrt{\lambda}\sinh\sqrt{\lambda}}, & x_0 < x < 1. \end{cases}$$

Using the calculus of residues, one obtains the following solution representation of the boundary value problem (4.233a,b) that is useful for large values of t:

$$u(x,t) = 2\sum_{n=1}^{\infty} e^{-n^2\pi^2 t} \sin n\pi x_0 \sin n\pi x. \tag{4.240}$$

Using the asymptotic expansion of $y(x;\lambda)$ valid for large values of $|\lambda|$ along the Bromwich contour of the inverse Laplace transform, one obtains the following solution representation of the boundary value problem (4.233a,b) that is useful for small values of t:

$$u(x,t) = \sum_{n=-\infty}^{\infty} [G(x - x_0 - 2n, t) - G(x + x_0 + 2n, t)],$$

where

$$G(x,t) = \frac{1}{\sqrt{4\pi t}} e^{-x^2/4t}.$$

In principle, the approach presented here applies to any boundary value problem for a linear PDE that has a nontrivial point symmetry admitted by the homogeneous PDE and its homogeneous boundary conditions with the symmetry also admitted by a boundary surface on which no boundary conditions are imposed. One proceeds by using canonical coordinates $r(x,y), s(x,y)$ associated with the point symmetry. Then $\partial/\partial s$ plays the role in solution of the transformed boundary value problem that $\partial/\partial t$ played in the solution of the boundary value problem (4.233a,b). The solution representation of the transformed boundary value problem would then be the inverse Laplace transform with s playing the role of t.

(ii) *Homogeneous Heat Equation with a Linear Nonhomogeneous Boundary Condition.* Consider the following boundary value problem for the linear heat equation:

$$u_t - u_{xx} = 0, \quad 0 < x < 1, \ t > 0, \tag{4.241a}$$

$$u(0,t) = u(1,t) = 0, \tag{4.241b}$$

$$u(x,0) = h(x). \tag{4.241c}$$

Clearly,

$$\mathrm{X}_\lambda = \frac{\partial}{\partial t} + \lambda u \frac{\partial}{\partial u}$$

is admitted by (4.241a,b). The invariant form for the related invariant solution is given by

$$u = \Theta(x,t;\lambda) = y(x;\lambda)e^{\lambda t}, \tag{4.242}$$

which satisfies (4.241a,b) if and only if

$$\lambda = \lambda_n = -n^2\pi^2,$$

with

$$y(x;\lambda_n) = a_n \sin n\pi x,$$

where a_n is an arbitrary constant, $n = 1,2,\ldots$. If $h(x) = \delta(x - x_0)$, then the superposition of invariant solutions

$$u(x,t) = \sum_{n=1}^{\infty} \Theta(x,t;\lambda_n)$$

satisfies the initial condition if $a_n = 2\sin n\pi x_0$. Of course, this is the solution representation (4.240) since the boundary value problems (4.233a,b) and (4.241a–c) are equivalent problems when $h(x) = \delta(x - x_0)$. Let

$$K(x,t;x_0) = 2\sum_{n=1}^{\infty} e^{-n^2\pi^2 t} \sin n\pi x_0 \sin n\pi x.$$

Then the solution of the boundary value problem (4.141a–c) is given by

$$u(x,t) = \int_0^1 h(x_0)K(x,t;x_0)\,dx_0.$$

(2) *An Inverse Stefan Problem*
A nontrivial example is illustrated by the inverse Stefan problem that is given by the boundary value problem

$$u_t = u_{xx}, \quad 0 < x < X(t), \ t > 0, \tag{4.243a}$$

$$u(X(t),t) = 0, \quad t > 0, \tag{4.243b}$$

$$u_x(0,t) = h_1(t), \quad t > 0, \tag{4.243c}$$

$$u(x,0) = h_2(x), \quad 0 < x < 1, \tag{4.243d}$$

$$h_3(t) = ku_x(X(t),t) - X'(t), \quad t > 0, \tag{4.243e}$$

where, for a prescribed moving boundary $X(t)$ with $X(0) = 1$, an arbitrary initial distribution $h_2(x)$, fixed constant *k*, and arbitrary flux $h_1(t)$, the aim is to determine $u(x,t)$ and the flux $h_3(t)$ so that the boundary value problem (4.243a–e) is solved.

Our strategy will be to first obtain a solution $u = \Theta_1(x,t)$ of (4.243a–c). Then we will solve (4.243a–d), with $h_1(t) \equiv 0$ and $u(x,0) = h_2(x) - \Theta_1(x,0)$, to obtain a second function $u = \Theta_2(x,t)$ that is the unique solution of this second boundary value problem. Consequently, the solution of the boundary value problem (4.243a–e) is given by $u = \Theta_1(x,t) + \Theta_2(x,t)$ with

$$h_3(t) = k\left[\frac{\partial\Theta_1}{\partial x}(X(t),t) + \frac{\partial\Theta_2}{\partial x}(X(t),t)\right] - X'(t).$$

Details of this example appear in Bluman and Cole (1974, pp. 213–219, 235–245) and Bluman (1974).

Consider the six-parameter Lie group of point transformations admitted by the heat equation (4.44) with its infinitesimals given by (4.49a–c). One can show that this group leaves invariant a fixed curve, representing a moving boundary, $x = X(t)$ with $X(0) = 1$, if and only if $\kappa = \delta = 0$. After solving the resulting ODE given by

$$\xi(X(t),t) = \tau(t)X'(t) \quad \text{with } X(0) = 1,$$

one can show that $X(t)$ must be of the form

$$X(t) = \sqrt{1 + 2\beta t + \gamma t^2},$$

for arbitrary constants β and γ. We now examine in detail the interesting subcase where $\gamma = \beta^2$. Here

$$X(t) = 1 - \frac{t}{T}, \quad T = -\beta^{-1}. \tag{4.244}$$

If $X(t)$ is of the form (4.244), then (4.243a,b) admits

$$X_1 = \beta x(1+\beta t)\frac{\partial}{\partial x} + (1+\beta t)^2\frac{\partial}{\partial t} - \beta^2(\tfrac{1}{4}x^2 + \tfrac{1}{2}t)u\frac{\partial}{\partial u}. \tag{4.245}$$

We consider the situation where $T > 0$, so that $0 < t < T$, i.e., $X'(t) < 0$, which corresponds to a "melting" situation with the melt completed when $t = T$. The corresponding similarity variable is given by

$$\zeta = \frac{x}{X(t)} = \frac{x}{1-\dfrac{t}{T}}, \quad 0 < \zeta < 1, \tag{4.246}$$

with $\zeta = 0$ corresponding to $x = 0$, and $\zeta = 1$ corresponding to $x = X(t) = 1 - t/T$.

The similarity curves (invariant curves) $\zeta = \text{const}$ are illustrated in Figure 4.1.

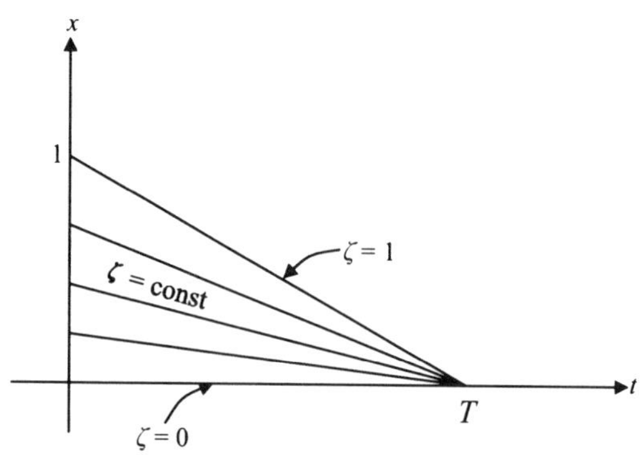

Figure 4.1. The invariant curves $\zeta = \text{const}$.

The invariant form corresponding to the infinitesimal generator $X_\lambda = X_1 + \lambda u\dfrac{\partial}{\partial u}$ is given by

$$u = \Phi(x,t;\nu) = \frac{1}{\sqrt{X(t)}}\exp\left[-\frac{\nu^2 T}{X(t)} + \frac{\zeta^2 X(t)}{4T}\right]y(\zeta;\nu), \tag{4.247}$$

with

$$\nu^2 = \lambda - \tfrac{1}{2}T^{-1}. \tag{4.248}$$

After substituting the invariant form (4.247) into the heat equation (4.243a), we find that $y(\zeta;\nu)$ satisfies the ODE

$$y_{\zeta\zeta} + \nu^2 y = 0. \tag{4.249}$$

The boundary condition (4.243b) yields

$$y(1;\nu) = 0, \tag{4.250}$$

so that

$$y(\zeta;\nu) = A(\nu)\ \sin\ \nu(\zeta - 1), \tag{4.251}$$

for an arbitrary constant $A(\nu)$. Hence any formal superposition of invariant solutions

$$u = \sum_{\nu} \frac{A(\nu)}{\sqrt{X(t)}} \exp\left[-\frac{\nu^2 T}{X(t)} + \frac{\zeta^2 X(t)}{4T}\right] \sin \nu(\zeta - 1) \tag{4.252}$$

solves the homogeneous problem (4.243a,b).

Now let

$$r = \frac{t}{X(t)}, \quad s = -\nu^2, \quad B(s) = \frac{2\pi i}{\sqrt{T}} e^{-\nu^2 T} A(\nu),$$

and replace $\sum_{\nu}$ by $\int_{\gamma-i\infty}^{\gamma+i\infty} ds$. Thus, formally, we obtain the following solution representation of the boundary value problem (4.243a–c) in terms of an inverse Laplace transform:

$$u = \Theta_1(x,t) = \sqrt{r+T} e^{\zeta^2/4(r+T)} \frac{1}{2\pi i} \int_{\gamma-i\infty}^{\gamma+i\infty} B(s) \sinh[\sqrt{s}(1-\zeta)] e^{sr}\, ds. \tag{4.253}$$

Letting

$$H(r) = h_1(t) = h_1\left(\frac{rT}{r+T}\right),$$

and then taking the inverse of (4.253) so that the boundary condition (4.243c) is satisfied, we find that

$$B(s) = -\frac{\beta}{\sqrt{s}\cosh\sqrt{s}} \int_0^\infty \frac{H(r)}{(r+T)^{3/2}} e^{-sr}\, dr. \tag{4.254}$$

Now we proceed to find $\Theta_2(x,t)$. Let

$$H_2(x) = h_2(x) - \Theta_1(x,0),$$

and consider the boundary value problem (4.243a–d) with $h_1(t) \equiv 0$ and $h_2(x)$ replaced by $H_2(x)$. In particular, we consider the boundary value problem

$$u_t = u_{xx}, \quad 0 < x < X(t), \quad t > 0, \tag{4.255a}$$

$$u(X(t),t) = 0, \quad t > 0, \tag{4.255b}$$

$$u_x(0,t) = h_1(t), \quad t > 0, \tag{4.255c}$$

$$u(x,0) = H_2(x), \quad 0 < x < 1, \tag{4.255d}$$

with $X(t)$ given by (4.244). It is easy to check that the point symmetry (4.245) is admitted by (4.255a–c). Consequently, one obtains the similarity variable (4.246). The infinitesimal generator $X_\lambda = X_1 + \lambda u \frac{\partial}{\partial u}$ yields the invariant form (4.247), (4.248). The substitution of (4.247) into the heat equation (4.255a) leads to ODE (4.249). The boundary conditions (4.255b,c) then yield the homogeneous boundary conditions (4.250) and

$$y_\zeta(0;\nu) = 0 \tag{4.256}$$

for ODE (4.249). Thus,

$$\nu = \nu_n = (n + \tfrac{1}{2})\pi$$

and

$$y(\zeta;\nu_n) = A_n \cos \nu_n \zeta,$$

where A_n is an arbitrary constant, $n = 0, 1, 2, \ldots$. The formal superposition of invariant solutions

$$u = \Theta_2(x,t) = \sum_{n=0}^{\infty} \frac{A_n}{\sqrt{X(t)}} \exp\left[-\frac{(\nu_n)^2 T}{X(t)} + \frac{\zeta^2 X(t)}{4T}\right] \cos \nu_n \zeta \tag{4.257}$$

satisfies (4.255a–c). The initial condition (4.255d) is satisfied by setting

$$H_2(x) = \sum_{n=0}^{\infty} A_n \exp\left[\frac{x^2}{4T} - (n + \tfrac{1}{2})^2 \pi^2 T\right] \cos(n + \tfrac{1}{2})\pi x.$$

Let $\psi_n(x) = \cos(n + \frac{1}{2})\pi x$, $n = 0, 1, 2, \ldots$. Then the set of eigenfunctions $\{\psi_n(x)\}$ form a complete orthogonal set of functions on the interval $[0,1]$ with $\int_0^1 \psi_n(x) \psi_m(x)\, dx = \frac{1}{2}\delta_{nm}$. Thus,

$$A_n = 2e^{(n+1/2)^2 \pi^2 T} \int_0^1 H_2(x) \psi_n(x) e^{-x^2/4T} dx, \quad n = 0, 1, 2, \ldots.$$

The above solutions have been used to develop a numerical procedure for solving the direct nonlinear Stefan problem described by boundary value problem (4.243a–e) where the aim is to find the unknown moving boundary $X(t)$ and the distribution $u(x,t)$ for arbitrary $h_1(x), h_2(t)$, $h_3(t)$, and constant k [Milinazzo (1974); Milinazzo and Bluman (1975)].

4.4.3 INCOMPLETE INVARIANCE FOR A LINEAR SYSTEM OF PDEs

As an example, consider the initial value problem for the linear system of wave equations

$$v_t = u_x, \tag{4.258a}$$

$$u_t = c^2(x)v_x, \tag{4.258b}$$

with

$$u(x,0) = U(x), \tag{4.258c}$$

$$v(x,0) = V(x), \tag{4.258d}$$

on the domain $-\infty < x < \infty, t > 0$. We consider the physically interesting wave speed $c(x)$ found in Section 4.3.4 that satisfies the ODE

$$c' = m\sin(\upsilon \log c), \quad \upsilon = \text{const.} \tag{4.259}$$

One can show that for any solution of ODE (4.259), the wave speed $c(x)$ is a monotonic function of x. In particular, for the corresponding wave equation (4.147), such a wave speed $c(x)$ describes wave propagation in two-layered media with smooth transitions, with the properties,

$$\lim_{x\to-\infty} c(x) = 1, \tag{4.260a}$$

$$\lim_{x\to\infty} c(x) = e^{\pi/\upsilon} = \gamma, \quad \gamma > 0, \tag{4.260b}$$

$$\max_{x\in(-\infty,\infty)} c'(x) = m \quad \text{when } m > 0, \tag{4.260c}$$

where γ, m are independent parameters with γ representing the ratio of asymptotic wave speeds. [One can easily adapt the results presented here to the situation where

$$\lim_{x\to-\infty} c(x) = c_1 > 0, \quad \lim_{x\to\infty} c(x) = c_2 > 0,$$

by appropriate scalings.] Note that, without loss of generality, from the invariance of ODE (4.259) under translations in x, we can set $c'(0) = m$. A typical profile for such a wave speed $c(x)$ is exhibited in Figure 4.2.

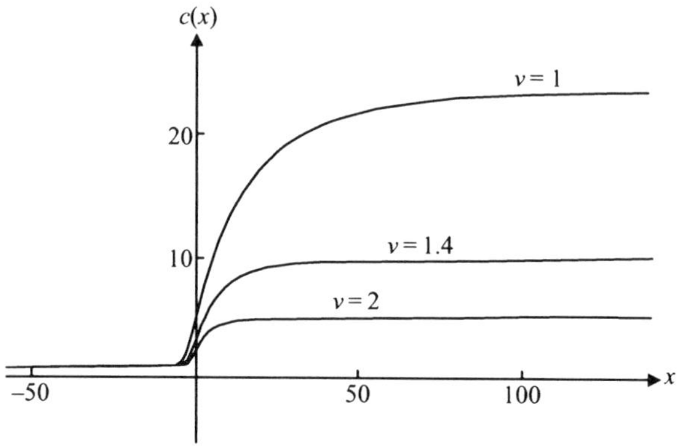

Figure 4.2. Typical profiles for the wave speed $c(x)$.

The four-parameter Lie group of point transformations (4.160) is admitted by the system of PDEs (4.258a,b) when $c(x)$ satisfies the first-order ODE (4.259). It is easy to see that the point symmetry $\mathrm{X} = \mathrm{X}_2 + \mathrm{X}_3$ leaves invariant the curve $t = 0$. One can show that the relevant invariant solutions of the given system of PDEs (4.258a,b) arise from its invariance under

$$\mathrm{X} + 4\upsilon n i\left[u\frac{\partial}{\partial u} + v\frac{\partial}{\partial v}\right] \tag{4.261}$$

for all integers n.

For $n = 0, 1, 2, \ldots$, these invariant solutions are given by

$$\begin{bmatrix} u(x,t) \\ v(x,t) \end{bmatrix} = \begin{bmatrix} u_n(x,t) \\ v_n(x,t) \end{bmatrix} = \sqrt{\sin y}e^{-2ni\arctan[\cot y\,\mathrm{sech}(m\upsilon t)]} \times \begin{bmatrix} c^{1/2}(x) & 0 \\ 0 & c^{-1/2}(x) \end{bmatrix}$$

$$\times \begin{bmatrix} \sqrt{\cosh(m\upsilon t) + \sinh(m\upsilon t)\cos y} & \sqrt{\cosh(m\upsilon t) - \sinh(m\upsilon t)\cos y} \\ \sqrt{\cosh(m\upsilon t) + \sinh(m\upsilon t)\cos y} & -\sqrt{\cosh(m\upsilon)t - \sinh(m\upsilon t)\cos y} \end{bmatrix} \times \begin{bmatrix} f_n(z) \\ g_n(z) \end{bmatrix}, \tag{4.262}$$

where

$$y = \upsilon \log c(x), \quad z = \sinh(m\upsilon t)\sin y, \quad \begin{bmatrix} f_n(z) \\ g_n(z) \end{bmatrix} = M_n(z)\begin{bmatrix} f_0(z) \\ g_0(z) \end{bmatrix}\begin{bmatrix} P_n \\ Q_n \end{bmatrix},$$

with

$$\begin{bmatrix} f_0(z) \\ g_0(z) \end{bmatrix} = (z^2+1)^{-1/2}\begin{bmatrix} \cos\psi(z) & \sin\psi(z) \\ -\sin\psi(z) & \cos\psi(z) \end{bmatrix}, \quad \psi(z) = \frac{1}{2\upsilon}\log(z + \sqrt{z^2+1}),$$

$$M_n(z) = R_n(z)\times R_{n-1}(z)\times\cdots\times R_1(z)\times R_0(z), \quad R_0(z)=\begin{bmatrix}1 & 0\\ 0 & 1\end{bmatrix},$$

and, for $n \geq 1$,

$$R_n(z)=\begin{bmatrix}(n^2-\tfrac{1}{4})\left(\dfrac{z-i}{z+i}\right)-\tfrac{1}{4}\upsilon^{-2} & \dfrac{i-2nz}{2\upsilon\sqrt{z^2+1}}\\ \dfrac{2nz+i}{2\upsilon\sqrt{z^2+1}} & (n^2-\tfrac{1}{4})\left(\dfrac{z+i}{z-i}\right)-\tfrac{1}{4}\upsilon^{-2}\end{bmatrix};$$

the constants P_n and Q_n are chosen separately for each invariant solution pair $(u_n(x,t), v_n(x,t))$ in terms of the initial data (4.258c,d).

For $n=-1,-2,\ldots,$ the corresponding invariant solutions can be expressed in terms of the invariant solutions (4.262) through

$$\begin{bmatrix}u(x,t)\\ v(x,t)\end{bmatrix}=\begin{bmatrix}u_n(x,t)\\ v_n(x,t)\end{bmatrix}=\begin{bmatrix}\overline{u}_{-n}(x,t)\\ \overline{v}_{-n}(x,t)\end{bmatrix},$$

where a bar denotes complex conjugation.

Consequently, the solution of the initial value problem (4.258a–d) can be represented formally in the form

$$\begin{bmatrix}u(x,t)\\ v(x,t)\end{bmatrix}=\sum_{n=-\infty}^{\infty}\begin{bmatrix}u_n(x,t)\\ v_n(x,t)\end{bmatrix}=2\,\mathrm{Re}\left(\sum_{n=1}^{\infty}\begin{bmatrix}u_n(x,t)\\ v_n(x,t)\end{bmatrix}\right)+\begin{bmatrix}u_0(x,t)\\ v_0(x,t)\end{bmatrix}.$$

The constants P_n and Q_n are now determined. First, note that

$$u_n(x,0)=(-1)^n(P_n+Q_n)\sqrt{c(x)\sin y}\,e^{i2ny}, \tag{4.263a}$$

$$v_n(x,0)=(-1)^n(P_n-Q_n)\sqrt{\frac{\sin y}{c(x)}}\,e^{i2ny}, \tag{4.263b}$$

$0<y<\pi$. Consequently, from the Fourier series representation (4.263a,b), we find that

$$P_n=\frac{(-1)^n}{2\pi}\int_0^{\pi}e^{-i2ny}(\sin y)^{-1/2}[e^{-y/2\upsilon}U(x(y))+e^{y/2\upsilon}V(x(y))]\,dy,$$

$$Q_n=\frac{(-1)^n}{2\pi}\int_0^{\pi}e^{-i2ny}(\sin y)^{-1/2}[e^{-y/2\upsilon}U(x(y))-e^{y/2\upsilon}V(x(y))]\,dy.$$

For a given initial value problem (4.258a–d), after determining the constants P_n and Q_n, for $n=1,2,\ldots,$ one can directly compute the solution for any time t, $0<t<\infty$. Note that no time-step marching is required as would be the case for numerical procedures based on the method of characteristics. Full details of the derivation of these solutions and their properties are found in Bluman and Kumei (1988).

EXERCISES 4.4

1. Prove Theorem 4.4.1-1.
2. Prove Theorem 4.4.1-2.
3. Prove Theorem 4.4.1-3.
4. Obtain the fundamental solution of the heat equation (4.47) for an infinite spatial domain $(a,b)=(-\infty,\infty)$ by using the invariant forms arising from the following combinations of the infinitesimal generators (4.185):
   (a) X_1, X_3; and
   (b) X_2, X_3.
5. The problem of finding the steady-state temperature distribution near the surface of the Earth due to a periodic temperature variation at the Earth's surface approximately reduces to finding the steady-state solution of the following boundary value problem:

$$u_t = u_{xx}, \quad 0<x<\infty, \ 0<t<\infty, \tag{4.264a}$$

$$u(x,0) = h(x), \tag{4.264b}$$

$$u(0,t) = A\cos \omega t, \tag{4.264c}$$

$$u(\infty,t) = 0. \tag{4.264d}$$

   (a) Show that the steady-state solution of the boundary value problem (4.264a–d) is independent of the initial distribution $h(x)$.
   (b) Let $v(x,t)$ solve

$$v_t = v_{xx}, \quad 0<x<\infty, \quad 0<t<\infty, \tag{4.265a}$$

$$v(0,t) = Ae^{i\omega t}, \tag{4.265b}$$

$$v(\infty,t) = 0. \tag{4.265c}$$

   Find a one-parameter Lie group of point transformations that is admitted by the boundary value problem (4.265a–c). Find the resulting invariant solution of (4.265a–c).
   (c) Find the steady-state solution of the boundary value problem (4.264a–d).
6. Find the fundamental solution (Riemann function) for the Euler–Poisson–Darboux equation, i.e., solve

$$u_{xx} + \frac{\lambda}{x}u_x - u_{tt} = \delta(x-x_0)\delta(t). \tag{4.266}$$

   This solution is the source solution for isentropic flow for a polytropic gas where x is the sound speed in the gas, t is the fluid velocity in some fixed direction, u is the time variable, and the constant λ is related to the ratio of specific heats of the gas.

(a) Show that (4.266) admits the point symmetry

$$\mathrm{X} = -\frac{2xt}{\lambda}\frac{\partial}{\partial x} + \left[\frac{(x_0)^2 - x^2 - t^2}{\lambda}\right]\frac{\partial}{\partial t} + tu\frac{\partial}{\partial u}. \tag{4.267}$$

(b) Show that the similarity variable arising from the point symmetry (4.267) is given by

$$\zeta = \frac{(x - x_0)^2 - t^2}{x}. \tag{4.268}$$

(c) Derive the invariant form for the resulting invariant solution.

(d) Show that the solution of PDE (4.266) is given by

$$u = u(x,t;x_0) = \frac{(2x_0)^{\lambda}}{[(x + x_0)^2 - t^2]^{\lambda/2}} F\left(\tfrac{1}{2}\lambda, \tfrac{1}{2}\lambda; 1; \frac{(x - x_0)^2 - t^2}{(x + x_0)^2 - t^2}\right),$$

where $F(a,b;c;z)$ is the hypergeometric function [Bluman (1967)].

7. Consider the boundary value problem for the response due to a unit impulse for a vibrating string with a nonlinear restoring force:

$$u_{tt} - u_{xx} + f(u) = \delta(x)\delta(t),$$

$$u \equiv 0 \quad \text{if } x > t,$$

with

$$f(u) = -f(-u), \quad f(u) > 0 \quad \text{for } u > 0.$$

(a) Find a point symmetry admitted by this boundary value problem.

(b) Find the invariant form for the resulting invariant solution.

(c) For the case $f(u) = ku^3$, study the solution in a suitable phase plane. What conditions must apply at the wavefront $x = t$?

8. Consider the Fokker–Planck equation

$$u_t = u_{xx} + (\phi(x)u)_x, \quad t > 0, \ 0 < x < \infty, \tag{4.269a}$$

with the drift

$$\phi(x) = \frac{\alpha}{x} + \beta x, \quad \alpha < 1, \ \beta > 0,$$

and the initial condition

$$u(x,0) = \delta(x - x_0), \quad 0 < x_0 < \infty. \tag{4.269b}$$

(a) Find a point symmetry X admitted by the boundary value problem (4.269a,b).

(b) Let $u(x,t;x_0)$ be the solution of the boundary value problem (4.269a,b). Since the conservation law equation

$$\int_0^\infty u(x,t;x_0)\,dx = 1 \tag{4.270}$$

must hold for any allowed values of t, x_0, it follows that (4.270) must admit the point symmetry X. Consequently, use the invariance of the conservation law (4.270) under the point symmetry X to show that the second moment of the solution $u(x,t;x_0)$ of the boundary value problem (4.269a,b) is given by

$$\langle x^2\rangle = \int_0^\infty x^2 u(x,t;x_0)\,dx = \left(\frac{1-\alpha}{\beta}\right)(1-e^{-2\beta t}) + (x_0)^2 e^{-2\beta t} \tag{4.271}$$

without explicitly determining $u(x,t;x_0)$.

(c) Show that the solution of the boundary value problem (4.269a,b) is given by the following invariant solution related to invariance under the point symmetry X:

$$u(x,t;x_0) = \beta\sqrt{\frac{x_0\zeta}{2\sinh\beta t}}\left(\frac{x_0}{x}\right)^{\alpha/2} \exp[\tfrac{1}{2}(1-\alpha)\beta t - \tfrac{1}{4}\beta(1+\coth\beta t)(x-x_0e^{-\beta t})^2]$$

$$\times I_{-(1/2)(1+\alpha)}(\beta x_0\zeta), \tag{4.272}$$

where the similarity variable is given by

$$\zeta = \frac{x}{2\sinh\beta t},$$

and $I_\nu(z)$ is a modified Bessel function.

(d) Use (4.270)–(4.272) and the fact that *all* moments must admit the point symmetry X to derive explicit expressions for definite integrals involving $I_\nu(z)$ [Bluman and Cole (1974)].

9. Use group methods to find the fundamental solution of the heat equation in $n \geq 2$ spatial dimensions, i.e., solve the initial value problem

$$\frac{\partial u}{\partial t} = \sum_{i=1}^{n}\frac{\partial^2 u}{\partial x_i^2}, \quad t>0,\ -\infty < x_i < \infty \text{ for } i = 1,2,\ldots,n,$$

with

$$u(x_1,x_2,\ldots,x_n,0) = \delta(x_1)\delta(x_2)\cdots\delta(x_n).$$

10. Consider the problem of finding the Green's function for an instantaneous line particle source diffusing in a gravitational field and under the influence of a linear shear wind [Neuringer (1968); Bluman and Cole (1974)]. This problem reduces to solving the initial value problem

$$u_t + yu_x - u_y - d(u_{xx} + u_{yy}) = 0, \tag{4.273a}$$

$$u(x,y,0) = \delta(x)\delta(y-y_0), \tag{4.273b}$$

on the domain $t > 0$, $-\infty < x < \infty$, $-\infty < y < \infty$ with $-\infty < y_0 < \infty$.

(a) Show that the initial value problem (4.273a,b) admits the point symmetries

$$\mathrm{X}_1 = t^2\frac{\partial}{\partial x} + 2t\frac{\partial}{\partial y} + d^{-1}(y_0 - y - t)u\frac{\partial}{\partial u},$$

$$\mathrm{X}_2 = (t^3 - 6t)\frac{\partial}{\partial x} + 3t^2\frac{\partial}{\partial y} + 3d^{-1}(x - \tfrac{1}{2}t^2 - yt)u\frac{\partial}{\partial u}. \tag{4.274}$$

(b) Find invariant forms for the solution of the initial value problem (4.273a,b) resulting from the infinitesimal generators X_1 and X_2, respectively. Then show that the solution of the initial value problem (4.273a,b) reduces to solving a first-order ODE.

(c) Show that the solution of the initial value problem (4.273a,b) is given by

$$u(x,y,t) = \frac{1}{4d\pi t\sqrt{1+\frac{1}{12}t^2}}\exp\left[-\frac{1}{16d}\left(\frac{[2x-(y+y_0)t]^2}{t[1+\frac{1}{12}t^2]} + 4\frac{(y-y_0+t)^2}{t}\right)\right]. \tag{4.275}$$

(d) Use the point symmetries (4.274) to compute directly the moments $\langle x\rangle, \langle y\rangle$, $\langle xy\rangle$, $\langle x^2\rangle$, $\langle y^2\rangle$, etc., *without using the explicit solution* (4.275).

11. The Poisson kernel is the solution of

$$\nabla^2 u = u_{rr} + \frac{1}{r}u_r + \frac{1}{r^2}u_{\theta\theta} = 0, \quad 0 \le r < 1, \ 0 \le \theta < 2\pi, \tag{4.276a}$$

with

$$u(1,\theta) = \delta(\theta). \tag{4.276b}$$

(a) Let $z = re^{i\theta}$. Show that the boundary value problem (4.276a,b) admits the infinite-parameter Lie group of point transformations corresponding to the infinitesimal generator

$$\mathrm{X}_\infty = rS(r,\theta)\frac{\partial}{\partial r} + T(r,\theta)\frac{\partial}{\partial \theta} + \lambda u\frac{\partial}{\partial u},$$

where

$$S(r,\theta) = \sum_{n=1}^{\infty}[a_n(z^n - \bar{z}^{-n}) + b_n(z^{-n} - \bar{z}^n)], \quad \bar{z} = re^{-i\theta},$$

$T(r,\theta)$ is the harmonic conjugate of $S(r,\theta)$, $T(1,0) = 0$, $\lambda = -T_\theta(1,0)$, and a_n, b_n are arbitrary complex parameters for $n = 1,2,\ldots$.

(b) Consider the subgroup for which $a_1 = a \ne 0, b_1 = b \ne 0, a_j = b_j = 0$ for $j \ne 1$. Show that the boundary value problem (4.276a,b) admits the two-parameter subgroup given by the infinitesimal generators

$$X_1 = (1-r^2)\sin\theta\frac{\partial}{\partial r} + \left[\left(r+\frac{1}{r}\right)\cos\theta - 2\right]\frac{\partial}{\partial\theta},$$

$$X_2 = (r^2-1)\cos\theta\frac{\partial}{\partial r} + \left(r+\frac{1}{r}\right)\sin\theta\frac{\partial}{\partial\theta} - 2u\frac{\partial}{\partial u}.$$

(c) Show that the invariant solution resulting from point symmetry X_1 has the invariant form $u = \Phi(\zeta)$, with the similarity variable given by

$$\zeta = \frac{1-r^2}{1-2r\cos\theta + r^2}.$$

(d) Use the invariant surface condition $X_2(u-\Phi(\zeta)) = 0$ when $u = \Phi(\zeta)$, to show that $\Phi(\zeta)$ satisfies the ODE $\zeta\Phi'(\zeta) - \Phi(\zeta) = 0$. Hence, derive the Poisson kernel

$$u(r,\theta) = \frac{1}{2\pi}\frac{1-r^2}{1-2r\cos\theta + r^2}.$$

[Bluman and Cole (1974).]

12. Consider a well-posed boundary value problem for a linear homogeneous PDE with independent variables t and $x = (x_1, x_2, \ldots, x_n)$:

$$\mathrm{L}u = 0, \quad t > 0, \ x \in \Omega, \tag{4.277a}$$

with $k-1$ linear homogeneous boundary conditions

$$\mathrm{L}_\alpha u = 0 \ \text{ on } \omega_\alpha(x) = 0, \quad \alpha = 1, 2, \ldots, k-1, \tag{4.277b}$$

and one linear nonhomogeneous initial condition

$$\mathrm{L}_k u = h(x) \quad \text{when } t = 0. \tag{4.277c}$$

If the associated homogeneous boundary value problem admits $X = \frac{\partial}{\partial t}$, show that the solution of the boundary value problem (4.277a–c) has the inverse Laplace transform representation

$$u(x,t) = \frac{1}{2\pi i}\int_{\gamma - i\infty}^{\gamma + i\infty} F(x;s)e^{st}\,ds, \tag{4.278}$$

where $F(x;s)$ is determined by substituting (4.278) into the boundary value problem (4.277a–c). What is the situation when the right-hand side of (4.277a) is $g(x)$ and the initial condition (4.277c) becomes homogeneous?

4.5 DISCUSSION

In this chapter, we showed how to:

(i) find the point symmetries admitted by a scalar PDE or system of PDEs;
(ii) use admitted point symmetries of PDEs to construct resulting invariant solutions (also called similarity solutions); and
(iii) find and use point symmetries admitted by a boundary value problem for a PDE to reduce the boundary value problem to one involving fewer independent variables.

Invariant solutions for scalar PDEs were discovered by Lie (1881). Such solutions for scalar PDEs or systems of PDEs can be determined from an admitted point symmetry in two ways:

(i) Using the Invariant Form Method, one first solves explicitly the characteristic equations arising from the invariant surface conditions to obtain the invariant form for resulting invariant solutions. The invariant solutions are then determined by substituting the invariant form into the given PDEs.

(ii) Using the Direct Substitution Method [Bluman and Kumei (1989b)], one first isolates a specific independent variable and treats it as a parameter. Then one substitutes the invariant surface conditions and necessary differential consequences into the given PDEs in order to eliminate all derivatives with respect to this isolated (parametric) independent variable.

The resulting invariant solutions are determined by solving the reduced PDEs (with one less independent variable than the given PDEs) and then substituting solutions of the reduced PDEs into either the invariant surface conditions or the given PDEs. Most important, the Direct Substitution Method allows one to construct invariant solutions without explicitly solving the characteristic equations corresponding to the invariant surface conditions. One can extend these two methods to obtain invariant solutions from an admitted multiparameter group of point symmetries [cf. Section 4.4.1].

A boundary value problem for a scalar PDE, or system of PDEs, admits a point symmetry if the symmetry separately leaves invariant the boundary, the boundary conditions, and the PDEs of the boundary value problem. If the boundary value problem is well-posed, then its solution is an invariant solution resulting from the admitted point symmetry. The construction of the solution of the boundary value problem further simplifies if the boundary value problem admits a multiparameter Lie group of point transformations.

When applying invariance under point symmetries to a linear boundary value problem, it is unnecessary to leave invariant the boundary conditions of the boundary value problem. Moreover, one only needs to leave invariant the associated homogeneous PDE of a nonhomogeneous PDE, since a homogeneous PDE always admits a uniform scaling of its dependent variables. Here a superposition of invariant solutions or invariant forms, arising from invariance of the associated homogeneous boundary value problem, can be used to solve the given boundary value problem.

Invariant solutions arising from a multiparameter group of point symmetries admitted by a system of PDEs are considered in Anderson, Fels, and Torre (2000) for the situation when the group does not admit sufficiently many independent invariants in relation to the number of independent and dependent variables. This requires a more complicated invariant form for finding invariant solutions than in the situation arising from invariance under a one-parameter group [cf. Section 4.3.1]. In particular, this is the case when seeking rotationally invariant solutions that arise from invariance under the rotation group SO(3) for systems of PDEs in three spatial dimensions, e.g., the Euler equations for fluid flow whose dependent variables include a vector function of its independent variables.

Often, the asymptotic solution of a boundary value problem for a nonlinear PDE is either an invariant solution of self-similar type (*self-similar* or *automodel solution*) arising from scaling invariance, or a traveling-wave solution arising from invariance under translations in space and time. Comprehensive reviews of self-similar asymptotics appear in Newman (1984) and Galaktionov, Dorodnitsyn, Elenin, Kurdyumov, and Samarskii (1988). For applications of self-similar and traveling-wave asymptotics to physical problems, see Barenblatt and Zel'dovich (1972), Barenblatt (1979, 1987, 1996), and Goldenfeld (1992). Barenblatt and Zel'dovich (1972) and Barenblatt (1979, 1987, 1996) consider examples of "intermediate asymptotics" where, in an intermediate space–time domain, the solution of a boundary value problem is approximated by a similarity solution which does not depend on the given boundary conditions—in such examples the similarity solution is not an equilibrium state. Kamin (1975) rigorously justified the evolution of the solution of a porous medium equation to a self-similar solution. For other papers that rigorously justify self-similar asymptotics, see Atkinson and Peletier (1974), Friedman and Kamin (1980), Galaktionov and Samarskii (1984), and Kamin (1975).

Point symmetries of a scalar PDE, or a system of PDEs, describe geometrical motions on its solution space. As is the situation for ODEs, such motions are naturally formulated in the jet space [cf. Section 2.8] associated to the PDE, or system of PDEs, with coordinates given by the independent variables, dependent variables and their partial derivatives up to a finite order. Here an admitted point symmetry geometrically represents the integral curve of a vector field that is tangent to the surface defined by a given scalar PDE, or simultaneously tangent to the set of surfaces defined by a given system of PDEs, and preserves the derivative relations (contact ideal) among the coordinates on the entire jet space. Such vector fields also arise naturally when one considers first-order and higher-order symmetries. Point symmetries in particular correspond to vector fields given by extensions (prolongations) of one-parameter Lie groups of transformations defined on the coordinates for the independent and dependent variables to transformations acting on the coordinates including all partial derivatives of dependent variables with respect to the independent variables [cf. Section 2.4] in jet space. All the point symmetries admitted by a scalar PDE, or a system of PDEs, form a group which has the structure of an abstract connected Lie group [cf. Section 2.8] whose Lie algebra is characterized by a Lie bracket which is isomorphic to the commutators of the vector fields representing the point symmetries on the jet space associated to the PDE, or the system of PDEs.

In a subsequent volume, we will consider many other topics related to the invariance of PDEs, including:

(i) the algorithmic computation of local conservation laws for PDEs, analogous to finding first integrals of ODEs [cf. Sections 3.6 and 3.7], [Olver (1986); Anco and Bluman (1997a, 2002a,b)];

(ii) the computation and use of higher-order symmetries (so-called Lie–Bäcklund symmetries) for PDEs, including recursion operators for linearization [Anderson, Kumei, and Wulfman (1972); Olver (1977, 1986); Bluman and Kumei (1980, 1989b); Mikhailov, Shabat, and Sokolov (1991); Krasil'shchik and Vinogradov (1989)];

(iii) the use of point symmetries and contact symmetries for linearizations of PDEs and to discover mappings relating PDEs [Kumei and Bluman (1982); Bluman and Kumei (1989b, 1990a); Bluman (1983b)];

(iv) the computation of nonlocal symmetries (including potential symmetries resulting from conservation laws) and their uses for finding invariant solutions, linearizations, and conservation laws [Bluman and Kumei (1987, 1988, 1989b, 1990b); Bluman, Kumei, and Reid (1988); Mikhailov, Shabat, and Sokolov (1991); Anco and Bluman (1996, 1997b)]; and

(v) the generalization of the method of finding invariant solutions resulting from invariance under point symmetries to the nonclassical method for finding solutions of PDEs [Bluman and Cole (1969)] which, in the case of a nonlinear PDE, allows one to find solutions that cannot be obtained as invariant solutions resulting from the point symmetries of the PDE [Levi and Winternitz (1989); Nucci and Clarkson (1992); Clarkson and Mansfield (1994a,b)].

References

Abraham-Shrauner, B., Leach, P.G.L., Govinder, K.S., and Ratcliff, G. (1995). Hidden and contact symmetries of ordinary differential equations. *J. Phys.* **A28**, 6707–6716.

Abramowitz, M. and Stegun, I.A. (eds.) (1970). *Handbook of Mathematical Functions.* Dover, New York, NY.

Aguirre, M. and Krause, J. (1985). Infinitesimal symmetry transformations, II. Some one-dimensional nonlinear systems. *J. Math. Phys.* **26**, 593–600.

Akhatov, I.S., Gazizov, R.K., and Ibragimov, N.H. (1988). Bäcklund transformations and nonlocal symmetries. *Soviet Math. Dokl.* **36**, 393–395. See also *CRC Handbook of Lie Group Analysis of Differential Equations,* Vol. I, 1994, Chap. 13, pp. 258–291 (N.H. Ibragimov, ed.). CRC Press, Boca Raton, FL.

Aksenov, A.V. (1995). Symmetries of linear partial differential equations and fundamental solutions. *Dokl. Math.* **51**, 329–331.

Ames, W.F., Lohner, R.J., and Adams, E. (1981). Group properties of $u_{tt} = [f(u)u_x]_x$. *Internat. J. Nonlinear Mech.* **16**, 439–447.

Anco, S.C. and Bluman, G.W. (1996). Derivation of conservation laws from nonlocal symmetries of differential equations. *J. Math. Phys.* **37**, 2361–2375.

Anco, S.C. and Bluman, G.W. (1997a). Direct construction of conservation laws from field equations. *Phys. Rev. Lett.* **78**, 2869–2873.

Anco, S.C. and Bluman, G.W. (1997b). Nonlocal symmetries and nonlocal conservation laws of Maxwell's equations. *J. Math. Phys.* **38**, 3508–3532.

Anco, S.C. and Bluman, G.W. (1998). Integrating factors and first integrals for ordinary differential equations. *European J. Appl. Math.* **9**, 245–259.

Anco, S.C. and Bluman, G.W. (2002a). Direct construction method for conservation laws of partial differential equations. Part I: Examples of conservation law classifications. To appear in *European J. Appl. Math.*

Anco, S.C. and Bluman, G.W. (2002b). Direct construction method for conservation laws of partial differential equations. Part II: General treatment. To appear in *European J. Appl. Math.*

Anderson, I.M., Fels, M., and Torre, C.G. (2000). Group invariant solutions without transversality. *Comm. Math. Phys.* **212**, 653-686.

Anderson, R.L., Kumei, S., and Wulfman, C.E. (1972). Generalization of the concept of invariance of differential equations. *Phys. Rev. Lett.* **28**, 988–991.

Atkinson, F.V. and Peletier, L.A. (1974). Similarity solutions of the nonlinear diffusion equation. *Arch. Rat. Mech. Anal.* **54**, 373–392.

Baikov, V.A., Gazizov, R.K., and Ibragimov, N.H. (1990). Classification of multi-dimensional wave equations with respect to exact and approximate symmetries. Preprint 51, Institute of Applied Mathematics, Academy of Sciences, USSR, Moscow. See also *CRC Handbook of Lie Group Analysis of Differential Equations,* Vol. I, 1994, Chap. 12, pp. 222–224 (N.H. Ibragimov, ed.). CRC Press, Boca Raton, FL.

Baikov, V.A., Gazizov, R.K., and Ibragimov, N.H. (1991). Approximate symmetries and conservation laws. In *Number Theory, Algebra, Analysis and Their Applications.* Proc. Steklov Math. Inst., Academy of Sciences, USSR, Vol. 200, Nauka, Moscow. See also *CRC Handbook of Lie Group Analysis of Differential Equations,* Vol. I, 1994, Chap. 12, pp. 222–224 (N.H. Ibragimov, ed.). CRC Press, Boca Raton, FL.

Barenblatt, G.I. (1979). *Similarity, Self-Similarity, and Intermediate Asymptotics.* Consultants Bureau, New York, NY.

Barenblatt, G.I. (1987). *Dimensional Analysis.* Gordon and Breach, New York, NY.

Barenblatt, G.I. (1996). *Scaling, Self-Similarity, and Intermediate Asymptotics.* Cambridge University Press, Cambridge, UK.

Barenblatt, G.I. and Zel'dovich, Ya.B. (1972). Self-similar solutions as intermediate asymptotics. *Ann. Rev. Fluid Mech.* **4**, 285–312.

Baumann, G. and Nonnenmacher, T.F. (1987). Lie transformations, similarity reduction, and solutions for the nonlinear Madelung fluid equations with external potential. *J. Math. Phys.* **28**, 1250–1260.

Becker, H.A. (1976). *Dimensionless Parameters: Theory and Method.* Wiley, New York, NY.

Bianchi, L. (1918). *Lezioni sulla Teoria dei Gruppi Continui Finiti di Transformazioni.* Enrico Spoerri, Pisa, Italy.

Birkhoff, G. (1950). *Hydrodynamics: A Study in Logic, Fact and Similitude.* Princeton University Press, Princeton, NJ.

Bluman, G.W. (1967). Construction of Solutions to Partial Differential Equations by the Use of Transformation Groups. Ph.D. Thesis, California Institute of Technology.

Bluman, G.W. (1971). Similarity solutions of the one-dimensional Fokker–Planck equation. *Internat. J. Nonlinear Mech.* **6**, 143–153.

Bluman, G.W. (1974). Applications of the general similarity solution of the heat equation to boundary value problems. *Quart. Appl. Math.* **31**, 403–415.

Bluman, G.W. (1980). On the transformation of diffusion processes into the Wiener process. *SIAM J. Appl. Math.* **39**, 238–247.

Bluman, G.W. (1983a). Dimensional analysis, modelling, and symmetry. *Internat. J. Math. Ed. Sci. Tech.* **14**, 259–272.

Bluman, G.W. (1983b). On mapping linear partial differential equations to constant coefficient equations. *SIAM J. Appl. Math.* **43**, 1259–1273.

Bluman, G.W. (1990a). Simplifying the form of Lie groups admitted by a given differential equation. *J. Math. Anal. Appl.* **145**, 52–62.

Bluman, G.W. (1990b). A reduction algorithm for an ordinary differential equation admitting a solvable Lie group. *SIAM J. Appl. Math.* **50**, 1689–1705.

Bluman, G.W. (1990c). Invariant solutions for ordinary differential equations. *SIAM J. Appl. Math.* **50**, 1706–1715.

Bluman, G.W. and Cole, J.D. (1969). The general similarity solution of the heat equation. *J. Math. Mech.* **18**, 1025–1042.

Bluman, G.W. and Cole, J.D. (1974). *Similarity Methods for Differential Equations.* Applied Mathematical Sciences, No. 13. Springer-Verlag, New York, NY.

Bluman, G.W. and Doran-Wu, P. (1995). The use of factors to discover potential systems or linearizations. *Acta Appl. Math.* **2**, 79–96.

Bluman, G.W. and Gregory, R.D. (1985). On transformations of the biharmonic equation. *Mathematika* **32**, 118–130.

Bluman, G.W. and Kumei, S. (1980). On the remarkable nonlinear diffusion equation

$$\frac{\partial}{\partial x}\left[a(u+b)^{-2}\frac{\partial u}{\partial x}\right]-\frac{\partial u}{\partial t}=0.$$

J. Math. Phys. **21**, 1019–1023.

Bluman, G.W. and Kumei, S. (1987). On invariance properties of the wave equation. *J. Math. Phys.* **28**, 307–318.

Bluman, G.W. and Kumei, S. (1988). Exact solutions for wave equations of two-layered media with smooth transition. *J. Math. Phys.* **29**, 86–96.

Bluman, G.W. and Kumei, S. (1989a). Use of group analysis in solving overdetermined systems of ordinary differential equations. *J. Math. Anal. Appl.* **138**, 95–105.

Bluman, G.W. and Kumei, S. (1989b). *Symmetries and Differential Equations.* Applied Mathematical Sciences, No. 81. Springer-Verlag, New York, NY.

Bluman, G.W. and Kumei, S. (1990a). Symmetry-based algorithms to relate partial differential equations, I. Local symmetries. *European J. Appl. Math.* **1**, 189–216.

Bluman, G.W. and Kumei, S. (1990b). Symmetry-based algorithms to relate partial differential equations, II. Linearization by nonlocal symmetries. *European J. Appl. Math.* **1**, 217–223.

Bluman, G.W., Kumei, S., and Reid, G.J. (1988). New classes of symmetries for partial differential equations. *J. Math. Phys.* **29**, 806–811; Erratum, *J. Math. Phys.* **29**, 2320.

Bluman, G.W. and Shtelen, V.M. (1998). Preprint. Nonlocal transformations of diffusion processes to Wiener processes. Department of Mathematics, University of British Columbia, BC.

Boisvert, R.E., Ames. W.F., and Srivastava, U.N. (1983). Group properties and new solutions of Navier-Stokes equations. *J. Engrg. Math.* **17**, 203–221.

Boyer, T.H. (1967). Continuous symmetries and conserved currents. *Ann. Physics* **42**, 445–466.

Bridgman, P.W. (1931). *Dimensional Analysis,* 2nd ed. Yale University Press, New Haven, CT.

Buckingham, E. (1914). On physically similar systems; illustrations of the use of dimension equations. *Phys. Rev.* **4**, 345–376.

Buckingham, E. (1915a). The principle of similitude. *Nature* **96**, 396–397.

Buckingham, E. (1915b). Model experiments and the forms of empirical equations. *Trans. ASME* **37**, 263–296.

Cantwell, B.J. (1978). Similarity transformations for the two-dimensional, unsteady, stream-function equation. *J. Fluid Mech.* **85**, 257–271.

Champagne, B., Hereman, W., and Winternitz, P. (1991). The computer calculation of Lie point symmetries of large systems of differential equations. *Comput. Phys. Comm.* **66**, 319–340.

Cheb-Terrab, E.S. and Roche, A.D. (1999). Integrating factors for second-order ODEs. *J. Symbolic Comput.* **27**, 501–519.

Cicogna, G. and Vitali, D. (1990). Classification of the extended symmetries of Fokker–Planck equations. *J. Phys.* **A23**, L85–L88.

Clarkson, P.A. and Mansfield, E.L. (1994a). Symmetry reductions and exact solutions of a class of nonlinear equations. *Physica* **D70**, 250–288.

Clarkson, P.A. and Mansfield, E.L. (1994b). Algorithms for the nonclassical method of symmetry reductions. *SIAM J. Appl. Math.* **54**, 1693–1719.

Coddington, E.A. (1961). *An Introduction to Ordinary Differential Equations.* Prentice-Hall, Englewood Cliffs, NJ. Reprinted (1989), Dover, New York, NY.

Cohen, A. (1911). *An Introduction to the Lie Theory of One-Parameter Groups, with Applications to the Solution of Differential Equations.* D.C. Heath, New York, NY.

Cohn, P.M. (1965). *Lie Groups.* Cambridge Tracts in Mathematics and Mathematical Physics, No. 46. Cambridge University Press, Cambridge, UK.

Cole, J.D. and Wagner, B. (1996). On self-similar solutions of Barenblatt's nonlinear filtration equation. *European J. Appl. Math.* **7**, 151–167.

Curtis, W.D., Logan, J.D., and Parker, W.A. (1982). Dimensional analysis and the pi theorem. *Linear Algebra Appl.* **47**, 117–126.

de Jong, F.J. (1967). *Dimensional Analysis for Economists.* North-Holland, Amsterdam.

Dickson, L.E. (1924). Differential equations from the group standpoint. *Ann. of Math.* **25**, 287–378.

Dresner, L. (1983). *Similarity Solutions of Nonlinear Partial Differential Equations.* Research Notes in Mathematics, No. 88. Pitman, Boston, MA.

Dresner, L. (1999). *Applications of Lie's Theory of Ordinary and Partial Differential Equations.* Institute of Physics, Bristol, UK.

Drew, M.S., Kloster, S.C., and Gegenberg, J.D. (1989). Lie group analysis and similarity solutions for the equation

$$\frac{\partial^2 u}{\partial x^2}+\frac{\partial^2 u}{\partial y^2}+\frac{\partial^2 (e^u)}{\partial z^2}=0.$$

Nonlinear Anal., Theory, Methods Appl. **13**, 489–505.

Edwards, M.P. and Broadbridge, P. (1995). Exceptional symmetry reductions of Burgers' equation in two and three spatial dimensions. *Z. Angew. Math. Phys.* **46**, 595–622.

Eisenhart, L.P. (1933). *Continuous Groups of Transformations.* Princeton University Press, Princeton, NJ.

Friedman, A. and Kamin, S. (1980). The asymptotic behavior of gas in an n-dimensional porous medium. *Trans. Amer. Math. Soc.* **262**, 551–563.

Galaktianov, V.A., Dorodnitsyn, V.A., Elenin, G.G., Kurdyumov, S.P., and Samarskii, A.A. (1988). A quasilinear heat equation with a source: Peaking, localization, symmetry, exact solutions, asymptotics, structures. *J. Soviet Math.* **41**, 1222–1292.

Galaktionov, V.A. and Samarskii, A.A. (1984). Methods of constructing approximate self-similar solutions of nonlinear heat equations, IV. *Math. USSR-Sb.* **49**, 125–150.

Gandarias, M.L. (1996). Classical point symmetries of a porous medium equation. *J. Phys.* **A29**, 607–633.

Gilmore, R. (1974). *Lie Groups, Lie Algebras, and Some of Their Applications.* Wiley, New York, NY.

Goldenfeld, N. (1992). *Lectures on Phase Transitions and the Renormalization Group.* Addison-Wesley, Reading, MA.

Gonzalez-Gascon, F. and Gonzalez-Lopez, A. (1983). Symmetries of differential equations, IV. *J. Math. Phys.* **24**, 2006–2021.

Gordon, T.J. (1986). On the symmetries and invariants of the harmonic oscillator. *J. Phys.* **A19**, 183–189.

Görtler, H. (1975). Zur Geschichte des π-Theorems. *Z. Angew. Math. Mech.* **55**, 3–8.

Greub, W.H. (1967). *Multilinear Algebra.* Springer-Verlag, New York, NY.

Hansen, A.G. (1964). *Similarity Analyses of Boundary Value Problems in Engineering.* Prentice-Hall, Englewood Cliffs, NJ.

Haynes, R. (1982). *An Introduction to Dimensional Analysis for Geographers.* Institute of British Geographers, London, UK.

Head, A.K. (1992). LIE: A muMATH Program for the Calculation of the LIE Algebra of Differential Equations. CSIRO Division of Material Sciences, Clayton, Australia.

Heredero, R.H., and Olver, P.J. (1996). Classification of invariant wave equations, *J. Math. Phys.* **37**, 6414–6438.

Hereman, W. (1996). *CRC Handbook of Lie Group Analysis of Differential Equations,* Vol. 3: *New Trends in Theoretical Developments and Computational Methods,* Chap. 13, pp. 367–413. (N.H. Ibragimov, ed.). CRC Press, Boca Raton, FL.

Holmes, M.H. (1984). Comparison theorems and similarity solution approximations for a nonlinear diffusion equation arising in the study of soft tissue. *SIAM J. Appl. Math.* **44**, 545–556.

Hydon, P.E. (2000). *Symmetry Methods for Differential Equations. A Beginner's Guide.* Cambridge University Press, Cambridge, UK.

Ibragimov, N.H. (1985). *Transformation Groups Applied to Mathematical Physics.* Reidel, Boston, MA. See also *CRC Handbook of Lie Group Analysis of Differential Equations,* Vol. 2, Chap. 11, pp. 272–273 (N.H. Ibragimov, ed.). CRC Press, Boca Raton, FL.

Ibragimov, N.H. (1995). *CRC Handbook of Lie Group Analysis of Differential Equations,* Vol. 2, Chap. 11, pp. 269–293 (N.H. Ibragimov, ed.). CRC Press, Boca Raton, FL.

Kamin, S. (1975). Similarity solutions and the asymptotics of infiltration eqautions. *Arch. Rational Mech. Anal.* **60**, 171–183.

Kamke, E. (1943). *Differentialgleichungen, Lösungsmethoden und Lösungen,* Vol. 1, 2nd ed. Akademische Verlagsgesellschaft, Leipzig.

Kaplan, W. (1958). *Ordinary Differential Equations.* Addison-Wesley, Reading, MA.

Kersten, P.H.M. (1987). *Infinitesimal Symmetries: a Computational Approach.* CWI Tract No. 34. Centrum voor Wiskunde en Informatica, Amsterdam.

King, J.R. (1989). Exact solutions to some nonlinear diffusion equations. *Quart. J. Mech. Appl. Math.* **42**, 537–552.

King, J.R. (1991). Exact results for the nonlinear diffusion equations

$$\frac{\partial u}{\partial t} = \frac{\partial}{\partial x}\left(u^{-4/3}\frac{\partial u}{\partial x}\right) \quad \text{and} \quad \frac{\partial u}{\partial t} = \frac{\partial}{\partial x}\left(u^{-2/3}\frac{\partial u}{\partial x}\right).$$

J. Phys. **A24**, 5721–5745.

Klamkin, M.S. (1962). On the transformation of a class of boundary value problems into initial value problems for ordinary differential equations. *SIAM Rev.* **4**, 43–47.

Krasil'shchik, I.S. and Vinogradov, A.M. (1989). Nonlocal trends in the geometry of differential equations: Symmetries, conservation laws, and Bäcklund transformations. *Acta Appl. Math.* **15**, 161–209.

Kumei, S. and Bluman, G.W. (1982). When nonlinear differential equations are equivalent to linear differential equations. *SIAM J. Appl. Math.* **42**, 1157–1173.

Kurth, K. (1972). *Dimensional Analysis and Group Theory in Astrophysics.* Pergamon Press, Oxford, UK.

Lefschetz, S. (1963). *Differential Equations: Geometric Theory,* 2nd ed. Interscience, New York, NY.

Levi, D. and Winternitz, P. (1989). Non-classical symmetry reduction: example of the Boussinesq equation. *J. Phys.* **A22**, 2915–2924.

Lie, S. (1881). Über die Integration durch bestimmte Integrale von einer Klasse linearer partieller Differentialgleichungen. *Arch. Math.* **6**, 328–368; also *Gesammelte Abhandlungen,* Vol. III, pp. 492–523. B.G. Teubner, Leipzig, 1922.

Lie, S. (1893). *Theorie der Transformationsgruppen,* Vol. III. B.G. Teubner, Leipzig.

Lighthill, M.J (1958). *Introduction to Fourier Analysis and Generalised Functions,* Cambridge University Press, Cambridge, UK.

Lisle, I. (1992). Equivalence Transformations for Classes of Differential Equations. Ph.D. Thesis, University of British Columbia, BC.

Liu, Q. and Fang, F. (1986). Symmetry and invariant solution of the Schlögl model. *Physica* **139A**, 543–552.

Mandelbrot, B.B. (1977). *Fractals: Form, Chance and Dimension.* W.H. Freeman, San Francisco, CA.

Mandelbrot, B.B. (1982). *The Fractal Geometry of Nature.* W.H. Freeman, New York, NY.

Mansfield, E.L. (1996). The differential algebra package diffgrob2. *Maple Tech.* **3**, 33–37.

Mansfield, E.L. and Clarkson, P.A. (1997). Applications of the differential algebra package diffgrob2 to classical symmetries of differential equations. *J. Symbolic Comput.* **23**, 517–533.

Marsden, J.E. and Ratiu, T.S. (1999). *Introduction to Mechanics and Symmetry.* Texts in Applied Mathematics, No. 17. Springer, New York, NY.

Mikhailov, A.V., Shabat, A.B., and Sokolov, V.V. (1991). The symmetry approach to classification of integrable equations. In *What Is Integrability*? (V.E. Zakharov, ed.). Springer-Verlag, Berlin, pp. 115–184.

Milinazzo, F. (1974). Numerical Algorithms for the Solution of a Single Phase One-Dimensional Stefan Problem. Ph.D. Thesis, University of British Columbia, BC.

Milinazzo, F. and Bluman, G.W. (1975). Numerical similarity solutions to Stefan problems. *Z. Angew. Math. Mech.* **55**, 423–429.

Miller, W., Jr. (1977). *Symmetry and Separation of Variables.* Addison-Wesley, Reading, MA.

Mimura, F. and Nôno, T. (1994). A new conservation law for a system of second-order differential equations. *Bull. Kyushu Inst. Tech.* **41**, 1–10.

Murota, K. (1985). Use of the concept of physical dimensions in the structural approach to systems analysis. *Japan. J. Appl. Math.* **2**, 471–494.

Na, T.Y. (1967). Transforming boundary conditions to initial conditions for ordinary differential equations. *SIAM Rev.* **9**, 204–210.

Na, T.Y. (1979). *Computational Methods in Engineering Boundary Value Problems.* Academic Press, New York, NY.

Neuringer, J.L. (1968). Green's function for an instantaneous line particle source diffusing in a gravitational field and under the influence of a linear shear wind. *SIAM J. Appl. Math.* **16**, 834–842.

Newman, W.I. (1984). A Lyapunov functional for the evolution of solutions to the porous medium equation to self-similarity, I. *J. Math. Phys.* **25**, 3120–3123.

Noether, E. (1918). Invariante Variationsprobleme. *Nachr. König. Gesell. Wissen. Göttingen, Math.-Phys. Kl.*, pp. 235–257.

Nucci, M.C. and Clarkson, P.A. (1992). The nonclassical method is more general than the direct method for symmetry reductions. An example of the Fitzhugh–Nagumo equation. *Phys. Lett.* **A164**, 49–56.

Olver, P.J. (1977). Evolution equations possessing infinitely many symmetries. *J. Math. Phys.* **18**, 1212–1215.

Olver, P.J. (1986). *Applications of Lie Groups to Differential Equations.* Graduate Texts in Mathematics, No. 107. Springer-Verlag, New York, NY.

Ovsiannikov, L.V. (1959). Group properties of the nonlinear heat conduction equation. *Dokl. Akad. Nauk. USSR* **125**, 492–495 (in Russian).

Ovsiannikov, L.V. (1962). *Group Properties of Differential Equations.* Nauka, Novosibirsk (in Russian).

Ovsiannikov, L.V. (1982). *Group Analysis of Differential Equations.* Academic Press, New York, NY.

Page, J.M. (1896). Note on singular solutions. *Amer. J. Math.* **XVIII**, 95–97.

Page, J.M. (1897). *Ordinary Differential Equations with an Introduction to Lie's Theory of the Group of One Parameter.* Macmillan, London, UK.

Reid, G.J. (1990). A triangularization algorithm which determines the Lie symmetry algebra of any system of PDEs. *J. Phys.* **A23**, L853–L859.

Reid, G.J. (1991). Algorithms for reducing a system of PDEs to standard form, determining the dimension of its solution space and calculating its Taylor series solution. *European J. Appl. Math.* **2**, 293–318.

Rosinger, E.E. and Walus, Y.E. (1994). Group invariance of generalized solutions obtained through the algebraic method. *Nonlinearity* **7**, 837–859.

Rudra, P. (1990). Symmetry classes of Fokker–Planck-type equations. *J. Phys.* **A23**, 1663–1670.

Sarlet, W., Cantrijn, F., and Crampin, M. (1987). Pseudo-symmetries, Noether's theorem and the adjoint equations. *J. Phys.* **A20**, 1365–1376.

Sarlet, W., Prince, G.E., and Crampin, M. (1990). Adjoint symmetries for time-independent second-order equations. *J. Phys.* **A23**, 1335–1347.

Schepartz, B. (1980). *Dimensional Analysis in the Biomedical Sciences.* C.C. Thomas, Springfield, IL.

Schlichting, H. (1955). *Boundary Layer Theory.* McGraw-Hill, New York, NY.

Schwarz, F. (1985). Automatically determining symmetries of partial differential equations. *Computing* **34**, 91–106.

Schwarz, F. (1988). Symmetries of differential equations: From Sophus Lie to computer algebra. *SIAM Rev.* **30**, 450–481.

Sedov, L.I. (1982). *Similarity and Dimensional Methods in Mechanics,* 9th ed. Mir, Moscow.

Senthilvelan, M. and Lakshmanan, M. (1995). Lie symmetries and infinite-dimensional Lie algebras of certain nonlinear dissipative systems. *J. Phys.* **A28**, 1929–1942.

Seshadri, R. and Na, T.Y. (1985). *Group Invariance in Engineering Boundary Value Problems.* Springer-Verlag, New York, NY.

Sheftel, M.B. (1997). *A Course on Group Analysis of Differential Equations.* Part II. *Ordinary Differential Equations.* Nauka, St. Petersburg (in Russian).

Sophocleous, C. (1992). On symmetries of radially symmetric nonlinear diffusion equations. *J. Math. Phys.* **33**, 3687–3693.

Stephani, H. (1989). *Differential Equations: Their Solution Using Symmetries.* Cambridge University Press, Cambridge, UK.

Stognii, V.I. and Shtelen, V.M. (1991). Symmetries and particular solutions of some Fokker–Planck equations. *Ukrainian Math. J.* **43**, 456–460 (in Russian).

Tajiri, M. (1983). Similarity reductions of the one- and two-dimensional nonlinear Schrödinger equations. *J. Phys. Soc. Japan* **52**, 1908–1917.

Tajiri, M. and Hagiwara, M. (1983). Similarity solutions for the two-dimensional coupled nonlinear Schrödinger equation. *J. Phys. Soc. Japan* **52**, 3727–3734.

Taylor, E.S. (1974). *Dimensional Analysis for Engineers.* Clarendon Press, Oxford, UK.

Taylor, Sir G.I. (1950). The formation of a blast wave by a very intense explosion, II. The atomic explosion of 1945. *Proc. Roy. Soc.* **A201**, 175–186.

Torrisi, M. and Valenti, A. (1985). Group properties and invariant solutions for infinitesimal transformations of a nonlinear wave equation. *Internat. J. Nonlinear Mech.* **20**, 135–144.

Venikov, V.A. (1969). *Theory of Similarity and Simulation.* MacDonald Technical and Scientific, London, UK.

Warner, F.J. (1983). *Foundations of Differentiable Manifolds and Lie Groups.* Graduate Texts in Mathematics, No. 94. Springer-Verlag, New York, NY.

Watson, G.N. (1922). *A Treatise on the Theory of Bessel Functions.* Cambridge University Press, Cambridge, UK.

Wolf, T. (2002a). Investigating differential equations with CRACK, LiePDE, Applysymm and ConLaw. To appear in *Handbook of Computer Algebra, Foundations, Applications, Systems* (J. Grabmeier, E. Kaltofen, and V. Weispfenning, eds.). Springer, New York, NY.

Wolf, T. (2002b). A comparison of four approaches to the calculation of conservation laws. To appear in *European J. Appl. Math.*

Wolf, T. and Brand, A. (1992). The computer algebra package CRACK for investigating PDEs. In: *Proceedings of ERCIM Advanced Course on Partial Differential Equations and Group Theory*, Bonn.

Wulfman, C.E. (1979). Limit cycles as invariant functions of Lie groups. *J. Phys.* **A12**, L73–L75.

Zel'dovich, Ya.B. (1956). The motion of a gas under the action of a short term pressure (shock). *Akust. Zh.* **2**, 28–38 (in Russian).

Zierep, J. (1971). *Similarity Laws and Modeling.* Marcel Dekker, New York, NY.

Author Index

Subject Index

Boldface indicates a key reference. References to exercises are bracketed.

F

M

Applied Mathematical Sciences

(continued from page ii)

60. *Ghil/Childress:* Topics in Geophysical Dynamics: Atmospheric Dynamics, Dynamo Theory and Climate Dynamics.
61. *Sattinger/Weaver:* Lie Groups and Algebras with Applications to Physics, Geometry, and Mechanics.
62. *LaSalle:* The Stability and Control of Discrete Processes.
63. *Grasman:* Asymptotic Methods of Relaxation Oscillations and Applications.
64. *Hsu:* Cell-to-Cell Mapping: A Method of Global Analysis for Nonlinear Systems.
65. *Rand/Armbruster:* Perturbation Methods, Bifurcation Theory and Computer Algebra.
66. *Hlavácek/Haslinger/Necasl/Lovísek:* Solution of Variational Inequalities in Mechanics.
67. *Cercignani:* The Boltzmann Equation and Its Applications.
68. *Temam:* Infinite-Dimensional Dynamical Systems in Mechanics and Physics, 2nd ed.
69. *Golubitsky/Stewart/Schaeffer:* Singularities and Groups in Bifurcation Theory, Vol. II.
70. *Constantin/Foias/Nicolaenko/Temam:* Integral Manifolds and Inertial Manifolds for Dissipative Partial Differential Equations.
71. *Catlin:* Estimation, Control, and the Discrete Kalman Filter.
72. *Lochak/Meunier:* Multiphase Averaging for Classical Systems.
73. *Wiggins:* Global Bifurcations and Chaos.
74. *Mawhin/Willem:* Critical Point Theory and Hamiltonian Systems.
75. *Abraham/Marsden/Ratiu:* Manifolds, Tensor Analysis, and Applications, 2nd ed.
76. *Lagerstrom:* Matched Asymptotic Expansions: Ideas and Techniques.
77. *Aldous:* Probability Approximations via the Poisson Clumping Heuristic.
78. *Dacorogna:* Direct Methods in the Calculus of Variations.
79. *Hernández-Lerma:* Adaptive Markov Processes.
80. *Lawden:* Elliptic Functions and Applications.
81. *Bluman/Kumei:* Symmetries and Differential Equations.
82. *Kress:* Linear Integral Equations, 2nd ed.
83. *Bebernes/Eberly:* Mathematical Problems from Combustion Theory.
84. *Joseph:* Fluid Dynamics of Viscoelastic Fluids.
85. *Yang:* Wave Packets and Their Bifurcations in Geophysical Fluid Dynamics.
86. *Dendrinos/Sonis:* Chaos and Socio-Spatial Dynamics.
87. *Weder:* Spectral and Scattering Theory for Wave Propagation in Perturbed Stratified Media.
88. *Bogaevski/Povzner:* Algebraic Methods in Nonlinear Perturbation Theory.
89. *O'Malley:* Singular Perturbation Methods for Ordinary Differential Equations.
90. *Meyer/Hall:* Introduction to Hamiltonian Dynamical Systems and the N-body Problem.
91. *Straughan:* The Energy Method, Stability, and Nonlinear Convection.
92. *Naber:* The Geometry of Minkowski Spacetime.
93. *Colton/Kress:* Inverse Acoustic and Electromagnetic Scattering Theory, 2nd ed.
94. *Hoppensteadt:* Analysis and Simulation of Chaotic Systems, 2nd ed.
95. *Hackbusch:* Iterative Solution of Large Sparse Systems of Equations.
96. *Marchioro/Pulvirenti:* Mathematical Theory of Incompressible Nonviscous Fluids.
97. *Lasota/Mackey:* Chaos, Fractals, and Noise: Stochastic Aspects of Dynamics, 2nd ed.
98. *de Boor/Höllig/Riemenschneider:* Box Splines.
99. *Hale/Lunel:* Introduction to Functional Differential Equations.
100. *Sirovich (ed):* Trends and Perspectives in Applied Mathematics.
101. *Nusse/Yorke:* Dynamics: Numerical Explorations, 2nd ed.
102. *Chossat/Iooss:* The Couette-Taylor Problem.
103. *Chorin:* Vorticity and Turbulence.
104. *Farkas:* Periodic Motions.
105. *Wiggins:* Normally Hyperbolic Invariant Manifolds in Dynamical Systems.
106. *Cercignani/Illner/Pulvirenti:* The Mathematical Theory of Dilute Gases.
107. *Antman:* Nonlinear Problems of Elasticity.
108. *Zeidler:* Applied Functional Analysis: Applications to Mathematical Physics.
109. *Zeidler:* Applied Functional Analysis: Main Principles and Their Applications.
110. *Diekmann/van Gils/Verduyn Lunel/Walther:* Delay Equations: Functional-, Complex-, and Nonlinear Analysis.
111. *Visintin:* Differential Models of Hysteresis.
112. *Kuznetsov:* Elements of Applied Bifurcation Theory, 2nd ed.
113. *Hislop/Sigal:* Introduction to Spectral Theory: With Applications to Schrödinger Operators.
114. *Kevorkian/Cole:* Multiple Scale and Singular Perturbation Methods.
115. *Taylor:* Partial Differential Equations I, Basic Theory.
116. *Taylor:* Partial Differential Equations II, Qualitative Studies of Linear Equations.

(continued on next page)

Applied Mathematical Sciences

(continued from previous page)